J. Patrick Powers
CONSTRUCTION DEWATERING: A GUIDE TO THEORY AND PRACTICE

Harold J. Rosen
CONSTRUCTION SPECIFICATIONS WRITING: PRINCIPLES AND PROCEDURES, Second Edition

Walter Podolny, Jr. and Jean M. Müller
CONSTRUCTION AND DESIGN OF PRESTRESSED CONCRETE SEGMENTAL BRIDGES

Ben C. Gerwick, Jr. and John C. Woolery
CONSTRUCTION AND ENGINEERING MARKETING FOR MAJOR PROJECT SERVICES

James E. Clyde
CONSTRUCTION INSPECTION: A FIELD GUIDE TO PRACTICE, Second Edition

Julian R. Panek and John Philip Cook
CONSTRUCTION SEALANTS AND ADHESIVES, Second Edition

Courtland A. Collier and Don A. Halperin
CONSTRUCTION FUNDING: WHERE THE MONEY COMES FROM, Second Edition

James B. Fullman
CONSTRUCTION SAFETY, SECURITY, AND LOSS PREVENTION

Harold J. Rosen
CONSTRUCTION MATERIALS FOR ARCHITECTURE

William B. Kays
CONSTRUCTION OF LININGS FOR RESERVOIRS, TANKS, AND POLLUTION CONTROL FACILITIES, Second Edition

Walter Podolny and John B. Scalzi
CONSTRUCTION OF CABLE-STAYED BRIDGES, Second Edition

Edward J. Monahan
CONSTRUCTION OF AND ON COMPACTED FILLS

Ben C. Gerwick, Jr.
CONSTRUCTION OF OFFSHORE STRUCTURES

David M. Greer and William S. Gardner
CONSTRUCTION OF DRILLED PIER FOUNDATIONS

*Construction
of Offshore Structures*

Copyright © 1986 by John Wiley & Sons, Inc.

All rights reserved. Published simultaneously in Canada.

Reproduction or translation of any part of this work beyond that permitted by Section 107 or 108 of the 1976 United States Copyright Act without the permission of the copyright owner is unlawful. Requests for permission or further information should be addressed to the Permissions Department, John Wiley & Sons, Inc.

Library of Congress Cataloging in Publication Data:

Gerwick, Ben C.
   Construction of offshore structures.

   (Wiley series of practical construction guides)
   "A Wiley-Interscience publication."
   Includes index.
   1. Offshore structures—Design and construction.
I. Title. II. Series.

TC1665.G47   1986   627'.98   86-7753
ISBN 0-471-29705-4

Printed in the United States of America

10 9 8 7 6 5 4 3 2

To my wife, Martelle, with warmest gratitude for her inspiration, combined with determination to keep me "a-going" during the years in which this book was conceived, nurtured, and written.

Series Preface

The Wiley Series of Practical Construction Guides provides the working constructor with up-to-date information that can help to increase the job profit margin. These guidebooks, which are scaled mainly for practice, but include the necessary theory and design, should aid a construction contractor in approaching work problems with more knowledgeable confidence. The guides should be useful also to engineers, architects, planners, specification writers, project managers, superintendents, materials and equipment manufacturers and, the source of all these callings, instructors and their students.

Construction in the United States alone will reach $250 billion a year in the early 1980s. In all nations, the business of building will continue to grow at a phenomenal rate, because the population proliferation demands new living, working, and recreational facilities. This construction will have to be more substantial, thus demanding a more professional performance from the contractor. Before science and technology had seriously affected the ideas, job plans, financing, and erection of structures, most contractors developed their know-how by field trial-and-error. Wheels, small and large, were constantly being reinvented in all sectors, because there was no interchange of knowledge. The current complexity of construction, even in more rural areas, has revealed a clear need for more proficient, professional methods and tools in both practice and learning.

Because construction is highly competitive, some practical technology is necessarily proprietary. But most practical day-to-day problems are common to the whole construction industry. These are the subjects for the Wiley Practical Construction Guides.

M.D. MORRIS, P.E.

Preface

This book has been conceived as a means of assembling into one volume the principal considerations for planning and executing offshore construction operations. Thus it is designed to serve as both a guide and a reference volume for practicing construction engineers and as a text for graduate students and others seeking to enter this new field of construction.

Until now, this information has only been available scattered among hundreds of technical papers, usually reporting on individual specific projects, and in abbreviated form in rules and recommendations of various technical organizations. A number of excellent books have been published dealing with the design of offshore structures, but they do not address construction operations and methods from the viewpoint of the offshore constructor.

This book has been almost 10 years in preparation. The author has drawn from over 30 years of his own experience offshore, first as a contractor and more recently as a construction consultant, and from his service on technical committees preparing codes and recommendations for offshore structures. He has extensively extracted information, data, and examples from the technical literature and has supplemented these by in-house reports made available by oil companies and contractors serving the offshore industry.

The text has been illustrated by many sketches and photographs of offshore construction operations. Many of the photos were taken by the author; others have been made available by individuals and firms engaged in offshore activity.

BEN C. GERWICK, JR.

Acknowledgments

My grateful appreciation goes to all who have contributed information, data, pictures, and drawings. Among those who have been especially helpful are the following:

ABAM Engineers, Inc.
American Petroleum Institute
H. V. Anderson Engineers
Ballast Nedam Group
Bechtel Corporation
Brian Watt Associates, Inc.
Brown and Root, Inc.
Chevron Petroleum
Coflexip
Det Norske Veritas
Dome Petroleum
C. G. Doris
The Dutch Consortia for the Construction of the Oosterschelde Storm Surge Barrier
Exxon Production Research, Inc.
Global Marine Development Corp.
Gulf Canada Resources Ltd.
Han–Padron Associates
Hendrik Boogaard BV
Honshu-Shikoku Bridge Authority
Howard–Doris
Imodco
Kaiser Steel Corporation
Kajima Corporation
London Offshore Consultants
J. Ray McDermott, Inc.
Marconaflo Corp.
Marine Board, National Research Council
Mobil Exploration Norway
Mobil Research and Development Corp.
Morrison–Knudsen Company, Inc.
Noble–Denton Associates
Norwegian Contractors
Ocean Beach Outfall Constructors
Ocean Engineering Interdisciplinary Program, University of California, Berkeley
Offshore
Offshore Engineer
Parsons, Brinckerhoff, Quade and Douglas, Inc.
Petroleum Engineer International
Phillips Petroleum
PMB Systems Engineering, Inc.
Rendel and Partners
Riedel International, Inc.
Santa Fe—International
Seaflo Systems, Inc.
Shell Oil Company
Sohio Petroleum
Soros Associates
Swan Wooster Engineering Co., Inc.
Tokola Offshore, Inc.
Union Oil Co.
Utah International
Veritec
Vetco
Woodside Petroleum
Wright–Forssen, Inc.

Special thanks are due to Mr. Adrian Cottrill, who reviewed the book and whose suggestions have been most helpful, and to Ms. Jenny Brown, who "word-processed" the manuscript.

BEN C. GERWICK, JR.

Contents

Introduction 1
    I.1 General 1
    I.2 Geography 2
    I.3 Ecological Environment 3
    I.4 Legal Jurisdiction 3
    I.5 Offshore Construction Relationships and Sequences 4
    I.6 Typical Offshore Structures 7
    I.7 Interaction of Design and Construction 7

1 Physical Environmental Aspects of Offshore Construction 12
    1.1 Distances and Depths 12
    1.2 Hydrostatic Pressure and Buoyancy 12
    1.3 Temperature 13
    1.4 Seawater and Sea–Air Interface Chemistry; Dissolved and Suspended Gases and Solids; Marine Organisms and Bacteria 15
    1.5 Currents 16
    1.6 Waves and Swells 20
    1.7 Winds and Storms 25
    1.8 Tides and Storm Surges 28
    1.9 Rain, Snow, Fog, Whiteout and Spray; Atmospheric Icing, Lightning 29
    1.10 Sea Ice and Icebergs 30
    1.11 Seismicity, Seaquakes, and Tsunamis 37

2 Geotechnical Aspects: Seafloor Soils 38
    2.1 General 38
    2.2 Dense Sands 39
    2.3 Calcareous Sands 40
    2.4 Boulders on and near the Seafloor Surface; Glacial Till 40
    2.5 Overconsolidated Silts 41
    2.6 Subsea Permafrost and Clathrates 41
    2.7 Weak Arctic Silts and Clays 41
    2.8 Ice Scour and Pingos 42
    2.9 Methane Gas 42
    2.10 Muds and Clays 42
    2.11 Coral and Similar Biogenic Soils; Cemented Soils 44
    2.12 Unconsolidated Sands 45
    2.13 Underwater Sand Dunes ("Megadunes") 46
    2.14 Rock Outcrops 46
    2.15 Cobbles 47
    2.16 Deep Gravel Deposits 47
    2.17 Seafloor Oozes 47
    2.18 Seafloor Instability and Slumping; Turbidity Currents 47
    2.19 General Remarks Concerning Seafloor Construction 48

3 Protection of the Natural and Built Environment; Constraints on Construction 50
    3.1 Ecological Considerations 50
    3.2 Protection of Existing Structures 53

4 Materials and Fabrication for Offshore Structures 55
    4.1 Steel Structures for the Offshore Environment 55
    4.2 Structural Concrete 64

xiv Contents

    4.3 Hybrid and Composite Steel–Concrete Structures 72
    4.4 Plastics and Synthetic Materials 75
    4.5 Titanium 76
    4.6 Rock, Sand, and Asphaltic–Bituminous Material 76

5 Offshore Construction Equipment 78
    5.1 Basic Motions in a Seaway 78
    5.2 Buoyancy, Draft, and Freeboard 80
    5.3 Stability 81
    5.4 Damage Control 83
    5.5 Barges 85
    5.6 Crane Barges 89
    5.7 Offshore Derrick Barges 93
    5.8 Semisubmersible Barges 96
    5.9 Jack-Up Construction Barges 99
    5.10 Launch Barges 101
    5.11 Offshore Dredges 103
    5.12 Pipelaying Barges 107
    5.13 Supply Boats 110
    5.14 Anchor-Handling Boats 111
    5.15 Towboats 111
    5.16 Drilling Vessels 111
    5.17 Crew Boats 112
    5.18 Special-Purpose Equipment 112

6 Marine Operations 113
    6.1 Towing 113
    6.2 Moorings and Anchors 119
    6.3 Handling Heavy Loads at Sea 127
    6.4 Personnel Transfer at Sea 132
    6.5 Diving and Underwater Work Systems 133
    6.6 Underwater Concreting and Grouting 137
    6.7 Offshore Surveying and Navigation 143
    6.8 Temporary Buoyancy Augmentation 147

7 Seafloor Modifications and Improvements 149
    7.1 General 149
    7.2 Types of Seafloor Modifications 150
    7.3 Controls for Grade and Position; Determination of Existing Conditions 150
    7.4 Seafloor Leveling and Obstruction Removal 151
    7.5 Dredging and Removal of Seafloor Sediments 153
    7.6 Dredging and Removal of Hard Material 156
    7.7 Placement of Underwater Fills 159
    7.8 Densification and Consolidation of Fills 162
    7.9 Consolidation and Strengthening of Weak Soils 163
    7.10 Prevention of Liquefaction 164
    7.11 Scour Protection 165
    7.12 Concluding Remarks 168

8 Installation of Offshore Piles 169
    8.1 General 169
    8.2 Pile Fabrication 169
    8.3 Transportation of Piling 171
    8.4 Installing Piles in Offshore Jackets 172
    8.5 Methods of Increasing Penetration 186
    8.6 Insert Piles 188
    8.7 Anchoring into Rock or Hardpan 189
    8.8 Prestressed Concrete Piles 189
    8.9 Handling and Positioning of Piles for Offshore Terminals 190
    8.10 Drilled and Grouted Piles 191
    8.11 Belled Footings 194
    8.12 Other Installation Methods and Practices 196
    8.13 Improving the Capacity of Offshore Piles 196

9 Offshore Platforms: Steel Jackets and Piles 199
    9.1 General 199
    9.2 Loadout and Tie-down 199
    9.3 Transport 207
    9.4 Removal of Jacket from Barge; Lifting; Launching 209

9.5 Upending of Jacket 217
9.6 Installation on the Seafloor 220
9.7 Pile and Conductor Installation 222
9.8 Deck Installation 224
9.9 Examples 227

10 Concrete Offshore Platforms (Gravity-Base Structures) 236

10.1 General 236
10.2 Construction Stages 237
10.3 Enhancing Caisson–Foundation Interaction 273
10.4 Sub-base Construction 277
10.5 Platform Removal 278

11 Other Applications of Offshore Structures 279

11.1 Offshore Terminals 279
11.2 Single-Point Moorings 292
11.3 Articulated Columns 297
11.4 Guyed Towers 306
11.5 Tension Leg Platforms 310
11.6 Seafloor Well Templates 313
11.7 Underwater Oil Storage Vessels 315
11.8 Bridge Piers in Offshore Areas 319
11.9 Cable Arrays, Moored Buoys, and Seafloor Deployment 322
11.10 Subaqueous Tunnels (Tubes): "Bottom-Founded" and "Submerged–Floating" 324
11.11 Ocean Thermal Energy Conversion (OTEC) Systems 326
11.12 Oosterschelde Storm Surge Barrier 329

12 Installation of Steel Submarine Pipelines 336

12.1 General 336
12.2 Lay Barge 338
12.3 Reel Barge 351
12.4 Bottom-Pull Method 353
12.5 Controlled Above-Bottom Pull 357
12.6 Flotation 358
12.7 Controlled Underwater Flotation (Controlled Subsurface Pull) 359
12.8 J-Tube Method: Single- and Double-Pull 360
12.9 S-Curve 360
12.10 Bundled Pipes 360
12.11 Laying under Ice 360
12.12 Burial of Pipelines 361
12.13 Support of Pipelines 367

13 Nonsteel Pipelines and Cables 368

13.1 Polyethylene Lines 368
13.2 Fiber-Reinforced Glass Pipes 368
13.3 Flexible Pipelines and Risers 369
13.4 Outfall and Intake Lines 369
13.5 Cable Laying 373

14 Topside Installation 376

14.1 General 376
14.2 Module Erection 376
14.3 Hook-up 378
14.4 Giant Modules and Transfer of Complete Deck 379

15 Underwater Repairs 381

15.1 General 381
15.2 Repairs to Steel Jacket–Type Structures 382
15.3 Repairs to Concrete Offshore Structures 384
15.4 Repairs to Foundations 386
15.5 Fire Damage 387
15.6 Pipeline Repairs 387

16 Strengthening Existing Platforms 390

16.1 General 390
16.2 Strengthening Existing Members or Assemblies 390
16.3 Increasing Capacity of Existing Piles for Axial Loads 391
16.4 Increasing Lateral Capacity of Piles and Structure in Their Interaction with the Near-Surface Soil 394
16.5 Strengthening of Concrete and Steel Gravity-Base Platforms 395

17 Salvage and Removal 398

- 17.1 General 398
- 17.2 Piled Structures (Terminals, Trestles, Shallow-Water Platforms) 399
- 17.3 Offshore Drilling and Production Platforms (Jackets with Skirt Piles) 399
- 17.4 Gravity-Base Platforms 401
- 17.5 New Developments in Salvage Techniques 403

18 Constructibility 404

- 18.1 General 404
- 18.2 Construction Stages 405
- 18.3 Principles of Construction 408
- 18.4 Facilities and Methods for Initial Fabrication and Launching 408
- 18.5 Assembly and Jointing Afloat 411
- 18.6 Material Selection and Procedures 412
- 18.7 Construction Procedures 412
- 18.8 Access 417
- 18.9 Tolerances 418
- 18.10 Survey Control 419
- 18.11 Quality Control and Assurance 420
- 18.12 Safety 421
- 18.13 Control of Construction: Feedback and Modification 421
- 18.14 Contingency Planning 422
- 18.15 Manuals 423
- 18.16 On-Site Instruction Sheets 424
- 18.17 Risk and Reliability Evaluation 425

19 Construction in the Deep Sea 427

- 19.1 General 427
- 19.2 Considerations and Phenomena Related to Deep-Sea Operations 428
- 19.3 Techniques for Deep-Sea Construction 428
- 19.4 Properties of Materials for Use in the Deep Sea 429
- 19.5 Construction in the Deep Sea 431

20 Arctic Marine Structures 440

- 20.1 General 440
- 20.2 Sea Ice and Icebergs 440
- 20.3 Atmospheric Conditions 445
- 20.4 Arctic Seafloor and Geotechnics 446
- 20.5 Oceanographic 449
- 20.6 Ecological Considerations 449
- 20.7 Logistics and Operations 450
- 20.8 Earthwork in the Arctic Offshore 451
- 20.9 Ice Structures 456
- 20.10 Steel and Concrete Structures for the Arctic 457
- 20.11 Deployment of Structures in the Arctic 463
- 20.12 Installation at Site 465
- 20.13 Ice Condition Survey and Ice Management 470
- 20.14 Durability 473
- 20.15 Constructibility 474
- 20.16 Pipeline Installation 474

Epilogue 477

Bibliography and Technical Literature of Special Relevance to Offshore Construction 479

Appendices

Appendix 1 Excerpts from Det Norske Veritas, *Rules for the Design, Construction, and Inspection of Offshore Structures* 481

Appendix 2 Excerpts from Fédération Internationale de la Précontrainte, *Recommendations for the Design and Construction of Concrete Sea Structures*, 4th ed. 497

Appendix 3 Excerpts from American Petroleum Institute, *API RP2A, Planning, Designing, and Construction of Fixed Offshore Platforms*, 15th ed. 508

Appendix 4 Excerpts from Det Norske Veritas, *Rules for Submarine Pipeline Systems* 523

Appendix 5 Excerpts from Det Norske Veritas, *Rules for the Design, Construction, and Inspection of Offshore Structures*, Appendix H, "Marine Operations" 537

Index 545

*Construction
of Offshore Structures*

Introduction

I.1 General

The oceans are the dominant features of Earth, comprising more than two-thirds of its surface, stabilizing its temperature so that life as we know it can exist, providing the water vapor which later falls as rain on the continental "islands," the original source of life and the ultimate collector or sink of all surficial matter, including waste. Oceans have been both a barrier and a conduit over which people and goods have moved with relative ease, spreading culture while garnering Earth's remote resources.

Yet the ocean is fiercely inhospitable, making us dependent on land bases for support. Storm waves have destroyed even the largest vessels, as well as the puny attempts of humans to protect the coastline from the oceans' attack. The northernmost ocean, the Arctic, is almost completely covered with perpetual sea ice, while the southern, the Antarctic, carries with it huge tabular icebergs that stretch beyond the horizon.

Opportunity and challenge, safety and terror, wealth and destruction: these are the paradoxes of the seas.

Since before recorded history, oceans have been used for transport, for food, for conquest, and for waste disposal. The Phoenicians sailed as far as Norway to the North and Capetown to the South, perhaps even on to South America; the Polynesians crossed the Pacific to where they sighted the great wall of the Andes, which to them marked the "end of the world," and to Japan and Indonesia; and navigators from Kerala reached Africa and Indonesia, completing early man's circumnavigation of the globe. Much later came the Arabian sailors whose sea empire extended from West Africa to the Philippines, the Vikings, who sailed to Venice and Canada, and eventually the Western European navigators of the Age of Exploration, who challenged the utmost corners of the globe, including both the Arctic and the Antarctic. Today, more than 30,000 ships ply the trade routes of the world.

Mahan's brilliant insights in *The Influence of Sea Power on the History of the World* (1890) demonstrated the decisive role which has been played by the navies of nations who strove either to dominate the world or repel the challenger. As Mahan points out, it was the Greek sea victory at Artemis which blunted the expansion of the Persians; the Roman domination of the Mediterranean which forced Hannibal to his audacious but futile march through Spain and across the Alps to try to break Rome's stranglehold on Carthaginian trade; Drake's defeat of the Spanish Armada which eventually led to Britain's worldwide empire; and the temporary repulse of the British fleet by the French navy which enabled Washington to force surrender of the food-and-munitions starved Cornwallis. Similarly, it was the United States' navy's destruction of the Japanese fleet that led to victory in the Pacific in World War II.

The oceans have been regarded, at least until recently, as an inexhaustible source of food; one need only be clever enough to trap the fish which roam along its coastlines and its great internal rivers where the cold water, rich in nutrients, intermixes with the warm. Fishermen have learned to survive the storm waves, hurricane winds, dense fogs, and "black" ice that have destroyed their less able predecessors.

As cities have grown, clustered mainly at the junctures where great rivers meet the ocean, so have the problems of waste disposal grown, whether it be sewage, the effluent from industrial processes, the runoff of waste oil from urban lands and of nutrients from farmlands, or the warm water discharges from power plants.

2 Introduction

The sea has been a compliant receiver, quick to disperse and dilute all but the most toxic wastes. The oils have been consumed by bacteria, and most of the excess minerals precipitated to the seafloor.

The above describes the state of the ocean until recent times, the latter half of the twentieth century. Now suddenly humankind has burst out with explosive force, increasing both population and human activities at an exponential rate. In the forefront of this revolutionary expansion, this "big bang" of cultural spreading, has been the technological exploitation of the oceans.

In the field of transport, we see new ship types and modes, from containerization to catamarans, hovercraft and VLCCs. In fishing, we see electronic search, sea ranching, and the beginning of exploitation of Antarctic krill, that tiny shrimp whose numbers render it the most abundant source of protein on Earth. Even though we live in the Space Age, it remains the seas in which military might dominates, for the nuclear-powered submarine with its awesome destructive power lurks almost undetectable in the ocean depths or underneath sea ice cover. Waste discharges continue, but now there is a global awareness of the need for at least primary treatment and mechanical dispersal so as to avoid overconcentrations along the vulnerable coasts.

The thermal attributes of the ocean, as a source of cooling water, a sink for warm water, and even a potential source of energy, have long been recognized. Although ocean thermal energy conversion (OTEC) projects are not currently economically viable, their technological feasibility has been demonstrated. In the long term, it will probably be the unlimited source of cooling water of the oceans combined with their capacity to accept discharges which will lead to seaborne industrial processing plants on a large scale.

It has only been in the latter half of the twentieth century that full recognition has been given to the oceans and their sediments as a major source of mineral wealth, both hard minerals and petroleum. Offshore oil and gas can potentially supply one-third of the world's energy needs: in fact, it has been stated by the U.S. Geological Survey that the offshore sedimentary basins within the U.S. Economic Zone hold forth the greatest potential for major new discoveries.

An immense amount of publicity has been given to the manganese nodules which cover large areas of tropical and subtropical seafloors. More recently the scientific world has been excited by the discovery of the thermal vents from seafloor rifts, with their strange new forms of life and their apparently rich deposits of polysulfide minerals. Extraction of soluble minerals from the sea has been carried out since prehistoric times: salt (sodium chloride) and, in modern times, magnesium and bromine.

Coastal sediments are also rich deposits of precious minerals such as gold, tin, and probably chromium and platinum. Seabed mining of such unsophisticated minerals as sand and gravel is of major importance in Japan, many European countries, and the Arctic.

However, because of the tremendous economic importance of offshore oil and gas and the concentrated development of technology for their exploitation, most of the recent offshore construction practice has been devoted to the installation of facilities to serve the needs of the petroleum industry, and hence this will serve as the focal subject matter of this book.

1.2 Geography

The great ocean basins, long thought to be relatively simple, with stable slopes and flat seafloors, have turned out to be exceedingly complex and dynamic. The study of plate tectonics has revealed the underlying mechanism by which seafloors spread and sediments are eventually subducted. The seafloor is marked by deep canyons and steep escarpments. Great volcanic mountains rise above the ocean floor far higher than Everest rises above its base. Seamounts which have not yet reached the surface or which have been eroded or submerged below it now sport crowns of coral.

Among these oceans float the continents, whose margins extend well beneath the sea. That which essentially extends the continents out under the sea is known as the continental shelf, an area rich in sediments washed off the continents and eroded from the shore to be deposited here in relatively concentrated zones. At the outer edge of the continents is the continental slope, dipping down to the abyssal plain (whose surface is often far from planar). Sediments accumulated on the shelves periodically flow down the slopes, to form giant fans at their base.

The shelves and slopes are then inherently unstable, geologically speaking, at least in their surficial deposits. Many of the most striking geological features have occurred during episodic events, as opposed to gradual continuous erosion and deposition: these episodic events include gigantic submarine landslides and turbidity current flows.

A basic property of the oceans, affecting all human activities thereon, is their vastness, their "illimitable expanse" which necessitates long-distance transport of all materials, structures, equipment, and personnel. There are no easy geographic reference points, no stable support for adjoining activity or storing of supplies. This problem of logistics dominates all considerations of construction activities and integrates construction with the transport functions upon which it so heavily depends.

1.3 Ecological Environment

Interest in the life of the sea has expanded in recent years beyond the fascination of *Moby Dick* and the locating of desirable schools of fish to a deep concern for all living creatures, especially those of the sea, which share our common source of life and which have evolved along parallel lines to relatively high levels of intelligence. As with all newly emerging concerns, there have perhaps been excesses of zeal, but the underlying recognition that the life of the sea must be protected from wholesale depredation has become a basic ethical tenet of our society.

Thus construction activities in the ocean, especially those in the coastal zones, must take cognizance of ecological and environmental constraints, whether these be limitations on noise generated in the water column by dredging, which may affect the navigational and communications abilities of marine mammals, or pollution by persistent chemical discharges. Paradoxically, the esthetically and legally unforgivable presence of a sheen of oil on the water may also be the least environmentally harmful, due to its rapid biodegradation into edible protein.

1.4 Legal Jurisdiction

The seas have long presented a curious contrast between freedom and domination. As a result, a patchwork set of arrangements have emerged. The recent Law of the Sea conferences were an attempt to establish a more logical and politically viable legal basis for the rapidly expanding development of the oceans.

Freedom of navigation, except in a narrow zone close to the shores, and freedom of innocent passage through straits have long been established and enforced by the world's great sea powers. More recently, freedom of scientific research has been promulgated, only to founder on the narrow distinctions between research, exploration for mineral resources, and military intelligence.

While the Law of the Sea treaty has been signed by several nations, it has been rejected by the United States because of inclusion of the concept of a supranational Seabed Authority, to be granted jurisdiction and concomitant expropriation-like rights over seabed mineral resources and the technology used to produce them. Despite the current lack of treaty ratification, however, most of the treaty's other provisions are coming into being as common law, through voluntary observance and through unilateral proclamations of clauses similar to those of the treaty, such as the establishment of 200-mile-wide economic zones.

The most immediate results pertain to control of fishing in these vastly expanded national jurisdictions. The United States, for example, has by one stroke increased by almost 25 percent the area of the globe over which it asserts jurisdiction.

Political jurisdiction in the Arctic remains confused, with some of the nations bordering the Arctic asserting the sector theory, that is, the extension of a meridian from their northernmost land boundary directly to the North Pole. The five nations bordering the Arctic Ocean are Greenland (which belongs to Denmark), Canada, the United States (in Alaska), Norway, and the USSR, the sector of the last extending almost halfway around the globe. The USSR claims the shallow waters of the Barents, Kara, and East Siberian Seas as territorial seas. Canada similarly claims that the channels between the Arctic islands are territorial waterways, while the United States asserts that they are international straits with free right of passage.

The Antarctic seas, south of the 60th parallel, remain an anomaly under a regional authority set up under U.N. provisions. The Antarctic Treaty Organization, originally formed to prohibit military use of Antarctica and to foster exchange of scientific information, has more recently been expanded by the establishment of a "Living Resources" regime, primarily to control the exploitation of krill in the cold, upwelling waters around the continent. A parallel "Mineral Resources" regime is currently under negotiation, aimed primarily at potential petroleum resources such as the vast submarine sediments of the Ross, Weddell, and Bellingshausen Seas. With a so far more practicable approach to rights and obligations, the protection of the operator, and the sharing of gain, this could eventually set a pattern for revision of the objectionable provisions of the Law of the Sea treaty.

Closer to the continental shores, within those areas of national jurisdiction, is a 12-mile zone under full control of the national state. In the United States this is further subdivided by a 3-mile zone under the jurisdiction of the adjoining state. This latter is administered under the provisions of the Coastal Zone Management Act, which takes into account the onshore impacts of offshore activities as well as the direct activities themselves.

The environmental impact statements, carried out under the laws governing these several zones surrounding the coasts, have resulted in a series of agreements concerning specific projects, including their construction. These constraints may affect the procedures, methods, and sequence of construction and carry the force of law.

1.5 Offshore Construction Relationships and Sequences

Many offshore projects relate to the development of an ocean resource, especially oil and gas. Therefore, the sequence and relationships of the parties involved will help the constructor in planning his operations.

Leases for specific offshore areas are granted by the sovereign nation having jurisdiction to a petroleum company, with provision for payments, royalties, taxes, and conduct of operations. These latter may involve specific agreements as to training and employment of nationals during the construction period, use of local contractors, fabricators, and suppliers, purchase of local materials, areas in which work will be carried out, and research and educational activities to be supported. Many of these act as constraints upon the construction contractor, who may be required to associate with a local partner or a national enterprise. The constructor may be subject to restrictions regarding the number of foreign workers, including skilled workers and supervision, that he may employ. Often he is required to build new construction equipment in the country or, if he brings in outside equipment, to do so under bond assuring its subsequent export.

Federal U.S. laws—for example, the Jones Act—prohibit the use of foreign-registered dredges within U.S. waters and restrict the use of foreign towboats.

Following lease arrangements, the oil company will carry out extensive geophysical investigations, including seismic surveys. At this time it may also get shallow core borings, bathymetric data, and environmental information.

Exploratory drilling is usually carried out by floating vessels; drill ships and semisubmersibles being the most used in deep offshore areas and jack-ups in more limited depths. In the Arctic and in shallow waters, exploratory drilling may be carried out from bottom-founded mobile structures. These vessels are collectively called "mobile offshore drilling vessels."

The petroleum company, having discovered a field, now carries out delineation drilling to determine reservoir characteristics. It also intensifies planning for offshore structures and development, carrying out feasibility studies and preliminary engineering so as to select a concept and contractors.

Most often, these studies are carried out by one or more engineering contractors or integrated engineering–construction firms. However, some of these may be carried out by in-house teams of the petroleum company with the aid of consultants.

At the same time, the company proceeds to obtain more accurate, site-specific geotechnical and environmental data for use in design.

Arrangements must be made for shore-based facilities. Environmental impact statements must be prepared. Financing arrangements are made.

During this entire period, the oil company operator will often have put together a consortium of companies to participate in the project. These may include the national oil company and from 1 to 20 other petroleum companies. Usually, but not always, the operator has the largest percentage, except for the national oil company, which typically may have the rights to 49 percent or 51 percent of the project.

The position of *operator* is much sought after by petroleum companies, as it confers control of the project, usually with substantial fees to cover management and overhead costs. It also enables the company to develop in-house engineering and management capabilities.

With the project approved, the operator now lets contracts for the offshore platform. In many cases these will be broken up into the following segments:

Design of the substructure
Design of the deck
Fabrication of substructure
Procurement of process equipment
Fabrication of deck and installation of equipment
Installation of platform
Offshore hookup
Production drilling

Several of these may be combined in logical groups and awarded to one contractor. Rarely, except for small structures, are they all combined in one package.

Pipelines are usually divided into segments such as the following:

Design of submarine pipeline
Procurement of pipe
Coating of pipe
Installation and trenching of pipe

To oversee and manage such a complex series of contracts, the operating oil company may set up its own management team or may engage a *construction manager*. In the latter case it may integrate its staff into the construction manager's activities.

Most nations which have major offshore oil activities in their economic zones have established regulatory agencies to control and supervise their development. These governmental agencies are typically assigned responsibility for ensuring safety during development and operation with respect to the following:

Prevention of pollution
Prevention of loss or waste of the resource
Prevention of injury and death to personnel working on or in conjunction with the development

Such national agencies (e.g., the Minerals Management Service of the U.S. government) have established rules for the conduct of offshore minerals development, including especially offshore oil and gas. These rules provide for a review of design, fabrication, and installation by an independent, third-party "verification agent." The constructor is intimately involved in the fabrication and installation phases and must therefore not only submit properly prepared and documented plans for approval, but must carry out the work in compliance with these documents and sound construction practice.

Provision must also be made for quality assurance during construction. This may or may not be part of the verification contracts but of course must be integrated with them.

Following completion of the platform, development drilling is carried out from the platform. In many cases, production will start after a few wells are completed, with concurrent drilling continuing.

After all wells are operating, workovers must be carried out from time to time in order to ensure the continuing productivity of the wells.

Water and gas injection may be instituted to enhance well productivity. Subsea satellite wells may be tied into the primary platform. On the platform, the gas and oil must be separated. Produced water and sand must be separated out, with the water requiring treatment prior to discharge into the sea.

In some cases oil is stored at the offshore platform, usually by the salt-water displacement system. Discharge ballast water must have the hydrocarbons removed by separators to the concentration permitted by regulation.

Most shipment of oil and essentially all shipment of gas is by pipeline to a shore terminal. Some gas is used on board to power the platform operation. Flaring of gas from the platform is usually limited by regulation to the initial period of operations and to emergencies.

Oil may also be shipped by tanker. In this case, a submarine pipeline from the platform runs to a loading buoy for direct transfer by means of a swivel head to the tanker.

During the operating life of the platform, maintenance must be carried out, repairs performed, and modifications made. While these are usually relatively small contracts in magnitude, they frequently are equally demanding in terms of technical skill and specialized equipment.

When the field has reached the end of its economic lifetime, usually after 20–30 years, regulations of most countries provide that all facilities be removed to several meters below the mudline. The

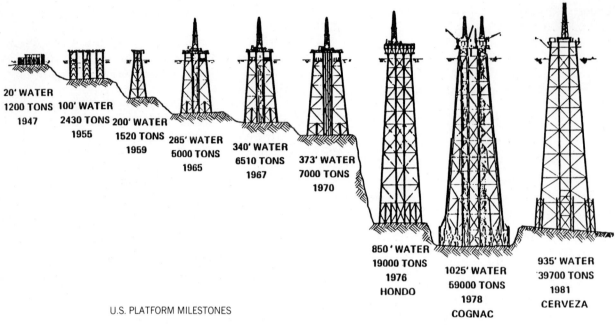

Figure I.6.1 Jacket and pin pile structures on U.S. continental shelf.

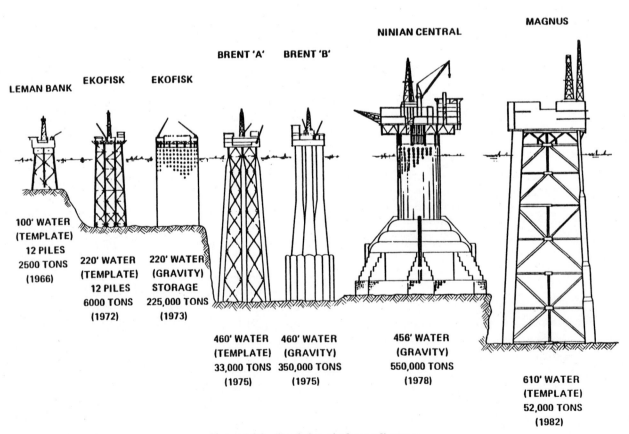

Figure I.6.2 North Sea platform milestones.

wells are capped and cemented off, and the topside equipment is removed. Piles and well casing may be cut off by special explosive charges. The platform is now dismantled and removed.

I.6 Typical Offshore Structures

The range of offshore structures is very great. A few of the more important types are illustrated in Figures I.6.1 to I.6.12.

More details, especially those relating to construction, are shown in the later chapters and sections devoted to specific operations.

I.7 Interaction of Design and Construction

This book is directed to the physical operations in the ocean: their conception, planning, preparation, and execution. Such operations, subsumed in the general category "construction," obviously involve many design-related activities, including engineering of a degree of sophistication appropriate to the activity. Once completed, the structure must perform satisfactorily under service conditions while safely enduring extreme environmental events and credible accidents. The structure must not suffer progressive collapse as a result of such extreme events as earthquake, iceberg impact, extreme storm, or even ship collision. It must withstand the

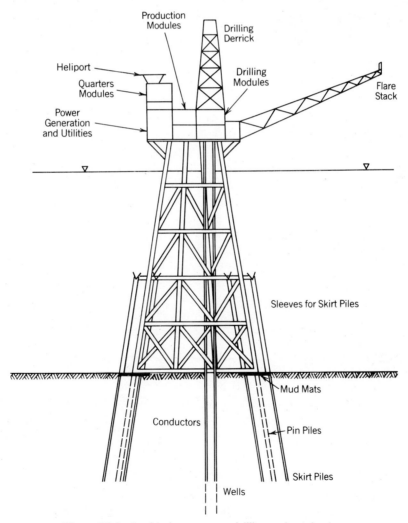

Figure I.6.3 Steel jacket structure drilling and production.

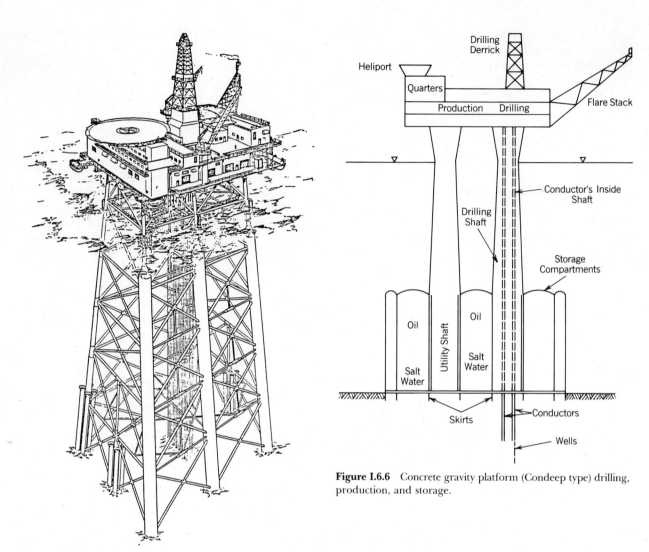

Figure I.6.4 Self-floating steel jacket and pile platform.

Figure I.6.6 Concrete gravity platform (Condeep type) drilling, production, and storage.

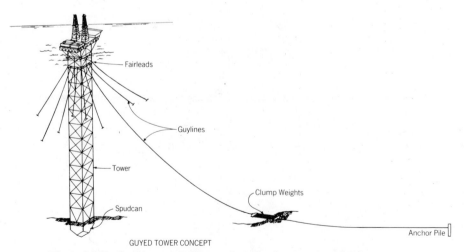

Figure I.6.5 Guyed-tower drilling and production structure for deep water.

8

I.7 Interaction of Design and Construction

Figure I.6.7 The shafts of the concrete gravity-based platform Statfjord C rise majestically above the waters of Stavangerfjord in Norway.

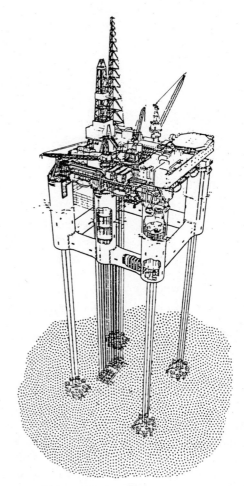

Figure I.6.8 Tension leg platform.

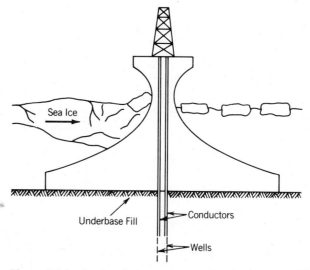

Figure I.6.9 Arctic caisson (cone) drilling and production platform.

repeated loads typical of the marine environment: an offshore platform, for example, may experience 2×10^8 cycles of wave loading during its design service life.

While broad references to the design of the structure cannot be avoided in a book on construction, the two aspects being inseparable, no attempt will be made to include detailed descriptions of analytical procedures for determining dynamic response, fatigue, soil consolidation, and so forth; instead, these will be briefly categorized and identified at the beginning of each section, leaving a description of appropriate analytical matters to other texts and technical papers, many of which are referenced in the Bibliography in the appendices to this book. Indeed, the technical literature in the field is extremely rich, consisting of the journals of professional engineering societies and the proceedings of conferences and symposia such as the Offshore Technology Conferences. Many of these are listed in the Bibliography.

The principal rules, standards, and recommendations for the design and construction of offshore structures are the following:

10 *Introduction*

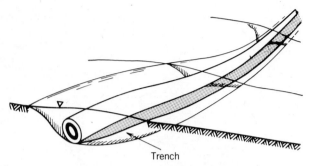

Figure I.6.10 Submarine pipeline.

1. API-RP2A, *Planning, Designing, and Constructing Fixed Offshore Structures*, 16th edi., American Petroleum Institute, Dallas, 1986.
2. DNV, *Rules for the Design, Construction, and Inspection of Offshore Structures, 1977*, repr. ed., (with Appendices A–I), Det Norske Veritas, Oslo, 1981.
3. British Standards Institute, *Code of Practice for Fixed Offshore Structures*, BS 6235, London, 1982.
4. ABS, *Rules for Building and Classing Offshore Installations*, Part I, *Structures*, New York, 1983.
5. Bureau Veritas, *Rules and Regulations for the Construction and Classification of Offshore Platforms*, Paris, 1975, with Amendments and Additions No. 1, August 1982. (Available in English.)
6. FIP, *Recommendations for the Design and Construction of Concrete Sea Structures*, 4th ed., Telford, London, 1985.

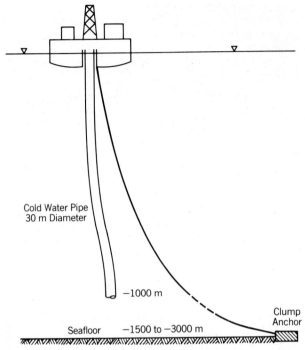

Figure I.6.12 Ocean thermal energy conversion (OTEC) plant.

7. API Bulletin 2N, *Planning, Designing, and Constructing Fixed Offshore Structures in Ice Environments*, American Petroleum Institute, Dallas, 1982.
8. DNV, *Rules for Submarine Pipeline Systems*, Det Norske Veritas, Oslo, 1981.

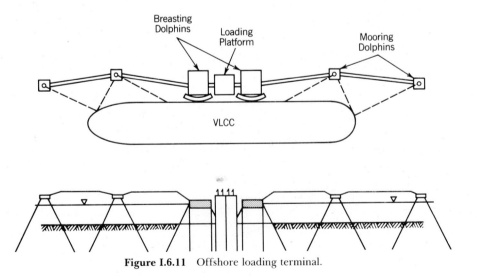

Figure I.6.11 Offshore loading terminal.

The Canadian Standards Institute is currently preparing a set of standards for offshore structures in *Frontier and Arctic Areas*.

Excerpts relating to construction have been reproduced in the appendices and quoted within the text, in each case by permission of the institute or society involved.

Consideration of the many demands of construction and their interaction with design, regulatory requirements, the environment, logistics, economics, schedule, risk, and reliability have led to the development of the concept of *constructibility*, a new term describing a process which has been evolving over many years. *Constructibility* denotes a process that has input to every phase of an offshore project, from conception to maintenance, repair, and eventual removal. It requires consideration of all the applicable provisions described in this book. Because of the growing importance of this concept, a special chapter (Chapter 18) has been devoted to the methodology.

Offshore engineering practice typically employs a mixture of both English units and SI units. In general, the system most commonly employed will be given first, followed by the equivalent value in the other system within parentheses.

Offshore construction, with its many current opportunities and its tremendous demands and challenges, is emerging as one of the most exciting fields of engineering practice, one which will test humankind's ability to rise to new heights of skill and courage.

Come, my friends
'Tis not too late to seek a newer world.
Push off, and sitting well in order, smite
The sounding furrows; for my purpose holds
To said beyond the sunset and the baths
Of all the western stars, until I die.

ALFRED, LORD TENNYSON, "ULYSSES"

1
Physical Environmental Aspects of Offshore Construction

The oceans present a unique set of environmental conditions which dominate the methods, equipment, support, and procedures to be employed in construction offshore. This same unique environment also of course dominates the design. Many books have therefore addressed the extreme environmental events and adverse exposures as they affect design. Unfortunately, relatively small attention has been given in published texts to the environment's influence on construction. Since the design of offshore structures is based to a substantial degree upon the ability to construct, there is an obvious need to understand and adapt to environmental aspects as they affect construction.

In this chapter, the principal environmental factors will be examined individually. As will be repeatedly emphasized elsewhere in this book, a typical construction project will be subjected to many of these concurrently, and it will be necessary to consider their interaction with each other and with the construction activity.

1.1 Distances and Depths

As noted in the Introduction, most offshore construction takes place at substantial distances from shore and even from other structures, often being out of sight over the horizon. Thus construction activities must be self-supporting insofar as possible, able to be manned and operated with a minimum dependency on a shore-based infrastructure.

Distance has a major impact upon the methods used for determining position and the practical accuracies obtainable. Distance affects communication. It necessitates arrangement of facilities to deliver fuel and spare parts and to transport personnel. Distance requires that supervisory personnel at the site be capable of interpreting and integrating all the many considerations and of making appropriate decisions. Distance also produces psychological effects: Men involved in offshore construction must be able to work together in harmony and to endure long hours under often miserable conditions.

Offshore regions extend from the coast to the deep ocean. Construction operations have been already carried out in over 300 meters water depth, exploratory oil drilling operations in 2000 meters, and offshore mining tests in even deeper water. The average depth of the ocean is 4000 meters, the maximum over 10,000 meters, deeper than Everest rises above sea level.

The ocean depths, even those in which work is currently carried out, are inhospitable and essentially dark, and thus require special equipment, tools, and procedures for location, control, operations, and communication. Amazing technological developments have arisen to meet these demands: the work submersible, remote-operated vehicles (ROVs), fiber optics, acoustic imaging, and special gases for diver operations. While some of these advances and their uses will be discussed in detail later, it is important to recognize the general limitations placed on construction operations by depth.

1.2 Hydrostatic Pressure and Buoyancy

The external pressure of seawater acting on a structure and all of its elements follows the simple hydraulic law that pressure is proportional to depth:

$$P = \gamma_w h$$

where h = depth, γ_w = density of seawater, and P = unit pressure. This can be very roughly expressed in the metric system as 1 ton per square meter per meter of depth. More accurately, for seawater, the density is 1026 kg/m^3. In the English system, density of seawater is 64 lb/ft^3.

Hydrostatic pressure acts uniformly in all directions, downward, sideways, and up. The depth is of course influenced by wave action: directly below the crest, the hydrostatic pressure is determined by the elevation of the crest and is therefore greater than that directly below the trough. This effect diminishes with depth, with differences due to moderate waves becoming negligible at 100 m and those due to storm waves becoming negligible at 200 m.

Hydrostatic pressure is also transmitted through channels below structures and channels (pores) in the soil. The difference in pressure causes flow. Flow is impeded by friction. The distribution of hydrostatic pressure in the pores of soils under wave action is thus determined by the water depth, wave length, wave height, and friction within the pores or channels.

Hydrostatic pressure is linked with the concept of buoyancy. Archimedes' principle is that a floating object displaces a weight of water equal to its own weight. From another viewpoint, it can be seen that the body sinks into the fluid—in this case, seawater—until its weight is balanced by the upward hydrostatic pressure.

In the case of a submerged object, its net weight in water (preponderance) can also be thought of as the air weight less either the displaced weight of water or the difference in hydrostatic pressures acting upon it. See Figure 1.2.1.

Hydrostatic pressure not only exerts a collapsing force on structures in total but also tends to compress the materials themselves. This latter can be significant at great depths, and even at shallower depths for materials containing closed pores, for example, plastic foam. This decreases the volume and buoyancy while increasing the density.

Hydrostatic pressure acts as a driving force to force water through permeable materials, membranes, cracks, and holes. In the cases of cracks and very small holes, flow is impeded by frictional forces. In the case of porous materials, capillary forces may augment the hydrostatic force.

It is important for the construction engineer to remember that full external hydrostatic pressure can be exerted in even a relatively small hole, for example, an open prestressing duct or duct left by removal of a slip-form climbing rod.

Hydrostatic pressure acting on gases or other fluids will transmit its pressure at the interface to the other substance. Thus where an air cushion is utilized to provide increased buoyancy to a structure, the pressure at the interface will be the hydrostatic pressure of the seawater.

The density of seawater increases slightly with depth. This can be important in determining net weight of objects at great depths.

The density of seawater also varies with temperature, salinity, and the presence of suspended solids such as silts. See Chapter 19, "Construction in the Deep Sea," in which the effects are quantified.

Special care must be taken during inshore or near-shore operations, where buoyancy, freeboard, and underkeel clearance are critical, where large masses of fresh water may be encountered, with their lowered density and consequent effect on draft. An example of such suddenly occuring events is the annual release of the lake behind St. George Glacier in Cook Inlet, or a flood on the Orinoco River, whose effects may extend almost to Trinidad. A more static situation exists north of Bahrain in the Arabian Gulf, where fresh water emerges from seafloor aquifers.

1.3 Temperature

The surface temperature in the seas varies widely from a low of $-2°C$ (28°F) to a high of 32°C (90°F). The higher temperatures decrease rapidly with depth, reaching a steady-state value of about 2°C (35°F) at a depth of 1000 m (3280 ft).

Temperatures of individual masses and strata of seawater are generally distinct, with abrupt changes across the thermal boundaries. This enables ready identification of global currents; for example, a rise in temperature of as much as 2°C may occur when entering the Gulf Stream.

While horizontal differentiation (on the surface) has long been known, vertical differentiation and upwelling have recently been determined as major phenomena in the circulation of the sea.

Temperature affects the growth of marine organisms, both directly and by its effect on the

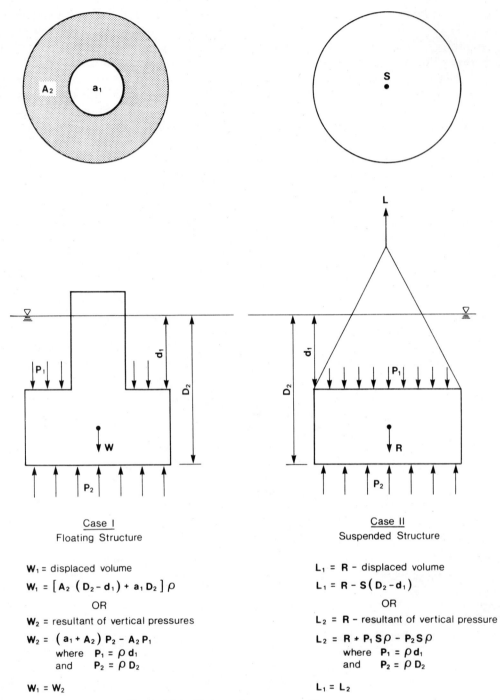

Figure 1.2.1 Hydrostatic pressure and buoyancy.

amount of dissolved oxygen in the water. Marine organisms are very sensitive to sudden changes in the temperature: a sudden rise or fall produces a severe shock that either inhibits growth or kills.

Air temperatures show much greater variation. In the tropics, day air temperatures may reach 40°C. In semi-enclosed areas such as the Arabian–Persian Gulf and the Arabian Sea, air temperatures may even reach 50°C. Humidity is extremely high in such areas, resulting in rapid evaporation, which can produce a "salt fog" in the mornings, causing saline condensation to form on metal surfaces.

The other extreme is the Arctic, where air temperatures over the ice may reach −40°C to −50°C. When the wind blows, air friction usually raises the temperature 10–20°C. However, the combination of low temperature and wind produces "wind chill," which severely affects ability of men to work. Wind may similarly remove heat from materials (e.g., weldments) far more rapidly than when the air is merely cold but still.

Air temperature in the temperate zones varies between these extremes. The ocean's thermal capacity, however, tends to moderate air temperatures from the extremes that occur over land. The rate of sound transmission varies with temperature.

The temperature of the surrounding seawater has an important effect on the behavior of material, since it may be below the transition temperature for many steels, leading to brittle failure under impact. Properties of many other materials, such as concrete, improve slightly at these lower temperatures. Chemical reactions take place more slowly at lower temperatures: this combined with the decrease in oxygen content with depth reduces greatly the rate of corrosion for fully submerged structures.

Temperature also has a major effect on the density (pressure) of enclosed fluids and gases which may be used to provide buoyancy and pressurization during construction. The steady temperature of the seawater will constantly tend to bring the enclosed fluid to the same temperature: where this enclosed fluid is subject to a transient phenomenon, density and thermal gradients will be set up in it.

The atmosphere immediately above seawater is greatly modified by the water temperature. Nevertheless, it can be substantially below freezing, as for example in the Arctic, or substantially above the water temperature, as in areas off Peru, where cold water contrasts with warm air. This produces a thermal gradient and thermal strains in structures which pierce the water plane. These above-water structures may also be directly heated by the sun. Thus there may be a significant expansion of the deck of a barge or pontoon, contrasting with the contraction of its bottom, so as to produce bending in the hull and deck and shear in the sides and longitudinal bulkheads.

Where the structure contains heated products (e.g., hot oil) or extremely cold products (such as liquefied natural gas, LNG) the thermal strains may be severe and require special attention, particularly at points of rigidity such as structural intersections and corners.

These thermal strains are discussed more fully in Chapter 4.

1.4 Seawater and Sea–Air Interface Chemistry; Dissolved and Suspended Gases and Solids; Marine Organisms and Bacteria

The dominant chemical characteristic of seawater is, of course, its dissolved salt, which typically constitutes 35 parts per thousand (3.5 percent) by weight. The principal ions are sodium, magnesium, choride, and sulfate.

These ions are of importance to the construction of structures in the ocean in many ways. Chloride (Cl^-) acts to reduce the protective oxidized coatings which form on steel and thus accelerates corrosion.

Magnesium (Mg^{2+}) will gradually replace the calcium in various chemical constituents of hardened concrete. Magnesium salts are very soft and tend to high permeability.

Sulfates (SO_4^{2-}) attack concrete, both the cement paste and the aggregates themselves, causing expansion and disintegration. Fortunately, the other constitutents of seawater tend to inhibit sulfate attack.

Oxygen is present in the air immediately adjacent to the seawater–air interface and is also present in the water in the form of entrapped air bubbles and dissolved oxygen. Oxygen plays an essential role in the corrosion of steel in the sea environment, whether the steel is exposed, coated, or encased in concrete.

Carbon dioxide (CO_2) and hydrogen sulfide (H_2S) are also dissolved in seawater in varying degrees depending on location and temperature. They lower the pH of seawater. In addition, H_2S may cause hydrogen embrittlement of steel.

Entrapped bubbles of water vapor, as in foam, may collapse suddenly, leading to cavitation which pits and erodes the surface of concrete structures.

Silt and clay are suspended in water, usually in colloidal form, as the result of river runoff and also as the result of bottom erosion and scour due to current and waves. Colloidal silt in fresh water will drop out of suspension upon encountering seawater: this, as well as reduced velocity, accounts for the formation of deltas. The zone or band where such deposition takes place is often very narrow, resulting in a disproportionate deposition and buildup in this zone.

Fine sand, silts and clays, and even gravel may

also be carried along with strong currents or wave action, to be deposited as soon as the velocity drops below critical for that particular particle size and density. This results in horizontal stratification of deposits.

The colloidal and suspended silts render vision and optics difficult due to their turbidity, which scatters light rays. Thus in many coastal areas, divers and submersible operations are limited in visibility with normal light spectra.

Moving silt, sand, and gravel may erode surfaces of paint, steel, and so on and remove the protective film of rust from steel, exposing fresh surfaces to corrosion.

Marine organisms have several effects upon sea structures. The first is the increase of drag due to the obstruction to the free flow of water past the surface of the structure. This is the "fouling" of ship bottoms. Mussels may clog intakes to power plants. Eels have entered circulating water systems and grown and plugged the system. Barnacles and algae increase the diameter of steel piles by several times.

Fouling increases the size of the member and, more importantly, increases the surface roughness. Because of this latter, the drag coefficient, C_D used in Morrison's equation, is usually increased by 10–20 percent.

Fortunately, most marine organisms have a specific gravity only slightly greater than that of the seawater itself; thus they do not add an appreciable mass. Fortunately, also, they tend to be fragile and are often torn or broken off by storm.

Barnacles and sea urchins secrete an acid which pits and erodes steel. Sea urchins are particularly active near the sand line and can attack the steel piling and jacket legs.

Mollusks secreting acids bore into rocks and soft concrete. Very aggressive mollusks exist in the Arabian–Persian Gulf. These bore into the hard limestone aggregate of high-strength concrete; they also can eat through bitumastic coatings protecting steel piles.

Of particular importance to the constructor is the attack of marine organisms on timber. Teredo enter into wood through a relatively small hole, eating out the heart. Limnoria attack the surface of the wood, generally in the tidal range. The action of teredo may be very rapid, especially in fast-flowing, clean seawater. Untreated timber piles have been eaten off within a period of 3 to 4 months.

Fishbite, attacking fiber mooring lines, is of increasing concern for deep-sea operations. Sharks apparently exercise their teeth on the lines, causing them to fray, which then attracts smaller fish. Fishbite is especially severe in the first month or two of exposure, apparently due to the curiosity of the sharks. Fishbite attacks occur to 1000 m in sub-Arctic waters and probably twice that depth in tropical waters.

Marine organisms play a major role in the soil formation on the seafloor and in disturbing and reworking the surficial soils. Walruses apparently plow up large areas of sub-Arctic seafloors in search of mollusks, leading to turbidity and erosion.

Marine growth is influenced by temperature, oxygen content, pH, salinity, current, turbidity, and light. While the majority of growth takes place in the upper 20 m, significant growth has occasionally been found at three times that depth.

Anaerobic sulfur-based bacteria are often trapped in the ancient sediments of the oil reservoir. Upon release to the salt water, they convert to sulfates, and upon subsequent contact with air they produce sulfides (H_2S). *Theobacillus concretivorous* bacteria attack weak and permeable concrete. Hydrogen sulfide causes pitting corrosion in steel. Even more serious, hydrogen sulfide is deadly poisonous and may be odorless. Hence entry to compartments previously filled with stored oil must be preceded by thorough purging not only of hydrocarbons but also of any hydrogen sulphide. These anaerobic bacteria may also react with each other to produce methane and hydrogen.

Bacteria in the Arabian Gulf have attacked polysulfide sealants, turning them into a spongy mass.

Within the sea itself, there is much horizontal and vertical stratification, with rather definite boundaries separating zones of slightly different temperature, chemistry, and density. These zones will have recognizably different acoustic and light transmission properties, and the boundaries may give reflections from sonic transmissions.

Seawater is subject to uptake into the atmosphere by the mechanisms of spray and evaporation. As noted earlier, fog formed over seawater may contain appreciable quantities of salt.

1.5 Currents

Currents, even when small in magnitude, have a significant effect on construction operations. They obviously have an influence on the movement of vessels and floating structures and on moorings. They change the characteristics of waves. They exert horizontal pressures against structural surfaces and, due to the Bernoulli effect, develop uplift or downdrag forces on horizontal surfaces. Currents create eddy patterns around structures, which may lead to scour and erosion of the soils.

Even before the start of construction, currents may have created scour channels and areas of deposition, thus creating surficial discontinuities at the construction site.

The vertical profile of currents is conventionally shown as decreasing with depth as a parabolic function. Recent studies in the ocean and on actual deep-water projects indicate, however, that in many cases, the steady-state current velocities just above the seafloor are almost as high as those nearer the surface.

There are several different types of currents: oceanic circulation, geostrophic, tidal, wind-driven, wave-induced, and density currents, as well as currents due to nearby river discharge. Some of these may be superimposed upon each other, with different phases. See Figure 1.5.1

The worldwide ocean circulatory system produces such currents as the Gulf Stream, with a relatively well-defined "channel" and direction and velocity of flow. Other major current systems exist but are often more diffuse, having a general trend but without the characteristics of a river. Thus the prevailing southeasterly trending current along the California and Oregon coasts gives an overall southward movement to sedimentary materials from river outflows.

These major currents may occasionally, often periodically, spin off eddies and branches; the lateral boundaries of the current are thus quite variable. Strong currents may thus occur many miles from the normal path of a current such as the Gulf Stream. Within local coastal configurations, a branch of the main current may sweep in toward shore or even eddy back close to shore.

Recent research has indicated that many of these current "streams" are fed by upwelling or downward movements of the waters and that there are substantial vertical components. These will become of importance as structures are planned and built in deeper waters and will require that accurate current measurements be taken at all depths.

Another major source of currents is that due to tidal changes. The stronger currents are usually in proximity to shore but may extend a considerable distance offshore where they are channeled by subsurface reefs or bathymetry. While they generally follow the tidal cycle, they frequently lag it by one-half to one hour; thus a tidal current may continue flooding on the surface for a short period after the tide has started to fall.

Actually tidal currents are often, if not usually, stratified vertically, so that the lower waters may be flowing in while the upper waters are flowing out. This is particularly noticeable where tidal currents are combined with river currents or where relatively fresh water of lower density overlies heavier salt water.

Since tidal currents are generally changing four times a day, it follows that their velocity and direction are constantly changing. Since the velocity head or pressure acting on a structure varies as the square of this current velocity, it can have a major effect on the mooring and control of structures during critical phases of installation. The current velocities are also superimposed on the orbital particle velocities of the waves, with the pressure and hence forces being proportional to the square of the vectorial addition.

While in regular harbor channels the tidal currents may move in and out along a single path, at most offshore sites the shoreline and subsurface configurations cause the directions to alter significantly, perhaps even rotate, during the tidal cycle.

Ebb currents may be directed not only 180° but often 150°, 120°, or even 90° from flood currents, and this variance itself may change periodically.

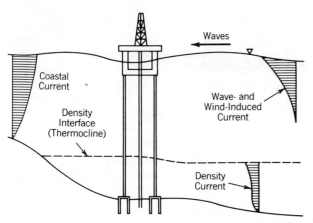

Figure 1.5.1 Wave–current flow field. (Adapted from Nabil Ismail, "Wave–Current Models for Design of Marine Structures" *Journal of Waterway, Port, Coastal, and Ocean Engineering*, ASCE, 1983.)

River currents, especially those of great rivers with large discharges, such as the Orinoco, extend far out to sea. Because the density of the water is less, and perhaps because of silt content, the masses of water tend to persist without mixing for a long period: thus substantial surface currents may reach to considerable distances from shore.

River currents may, as indicated earlier, combine with tidal currents to produce much higher velocities on ebb and reduced velocities on flood.

Wind persisting for a long period of time causes a movement of the surface water that is particularly pronounced adjacent to shallow coasts. This may augment, modify, or reverse coastal currents due to other causes.

Deep-water waves create oscillatory currents on the seafloor, so that there is little net translation of soil particles due to waves alone. When, however, a steady-state current is superimposed upon currents due to waves, the sediment transport is noticeably increased, since its magnitude varies as the cube of the instantaneous current velocity.

Adjacent to the shore, the translational movement of the waves produces definite currents, with water flowing in on top and out either underneath or in channels. Thus a typical pattern of the sea will be to build up an offshore bar, over which the waves move shoreward. Inshore of the bar, the current is directed laterally, with of course a prevailing tendency depending on the angle of incidence of the waves and wind. At intervals, the excess water escapes out to sea, through channels that this outflowing current cuts in the offshore bar. These are the infamous "undertows." These lateral- and seaward-flowing currents may prove to be a major hazard or may be taken advantage of to keep a dredged channel clear.

In the deeper ocean, currents are generated by internal waves, by geostrophic forces, and by deeply promulgated eddies from major ocean streams such as the Gulf Stream. It appears that currents of magnitudes up to 0.5 knot exist on the continental shelf and slope and that currents up to 2.6 knots (1.3 m/s) can be found in the deep ocean.

It is important to note that currents may have vertical as well as horizontal components. The direction of currents often varies with depth. At different times, the direction of currents may vary from their norm.

Strong currents can cause vortex shedding on risers and piles, and vibration of wire lines and pipelines. Vortex shedding can result in scour in shallow water. It can also result in cyclic transverse forces which in extreme cases can lead to fatigue.

As mentioned earlier, water moving over a submerged surface or under a structure's base produces a vertical pressure (uplift or downdrag) in accordance with Bernoulli's theorem. This can cause significant constructional problems, of which the following examples may be given:

1. A large concrete tank being submerged in the Bay of Biscay had its compartments accurately sized for filling, so as to create a known preponderance for controlled sinking, without free surface. When filled and submerged a few meters, the waves moving over the top had their oscillatory motion changed to a translatory current, thus creating an uplift force. This has been called the "beach effect." The tank would sink no further.

    When by emergency measures, additional ballast was added to cause sinking to continue, then at a depth of some 30 ms, the current effect was reduced and the uplift force diminished. The tank was now too heavy and plunged. See Figure 1.5.2.

2. A caisson being submerged to the seafloor behaves normally until close to the bottom, when the current is trapped beneath the base and its velocity increases. This "pulls" the caisson down, while at the same time tending to scour a previously prepared gravel base.

3. A pipeline set on the seafloor is subjected to a strong current which erodes the sand backfill from around it. The pipeline is now subject to uplift (from the increased current flowing over it) and raises off the bottom. The current now can flow underneath; this pulls the pipeline back to the seafloor, where the process can be repeated. Eventually the pipeline may fail in fatigue. See Figure 1.5.3.

Currents produce both scour and deposition. It is important to note that eddies formed at the upstream and downstream corners of structures, such as those of a rectangular caisson, produce deep holes, whereas deposition may occur at the frontal and rear faces.

Scour is extremely difficult to predict. Model studies indicate tendencies and critical location but cannot be quantitatively accurate because of the

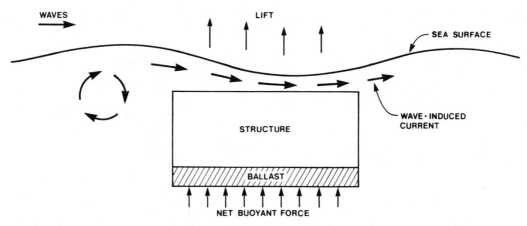

Figure 1.5.2 The "beach effect"—uplift forces on shallowly immersed structure due to wave-induced current.

impossibility of modeling the viscosity of water and the grain size and density, the effect of pore pressures, and so on. However, models can be effectively utilized to predict how the currents will be modified around a particular structure.

Another phenomenon due to currents which occurs when they reach critical velocity is that of vortices. These spin off in a regular pattern downstream of a structure, creating zones of low pressure, which may lead to uplift of soil and scour. For most sea structures, this critical velocity is about 3 knots. Vortices can also act on large-diameter cylinders such as cylinder piles and legs of platforms, causing an oscillatory movement of the cylinder as a whole and ovaling effects on the cylinder's walls.

The effect of a current on a structure is determined by the shape, dimensions, and surface conditions of the structure. While Morrison's equation can normally be applied, care has to be taken in assessing the coefficients. Marine growth, for example, not only increases the dimensions but also the surface roughness and hence the drag coefficient.

Currents have a significant effect on the wave profile. A following current will lengthen the apparent wave length and flatten the wave out, so that its slopes are much less steep. Conversely, an opposing current will shorten the wave length, increasing the height and steepness. Thus at an ocean site affected by strong tidal currents, the same incident waves will have quite different effects on the construction operations, depending on the phases of the tidal cycle. See Figure 1.5.4.

Currents have a serious effect on towing speed and time, a following current increasing the effective speed, an opposing current decreasing it. Translated into time, the decrease in time for a tow of a given distance is only marginally improved by a following current, whereas an opposing current may drastically increase the time required.

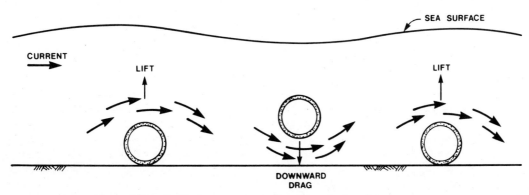

Figure 1.5.3 Oscillating movement of seafloor pipeline due to current.

20 Physical Environmental Aspects of Offshore Construction

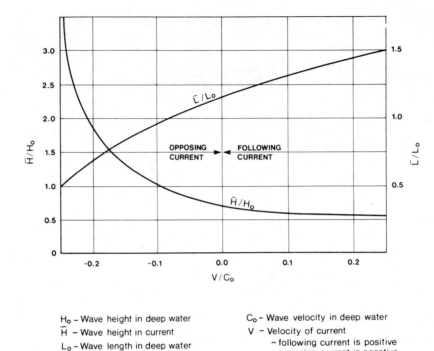

Figure 1.5.4 Changes in wave height and length in an opposing or following current.

Example

A boat can tow a barge 60 nautical miles at 6 knots in still water, requiring 10 hours. With a following current of 2 knots, the tow will take 60/(6 + 2) or 7½ hours, a saving of 25 percent or 2½ hours. With an opposing current of 2 knots, the tow will take 60/(6 − 2) or 15 hours, an increase of 50 percent or 5 hours.

Obviously, when towing a large caisson or similar large structure at an inherently lower speed, these effects can be even more significant.

1.6 Waves and Swells

Waves are perhaps the most obvious environmental concern for operations offshore. They cause a floating structure or vessel to respond in 6 degrees of freedom: heave, pitch, roll, sway, surge, and yaw. They constitute the primary cause of downtime and reduced operating efficiency. The forces exerted by waves are usually the dominant design criterion affecting fixed structures. See Figure 1.6.1.

Waves are primarily caused by the action of wind on water, which through friction transmits energy from the wind into wave energy. Waves which are still under the action of the wind and hence developing are called *waves*, whereas when these same waves have been transmitted beyond the wind-affected zone by distance or time, they are called *swells*.

Water waves can also be generated by other phenomena, such as high currents, landslides, ex-

Figure 1.6.1 North Sea storm waves pound Ekofisk Oil storage caisson.

plosions, and earthquakes. Those associated with earthquakes (e.g., tsunamis) will be dealt with in Section 1.11.

A wave is a traveling disturbance of the sea's surface. The disturbance travels, but the water particles within the wave move in a nearly closed circular orbit, with little net forward motion. See Figure 1.6.2.

Wave and swell conditions can now be predicted from a knowledge of the over-ocean winds. Routine forecasts are now available for a number of offshore operating areas. They are provided by governmental services such as the U.S. Naval Fleet Numerical Weather Control at Monterey, California. Many private companies now offer similar services. These forecasts are generally based on a very coarse grid, which unfortunately misses local storms such as tropical cyclones.

The height of a wave is governed by its speed, duration, and fetch (the distance that the wind blows over open water). The wave height can be roughly calculated by the formula

$$H_s = \frac{\sqrt{WF}}{30}$$

where H_s is the significant wave height in meters, W is wind velocity in km/hr, and F is fetch in kilometers. In the English system,

$$H_s = \frac{\sqrt{WF}}{6}$$

where H_s is in feet, W in statute miles per hour, and F in statute miles.

Deep-water wave forecasting curves can be prepared as a guide. See Figure 1.6.3. These values are modified slightly by temperature; for example, if the air is 20°F colder than the sea, the waves will be 20 percent higher, due to the greater density and hence energy in the wind. This can be significant in the sub-Arctic and Arctic.

Some interesting ratios can be deduced from Figure 1.6.3:

1. A 10-fold increase in fetch increases the wave height 2.5 times.
2. A 5-fold increase in wind velocity increases the wave height 13 times.
3. The minimum-duration curves indicate the duration which the wind must blow in order for the waves to reach their maximum height. The stronger the wind, the less time required.

The total energy in a wave is proportional to the square of the wave height.

While wave height is obviously an important parameter, wave period may be of equal concern to the constructor. Figure 1.6.3 gives the typical period associated with a fully developed wave in deep water. Long-period waves have great energy. When a vessel's length is less than one-half the wave length, it will see greatly increased dynamic surge forces.

Waves vary markedly within a site, of course, even at the same time. Therefore, they are generally characterized by their *significant height* and *significant period*. The significant height of a wave is the average of the highest one-third of the waves. It has been found from experience that this is what an experienced seaman will report as being the height of the waves in a storm.

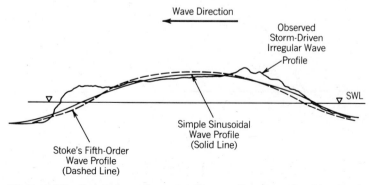

Figure 1.6.2 Comparison of water surface profiles as predicted by theories and as observed. (Adapted from "Production Risers," *Petroleum Engineer International*, Oct. 1984.)

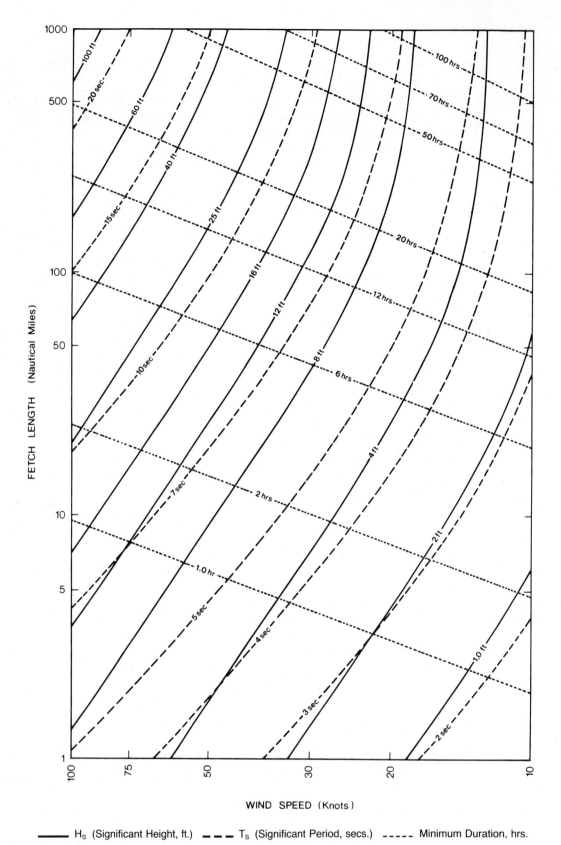

Figure 1.6.3 Deep-water wave forecasting curve. (Adapted from Shore Protection Manual, US Army Corps of Engineers, Coastal Engineering Research Center, Washington, D.C. 1984.)

If the duration of the strong wind is limited to less than the minimum duration, then the wave height will be proportional to the square root of the duration. A sudden squall will not be able to kick up much of a sea.

The majority of waves are generated by cyclonic storms, which rotate counterclockwise in the northern hemisphere and clockwise in the southern hemisphere. The storm itself moves rather slowly as compared to the waves themselves. Thus the waves travel out ahead of the generating area. Waves within the generating area are termed *seas*, those which move out ahead are termed *swells*. Swells can reach for hundreds and even thousands of miles.

The Antarctic continent is completely surrounded by open water. It is an area of intense cyclonic activity. Storms travel all the way around the continent, sending out swells that reach to the equator and beyond. The west coast of Africa, from southwest Africa to Nigeria and the Ivory Coast, and the west coast of Tasmania are notorious for their long swells that arrive from Antarctica. The long, high-energy swells that arrive at the coast of southern California in May are generated by tropical hurricanes in the South Pacific.

The swells eventually decay. Energy is lost due to internal friction and friction with the air. The shorter-period (high-frequency) waves are filtered out first, so that it is the longest of the long-period swells which reach furthest. Therefore swells tend to be more regular, each similar to the other, than waves. Whereas waves typically have significant periods of 5 to 15 seconds, swells may develop periods as great as 30 seconds or more.

The energy in swells is proportional to their length. Thus even relatively low swells can cause severe forces on moored vessels and structures.

Deep-sea waves tend to travel in groups, with a series of higher waves followed by a series of lower waves. The velocity of the group of waves is about half the velocity of the individual waves. This of course gives the opportunity to wait until a period of successive low waves arrives before carrying out certain critical construction operations of short duration, for example the setting of a load on a platform deck or the stabbing of a pile.

The average wave is about $0.63\ H_s$. Only 10 percent of the waves are higher than H_s. One wave in 1000 is 1.86 times higher than H_s; this is often considered to be the "maximum" wave, but more recent studies show that the value may be closer to 2.

Wave height H is the vertical distance from trough to crest, and T is the period, the elapsed time between the passage of two crests past a point. Wave length L is the horizontal distance between two crests. Velocity V, often termed celerity C, is the speed of propagation of the wave. Rough rule-of-thumb relationships exist between several of these factors.

In the metric system:

$$L = \frac{3}{2} T^2 = \frac{2}{3} V^2$$

$$T = 25\sqrt{L} = \frac{2}{3} V$$

$$V = 1.1\sqrt{L} = \frac{3}{2} T = \frac{L}{T}$$

where L is in meters, T in seconds, and V in m/s.

In the English system:

$$L = 5T^2 = \frac{1}{5} V^2$$

$$T = \frac{1}{2}\sqrt{L} = \frac{V}{5}$$

$$V = 2\sqrt{L} = 5T = \frac{L}{T}$$

where L is in feet, T in seconds, and V in m/s.

As noted in Section 1.5, currents have a significant effect upon wave length, steepness, and height. A following current increases the length and decreases the height, whereas an opposing current decreases the wave length and increases the height, thus significantly increasing the wave steepness. (See Figure 1.5.4.) Note that the influence of an opposing current is much more pronounced than that of a following current. Note also that the wave period remains constant.

When seas or swells meet a strong current at an angle, very confused seas result, with the wave crests becoming shorter, steeper, and sharper and thus hazardous for offshore operations.

Seas are often a combination of local wind waves from one direction and swells from another. Waves from a storm at the site may be superimposed upon the swells running out ahead of a second storm that is still hundreds of miles away. The result will be confused seas with occasional pyramidal waves and troughs. Major construction equipment and large

24 Physical Environmental Aspects of Offshore Construction

floating structures will respond more to the swells than the waves during such a period.

Waves are not "long-crested"; rather the length of the crest is limited. The crest length of wind waves averages 1.5 to 2.0 times the wave length. The crest length of swells averages 3 to 4 times the wave length. These crests are not all oriented parallel to one another but have a directional spread. Wind waves have more spread than swells. From a practical operational point of view, the majority of swells tend to be oriented within ±15°, whereas wind waves may have a ±25° spread. See Figure 1.6.4.

When waves in deep water reach a steepness greater than 1 in 13, they break. When these breaking waves impact against the side of a vessel or structure, they exert a very high local force, which in extreme cases may reach 30 tons/m² (0.3 MPa), or 40 psi. The areas subjected to such intense forces are limited, and the impact itself is of very short duration; however, these wave impact forces are similar to the slamming forces on the bow of a ship and thus may control the local design of floating construction equipment.

Data on wave climates for the various oceans are published by a number of governmental organizations. The U.S. National Oceanic and Atmospheric Administration (NOAA) publishes very complete sets of weather condition tables entitled "Summaries of Synoptic Meteorological Observations" (SSMOs), based on data compiled from ship observations and ocean data buoys. The published tables as of 1984 tend to underestimate the wave heights and period in the Pacific; recent data for the Pacific indicates that there is significant wave energy in longer periods (e.g., 20 to 22 seconds) during severe storms. The swells from such storms may affect operations even at a distance of many hundreds of miles.

The "persistence" of wave environmental conditions is of great importance to construction operations. Persistence is an indication of the number of successive days of low sea states one may expect to experience at a given site and season. To the offshore constructor, persistence is quite a different thing than a percentage chart of sea states exceeding various heights.

For example, assume that the limiting sea state for a particular piece of construction equipment is 2½ m. The percentage exceedance chart may show that seas greater than 2½ meters occur 20 percent of the month in question. This could consist of two storms of 3 days' duration each, interspersed between two 12-day periods of calm. Such a wave climate would allow efficient construction operations. Alternatively, this 20 percent exceedance could consist of 10 hours of high waves every other day, as typically occurs in the Bass Straits between Australia and Tasmania. Such a wave climate is essentially unworkable with the equipment postulated.

Typical persistence charts are shown in Figures 1.6.5 and 1.6.6. Further discussion on persistence is found in Section 1.7.

As swells and waves approach the land or shoal

Figure 1.6.4 Winter storm in the Pacific. In this case, swells ($H_s = 7$ m, $T_z = 18$–20 sec) were from 250 dg; while local wind waves ($H_s = 4$ m, $T_z = 12$ sec) were from 200 dg.

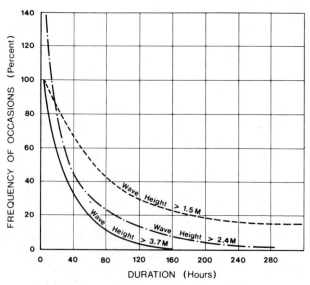

Figure 1.6.5 Persistence of unfavorable seas.

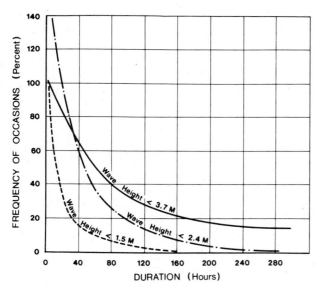
Figure 1.6.6 Persistence of favorable seas.

areas, the bottom friction causes them to slow down; thus the wave front will refract around toward normal with the shore. This is why waves always break on shore even though the winds may be blowing parallel to it.

Multiple refractions can create confused seas and make it difficult to orient a construction barge or vessel for optimum operational efficiency. At some locations, two refraction patterns will superimpose, increasing the wave height and steepness.

Submerged natural shoals and artificial berms increase the wave height and focus the wave energy toward the center. Waves running around a small island, natural or artificial, not only refract so as to converge their energy on the central portion but run around the island to meet in the rear in a series of pyramidal peaks and troughs. Such amplification of waves and the resultant confusion of the sea surface may make normal construction operations almost impossible.

Running along or around the vertical face of a caisson, waves will build up in an effect known as "mach stem" and spill over onto the island, but without radial impact.

Waves approaching a shore having a deep inlet or trench through the surf zone will refract away from the inlet leaving it relatively calm, while increasing the wave energy breaking in the shallow water on either side.

As waves and swells move from deep water into shallow water, their characteristics change dramatically. Only their period remains essentially the same. The wave length shortens and the height increases. This of course leads to steepening of the wave, until it eventually breaks.

While the above may be thought of as a typical description of coastal phenomena, the same can occur in the open ocean whenever shoaling is encountered. In the Bering Sea, for example, major storms generate high waves that spread into Norton Sound, with its relatively shallow depth over large areas. This results in an extremely difficult construction environment, with high, steep waves breaking and reforming over the shoal areas.

A deep-water wave is one for which the seafloor (the bottom), has essentially no effect. Since waves generate significant orbital motion of the water particles to a depth of about half a wave length, this has led to the adoption of a 200-m water depth as being "deep water." Most offshore construction to date has taken place in lesser depths and hence has been subject to the shallow-water effects.

Waves and the associated wave-generated currents may cause major sediment transport, eroding sand embankments under construction and causing rapid scour around newly placed structures, especially if the wave effects are superimposed upon currents.

Internal waves are waves which propagate beneath the surface, typically acting along the boundary line (thermocline) between the warm upper ocean waters and the colder, more saline waters which have greater density. This is usually located 100 to 200 m below the surface. Internal waves have been measured at a water depth of 1000 m with an internal height of 60 m. These can generate "density currents" as high as 2.6 knots (1.3 m/s) and thus are important for deep-sea operations.

1.7 Winds and Storms

The predominate patterns of ocean winds is their circulation around the permanent high-pressure areas which cover the ocean, clockwise in the northern hemisphere, counterclockwise in the southern hemisphere. In the region south of the major continents, that is, the Indian Ocean and the South Pacific, the pressures are lower and the strong westerlies prevail.

In the tropical and subtropical zones, the extreme heat and the interface between atmosphere and ocean create deep lows, which spawn the violent storms known as tropical cyclones in the Indian

Ocean, Arabian Sea, and offshore Australia; as hurricanes in the Atlantic and South Pacific; and as typhoons in the western Pacific. The occurrence of such storms in the subtropical and temperate zones is seasonal, late summer to early fall, and is fortunately somewhat infrequent. However, while they are easily spotted by satellite as well as observations of opportunity, prediction as to their route is still highly inaccurate. Thus there may be several alerts per year at a site, necessitating adoption of storm procedures. Most of these will turn out to be false alarms, but significant delays will nevertheless occur.

Typical of the violent cyclonic storms are the typhoons of the western Pacific. See Figure 1.7.1. These occur from May to December and have their source in the Pacific Ocean in the vicinity of the Caroline Islands. These storms travel westward, eventually dissipating over the Philippines, the China coast, or Japan. The diameters of these typhoons range from a maximum of 1000 miles (40 mph wind circle) to 100 miles or less. In most cases, the most severe wind, tide, and wave action occurs in a band about 50–100 miles wide. The storm as a whole can move in any direction, even circling back upon its previous track. It normally moves at a speed of less than 20 mph.

In the Indian Ocean, east of the Indian subcontinent, tropical cyclones can occur during two seasons of the year, one the southern hemisphere cyclone season, the other the northern hemisphere cyclone season.

There are many seas in which tropical cyclones were not thought to be present, primarily because there was so little human activity in the area. An example is the Arafura Sea, between Australia and the island of New Guinea. Not only do cyclones exist in this offshore region, but in 1982 one struck Darwin with devastating fury.

In typical offshore operations off the west coast of Australia, there will be a dozen or more cyclones spotted each season. See Figure 1.7.2. Several, perhaps four, will be close enough to cause a cyclone alert, with all operations shut down and equipment deployed in accordance with the prearranged procedures, either towed to the safety of a harbor or, in the case of larger vessels, towed clear to ride out the storm at sea. However, usually only one, sometimes none, of these cyclones will actually hit the construction site. Nothing more serious than long swells may arrive. Thus it often is the interruption to operations rather than actual storm damage that causes additional costs and delays.

Cyclonic storms can also spin off the interface between Arctic (and Antarctic) cold air masses with the warmer air of the temperate zones; these produce the typical winter storms of the North Atlantic, Gulf of Mexico, and North Pacific, and those sur-

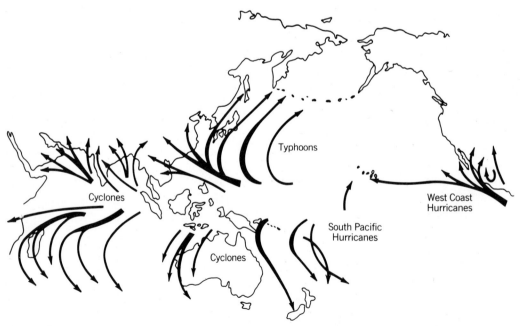

Figure 1.7.1 Major cyclonic storm tracks in Pacific and Indian oceans.

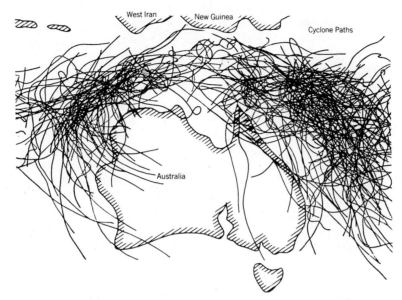

Figure 1.7.2 Typical cyclone paths around the Australian continent.

rounding the Antarctic continent. These storms tend to act over wide areas (several hundred miles) and persist for two to three days.

As noted in Section 1.6, patterns do exist for specific areas; for example, on the Pacific coast of the United States, six or seven days of storms will be followed by six or seven days of good weather. In the North Sea, in February, if there is a "good" day, then the probability of getting a "good" day the next day is 65 percent. The probability for the next two days is 40 percent, for the next three days 25 percent. Conversely, if today is a "bad" day, the probability of a "good" day tomorrow is only 15 percent.

Recent activity in the Arctic Ocean has shown that during the brief summer, the interaction between the cold air mass over the ice pack and the warmer air of the adjacent land masses can similarly spin off a series of intense but very local cyclonic storms, creating strong winds and short steep waves in the open water.

Storm procedures for the constructor involve a sequence of steps:—first to stop operations and secure them against the weather, second to move off site, as applicable, to a more protected location in which to moor. Alternatively, the floating construction equipment may be moored on a survival mooring or even towed clear to ride out the storms at sea. These procedures will all be dealt with in later sections of this book.

It will be noted that swells normally arrive prior to the winds. A seasoned mariner will note the sky: cirrus clouds indicate a low-pressure area; they usually radiate from the storm center. The initial event of an approaching storm is often a cessation of the normal high-pressure winds, "the calm before the storm," followed by winds shifting to the south or southeast and picking up as a fresh, intermittent breeze. This is in the northern hemisphere; in the southern hemisphere, as the cyclonic storm approaches, the winds will shift to the north or northeast.

It is often more dangerous to try to enter a harbor during the onset of a storm than to ride it out at sea. The harbor poses the additional problems of tidal currents and their effect on the waves, of nearby shores, and of winds often blowing across the channel.

Nautical terminology for wind directions is somewhat contrary. When a seaman speaks of a north wind, it means that the wind is coming from the north. When he speaks of an onshore wind, it is blowing from sea toward land. An offshore wind blows from land toward sea.

The term *lee* is the sheltered side of a vessel away from that from which the wind is blowing. However a *lee shore* is not a sheltered shore but the shore toward which the wind is blowing and hence a dangerous shore. The *windward* side of a vessel is the side against which the wind is blowing.

Wind speed increases as one goes up from sea level, for example, to the deck of a platform. For instance, the wind at a height of 20 m may be 10 percent greater than at 10 m, the usual reference height. Near sea level the friction of the waves decreases the speed significantly. Winds are of course not steady but blow in gusts: The 3-second gust, for example, may have one-third to one-half greater speed than the same storm averaged over one hour.

Wind storms may be of several types. There are the anticyclonic winds sweeping clockwise around the high-pressure areas over the major oceans of the northern hemisphere. Then there are the low-pressure cyclonic storms circulating counterclockwise around deep atmospheric depressions—the winter storms and the tropical cyclones.

In the tropics and subtropics, intense but very localized low-pressure zones result in sudden rain squalls which occasionally grow into waterspouts. Although high winds of short duration are associated with these phenomena, serious tornado and waterspout activity appears to be extremely rare. However, the squall winds, coming rather suddenly, can damage booms or interfere with offshore operations. Since a weather period of several days' duration may have frequent squall activity, this may require extra precautions during that season.

In many areas of the world adjacent to land masses, high-pressure cold air masses may build up over the land, then suddenly swoop down over the sea. They have many local names, such as "williwaw"; they are associated with clear visibility, usually a cloudless sky, and hence give little advance notice prior to the sighting of whitecaps advancing across the sea. Wind velocities may reach 100 km/hr (60–70 mph) or more. Because of their short duration, usually only a few hours, the seas are not fully developed; hence waves will be short and steep. They can catch construction operations unawares and do significant damage. Other offshore winds of longer duration are the *Schmall* of the Arabian and Persian Gulf and the dense air masses flowing down the slopes of Antartica and out to sea.

Offshore winds can also blow off the desert; the infamous *Santa Ana* of Southern California forms as an intense cyclonic storm over the hot desert interiors, and its intense winds sweep out over the adjacent ocean. Because of lack of moisture, these storms are usually cloudless but full of sand and dust. Similar storms reportedly occur off the west coast of North Africa.

Milder but prevalent types of winds are the onshore–offshore winds typical of all coasts. The land heats up during the day, the air rises, and the wind blows in from the sea during the afternoon. To a lesser extent, the process occurs in the opposite direction in early morning. The onshore afternoon winds can occasionally reach 30 knots or more. These onshore winds may occur almost every afternoon on the U.S. Pacific coast from June through August. Thus they pose a serious impediment to offshore operations.

Winds from two or more sources may be superimposed. The two most common are the high-pressure cyclonic winds plus the onshore afternoon winds. In extreme cases, these combined winds have reached 60 knots.

Storm forecasting services are available for all principal offshore operating areas. New areas, however, may suffer from lack of observation stations, especially in the southern hemisphere. Forecasting services also are not able to forecast local storms very well; all they can do is to warn when barometric pressures and air mass temperatures are right for such local storms to develop.

In the planning of offshore operations, every effort should be made to schedule critical operations for periods with low probability of storms or when those storms which may occur are of minimum intensity. Often there is a strong temptation to start early in the season or to work late just to finish up the project; often these are the times when storms occur and do damage that delays the project far more than a more prudent suspension of operations would entail.

It is customary in planning construction operations to (a) select the period during which the operations will be performed; (b) determine the maximum storms on a return period of at least 5 times the duration of work at the site (e.g., a 5-year return storm for that time of year); and (c) develop procedures and plans and select equipment capable of riding out such a storm without significant damage.

1.8 Tides and Storm Surges

Tides result from the gravitational pull of the moon and the sun. Due to the relative masses and distances, the sun exerts only half the influence on the tides as the moon.

During new and full moons, when the sun, the earth, and the moon are in approximate line, the highest tide ranges occur; these are called *spring tides*. When the sun and moon are approximately 90° apart, that is, at the first and third quarter of the moon, the ranges are lower; these are called *neap tides*.

The depth of the sea as shown on charts usually refers to MLLW (mean lower low water), which is the average of the low-water elevations during spring tides. Some authorities use LAT (lowest astronomical tide) as the reference datum.

Since the lunar month is one day shorter than the solar month, the times of tidal events are constantly changing. Normally, the tidal cycle moves back (later by clock time) by about 50 minutes each day. For example, the time of high tide will be 50 minutes later tomorrow than it is today.

Typically there are two tidal cycles each day. One of these tidal cycles will have significantly greater range (higher high tide, lower low tide) than the other. Some areas in the South Pacific (e.g., the Phillipines and West Irian) have a prolonged high tide once each day, followed by a low tide 12 hours later. These tidal cycles appear to follow the sun; hence the peaks occur about the same time each day.

The times of tide and their height for the reference station are tabulated and published one or more years in advance. However, the time of tidal peaks at a particular location depends not only on astronomical conditions but also on the local bathymetry. Tidal tables are published for most coastal areas of the world, showing the time differences and height differences from the reference station for each locality.

The tidal range in the deep ocean is relatively minor, usually less than 1 m. However, as one approaches the continental coasts, tidal ranges may increase radically, even though the site is still many kilometers offshore. This is especially noticeable off the west coast of North Africa and in the Bay of Biscay and on the northwest shelf of Australia.

Tidal ranges vary from approximately 0.5 in some locations to as much as 10 m in others. Here are some typical values (approximate):

Location	Range (m)
Boston, Massachusetts	2
Jacksonville, Florida	1
San Francisco, California	2.5
Puget Sound, Washington	5
Balboa, Panama (Pacific side)	4
Cristobal, Panama (Atlantic side)	0.2
Iceland	4
Mariana Islands, Pacific	0.6
Cook Inlet, Alaska	10
Prudhoe Bay, Alaska	0.5

It can thus be seen that the tidal ranges vary significantly depending on location.

The tidal cycles produce currents which are discussed in more detail in Section 1.5. The flood stage is that with significant flow on a rising tide; the ebb is the flow associated with a falling tide. Periods of little or no flow are called "slack water." The times of slack water do not coincide exactly with the peaks of high and low water, since the water continues to flow for some period after a peak has been reached. Tidal currents are often stratified, with the current at the surface in a different direction from that at some depth.

Storm surges are changes in the level of the sea which are superimposed upon the tide. They are caused primarily by the effect of the wind blowing for a long period in the same direction. A secondary cause is that of different barometric pressure. When low pressure exists at a site, the water level will rise to balance. Storm surges can be 1–3 m in height. They can also be negative, with a combination of offshore winds and high pressure depressing the surface of the sea.

1.9 Rain, Snow, Fog, Whiteout, and Spray; Atmospheric Icing, Lightning

Rain, snow, and fog are primarily a hazard to offshore operations because of their limitations on visibility. Fortunately, with the advent of radar and other sophisticated instrumentation, these no longer constitute as serious a constraint as in the past.

Rain can occur during general storms and also during the tropical rain squall, which moves through quickly and intensely. Appreciable amounts of water can enter through open hatches or other openings in the deck; if unattended over a period of time, they can adversely affect stability due to the free-surface effect. Proper drainage must be provided, adequate to remove the rainfall at its maximum rate.

Fog is of two types. The summer-type fog occurs when warm air passes over a colder ocean. Moisture condenses to form low stratus clouds. Usually there is clear visibility at the water surface. The second type is the winter fog, with cold air over warmer water. Here the fog is formed at the surface. Yet 20 or 30 m in the air it may be bright sunshine.

Rain, fog, and snow will affect helicopter operation, which requires visual observation for landing. Where fogs are prevalent, as for example the winter type off the east coast of Canada, then the helideck should be elevated as high as possible.

In the Arctic and sub-Arctic, there is almost always dense low fog at the ice edge, where the cold air above the ice condenses the warmer moisture evaporating from the open sea.

Snow presents the additional problem of removal; else it will accumulate and freeze. When the air temperature is at or above freezing, salt water can be used to wash the snow away, melting it as well as removing it by jet action. When the air temperature is more than a few degrees below freezing, mechanical means of snow removal may be necessary.

"Whiteout" is an atmospheric condition which occurs in the Arctic, in which the entire environment becomes white: sea, ice surface, land surface, and atmosphere alike. All perceptions of distance and perspective are lost and vertigo may occur. Obviously this is a serious problem for helicopter operation.

Spray is created by a combination of waves and wind. Waves breaking against a vessel or a structure hurl the spray into the air where it is accelerated by the wind. Many tons of water per hour can thus be dumped onto a structure or vessel.

Drains are often inadequate, having been sized for rainfall, not spray. In the Arctic and sub-Arctic, drains may become plugged by freezing.

On the Tarsiut Island Caisson Retained Island, an exploratory drilling structure in the Beaufort Sea north of Canada, the wave energy was concentrated against the caissons by the submerged berm. Waves reflecting from the vertical wall formed a clapotis peak (standing wave) which the wind picked up as spray. Also, waves running around the island were concentrated at discontinuities into a vertical plume, which the wind then hurled onto the island. Several hundred tons of water could thus be dumped onto the island during a 6-hour storm. The resulting runoff eroded the surface. Some equipment operation was impaired. The spray plumes reached 30 m or more into the air, endangering helicopters engaged in evacuation.

Spray in more serious quantities than rain can flood hatches and even minor deck openings of barges—for example, penetrations which were not properly seal-welded—leading to free-surface effects and loss of stability. Spray can also be so severe as to prevent men from working on deck.

It has been recently recognized that spray has been given inadequate attention in offshore operations, especially in the Arctic. Spray during those transition periods when the sea is still largely free from ice but the air temperature is well below freezing can produce dangerous accumulations of ice on vessels, booms, masts, antennas, and so forth.

Atmospheric icing ("black ice") is a notorious condition occurring in sub-Arctic regions, where the air is moist but the temperature low. The northern North Sea and the southern Bering Sea are especially prone to this phenomenon. Icing accumulates very rapidly on the exposed portions of the vessel. As it builds up, it adds topside weight and increases wind drag. Small vessels and boats are especially vulnerable. See Figure 1.9.1.

Atmospheric icing can be minimized by the application of special low-friction coatings. In the design of booms, masts, and spars of construction equipment which may see service in the sub-Arctic areas, use of a minimum number of widely spaced, round structural members is preferable to a larger number of smaller, closely spaced latticed shapes. Some specialized equipment is even provided with heat tracers to keep the surfaces warm and thus prevent icing.

Note that icing and frozen spray may occur in the same regions and under the same circumstances. Icing, if it occurs, should be removed promptly, by mechanical means and/or salt-water jetting. This latter can obviously only be used if it is free to drain.

Lightning is associated with storms, especially the tropical rain squall. The typical construction vessel is well equipped to discharge from its masts and ground through its hull. However, men should not be exposed on high decks such as helidecks or platform decks under construction during a lightning storm. Concrete offshore platforms should be grounded by providing an electrical connection from the deck to a conductor.

1.10 Sea Ice and Icebergs

Sea ice is found year round in the Arctic and during the winter and spring in the sub-Arctic oceans such as the east coast of Canada, Bering Sea, and Greenland Sea. Maximum southerly extent of sea ice generally occurs in February; the minimum extent, the retreat to the polar pack, in late August.

Sea ice is a unique substance, being a material that is near its melting point. It is saline, thus freezing at about $-2°C$. It occurs in many forms including the following. (Reportedly, the Inuit language contains more than 20 different words to describe the different types of sea ice.)

1.10 Sea Ice and Icebergs

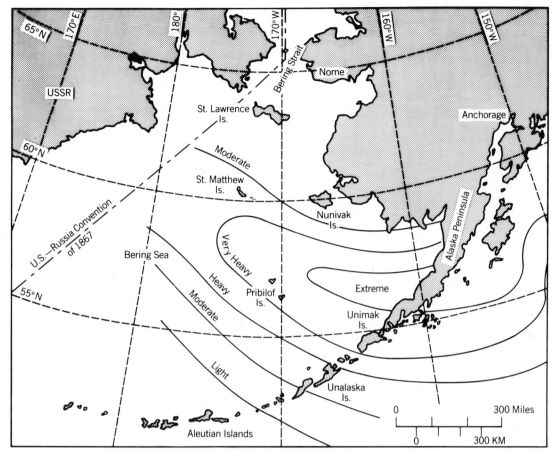

Figure 1.9.1 Atmospheric icing probabilities in the Bering Sea. Courtesy of Han-Padron Associates.

1. *Frazil ice* is ice that forms within supercooled water, especially when that water comes in contact with foreign bodies that enable amorphous ice to congeal. Thus it can clog intakes. It also increases the frictional effect on the hull of a vessel. It may act to reduce the impact force from ice floes and icebergs.
2. *Sheet ice* is the horizontal layer of sea ice that forms on relatively calm water, freezing from the top down.
3. *Leads* are formed when the thermal contraction of the ice causes ruptures in the sheet ice. Leads may also be formed by currents and wind.
4. *Rafting* occurs when one sheet of ice is driven up over the top of another.
5. *Ridges* (pressure ridges) are formed by a combination of refreezing of water in an open lead, then closure of the lead with rafting and crushing. See Figure 1.10.1.
6. A *compressive ridge* is formed when one sheet is forced normal to the face of the other, that is, normal to the lead. Rafting and crushing occurs, forming randomly oriented blocks of ice. The typical shape of a compressive ridge is shown in Figure 1.10.2. Such a ridge may be from 100 to 500 m in length, but will average about 200 m long. Ridges often follow each other in closely-spaced increments of 100 to 200 m. See Figure 1.10.3.
7. A *shear ridge* is formed when two sheets interact laterally along the lead, crushing up a more rectangular pile of lesser size.
8. *First year* or *annual ridge* denotes a ridge that has been recently formed and is less than one year old.
9. A *multiyear ridge* is one that has survived one summer and has had time for the meltwater from the sail to drain down through the bulk of the ridge, refreezing the broken blocks into

32 *Physical Environmental Aspects of Offshore Construction*

Figure 1.10.1 First-year pressure ridge I.

a more or less solid block of ice. See Figure 1.10.4. Multiyear ridges can be distinguished from first-year ridges by the fact that their blocks are rounded and refrozen together instead of just randomly heaped individual blocks.

10. *Sail* is the part of the ridge that extends above water. Maximum sail height is 10 m or less.
11. *Keel* is the deepest portion of the ridge. First-year keels may be hard ice and extend to almost 50 m depth: multiyear keels usually have a layer of weak ice (due to contact with the water) and extend to about 30 m maximum depth.
12. *Rubble piles* are found when sheet ice pushes up against grounded ridges. These rubble piles may reach 15 m in height and be quite extensive. They consist largely of broken blocks of ice.

When storm surges occur during the subsequent summer these large rubble piles, by now partially consolidated, float off, where they become "floe bergs," presenting many of the same hazards to navigation as small icebergs. Floe bergs are especially found in the northern Bering Sea. See Figure 1.10.5.

Rubble piles in the high Arctic may be formed by sheets being driven against ridges. In a *compressive field,* that is, a region in which the ice pack is exerting strong forces on the sheet and its ridges, a large rubble pile or series of rubble piles may be formed. These are known as a *hummock* or *hummock field*.

The polar pack itself consists primarily of multiyear ice. As it rotates slowly clockwise around the pole* (as seen from above) it impinges against the shallow-water *annual ice* which forms each winter. This annual ice is relatively stationary and is called *fast ice* or *shorefast ice*. At the boundary between the fast ice and the polar pack, a shear zone is set up which is highly dynamic, with much ridging and rubbing. The *shear zone* or *stahmuki zone* is usually located in 20–50 m of water depth, which is un-

---

* Actually, the center of rotation of the Arctic polar gyre is about 80°N, 150°W.

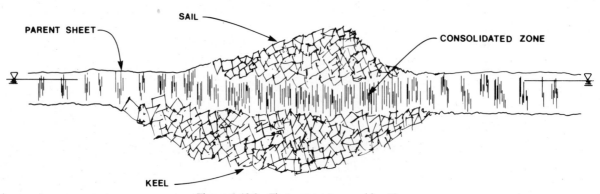

Figure 1.10.2 First-year pressure ridge II.

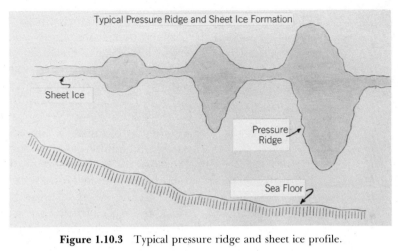

Figure 1.10.3 Typical pressure ridge and sheet ice profile.

Figure 1.10.4 Grounded multiyear ridge in shallow waters of Beaufort Sea, Alaska.

fortunately a region of high prospective activity related to the current development of petroleum resources. See Figure 1.10.6.

Large blocks of polar pack ice with embedded ridges may break off from the pack, becoming floes. *Multiyear floes* may be very large, several thousand meters in diameter, but usually are only 100 to 300 m in size. Their masses are very large. Since such floes may move with the general circulatory pattern, they may achieve velocities of 0.5–1 m/s. See Figure 1.10.7.

Finally, there are the *ice islands*, which are tabular bergs spawned off the Ward Hunt glacier on Ellesmere Island. See Figure 1.10.8. Caught up in the polar pack, they may circulate for many years, eventually grounding and breaking up into fragments. While the probability of encountering an ice island is extremely remote, there is a finite probability of encountering a fragment perhaps 100 m in diameter and 40 m thick.

Figure 1.10.5 Floe berg grounded in shallow water of Norton Sound, Bering Sea, Alaska.

Figure 1.10.6 Compressive hummock field closes leads opened by icebreaker. Chukchi Sea, off coast of Alaska.

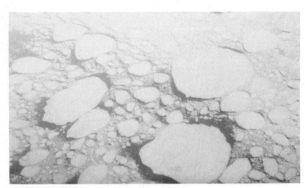

Figure 1.10.7 Multiyear ice floes as seen from 10,000-m height off northern Greenland. The larger floes are 10–20 km in diameter and 8–20 m thick.

The Arctic year can be divided into four seasons: the winter when ice covers the entire sea, the summer open-water season, the breakup, and the freezeup. Within 20–30 km of shore, depending on the water depth, the winter ice is fast ice and hence can be used for access and transport. During the breakup, which may occur in June and July, the sea is filled with remnant ice floes and bits of ice. The percentage of ice coverage is then given either in tenths or octas, an octa being 1/8. Maximum local pressure is usually about 4 MPa (600 psi), which is a value which is often used for design of barge and work boat hulls.

The summer open water varies with location and distance offshore. It may extend 300 km at times, but elsewhere may be only 20 km. Open water may last from as little as 11 days to 90 days or so. Some years there is no open water at critical locations; this occurred in 1975 at Point Barrow.

Figure 1.10.8 Huge tabular bergs break from floating ice sheets such as those of Ellesmere Island, northern Canada, to form "ice islands."

The fall freezeup consists of thin surface ice forming, which initially is relatively easily broken up by icebreakers. Thus Canmar Drilling Company of Canada has been able to extend its working season for floating drilling and construction equipment in the Canadian Beaufort Sea to November.

During the winter season, when the fast ice has reached a thickness of 2 m, ice roads can be built, enabling hauling of sand, gravel, and equipment. Airstrips can also be built on the fast ice. Dikes are built of snow and ice, and water is pumped up from holes in the ice and sprayed or flooded so as to thicken the ice. In the fast ice areas between the Arctic islands, insulating blankets of polyurethane have been placed on the early ice sheet, then water pumped on top to progressively freeze.

Early in the summer, when the rivers thaw and the snow in the mountains melts, the fresh water floods down over the shore fast ice. Flooded areas may be many miles in extent. Eventually the water thaws a hole in the ice sheet and large quantities of water pour through, eroding a cone in the seafloor to a depth of 7–10 m. This phenomenon is call *strudl-scour*.

During the open-water season, floating equipment may be employed. Some ice fragments will still be encountered and present the hazard of holing of the hull of vessels. These ice fragments may have local "hard spots" for which their crushing strength exceeds the average values by a factor of 2 or more. Measurements on icebreakers indicate that the usual maximum local pressure is 4MPa (600 psi), although recently, extreme values of 6MPa (800 psi) have been recorded. During this summer open-water season, one or more pack ice invasions may occur with multiyear ice floes moving at speeds of 0.5 to 1 m/s. At such times, all floating equipment must be towed to shallow water for protection.

The Arctic pack shields the open water from storms. The fetch in the north–south direction is very limited. However, the fetch in the east–west direction can be several hundred kilometers. Thus the short violent cyclonic storms that spin off of the ice pack can kick up considerable seas, with an H_s of 2–3 ms. Such storms are accompanied by storm surges of $+2$ m to -1 m.

In sub-Arctic areas the ice duration is much shorter. The ice itself is less strong, due to the warmer sea temperatures. However, some areas may still experience severe sea ice problems. One such is Norton Sound, where during the winter, north winds blow away huge sheets of recently frozen ice from the area south of the Seward Penin-

1.10 Sea Ice and Icebergs

Figure 1.10.9 Beautiful but dangerous icebergs move during the open-water season under the influence of wind and current.

sula. New ice forms and is in turn driven southward, to raft on the previous sheets.

The eastern sub-Arctic regions of Davis Strait and the Greenland Sea are characterized by annual sea ice but, more dramatically, by the thousands of icebergs which are calved off as the glaciers discharge into the sea. These are blocky bergs, not tabular bergs. As they melt, they assume the dramatic shapes often seen in pictures, with pinnacles and saddles. Such icebergs may range up to 10 million tons: however, the normal maxima are about 1 million tons. About 70–80 percent of the berg's mass lies below water and may extend out from the visible above-water portion, a phenomenon which led to the loss of the *Titanic*. See Figure 1.10.9. Bergs can also roll over whenever their centers of gravity and buoyancy, related to the waterplane area, result in a loss of stability.

Bergs are driven by a combination of wind, current, and wave drift force, and to a lesser extent by the Coriolis force. In their southernmost areas, especially, their paths are very erratic. Bergs may travel 20–40 km per day. They may safely pass a structure, only to return the next day. See Figure 1.10.10.

Icebergs have been successfully lassoed and/or harpooned, using two tugs and Kevlar or polypropylene lines to tow them clear of an operating site. Considerable mathematical effort has been expended in developing computer programs to try to predict motions and to determine the optimum direction and extent of towing force to apply. The towing operation is often complicated by the fog which frequently envelops the berg and by uncertainties as to its underwater profile.

Many berg fragments also clutter the water, especially in fall. *Bergy bits* are iceberg fragments from

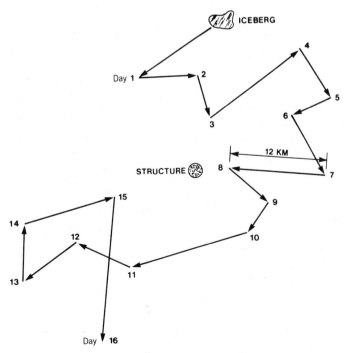

Figure 1.10.10 Typical trajectory of iceberg showing path on successive days.

120 to 540 tons in size. They generally float with less than 5 m of ice above the water, extending over 100–300 m² in area. They are small enough to be accelerated by storm waves in the open sea, with the smaller bergy bits reportedly reaching instantaneous velocities up to 4.5 m/s.

Growlers are smaller pieces of ice than bergy bits. They are often transparent and usually extend only 1 m above the surface. They range in mass up to 120 tons. These smaller ice features are also driven by the waves and may follow their orbital motion just as might a similar-sized vessel.

One major problem when working in an iceberg-prone area is the ability to detect them. As noted above, visual observation is often limited by fog. Radar is unreliable: it may "see" a berg at a distance, then suddenly lose it on the scope. Smaller features, largely submerged or even awash, may be missed altogether.

In winter the icebergs and fragments which remain unmelted are caught up in the sea ice, thus moving very much more slowly, but driven by the moving pack ice behind. Data is slowly being accumulated concerning iceberg types, masses, drafts and profiles, so as to enable a probabalistic approach for design. See Figures 1.10.11–1.10.13.

Tabular bergs are icebergs which break off a floating ice sheet extending out from a glacier over the water. They are the common icebergs of the Antarctic, where they may be 100–300 m thick and 100 km in diameter. As noted earlier, smaller tabular bergs break off the Ward Hunt Glacier on Ellesmere Island in the Arctic, where they become ice islands, trapped in the polar pack.

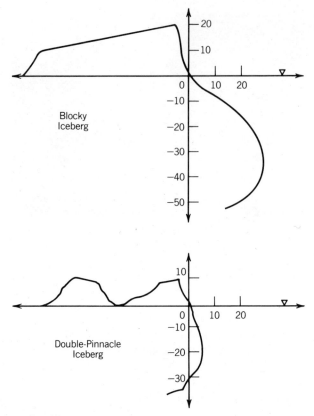

Figure 1.10.12 Typical underwater profiles of icebergs.

Iceberg ice is freshwater ice, as opposed to the saline ice from the sea. Iceberg ice reportedly contains a significant amount of entrapped air. For icebergs which have drifted near their southern limits, their mass density is about 0.9 and their near-surface ice-crushing strength is 4–7 MPa.

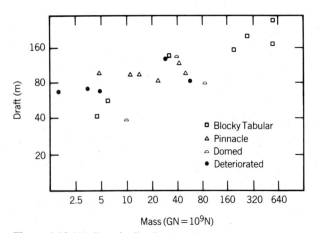

Figure 1.10.11 Sample distribution of iceberg types, masses, and drafts off Labrador. (From J. K. C. Lewis and C. P. Benedict, "Burial Parameters: An Integrated Approach to Limit Design."

Figure 1.10.13 "Blocky" berg in Davis Strait, between Canada and Greenland.

1.11 Seismicity, Seaquakes, and Tsunamis

Earthquakes, although a major feature of the design of offshore structures, are not normally a consideration for construction because of their infrequent occurrence. The most serious earthquakes occur on active tectonic plate margins such as the ring which surrounds the Pacific. These vary in intensity up to such cataclysmic events as the Alaskan earthquake of 1964. The effect of these stretches over 100 km or even more.

Earthquakes also occur within plates; these typically have a long return period, so that the area may have been popularly considered to be free from earthquakes. Their magnitude is usually limited to force 6 (very rarely force 7) on the Richter scale, but their effects spread for very long distances. Finally, local earthquakes of moderate intensity (force 4 to 5) are associated with vulcanism, both on adjacent islands and subsea.

Earthquake accelerations are modified with distance, the higher frequency components being filtered out, so that only long-period energy reaches the peripheral areas. Many of the structures built offshore, especially those in deeper water, and on weaker soils have long natural periods (2 to 4 seconds). However, the amount of energy in this frequency range is usually well below the peak energy in an active-margin-type earthquake. In evaluating the effect of earthquakes on structures, the designer will consider added mass effects and the nonlinear soil–structure interaction.

Although the above paragraphs would seem to apply primarily to the designer, they are of interest to the constructor because of three associated phenomena. The first is that of the seaquake, which is an intense pressure wave generated by the interaction of the seafloor and the water mass above. These short-period, high-intensity waves can be felt by ships, barges, and offshore structures for many hundreds of miles from the epicenter and usually are felt on the vessel as a shock similar to running aground or a collision. The overpressure generated is about equal to the hydrostatic pressure up to a depth of 100 m or so, then remains essentially constant below that depth.

It is now being recognized that previously unexplained cases of ship damage may have been due to these seaquakes. While not currently considered for marine operations, they may have to be considered for deep-water construction in seismically active areas in the future.

Earthquakes also generate very-long-period water waves, tsunamis, often mislabeled "tidal waves." The tsunami is so long and of such low height that it is rarely noticed on the open sea. When the energy is focused by the bathymetry and shoreline configuration, however, the kinetic energy in the wave is transformed into potential energy, resulting in disastrous lowering and raising of the water level. Such effects can spread completely across the ocean. The tsunami resulting from the Alaskan earthquake of 1964 did major damage in the Hawaiian Islands and California. South Pacific events have caused tsunamis in Chile and Japan.

Certain areas are known to be focal zones for tsunamis. While most of these are well inshore, others are in the areas currently considered "offshore" from a construction viewpoint.

Earthquakes have generated huge underwater landslides and turbidity currents, usually due to high-pore-pressure buildup and liquefaction. Tsunamis also generate landslides near shore by the sudden lowering of the sea level. Earthquakes also generate landslides and rockslides on land, which, in places like southeastern Alaska and western Canada, may slide into the sea, generating a huge water wave that may travel many kilometers.

Liquefaction and turbidity currents are dealt with in the following chapter.

I must go down to the seas again,
to the lonely sea and the sky,
And all I ask is a tall ship
and a star to steer her by;
And the wheel's kick and the wind's song
and the white sail's shaking,
And a grey mist on the sea's face
and a grey dawn breaking.

JOHN MASEFIELD, "SEA FEVER"

2

Geotechnical Aspects: Seafloor Soils

2.1 General

The seafloor of the ocean is highly complex due to its geological history and the action of the various elements, especially in the relatively shallow waters of the continental shelves. These shelves vary in extent depending on whether the margin is rising or slowly subsiding. Thus the east coast of the United States has a very wide continental shelf, whereas that on the Pacific coast of South America is very narrow. See Figure 2.1.1. Beyond the continental shelves are the slopes, averaging 4° of slope down to the abyssal plain. Submarine canyons, which cut through the shelf and slope may have side slopes as great as 30°. They usually terminate in a fan on the deep seafloor.

The ice ages have had a very dramatic influence on the shelf areas. When the Wisconsin ice age was at its peak about 20,000 years ago, immense quantities of water were withdrawn from the sea, lowering the sea level as much as 100 ms. This meant that the shelves were exposed to this contour: beyond the shoreline of that date, seas were shallower than at present. Rivers discharging from land cut troughs into that contour and a little deeper, which is why so many entrances to sounds and other large bodies of inland water are now approximately 100 m deep. On these coastal shelves, land erosion processes took place. The rivers were steeper and velocities higher, and hence sedimentary deposits were coarser. As the oceans have risen, the velocities have been reduced, and finer sediments—sands and silts—have been deposited on the shelves opposite large rivers.

During this ice age, glaciers extended far out into what are now ocean areas, carving deep trenches such as the Norwegian Trench, Cook Inlet, and the Straits of San Juan de Fuca. With the warming period in which the world now finds itself, the sea levels have been rising, slowly but inexorably flooding coastal areas, changing drainage patterns, creating shoreline features.

Glaciers retreated, leaving morainal deposits. Shallow freshwater lakes were eventually inundated, trapping their land sediments. Rivers dropped their sediments sooner, creating deltas through which new channels were cut. Volcanic ash fell through the shallow waters, as did windblown (aeolian) sand.

The corals were actively extending or building new reefs as old ones were flooded. The skeletons and shells of billions of marine organisms slowly sank to the seafloor, to be trapped in turn under new deposits.

Sand dunes formed, migrated shoreward, were eroded, and were covered with the rising water. Major sand dunes in the southern North Sea and along the coasts of Holland have been submerged by the sea and now move back and forth as underwater sand dunes known as "megadunes."

The shallow-water deposits have been acted upon by waves, reworking and densifying the sands and silts. In the Arctic regions, near-shore silts have been subjected to cyclic freezing and thawing and to scouring by the keels of sea ice ridges. Off Greenland and Labrador, icebergs have reached down hundreds of meters to scour both sediments and rocks. Many such deep scours in rock can be seen on detailed underwater photos and acoustic images of the Straits of Belle Isle, between Labrador and Newfoundland. Fault scars have been identified with 5-m-high scarps in numerous offshore areas and many more may exist, partially covered over by subsequent sediments.

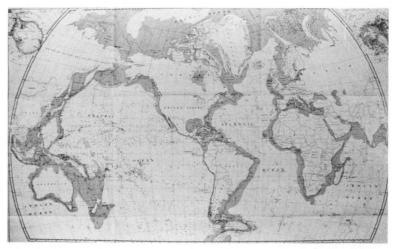
Figure 2.1.1 Rich resources abound on the continental shelves of the world.

Even today an interesting process can be seen in Cook Inlet, Alaska. The steep fjordlike walls of Turnagain Arm have rock falls which deposit large boulders on the tidal flats. In winter the high tide freezes so that cakes of ice form around the boulders. On a subsequent high tide, the ice raft floats away, carrying the boulder with it. As the raft moves into the southern portion of Cook Inlet, the increased salinity and warmer water melts the ice, dropping the boulder. This is one of the processes felt to be responsible for the many boulders found on the floor of the North Sea and Gulf of Alaska.

While the above is only a partial description of the many complex and interactive processes which have and continue to shape the sea floor, there are certain specific problem conditions which have caused many constructional problems. These will be addressed in sections below.

Throughout these subsections, reference will frequently be made to the difficulties which geotechnical engineers have in obtaining proper samples and data in critical seafloor soils. While great progress continues to be made in improving sampling methods and in applying new techniqes such as electrical resistivity measurements, many types of seafloor soils continue to give difficulty. In many of these cases, the in-place strength will be greater than indicated by conventional methods. The construction engineer needs to recognize these problems, so that he may adequately interpret the geotechnical reports and logs of borings and make appropriate decisions regarding his construction methods, equipment, and procedures. Failure to recognize these potential problem areas has led to a substantial number of cases of serious cost overruns and delays.

Most structures in the ocean extend over substantial areas. There may be significant variations in soil properties over this extent. Because of the cost and time required, it may not be possible to obtain a sufficient number of borings to show the true situation with its variations. There is a tendency to place undue emphasis on the few borings that may be available. Geophysical methods such as "sparker surveys" and a study of the site geology may help to alert the constructor to the range of soil properties that may be encountered.

2.2 Dense Sands

Sand deposits in the North Sea and off Newfoundland have been subjected to continuous pounding by the storm waves above. "Pounding" is perhaps an inaccurate term. What does happen is that the internal pore pressure in the upper layers of the sand is alternatively raised, then drained, only to be raised again. Pore pressure variations of 3.5 T/m^2 (35 KPa; 5 psi) have been measured. After millions of cycles, the sand becomes extremely dense, often with consolidation higher than can be reconstituted in the laboratory. Friction angles in excess of 40° may be found.

When sampled by conventional techniques, the

sands are automatically disturbed; hence laboratory tests will usually under-report their density and strength.

2.3 Calcareous Sands

Calcareous sands exist through much of the warmer seas of the world—for example, Australia's south and west coasts, the eastern Mediterranean, and Brazil. They are sandlike deposits formed by the minute shells of microorganisms. In laboratory tests, they typically give indications of relatively high friction angles. Yet their field behavior is far different than that of sands. High bearing values can be developed but only with substantial deformation. The friction on piling, however, may drop close to zero. The tiny shells crush and thus exert almost no effective pressure against the pile wall. Piles thus can be driven very easily, but develop little capacity in either downward bearing or uplift. In one extreme case, the measured force to extract a pile driven 60 m into these calcareous sands was little more than the weight of the pile!

Calcarenite, which is a calcareous "sandstone," and calcirubrite, which is calcareous conglomerate, while having initial rigidity and strength, are also subject to crushing of their constituent calcareous sands, resulting in sudden loss of strength and large deformations. These calcareous soils are relatively impermeable, so that the potential for liquefaction exists.

Sampling techniques inevitable crush some of the grains. Scale effects appear to be important in any testing and evaluation. Methods of constructing suitable foundations in such soils are given in Chapter 8.

2.4 Boulders on and near the Seafloor Surface; Glacial Till

Boulders on and near the seafloor are typically found in sub-Arctic areas where they may have been deposited by ice rafting. Another process, which has also been found in subtropical areas, is that of erosion, occurring when the sea was shallower than at present. The weaker deposits eroded away, dropping the boulders down and concentrating them. A third process occurs in granitic soils such as those of the east coast of Brazil and west coast of Africa, as well as in Hong Kong. As these deposits have weathered into residual soils, resistant cores have remained firm, thus becoming "boulders" formed in place.

Borings made in such residual soils usually miss most of the boulders. When they do encounter one, they frequently report it as "bedrock."

Boulders also exist in clay deposits. Some of these arose as morainal deposits from glaciers, discharging their bed load into shallow water muds, which have since been overconsolidated by subsequent advances of the glaciers. These are the boulder clays of the North Sea.

Glacial till is a term used to describe these unstratified conglomerate deposits of clay, gravel, cobbles, and boulders found in many Arctic and sub-Arctic regions. The term is very nonspecific: some glacial tills have little binder and may be largely composed of gravel, whereas others may contain large boulders. Perhaps the most difficult are the well-graded tills, with all the interstices filled with silts and clays, so that there is a very low percentage of voids. These deposits are usually heavily overconsolidated, resulting in a high unit weight and a structure superficially resembling that of concrete. Thus unit weights have reached 2200–2500 kg/m^3.

Geotechnical explorations, unless very carefully planned and carried out, often find only the finer sediments, and the samples will have been highly disturbed, resulting in the indication of much weaker material than that actually encountered. Such tills are hard to drill: the material is very abrasive and hard, yet the bond is weak. However, high-pressure jets have proven effective in penetrating these tills. If relief can be provided in the form of holes or exposed faces so that the overconsolidation pressure may be released, then more normal construction activities such as dredging and pile driving may be carried out effectively.

Individual boulders have not proven to be as difficult a construction problem as originally feared. A heavy-walled steel pile will usually displace them sideways through the soil. The same will occur when a large caisson with heavy steel or concrete skirts lands on a boulder.

Clusters of boulders are a much more difficult problem. Where suspected, efforts should be made to locate them and relocate the structure accordingly.

Large boulders underneath the base of a caisson,

for example, could exert a high concentrated local force on that base. Large seafloor boulders have been successfully removed from platform sites in the North Sea by using trawler techniques, dragging the boulders clear. Another means used has been to place shaped charges and break them into smaller pieces. These boulders do not show up well on side scan sonar or acoustic imaging. Work submarines (submersibles) taking video pictures, using special lighting, have been the most effective in determining the presence and size of seafloor boulders. Below-surface boulders can be sometimes located by sparker survey and, in shallower waters, by jet probes.

2.5 Overconsolidated Silts

Silts constitute one of the least-known types of soils, lying as they do in size range between sands and clays and exhibiting properties different from both. Silts are typical of Arctic and sub-Arctic regions, although they may also exist elsewhere. They have been encountered in the Beaufort Sea, in Cook Inlet, Alaska, and in the St. Lawrence Seaway.

One unique property is their overconsolidation. The overconsolidation of clays, for example, is usually due to their having been subjected to intense loads from overburden or ice (glaciers) which have subsequently been removed by erosional processes or melting. Silts, on the other hand, have frequently been found to be overconsolidated even when there is no prior geologic history of burial. Various hypotheses to account for this include freeze–thaw cycling in shallow water and wave action. Regardless of cause, these overconsolidated silts are extremely dense and resistant to penetration, pile driving, and dredging.

Sampling and even in situ vane shear tests always disturb these silts. In many conventional borings, the silt will be reported as "mud." These silts resist normal means of dredging and pile driving. They are very abrasive to drills. Yet they break up readily under the action of high-pressure water jets. Paradoxically, some of these silts remain in suspension for long periods, but when they do settle out, they become very dense once again. To the contractor they pose problems similar to a very soft rock. Jetting and sometimes explosives are necessary to break them into workable soils.

2.6 Subsea Permafrost and Clathrates

It is now known that relic permafrost extends out under the Arctic seas. It has been trapped there since the ice ages and is now covered by more recent sediments and seawater. The Arctic seawater, with its shallow-water temperatures ranging from $-2°C$ to $+8°C$ in summer, has been slow to thaw the ice from the top. The overlying sediments, with temperatures about $-1°C$, act as effective insulation to the permafrost.

The top zones of subsea permafrost may be sands. These may be only partially ice-bonded, due to the gradual thawing process. In other silt and clay deposits, ice lenses and frozen silt lenses may occur. Deeper down, fully bonded permafrost may be encountered. Typical depths below seafloor to the upper level of permafrost are 5–20 m. Permafrost is generally not continuous in areal extent.

Permafrost presents obvious problems in construction. Steam jets and high-pressure salt water jets have been used to aid pile driving and excavating. When permafrost thaws, consolidation settlements occur.

Clathrates are methane hydrates, which are a loose crystalline bond of ice and methane, stable under appropriate combinations of temperature and pressure. They exist under the seafloor in the deep ocean of temperate zones and at moderate depths of several hundred meters in the Arctic. When penetrated, they turn to their gaseous phase with a 500-fold expansion. They represent a potential problem for drilling of oil wells and for well casing but are generally too deep to affect construction.

2.7 Weak Arctic Silts and Clays

One of the worrisome problems of Arctic offshore construction is the presence of strata of apparently very weak silts and silty clays. Near the seafloor these are probably due to their recent deposition and the constant ploughing by sea ice keels. Much harder to explain are very low shear-strength measurements under 5–20 m of overlying stronger material.

It is now known that Arctic silts are anisotropic, having much greater strength for bearing than shear. One potential explanation for the extremely low strengths is that the released water and methane gas from the thawing of subsea permafrost zones have been trapped under the surficial layer of impervious silts, creating a high internal pore pressure, which has broken down the silt's structure. This phenomenon can also lead to serious sample disturbance.

2.8 Ice Scour and Pingos

The Arctic and much of the sub-Arctic seafloor has been scoured by sea ice ridge keels and by icebergs. Turning first to the Arctic, these scours occur as a regular yearly event in water depths from 10 to 50 m. The keels plow erratic furrows, perhaps a kilometer or more in length, to a depth of 2 m normally, up to 7 m at the extreme. While at any one time these scours are directional, the directions change, depending on currents and winds. Thus the entire seafloor in the applicable depth range appears to have been regularly reworked. New furrows are typically 2 m deep, perhaps 8 m wide, with small ridges pushed up at the top of the slope. Furrows rapidly fill in with soft unconsolidated sediments, many of them derived from the ploughing action.

There is a debate concerning the ice scour marks in deeper waters as to whether they are current or relics from a time when the seafloor was lower.

Icebergs can also scour the seafloor but at much greater depths. The individual scour marks are deeper and longer due to the greater energy in the berg. Icebergs can even produce surficial scour on exposed bedrock.

Another seafloor phenomenon occasionally encountered in the Arctic is that of "pingos," which are hillocks raised in silty clay soils by progressive frost heave. They are a common feature of the onshore coastal landscape. When encountered offshore, in shallow water, they are believed to be relics from a time when the sea level was lowered during the ice ages. When ancient pingos have thawed and collapsed, they have left small craters in the seafloor.

2.9 Methane Gas

Methane gas occurs at shallow depths in deltaic sediments which contained organic matter. This can be released by geotechnical borings or even by pile driving and has been known to cause minor explosions and injury to personnel. Methane gas also occurs at shallow depths in Arctic silts, which are not organic. It has been postulated that this gas was originally bound up in the subsea permafrost, perhaps as methane hydrates (clathrates) and has been gradually released as the permafrost warmed. The released gas has migrated upward and been trapped beneath the near-surface silts.

2.10 Muds and Clays

The end weathering process of many rocks results in the formation of clays, which constitute the large bulk of many deltas. These clays consolidate under the overburden of later deposits. These clays are highly impermeable and cohesive. They are often anisotropic, with greater horizontal permeability than vertical. Often there are thin lenses or strata of silts and sands embedded in the predominant clay body. Clays usually contain organic material.

The behavior of clays is determined by their particle shape, mineralogical composition, and water content. Thin flat plates similar to montmorillonite possess dynamic lubricating qualities. Other types of clays may be sticky, "gumbo," plastic, or firm.

Mud is a term used to denote very soft, highly plastic, recently deposited clays. Typically they show shear strengths ranging from 35 KPa (700 psf) down to 14 KPa (300 psf), although some surficial muds may have half that. These qualities of clay and mud present a number of problems to the constructor. Among them, the following deserve special attention.

2.10.1 Underwater Slopes

Clays tend initially to stand at relatively steep slopes when the excavation depth is limited. The buoyant weight of submerged clay is much less than the air weight; hence the driving force leading toward failure is much reduced compared to the same excavation in the same clay above water. With time, however, they strain (creep) and lose strength, failing in a typical curved shear plane. Thus their stable slope underwater may range from 1 horizontal to 1 vertical to as flat as 5 to 1. Increased depth of excavation increases the tendency for slope failure; thus deep cuts are often stepped down with one or more berms.

Surcharge on the dredged slope increases the driving force and may lead to sudden, large-scale failure. In practice, surcharge often arises as dredge disposal, especially when the excavation is performed by a bucket-type dredge. If the spoil pile is located so that its toe is farther back from the edge of the trench than the depth of the trench, the surcharge effect is usually negligible.

Clays are very sensitive to shock. The dropping of a large bucketload of dredged material onto the top of a trench-edge spoil pile may trigger slope failure.

Table 2.10.1 is presented as a rough guide to underwater slopes in clays. In this table, SPT refers to the standard penetration test, with N_{SPT} being the number of blows per foot for a hammer weighing 140 lb falling 30 in. When blow counts are recorded underwater, the value must be corrected for the lesser weight of hammer. The factor is approximately 1.15.

Waves will also create the same effect as a surcharge. An approximate method is to convert the wave height into an equivalent height of dredge spoil surcharge, $E = H_s/2$, where E = equivalent height and H_s = the significant wave height.

Strong continued wave action may cause cumulative strain in the clays, a process similar to fatigue, leading to a reduction in shear strength of up to 25 percent.

2.10.2 Pile Driving

Clays typically are penetrated rather readily by a pile under the dynamic blows of an impact hammer. Their short-term cohesion against the side of a pile is low. However, with a short period of rest, the soil will bind to the pile with its full cohesion, a process called "setup" or "set." Thus in driving piling in clays, when a stop is made to splice on another section, the blow count will jump up considerably when driving is recommenced. On occasion, it will not be possible to get the pile moving again.

2.10.3. Short-Term Bearing Strength

The constructor is often interested in the short-term bearing strength of clays, for example, their ability to support a jacket on mud mats. Bearing failures in cohesive materials such as clays are usually shallow shear failures. In the typical deep-clay deposit, the unit bearing value may range from 5 to 9 times the unit shear strength.

When thin layers of clay overlie much stronger soils, then the shear failure mechanism is inhibited and higher values may be sustained. This is also true when the base size is large, whereas a concentrated local load will penetrate.

Bearing strength may be increased by placing surcharge material around the loaded area, so as

TABLE 2.10.1 Correlation between SPT and Stable Dredged Slope in Cohesive Soils (Clays)

Soil Consistency	Unit	Very Soft to Soft	Soft to Medium Stiff	Medium Stiff to Stiff	Stiff to Very Stiff	Hard
N_{SPT}	bpf	0–2	2–4	4–8	8–16	16–32
Typical depth[a]	ft	0.1–10	15–25	25–40	40–80	80–100
Shear Strength From N_{SPT} or from unconsolidated, undrained laboratory tests or from field vane shear tests.	ksf	.25 (250 psf)	.5 (500 psf)	1.0 (1000 psf)	2.0 (2000 psf)	4.0 (4000 psf)
Stable slope[b]		Requires special consideration	4:1	1½:1	1:1	¾:1
Need to consider surcharge?		Yes	Yes	Possibly	Normally not required	

[a]The depth of normally consolidated clay associated with the shear strengths shown.
[b]This is the ratio of horizontal distance to vertical height.

to confine the material and force a larger area of clay to resist the shearing forces.

2.10.4. Dredging

Clays may present problems in dredging, due to their cohesive nature. In hydraulic dredging, clay balls may form. Flow will not be uniform. In bucket dredging, the clay may stick to the bucket and not discharge readily.

Once clay is in suspension, it becomes colloidal in behavior, and the discharge will be highly turbid. Where permitted, chemical flocculants can be used to cause more rapid flocculation and dropping out of suspension.

2.10.5. Sampling

The sampling of muds and clays presents special problems to the geotechnical engineer. Physical disturbance can lead to remolding and a temporary loss of strength. Thus sample disturbance is often responsible for lower strengths being indicated than exist in place.

Reference was made to the standard penetration test in Section 2.10.1. Other tests are the field vane shear test and cone penetrometer test. All such tests require correction factors for depth, strain rate, and the anisotropy of the clay.

2.10.6. Penetration

Structures designed to sit on the seafloor often have dowels or skirts which are required to penetrate into the soil. Caissons of the bridge pier type and large-diameter cylinders also must penetrate to prescribed levels. The resistance to such penetration is a combination of bearing failure under the point or edge and side shear, the latter dominating.

In some clays, an enlarged tip may create a temporary annulus around the outside and reduce the side shear resistance and hence the total resistance, even though the tip-bearing area is increased. Short-term cohesion is often lower than long-term, and any dynamic process usually results in local remolding and reduction in shear strength.

2.10.7. Consolidation of Clays; Improvement in Strength

Clays may undergo significant improvements in strength if they are drained. The reduction in water content causes consolidation, increase in shear strength, and generally beneficial gain in all properties. Such consolidation may be accomplished by surcharge (overburden), by provision of drainage (wick drains or sand drains), and by time. Due to this consolidation process, a heavy structure seated on clay soils will experience a gradual improvement in strength of the foundation.

2.10.8 Scour

Clays are normally quite resistant to scour. However, where strong bottom currents exist or where layered soils permit cyclic pore pressures to build up under overlying impervious strata, scour must be considered. Scour pockets in clays do not continuously refill as they do in sand.

2.11 Coral and Similar Biogenic Soils; Cemented Soils

As skeletons of dead marine animals who live and extract calcium carbonate from the sea, coral and other calcareous deposits initially have a complex structure which hardens through age. Storms wear off the weaker portions, exposing the older, harder portions upon which new coral can grow and may become mixed with the coral and embedded in it. New sediments may interbed with the coral. These may either be calcareous sand or silicous sand deposited by wind or shoreline transport.

Such rocks are usually highly, if irregularly, stratified. They are usually very hard, almost flint-like in small pieces, but having large voids, formed in many cases by seashells, so that the rock is very brittle.

Cap rock exists in many tropical and subtropical areas: it is a recent near-surface coral deposit, usually containing embedded sand grains. It is called by many local names; for example, in Kuwait it is known as *"gatch."*

Other coral layers have been submerged under the rising sea, so that it is not unusual to encounter numerous strata ranging from a recent coral reef or cap rock on top, down through various strata of sands, coral, calcareous silt, and limestone. These soil profiles are highly variable from one location to another. It is not unusual to find a very weak layer, almost fluid, beneath a 1-m-thick limestone stratum.

Coral can be dredged mechanically with very large heavy equipment, especially if it is possible to break through the overlying hard stratum. In other areas, the cap rock is too hard and thick. Off the east coast of Saudi Arabia, rock-breaking chisels are necessary. Off Hawaii, surface blasting, using either shaped charges or large charges of powder, will frequently be sufficient to break the surface stratum downward into the unconsolidated sands below.

Drilling and blasting are usually only effective if the charge can be located in the hard stratum. If below, it will result only in forming large, undredgeable slabs; if above, it may not break the stratum at all, but just blow sand and water into the air. Drilling of holes (e.g., for piling) may present major difficulties due to loss of drilling fluid in weak strata or voids.

Cemented and partially cemented sands are quite commonly found in subtropical and tropical seafloors. The cementing material is the calcareous deposits from marine organisms. These typically are highly irregular in their degree of cementation. They are often roughly stratified.

Since the cementation process has usually occurred over long periods, the cemented zones may have been exposed to erosion when the sea level lowered, to be later filled with looser, uncemented deposits as the sea again rose.

2.12 Unconsolidated Sands

In many offshore areas there are very extensive accumulations of sands, in some cases due to longshore sediment transport of sand discharged from rivers, in other cases due to ancient sand dunes, such as are found in the southern North Sea. Sands are extremely hard to sample during geotechnical investigations: they will almost always be highly disturbed by the sampling process. Thus the geotechnical reports must be very carefully evaluated. These cohesionless materials are very mobile and sensitive to disturbance by construction activity.

Under earthquake, storm wave pounding, or dynamic construction activity such as pile driving, the sands may locally liquefy, turning momentarily into a heavy fluid.

Surface sands are readily disturbed by waves and eddies, which increase the internal pore pressures, causing the sand grains to tend to rise up, thus ready for easy transport by currents. It is this that makes them so susceptible to scour and erosion.

Scour from wave action can occur when the water depth is less than half the significant wave length. Whenever the depth is less than one-fourth the wave length, substantial scour in sands is likely.

Bottom currents, whether wave-induced or from other causes (see Section 1.5) can move sand, especially if it is periodically raised by the pore pressure gradients induced by waves. The resultant scour holes tend eventually to stabilize at a condition where the rate of sand infill matches the erosion rate.

The excavation of underwater trenches in sands is typically rather difficult and complex due to the varying densities of the sands. However, stable side slopes can be achieved in the absence of severe currents and wave action at shallow depths. SPT (N) values provide a guide. SPT values obtained underwater must first be multiplied by 1.12 to account for the effect of submergence. In cohesionless sands, SPT values must also be corrected for depth. See Table 2.12.1.

With the N_{SPT} values so corrected, Table 2.12.2 can be used to evaluate the approximate behavior of underwater dredged slopes in sand. Where loose surficial sands overlie either clay or denser sands, the joint action of the waves and bottom currents may readily move these sands into the trench, infilling it.

Sands from the side slopes, especially the surficial loose sands, tend to move downward into an underwater excavation. If there is a prevailing current, the excavation will tend to fill in on the upstream side and erode on the downstream side. Thus a trench transverse to a current will tend to "migrate" downstream.

Trenches and other underwater excavations in relatively shallow ocean waters (50 m or so) have a

TABLE 2.12.1 Correction Multiplier to Apply to Measured SPT Valves (as Adjusted for Submergence) to Account for Overburden Pressure at Various Depths

Depth below Seafloor (ft)	Correction Multiplier to Give Value under Standard Confinement Pressure
2	2.3
5	2.0
10	1.8
15	1.5
20	1.3

46 Geotechnical Aspects: Seafloor Soils

TABLE 2.12.2 Correlation between SPT and Stable Dredged Slopes in Cohesionless Materials (e.g., Sand)

	Very Loose to Loose	Loose to Medium	Medium to Dense	Dense to Very Dense
Corrected N_{SPT}	0–4	4–10	10–30	30–50+
Relative density	.15	.35	.65	.85–1.0
Moist unit weight (lb/ft³)	70–100	90–120	110–130	120–140
Stable slope	4:1	2.25:1	1.75:1	1.5:1

diffraction effect on the waves, causing them to refract toward the edges and leaving the center of the excavation more calm. A long trench normal to the shore will not only be more calm, the waves refracting to either side, but will form a channel for a return current, tending to clean loose sediments from the trench.

Sands also migrate along the shore, in the direction of the overall net current. This is the well-known longshore transport. During summer, when wave energy is reduced, sand accumulates near shore. In winter, when wave energy increases, the sands move further offshore.

In deeper waters, such as those of the northern North Sea, relatively thin lenses of surface sand occur over the basically stiff clay seafloor. Thin lenses of sand, although unconsolidated, form zones of high resistance, and therefore of high local bearing pressure, on the flat base of any large structure. These are the "hard points" which may develop local pressures as great as 300 T/m² (3 MPa) (66 ksf). It is partly because of these hard points that a detailed bathymetry is required at the site of such a structure.

The construction of a structure on sands modifies their behavior. Wave energy acting on the structure is transmitted to the foundation, increasing the pore pressure. Due to its cyclic nature, the pore pressure may be progressively built up until local liquefaction occurs under the edge. This eventually leads to a loss of material under the edge and the tendency for the structure to rock, aggravating the problem. This is why concrete caissons usually fail outward under wave attack: they have lost the sand under the toe.

2.13. Underwater Sand Dunes ("Megadunes")

Under the action of strong currents such as those found in the southern North Sea and at the mouth of major rivers of South America and Southeast Asia, the sand bed may be formed into waves, that is, sand dunes. These dunes move just as their land-based counterparts do, eroding from the back, redepositing on the front. Typical maxima depths are 10 m. Thus in planning installations in such areas, it may be necessary to dredge the dunes down to or below the level of the troughs; otherwise the structure or pipeline may end up exposed above the seafloor.

The structure, if large, may interfere with subsequent movements. Large-scale temporary installations may cause large quantities of sand to deposit on one side and erode from the other.

2.14 Rock Outcrops

Bedrock outcrops present a highly site-specific problem. In deep water or when the outcrop is partially covered with sand, such outcrops pose the problem of irregularities and hard points.

There are several alternative methods by which such outcrops may be treated:

1. Softer rocks may be dredged so as to remove all significant irregularities, then backfilled with properly graded gravels or crushed rock. Alternatively, a pipeline plow may be used to rip the irregularities.
2. Harder rocks may require underwater drilling and blasting to enable them to be dredged.
3. In the case of isolated high points of rock, shaped charges may be used.
4. A blanket of rock (underwater embankment or berm) may be placed so as to cover all irregularities to a sufficient depth (e.g., 5–10 m) so as to give a uniform bearing.
5. Pipelines and structures may be designed to span between irregularities. Suitable anchors may be required to prevent movement, abrasion, pounding, and so forth, under the action of waves and currents.

Drilling of piling and the like into rock outcrops may present difficulties in getting the hole started, especially if the rock is highly irregular or covered with fractured material.

Rock outcrops which have been exposed in ancient geological times will have been weathered. Weathering may have been quite variable within the formation, proceeding down fractures and other discontinuities. Thus drilling may progress through hard rock into softer material below. Drilling may also encounter difficulties with loss of fluid into fractures or cavities, the latter being a special problem with coral and limestone deposits.

In some weak rocks, such as mudstones, it may be possible to install piling by driving if there is a relief of the confinement. For example, a pilot hole or holes of relatively small diameter may permit a larger-diameter pile to be driven.

2.15 Cobbles

Some seafloor areas which have been subjected to high currents or wave action are "paved" with cobbles, closely packed, with or without sand in the interstices. These cobbled areas are difficult to excavate because most conventional equipment has difficulty getting a "bite" into the material. Once a trench or hole is begun, the slopes may become very loose and unstable, taking a rather flat angle of repose, for example, 2:1 or even flatter, depending on the current. Their rounded surfaces give a low angle of friction.

2.16 Deep Gravel Deposits

In sub-Arctic areas, deep deposits of gravels are found. These have been eroded by glaciers and rivers from the mountains and discharged into the sea so as to have a minimum of stratification and fines. Although composed of good material, they often develop very low frictional resistance, due to their rounded surfaces and high void ratio. Thus they are quite unstable and dredged slopes will be very flat, for example, 3:1.

Piling driven into such material has typically failed to develop the desired skin friction. End bearing may also be less than expected, due to the large void ratio. It has proven necessary in many cases to provide enlarged tips of piles in order to obtain adequate bearing. Such gravel deposits are inherently difficult to sample by conventional means: any samples are more or less completely disturbed, and it is difficult to determine the degree of packing (consolidation).

2.17 Seafloor Oozes

Seafloor oozes are relatively thin, flocculant layers, usually of organic sediments lying on the surface of the seafloor in many of the deep basins of the seas. They are readily displaced, so that conventional objects sink readily through them. They create turbidity clouds upon the slightest disturbance and thus interfere with positioning and control. In many cases, they are so soft that a diver or an object (e.g., an ROV) will sink through them just as though they were a fluid of slightly heavier density than water.

Seafloor oozes do not normally show up on echo-sounding or sparker surveys and may easily wash out of sampling tubes; hence their presence may not always have been ascertained prior to construction. For example, a 25-m layer of silty ooze was discovered on the floor of a Norwegian fjord by grab, whereas acoustic soundings had penetrated through it without reflection.

2.18 Seafloor Instability and Slumping; Turbidity Currents

The offshore areas of principal interest to the petroleum industry are great sedimentary basins. While many are ancient deposits and relatively stable, others are still active deltaic areas. The great freshwater rivers—the Mississippi, the Amazon, the Orinoco, the Congo, and so on—are transporting huge quantities of silts and clays in colloidal suspension. Contact with the salt water causes flocculation, and the highly dispersed soil particles settle to the seafloor. Periodically, huge blocks of these recent sediments slump off and flow seaward. This process is prevalent even far out to sea, on the sides of submarine canyons such as the Baltimore Canyon of the U.S. East Coast and the outer edge of the continental shelves. Evidence of widespread slumping exists near the break of the continental shelf of the Alaskan Beaufort Sea. See Figure 2.18.1.

Underwater sand deposits may also be very loosely consolidated. If the internal pore pressure

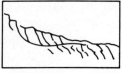

Subaqueous Slumping and Sliding   Subaqueous Mass Flow (Low-Density, Low-Velocity)

Turbidity Current Flow (High-Density, High-Velocity-up to 50 K)

Figure 2.18.1 Seafloor instability phenomenon. (From U.S. Naval Civil Engineering Laboratory Report R744-1971, Port Hueneme, Ca. 1971.)

is raised by any of several mechanisms, the sand turns into a heavy liquid. These underwater flows and turbidity currents have occurred in both sands and clayey silts. Some of these are going on frequently, where they are the mechanism by which shore sands are fed down submarine canyons, to be deposited in the fan at the bottom of the continental slope. During this downward flow, they erode the canyon itself.

Others are more infrequent, being triggered by an intense storm or by earthquake. In clay areas, these slides are often felt to be aggravated by entrapped methane gas in the silty clays. These failures often occur on very flat slopes, which superficially would appear to be stable.

While the natural occurrence of these would appear to be, like earthquakes, important to the designer but not to the constructor, there is one important difference: They may be and have been triggered by the construction operation itself. Pile driving, dredging, dynamic compaction, and underwater rock dumping have triggered flows in the Norwegian fjords which have involved millions of cubic meters of soil and have extended out into the offshore areas, eventually pouring into the Norwegian Trench.

Temporary structures constructed off the coast of California have transmitted wave energy into the sands, triggering slides 2 km wide, again involving millions of cubic meters of sand which were transported down into the Monterey submarine canyon. In one case an existing pier "disappeared"; in another the sheet pile cofferdam was lost. Turbidity currents off the U.S. East Coast have been known to sever submarine cables many kilometers offshore. These submarine slides and turbidity currents are typically limited as to the depth of soil involved to about 20 m.

As petroleum exploration and hence offshore construction extends out to the edge of the continental shelf and beyond to the slope and eventually the fans themselves, these submarine flows will become of increasing concern to the construction operations as well as the design.

2.19. General Remarks Concerning Seafloor Construction

Bathymetry at and adjacent to the construction site is extremely important, as it affects initial setdown of jackets and caissons as well as the landing of more flexible structures such as pipelines. Adequate surveying methods and locating systems must be employed to ensure that the bathymetry, geotechnical borings, and actual installation are all controlled to the same positioning relative to each other. Because of tolerances and systematic errors (e.g., night effect) on many electronic survey systems, a method which marks the site—for example, acoustic transponders or articulated spar buoys—will usually be found desirable.

The site survey should also identify carefully the relative position of any foreign or artificial objects—pipelines, abandoned anchors, dropped casing, and the like—which often litter the seafloor in the vicinity of an offshore site due to the prior conduct of exploratory drilling operations.

Scour and erosion are addressed in specific sections where their occurrence is most likely. The potential for scour exists at depths up to 100 m and even greater, where wave action may build up internal pore pressures, and eddy currents may have vertical as well as horizontal components. Recent studies indicate that scour potential may even exist in certain areas of the deep ocean where periodic eddy currents are superimposed upon steady-state unidirectional currents.

Since scour may occur during or immediately after the installation of a structure, it is essential that monitoring is carried out and adequate protection placed as soon after landing as possible. Scour protection methods are described in Chapter 7.

In some cases it may be necessary or desirable

to place scour protection prior to the installation. When a large caisson is being seated on the seafloor, for example, the water trapped underneath must escape, thus creating a scouring action. At this same stage of installation, when there is a relatively small gap under the structure, current or wave action may induce a high flow rate under the structure, causing scour to progress in susceptible soils almost as rapidly as the caisson descends.

A similar phenomenon is that of piping. Piping is the formation of a channel or tunnel under a structure due to pressure gradients which erode the soil locally. Such piping not only weakens the foundation but may prevent subsequent construction operations which require the maintenance of an overpressure (e.g., for removal of the structure) or underpressure (e.g., for aiding penetration of skirts).

Certain soils may degrade and soften when exposed to the changed conditions they experience during construction, especially if they have been previously blanketed by impervious material or were loosely cemented.

In summary, it may be fairly said that seafloor geotechnics represents the area of greatest concern to and difficulty for the constructor. Problems of instability, inability to penetrate, and slope failure continue to plague offshore construction activities. Overconsolidated silts in the Arctic and calcareous sands in the subtropics are perhaps the most demanding problems facing the constructor in today's offshore operations. Thus a closer relationship between the geotechnical and construction engineers should lead to more effective and economical offshore construction.

And I have loved thee, Ocean! and my joy
Of youthful sports was on thy breast to be
Borne, like thy bubbles, onward: from a boy
I wanton'd with thy breakers—they to me
Were a delight; and if the freshening sea
Made them a terror—'twas a pleasing fear
For I was as it were, a child of thee
And trusted to thy billows far and near,
And laid my hand upon thy mane—
   As I do here.

LORD BYRON, "CHILDE HAROLD'S PILGRIMAGE"

3

Protection of the Natural and Built Environment: Constraints on Construction

The offshore construction process is obviously not carried on in a sterilized and isolated sea, but in one which teems with life and in which many other processes are going on. So it is necessary at the commencement of planning of offshore construction that consideration be given to the existing environment, the natural ecology, and the man-built structures and activities that might be affected.

Too often in the past, these considerations have been addressed only after plans have been adopted. The approach has been one of amelioration of problems rather than integration. Early consideration facilitates permitting and minimizes the probability that external authorities will later impose excessive restrictions or order delays during the construction.

3.1 Ecological Considerations

The concept of ecology is one of a living system, ranging from microorganisms up to the whales, forming part of a continuous food chain. Any disruption at one point in this chain may have large-scale effects. Some of the more common disruptions occasioned by construction activities are addressed below.

3.1.1 Turbidity

Dredging, filling, and trenching are typical construction activities which churn up the sediments and cause the finer particles to go into colloidal suspension. The resulting turbidity may be quite persistent and locally affect the microorganisms' growth and reproductivity. Larger particles, fine sands, and silts may have a disastrous effect upon oyster beds, both natural and artificial.

Various procedures have been adopted to cope with turbidity. Suction dredging generally causes less turbidity than bucket dredging. Large clamshell buckets cause less disturbance than a bucket ladder dredge.

Discharge at the seafloor causes less widespread turbidity than discharge at the surface of the sea. Hence discharges from pipelines may employ a discharge pipe, suspended from a pontoon. Special devices may be placed at the tip to cause the fine material to be concentrated and drop out of suspension more rapidly.

Buckets may be required to lower the material to the seafloor before opening or else to discharge through tremie pipes. Silt curtains may be suspended from floats around a higher-level discharge, separating the water masses. These have been quite widely and successfully employed.

Turbidity may also be remedied by chemical treatment. However, the chemicals used must be selected with care to ensure that they are not toxic. The salinity of the natural seawater itself aids in this process.

While turbidity is harmful to some microorganisms, it has beneficial effects in cleansing the water column. Heavy metal ions are reportedly picked up by the fine particles and precipitated and then blanketed. Turbidity and discharge of fine sediments may also be beneficial to shrimp. Conversely, the dredging process may stir up and release such unwanted ions into the water column—hence the prior recommendation for minimizing disturbance by use of suction dredging or large buckets.

The U.S. Army Corps of Engineers carried out a very extensive study on discharge of dredged spoil. While these tests and investigations were primarily conducted in estuarine waters, some were in coastal waters as well. They showed that dumping from bottom-dump barges typically descended as a mass. As the mass hit the seafloor, it spread as a lateral wave. Very little "rebounded" upward to any distance, and the total extent of the lateral dispersion was of the order of two to four times the depth, depending on the material and bottom currents.

When discharging material such as graded crushed rock which has been loaded on or into the barge "in the dry," entrapped air causes dispersion and turbidity as the mass falls through the water column. This can be minimized by saturating the material just prior to discharge.

The operation of an airlift pump, cleaning the silt out of a cylinder pile, for example, can cause a highly visible "plume" if the discharge is above the sea surface. By discharging below water by use of an elbow, or by just locating the discharge below the surface, this plume can be minimized.

3.1.2 Noise

High-speed outboard motors, helicopters, low-flying aircraft, discharge of dredged gravel through submarine or floating pipelines, pile driving, drilling, sparker and seismic surveys, and even echo sounding are examples of construction operations which create noise in the water column.

Noise appears both to attract and repel sea animals. Low-frequency noises travel further in water. Concern has been expressed that wide-band noise spectra may interfere with the navigation used by the bowhead whale. There is concern among the Inuit hunters that the noise may drive the whales and seals further offshore, to the edge of the polar pack, where hunting is more difficult and dangerous. Several experts believe that the distance over which construction and drilling noise will have a significant effect is of the order of 1000 m.

Noise may be isolated by an air gap, such as that created by intense bubbling of air around the resonator in contact with the water. Ma, Veradan, and Veradan in OTC Paper 4506, (Offshore Technology Conference Preprints, 1985) have shown that gas or air bubbles in water and sediments can attenuate long-range, low-frequency underwater sound propagation very effectively.

Many larger animals (caribou, geese, ducks, etc.) appear to become accustomed to the noise of helicopters if they are not too close, although it is generally believed that loud noise such as that from low-flying aircraft is injurious to breeding birds.

3.1.3 Open Leads in Sea Ice

The use of icebreakers and the traffic of construction vessels tend of course to open more leads than are natural to the area and season, especially during the period of freezeup. Operations into the late fall and winter will also open leads which will, however, quickly freeze behind the vessel. Tests by Polar Gas Company and others have reported that seals and whales appear to distinguish between artificial and natural leads; in any event, there appears to be no harmful effect. Opening such leads in the fast ice may present new problems to the native hunters and impede their ability to travel across the ice in the late fall. Exercise of judgment appears warranted in the vicinity of native villages.

3.1.4 Explosion

The use of explosive charges for underwater dredging and demolition can cause death and damage to the fish in the immediate vicinity. For major undertakings, therefore, the placing of charges in drilled holes beneath the overburden will be found not only more effective but less damaging to fish, since the overburden cushions and attenuates the shock. Where charges are placed on the seabed, shaped charges are less damaging than omnidirectional charges.

3.1.5 Dredging

Dredging of the seafloor of course causes some turbidity, as noted earlier. Dredging causes trenches, pockets, and holes in the seafloor, which generally have a long-term beneficial effect on shrimp and other bottom dwellers. Dredging will of course probably destroy most of the seafloor community in the immediate vicinity; however, the effects are generally quite limited in extent.

As noted under Section 3.1.2, "Noise," the pumping of gravel and cobbles through a pipeline creates a resonate noise in the line.

3.1.6 Deposition: Spoil Disposal and Placing of Fill through Water

The dumping of material through the water column has been previously addressed in Section 3.1.1, "Turbidity," since this is the major adverse effect of deposition. Also noted there was the "cleansing effect" that deposition of finely divided particles such as clay and silt may have in removing heavy metal ions and contaminants from the water column. When large embankments are constructed underwater, they may affect the bottom currents, leading to deposition on the upstream edge and erosion on the downstream edge. Shoaling may be caused beyond that intended.

Construction of underwater rock embankments may significantly increase the benthic and pelagic populations by providing crevices in which the smaller life forms may hide from predators as well as protection for the eggs during storms. Rincon Island, an offshore rock and sandfill island off the coast of Southern California, for example, has developed a rich marine life in an area which had prior to construction been termed an "ecological desert."

3.1.7 Oil Spill

The construction contractor is generally not involved with drilling for oil but may very well be conducting operations in the vicinity of live oil lines. Thus he may well have the potential for causing an oil spill. The constructor's operations themselves involve the use of fuel oils (diesel, gasoline, etc.) and lubricants. Leaking equipment, errors in transfer of fuel, failure to close and seal valves, and so on may all produce the "sheen of oil on the surface" which is prohibited by regulations in many coastal waters and is unfortunately highly visible from the air. The amount of oil that is tolerable is of course the subject of highly emotional debates. However, there is no question but that this is a matter to which the construction contractor must give attention and that he must take active steps to prevent oil spills.

The most harmful immediate effects of oil are the contamination of the feathers of seabirds. Oil may travel long distances, eventually ending up on a beach where it has serious esthetic as well as some harmful biological effects. Fortunately, these latter do not seem to persist for long on active shoreline beaches. Oil in estuaries, marshes, and the like appears more harmful.

Gasoline and diesel oil are more toxic than crude oil. Oil, being an organic substance, bio-degrades in the open water, due to a combination of bacterial activity, oxygenation, and sunlight.

For the Arctic, there has been widespread concern over the consequences of an offshore oil spill in and under the ice. A number of tests have been run by the Canadian Offshore Oil Spill Research Association (COOSRA) organization as well as the Alaska Oil and Gas Association (AOGA) and the U.S. Coast Guard.

In the winter, when there is full ice coverage, the spilled oil tends to collect under the ice. Its spread is limited by the keels of ridges. It tends to coagulate. Some balls drop to the seafloor. Being thus naturally contained, the oil can eventually be disposed of.

In the open-water season, an oil spill in the Arctic is similar to that in the more temperate zones. The effect if it reaches the beaches is considered more serious because they are low-lying and heavily populated by breeding birds.

It is during spring breakup when the effects of an oil spill are potentially the most uncertain and inherently serious. Cleanup operations are impeded by the ice floes. The oil tends to concentrate in the open-water leads, which are the areas in which photoplankton growth normally starts earliest and which the sea mammals use as entry routes. It also migrates up through brine channels in the ice to form melt pools which can then be burnt on the ice surface. Considerable effort is currently being expended to develop effective oil spill cleanup capability under broken ice conditions.

3.1.8 Toxic Chemicals

Strict prohibitions are placed by international law on the disposal of toxic chemicals at sea. The constructor would rarely become involved in such a situation, and then more or less inadvertently if he were to have surplus or waste chemicals such as coal tar epoxy or solvents. Arrangements must be made for their containment and return to shore for disposal in accordance with regulations.

3.1.9 Shoreline or Critical Area Disturbance

In many areas, fish migration is very close to shore, in relatively shallow and protected waters. An example is along the west and north coasts of Alaska. In other areas, fish and migrating mammals often use a relatively narrow channel.

For such limited and constricted areas, even relatively minor disturbances can have important consequences. For example, an offshore jetty would obviously interfere, but so also might an overhead bridge, since many fish, such as salmon, reportedly are reluctant to swim under a shadow.

3.1.10 Sediment Transport

In most shallow-water and coastal areas, sediments are in constant movement. In addition to the local response to the orbital motion of waves and wave-induced currents, there is the offshore displacement due to storms, alternating with the onshore replacement during calm periods. More important is the general longshore transport due to the net current, which moves vast quantities of sand along the coasts. The disruption caused by major works such as breakwaters has long been recognized in civil engineering practice. The sand tends to build up on the "upstream" side and erode from the "downstream" side.

The constructor can create similar if shorter-term changes through the construction of trestles, cofferdams, or jetties for service boats or to facilitate the pullout of submarine pipelines. The design of such temporary facilities can mitigate this problem by providing large, relatively free openings for the free movement of the sand. If a structure such as an isolated platform, dolphin, or the like is constructed in the zone of coastal sediments, the focusing of wave energy will cause a tombolo or spit to form from shore out toward the structure. Thus over a period of time, such a spit may itself become an obstruction to coastal sand transport.

3.1.11 Summary

It is obvious from the discussions of the items of serious environmental impact that most of these relate to the coastal areas rather than the deep offshore. The offshore constructor is nevertheless dependent on the shore and the coasts for support and access and hence his offshore activities may involve significant coastal impact.

Most countries today require the filing of an environmental impact statement or report prior to the undertaking of a major offshore project. This will usually have been filed by the client. Included will be sections dealing with the impacts during the construction phase and the coastal and onshore impacts of offshore operations. It is important for the offshore construction contractor to become familiar with these documents and the constraints, restrictions, and mitigating procedures set forth in them. Compliance is not only legally required but, as a practical matter, important to assure that construction operations may proceed without interruption or delay. Lack of strict compliance may involve the constructor in legal disputes, even criminal charges, and in today's social and political environment may stir up public opposition to and interference in his operations.

3.2 Protection of Existing Structures

Many offshore construction operations must be carried out in the vicinity of existing structures and facilities. For example, it is increasingly becoming the practice for the oil company to have wells drilled prior to the installation of the jacket and platform. Subsea satellite wells may similarly have been completed ahead of platform construction. Flow lines and pipelines may be in the vicinity. It is of course essential that these not be damaged by the construction contractor through carelessness such as allowing an anchor line to be wrapped around a subsea well completion or an anchor to be dropped onto an existing pipeline. These have occurred, with serious financial cost for repairs. There is always the possibility that oil will be released to the sea.

Pipelines and moorings laid in an active bottom-fishing area have to cope with trawl boards and nets. While these can damage the line, most often it is the fishing gear which is lost (or claimed to be lost), with resultant claims for reimbursement.

In the vicinity of salt-water intakes for onshore facilities such as LNG plants, power plants, and so on, sediments, especially sand, may be a hazard. Sand particles, for example, swept into suspension in an intake may clog spray nozzles in the plant. Operations therefore will have to be planned so as to minimize stirring up of the seafloor sediments. In extreme cases it may be necessary to install barriers on the seafloor (e.g., steel frames with filter fabric curtains) to prevent sand movement just above the seafloor.

Considerations of existing installations require

that very careful surveys be made prior to the start of operations and that their relative position be tied in to visible structures, acoustic tranponders, or the like so that they may be a guide to subsequent operations. Side-scan sonar or more sophisticated profiling systems are the usual means for location. On occasion they may need to be supplemented by underwater visual or video means using a submersible or ROV or by diver surveys.

Careful records should be kept of all anchor locations and the survey plots used in setting them to provide verification of the contractor's work and to protect the contractor from claims for damage that may have been caused by others also working in the vicinity.

The client or others may be carrying on other operations in the vicinity of the new construction: tanker loading, offshore supply, drilling, and other construction. One common source of problems is anchor line interference. In such cases, carefully planned schedules and layouts should be agreed upon by all parties and then adhered to. If it becomes necessary to adjust them, all parties should be notified promptly, as more than one may have planned operations in the same "weather window" and space.

When working in the vicinity of operating facilities, especially offshore terminals, there may be very strict limitations on operations (e.g., welding) involving the possibility of fire or explosion. These may be conditioned upon the direction of the wind. Similarly, work near an operating flare may be dependent on wind direction in order to avoid excessive radiant heat.

Ships and vessels not under the control of either the client nor the constructor may be operating in close vicinity to the construction site. A recent study identified structures built in the vicinity of shipping lanes as presenting a major collision hazard. Many ships do not have a deck officer actually on the bridge at all times; he may also be carrying on other activities while on watch, leaving the ship on automatic pilot. The mandatory lookout may or may not be effective. Fog, rain, and snow obscure visibility. Radar may give false echoes from offshore platforms and confused echoes if reflecting from both a platform and construction equipment in the vicinity.

The constructor, recognizing these problems, should take special steps. He can arrange to have a "Notice to Mariners" issued by the proper regulatory agency (e.g., the U.S. Coast Guard). For operations which will go on for a long period, he can arrange for a danger area to be marked on the navigational charts. However, many ships keep neither their Notices to Mariners nor their charts up to date.

Provision of bright floodlights and/or horns on the platform have been found effective, but these must be cleared by the local regulatory agency to ensure that these do not conflict with navigational aids. Otherwise, the contractor may increase his potential liability.

Protected zones, usually 1000 m in diameter, have been established around offshore platforms. All vessels, large and small, are required to keep clear. In areas of potential hazard, the contractor may keep a boat full time to warn fishing boats and sightseeing boats away from the platform. However, these are of little use in event of a ship off track in bad weather. For these cases, the contractor may operate his own radar and use voice, radio, bullhorn, whistles, or signal lights to attract the attention of an approaching vessel.

The sightseeing boats referred to above present a special risk. Some enterprising boatowners advertise when a particular offshore operation is to take place and run tours. The potential liability of a contractor is enormous, as well as the more likely interference with his operations. Close liaison, scheduling, and personal communication with sightseeing boat operators is essential. The local harbor police or coast guard may be willing to aid in the interests of safety.

For all at last returns to the sea—to Oceanus, the ocean river, like the ever-flowing stream of time, the beginning and the end.

RACHEL CARSON, *The Sea Around Us*

4

Materials and Fabrication for Offshore Structures

The principal materials for offshore structures are steel and concrete. The fabrication and/or construction contractor is generally responsible for their procurement and quality control, although in some cases, especially pipeline steel, the basic material may be separately purchased by the client (operator) and made available to the constructor.

These materials must perform in a harsh environment, subject to the many corrosive and erosive actions of the sea, under dynamic cyclic and impact conditions over a wide range of temperatures. Thus special criteria and requirements are imposed on the material qualities and their control.

Fabrication is especially critical for both steel and concrete in order to assure that the structure will perform properly under both service and extreme loads. The cyclic nature of the loading combined with the corrosive environment tends to propagate cracks; hence improper fabrication details may grow into serious problems. Fabrication is also rendered more difficult because of the large sizes of offshore structures: Spatial dimensions are difficult to measure and maintain, and thermal strains cause significant temporary distortions.

4.1 Steel Structures for the Offshore Environment

4.1.1 Steel Materials for Structures

American Petroleum Institute standard API-RP2A, sec. 2.9.1, designates structural steels by groups, according to their strength, and by classes, according to their notch toughness characteristics and suitability for service in specific applications and temperatures.

Structural Steel Group I designates mild steels, with specified minimum yield strengths of 280 MPa (40,000 psi) or less, with carbon equivalent .40 percent or less. Carbon equivalent CE is defined as:

$$CE = C + \frac{Mn}{6} + \frac{Ni + Cu}{15} + \frac{Cr + Mo + V}{5}$$

Group I structural steels may be welded by any recognized welding process, such as those described in AWS D1.1.

Group II designates intermediate-strength steels with specified minimum yield strengths of 280–360 MPa (40,000–52,000 psi), having carbon equivalents up to .45 percent and higher. Low hydrogen welding processes are required.

Group III designates high-strength steels with specified minimum yield strengths in excess of 360 MPa (52,000 psi). When such steels are used, they must be investigated with regard to weldability, the special welding procedures required, fatigue problems resulting from working at high stresses, and notch toughness in relation to other aspects of fracture control such as fabrication, inspection procedures, service stress, and temperature environment.

The classes of structural steel are as follows:

Class C steels are those which have had a history of successful application in welded structures at temperatures above freezing, but impact tests are not required. These steels are applicable to primary members such as piling, jacket braces and legs, deck beams and legs, and so on which are of limited thickness, moderate forming, low restraint, modest stress concentration, quasi-static loading rate (having a rise time of 1 second or more), and structural redundancy such that an isolated fracture would not lead to progressive collapse.

Class B steels are for use where the conditions of thickness, cold work, restraint, stress concentration, impact loading, and/or structural redundancy indicate the need for improved notch toughness. Class B steels should meet the following Charpy V-notch impact test values at temperatures between 0 and 10°C:

For Group I steels—20 J (15 ft-lb)
For Group II steels—34 J (25 ft-lb)

Class A steels are suitable for use at subzero temperatures and for critical applications involving adverse combinations of the factors cited above. For critical applications, the Charpy test values cited for Class B steels, Groups I and II, should be met at temperatures 20–30°C below the lowest anticipated service temperature. The purpose is to prevent propagation of brittle fractures from large flaws.

The DNV (Det Norske Veritas) Rules divide structural steels in three "types" depending on usage in the structure. Special structural steel is for members essential to the overall integrity of the structure and subject to particularly arduous stress conditions, including can sections of the jacket, legs, deck-to-leg and deck-to-column connections, cruciform joints of main girders of the deck, and so on.

Primary structural steel is for members participating in the overall integrity of the structure and for members of importance for operational safety. This type includes jacket legs, piles, bracings, main girders, and beams in the deck, riser support frames, helicopter deck, module support structures, and cranes.

Secondary structural steel is for all other structural members.

Maxima carbon equivalents are .45 percent for special steels, .47 percent for primary steels, and .50 percent for secondary steels. Additional detailed limits are placed on specific alloy elements.

"Normal-strength" steel is defined as having a minimum specified yield strength of 300 MPa (42,000 psi), and "higher-strength" steel is that with a greater specified minimum yield strength. Minimum elongation percentages are required for the two strength classifications, ranging from 14 to 22 percent depending on the thickness.

The actual yield strengths of steel, as furnished, typically exceed the specified minimum values by 10–15 percent. Care must be exercised to ensure that steels do not appreciably exceed these over-yield-strength values since this may lead to poorer weldability and increased susceptibility to fracture propagation. DNV rules attempt to prevent excessive yield strength values by specifying that the ultimate strength shall be at least 18 percent greater than the yield.

To screen out laminated plates, special and primary structural steel is to be subjected to ultrasonic testing.

Primary structural steel which will be significantly strained in the thickness direction during welding (or service) is to be tested for through-thickness ductility against lamellar tearing. This is especially critical where one member is welded at a right angle to the thick flange of another member.

DNV rules specify minimum Charpy V-notch energy absorption for both the longitudinal and transverse directions (parallel and at right angles to the principal working direction) according to yield strength, thickness, and working temperature. See Figure 4.1.1.1.

Thicker members require the required energy absorption to be attained at lower temperatures than the design temperature. This latter is specified to be 5°C below the most probable monthly mean temperature of the ambient air or seawater in which the member will be used.

Led by the Japanese steel companies, new steels have recently been developed which have high yield strength (up to 700 and even 800 N/mm^2), good fatigue resistance in salt water, adequate toughness to resist impact at temperatures of $-50°C$ and even $-60°C$, and resistance to hydrogen-induced cracking, and which are weldable without preheating, while still retaining their toughness properties. Available plate thicknesses range from 60 mm for a yield strength of 800 N/mm^2 to 100 mm for a yield strength of 600 N/mm^2, 150 mm for a yield strength of 500 N/mm^2, and 200 mm for 400 N/mm^2. Thus steel materials are being designed to provide the special properties required for offshore and Arctic service.

API-RP2A also contains specifications for seamless or welded pipe used as a structural member. API-RP2A calls special attention to requirements for tubular joints, that is the intersections of tubular members at nodes of jacket frames. These joints are subjected to local stress concentrations which may lead to local yielding and plastic strains under the design load and to fatigue cracking under long-term cyclic loading. These in turn will place addi-

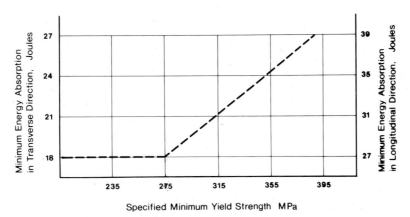
Figure 4.1.1.1 Charpy V-notch energy absorption.

tional demands on the ductility of the steel under dynamic loads. Heavy-wall joint cans (nodes) which must resist punching shear are of particular concern. The brace ends at tubular joints are also subject to high stress concentrations.

For such tubular joints, both above and below water, API-RP2A gives special guidance for Charpy testing, the use of the tougher Class A steels, and the use of steels having improved through-thickness properties. Because of the stress concentrations at such nodes, welds should achieve as full a joint penetration as practicable and the external weld profile should merge smoothly with the base metal on each side.

Welding materials are equally critical in assuring proper strength and ductility in service. The weld metal is to be compatible with the base material as regards heat treatment and corrosion. Crack-opening displacement tests or other fracture mechanics tests are normally to be conducted for selection of the welding consumables.

High-strength bolts and nuts, when used as structural elements, should have Charpy V-notch toughness values as required for the structural steel members being connected.

4.1.2 Fabrication and Welding

Welding procedures should be prepared, detailing steel grades, joint/groove design, thickness range, welding process, welding consumables, welding parameters, principal welding position, preheating/ working temperature, and postweld heat treatment. Stress-relieving is normally not required for the range of wall thicknesses normally used in the jackets and piles of offshore jackets in moderate environments such as the Gulf of Mexico, but is frequently required for the thicker members of large deck structures and for the joints (nodes) of the thicker-walled jackets of North Sea platforms.

The qualification of welding procedures is based on nondestructive testing and mechanical testing. These latter include tensile tests, bend tests, Charpy V-notch tests, and hardness tests. A macrosection cut through the weld should show a regular profile, with smooth transitions to the base material and without significant undercuts or excessive reinforcement. Cracks and cold lap (lack of fusion) are not acceptable. Porosity and slag inclusions are limited. Fracture mechanics toughness of heavy welded joints should be verified by crack-opening displacement tests.

Nondestructive testing may include X-ray (radiographic testing); ultrasonic testing (UT), and magnetic particle (MP) testing. Both the weld itself and the heat-affected zone should have notch toughness properties equal to those specified for the members.

Re-qualification is required for a welder who has interrupted his welding work for over 6 months.

Manual welding of all higher-strength steels and of normal-strength steel having a carbon equivalent greater than .41 should be carried out with low-hydrogen electrodes. For "special structural steels" and for all repair welding, DNV requires the use of extra-low-hydrogen electrodes. It is recommended by this author that all piling be welded with low-hydrogen electrodes in order to prevent fracture under impact.

Welding consumables should be kept in sealed moisture-proof containers at 20–30°C, but in any event at least 5°C above ambient. Opened containers should be stored at 70–150°C, depending on type of electrode. When electrodes are withdrawn for use, they should be kept in heated containers and used within 2 hours.

Consumables which have been contaminated by moisture, rust, oil, grease, or dirt should be discarded. Surfaces to be welded should be free from mill scale, slag, rust, grease, and paint. Edges should have a smooth and uniform surface.

No welding should be performed when surfaces are humid or damp. Suitable protection should be arranged when welding is performed under inclement weather conditions. Heating of the enclosed space can be used to raise the temperature above the dew point.

The groove should be dry at the time of welding. Moisture should be removed by preheating. The joint should be at a temperature of at least 5°C.

Fit-up should be checked before welding. Misalignment between parallel members should not exceed 10 percent of the thickness nor 3 mm. If the thicknesses of abutting members differ by more than 3 mm (⅛ in.), the thicker member should be tapered by grinding or machining to a slope of 1:4 or flatter. See Figure 4.1.2.1.

Each welding pass and the final weld are to be de-slagged and thoroughly cleaned.

Certain completed welds which are critical for fatigue endurance may be required to be ground to a smooth curve. This also reduces the probability of brittle fracture.

Welds which are essentially perpendicular to the direction of applied fluctuating stresses in members important to the structural integrity are normally to be of the full-penetration type and if possible welded from both sides.

Intersecting and abutting members for which the welding details have not been specified in the design should be joined by complete-penetration groove welds. This requirement includes "hidden" intersections, such as may occur in overlapped braces and pass-through stiffeners.

The construction contractor should detail all lifting plates, pad eyes, and so forth which are subject to dynamic impact stress, so that welds are not perpendicular to the principal tension. Welds acting in shear are much less susceptible to cracking than welds in tension. Where this is not practicable, then full-penetration welds must be used.

All temporary plates and fittings should be subjected to the same requirements for welding procedures and testing as the material of the member to which they are affixed. Lest this seem unduly conservative, remember the case of the *Alexander Kjelland* which capsized due to a fatigue crack initiated at the attachment of a sonar device to a principal structural member.

Permanent steel backing strips are permitted when properly accounted for in the design analysis. These are especially useful for piling and other members which are to be welded in the field and which are not accessible from both sides. Temporary backing can be provided by internal lineup clamps. Special skill is required for single-side welding of complete joint penetration tubular welds without backing. The interference of these backing strips with other operations such as drilling must be considered.

Temporary cutouts should be of sufficient size to allow sound replacement. Corners should be rounded so as to minimize stress concentrations.

Fillet welds for sealing purposes are required by DNV to have a leg length of at least 5 mm (¼ in.), whereas API-RP2A requires only 3 mm (⅛ in.). If such welds are perpendicular to the principal tension of a member subjected to dynamic impact, then great care must be taken to avoid undercutting.

Where welds are found to be defective, they should be rectified by grinding, machining, or welding as required. Welds of insufficient strength, ductility, or notch toughness should be completely removed prior to repair.

If arc-air gouging is used to remove a defective weld, it should be followed by grinding.

Whenever a discontinuity is removed, the gouged and ground area should be examined by

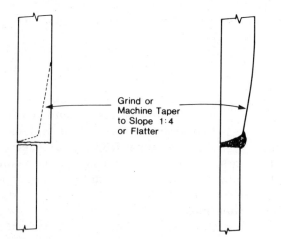

Figure 4.1.2.1 Tapering of thicker member for a full-penetration butt weld.

MP testing or other suitable methods to verify complete removal.

Repair welding should use extra-low-hydrogen electrodes and an appropriate preheating temperature, usually 25°C above the level used for production welding and at least 100°C.

All welds should be subjected to both visual and nondestructive testing as required by the specifications as fabrication and construction proceeds. All nondestructive testing should be properly documented and identified so that the tested areas may be readily retraced during fabrication and construction and after completed installation of the structure.

Increasing use is being made of computer-controlled cutting and beveling which ensures that all intersecting tubulars will fit properly. In many cases, the welding can then be carried out by semi-automatic welding equipment.

Because of the growing importance of documentation and the political sensitivity of many offshore structures, the contractor should make special effort to set up a quality assurance system that will ensure proper records of all testing.

Welding machines must be properly grounded to prevent underwater corrosion damage. Since welding machines are normally DC, the discharge to ground would otherwise occur underwater at piping penetrations or other similar points of concentration.

Fabrication of offshore steel structures should follow applicable provisions of codes for the fabrication and erection of structural steel for buildings, for example, the AISC specification for the design, fabrication, and erection of structural steel for buildings. Additional requirements are given in API-RP2A. These include the following.

Beams, whether of rolled shapes, tubulars, plate, or box girders, may be spliced. In cantilever beams, there should be no splice located closer to the point of support than one-half the cantilevered length. For beams within a span (continuous span) there should be no splice in the middle one-fourth of the span nor in the eighth of the span nearest a support, nor over a support.

The fabrication of an X-joint of two or more tubulars is especially difficult. In most normal practice, the larger-diameter and thicker member should continue through the joint and the smaller member frame into it. In a number of large and important recent jackets, the intersection node is specially fabricated, so that several or all intersecting members are continuous through the joint. In this case, the node is fabricated separately, so that it can be properly treated in the shop, and the members framing it are joined to the node by simple full-penetration butt welds. See Figures 4.1.2.2 and 4.1.2.3.

This same procedure has been employed quite effectively for jackets which have to be completed

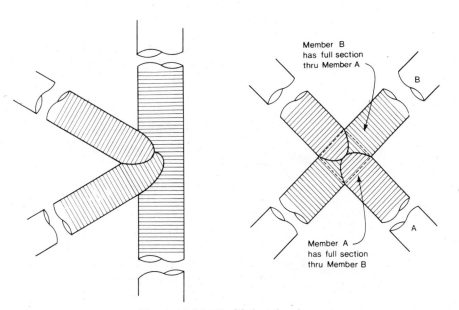

Figure 4.1.2.2 Prefabricated nodes.

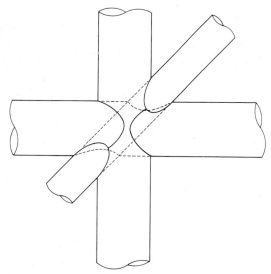

Figure 4.1.2.3 Intersecting joint of Hondo platform: full sections carried through joint.

at remote areas. The nodes are fabricated separately and shipped to the site. Then the main legs and braces are joined by butt welds.

Cast steel nodes are being used increasingly, in order to eliminate the critical welding details.

Typical details showing the proper beveling and weld for tubular members framing into or overlapping another member are shown in Figure 4.1.2.4.

For plate girders, the web-to-flange connection may usually consist of continuous double-fillet welds. Welds should have a concave profile and a smooth transition into flange and web. The connection between flanges and plates for stiffening the flanges should be a full-penetration weld made from both sides.

Stiffener plate-to-web connections may usually be continuous double-fillet welds. Weld metal and heat-affected zone notch toughness should not be less than the minimum toughness requirements for the girder.

Figure 4.1.2.5 illustrates some of the newly developed weld details of the "Subnap" process used to facilitate fabrication of offshore structures. Grinding of the weld's external profile may be required in order to improve the fatigue endurance. See Figure 4.1.2.6.

High-strength bolts may be effectively employed in temporary construction and in many cases of offshore construction such as the decks of offshore terminals. They are especially suitable for field connections, where spray and wave-induced vibration make it difficult to obtain high-quality welds. They are also an effective means for making connections under cold conditions.

The "turn-of-the-nut" method appears to be the most reliable method of ensuring that adequate torque has been applied. In a large joint, with multiple bolts, either the abutting plates should be premilled or shims should be employed to ensure a tight fit.

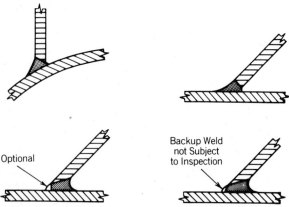

Figure 4.1.2.4 Welded tubular connections for shielded metal arc welding. (Adapted from API-RP2A, American Petroleum Institute, Dallas, Tx., 1984.)

Method	Edge Preparation and Pass Sequence	Applicable Plate Thickness	Features
Small Angular Groove Method	30–40° (a) 30–40° (b)	80 mm	Saves Welding Time and Materials Suitable for no Back-Gouging (b)
Single-Center-Pass Method	3°	150 mm	Saves Welding Time and Materials for Plates over 70 mm Thick
Two-pass Layer Method	2°	300 mm	Applicable to Thick Plates. Easy Slag Removal Low Heat Input Excellent Toughness

Figure 4.1.2.5 Japanese-developed "subnap" process. (From K. Itoh, "The Trend of Development of Steels used in the Arctic Ocean Field," *Arctic Ocean Engineering for the 21st Century*, Marine Technology Society, 1985.)

Figure 4.1.2.6 Intersection of bracing with main jacket leg of Hondo platform. Note that weld on left has been ground to smooth profile in order to enhance fatigue endurance.

4.1.3 Erection of Structural Steel

The spatial relationship of structural elements is critical for offshore structures which are to be assembled in the field or where major components are to be mated.

API-RP2A provides specific tolerances for final fabrication. For jacket and deck section columns, in any plane critical to field assembly, the horizontal distance from the center line of the adjacent columns should be within 6 mm (¼ in.) of the design dimension. The same tolerance should be applied in the other planes to working points on the outside of the columns.

Angles between corner columns should be within 1 minute of the design angle. Diagonals of rectangular layouts should be within 18 mm (¾ in.) of each other. Alignment of jacket columns should be maintained within 6 mm (¼ in.).

For jacket and deck section bracing, all braces should be within 12 mm (½ in.) of the design dimension.

The deck beams and cap beams at their ends should be within 12 mm (½ in.) of design position.

The assembly of a jacket frame, often having a spread at the base of 50 m or more, places severe demands on field layout and survey and on temporary support and adjustment bracing. A tunnel laser is being utilized to provide accurate levels and alignment on some recent platforms. Such large dimensions mean that thermal changes will be significant. Temperature differences may easily be as great as 20–30°C from before dawn to afternoon, and half that between various parts of the structure, resulting in several centimeters' distortion. On platform Cerveza, members were cut and shaped to the cold-side dimensions in the morning and then held for the midday sun to tightly fit metal to metal for welding.

Elastic deflections are also a source of difficulty in maintaining tolerances in the location of nodes. Foundation displacements under the skid beams and temporary erection skids must be carefully calculated and monitored.

Jacket frames are typically laid out flat and then rolled up by the use of multiple crawler cranes. See Figures 4.1.3.1 and 4.1.3.2. Because of the great distances and heights involved, some of the cranes may have to walk with their load. To coordinate such a rigging and lifting operation requires:

1. Thoroughly developed three-dimensional layouts
2. Firm, level foundations for the cranes
3. Trained and rehearsed operators
4. Proper communications

Twenty-four cranes were involved in the two major side-frame lifts during the erection of platform Cerveza, which was 1000 ft (300 m) in total length.

For the Magnus platform, a different procedure known as "toast rack" was employed. The jacket (in its final vertical position) was cut by horizontal planes so as to form four or five stages. Sub-assemblies weighing up to 1150 tons were erected to complete each stage.

On other projects the "slices of toast" have been completely fabricated off-site, then barged to the erection ways, skidded ashore, and joined to their neighbor.

For the Bullwinkle jacket, one of the world's largest, sections of the jacket are being fabricated in Japan, transported by barge to Texas, and assembled by the use of jacking towers, which roll up the sections to heights as great as 460 ft. (140m).

After the jacket frames have been rolled up, the final assembly will require staging and support high in the air. Safety of workers is of paramount importance. Often carefully planned scaffolding and staging can be attached prior to rollup. In other cases, attachment plates may be provided, then the staging set as a unit.

Adequate wind bracing must be provided. Usually this is done by means of guy wires (stays) and

62 *Materials and Fabrication for Offshore Structures*

Figure 4.1.3.1 "Rollup" of jacket framing for platform Eureka. (Courtesy Shell Oil Co.)

Figure 4.1.3.2 "Rollup" of side frames of platform Eureka. (Courtesy of Shell Oil Co.)

turnbuckles secured to deadmen or the skidways.

Communication can be provided by voice radio. Tools and supplies should be prepackaged and hoisted up as a unit. Power cables should be laid out so as to avoid interference with other operations and to minimize chances for snagging.

For jackets destined for shallow water, where the height is of the same order as or less than the plan dimensions, erection is usually carried out vertically, that is, in the same attitude as the final installation. Such jackets may be lifted onto the barge, if within the capacity of the cranes, or skidded out. In this latter case, adequate temporary pads and braces must be provided under the columns to distribute the loads for skidding.

Jackets destined for deeper water, in which the height is significantly greater than the plan dimensions, are usually erected on their side. Such jackets are loaded by skidding out onto a barge.

Another method used for very large jackets and self-floating jackets is to assemble them in a graving dock, on blocks, similar to shipbuilding practice. The loadout operations themselves are described in Chapter 9.

Large-diameter piles are fabricated from pile segments (cans) of rolled plate. The length of the can should be 1.5 m (5 ft) or more. The longitudinal seams of two adjacent cans should be at least 90°

apart. The pile should be straight, with a tolerance not greater than 3 mm (⅛ in.) in 3 m (10 ft), nor 12 mm (½ in.) in 12 m (40 ft) or more.

Out-of-roundness is often a problem with pile segments. The cans may have to be rotated and/or selected so as to match properly for welding. Outside diameter (O.D.) and out-of-roundness tolerances for adjacent segments should meet the requirements of the API specification 2B (*Specification for Fabricated Structural Steel Pipe*). As a general statement, the inside circumferences should match within 15 percent of the thickness of the thinner wall.

For joining pile segments of different wall thickness, if the thicker wall is more than 3 mm (⅛ in.) thicker than the thinner wall, the thicker wall should be tapered as shown in Figure 4.1.2.1.

Steel surfaces of piles and the inside of skirts or jacket legs where the connection is to be made by grout bond should be free of mill scale or varnish. Mechanical bond transfer devices such as weld beads or shear rings may be fabricated so as to enhance the effective bond shear.

4.1.4 Coatings and Corrosion Protection

Painting and coating of the steel members, where specified, should be carried out as far as practicable in the shop, under appropriate conditions of humidity and protection from extremes of weather. The joint surfaces should of course be masked so as to permit welding. Field coating of the joints and touchup of shop coats should be done only when the surfaces are dry and at the proper temperature. In some locations, portable tents or other protection will have to be provided. Heaters and/or dehumidifiers may be required.

It is extremely important that surface preparation be thorough and in accord with the specified requirements. The offshore environment will quickly degrade any coatings placed on damp steel, over mill scale, and the like.

DNV rules require that the provisions for coating include:

A description of general application conditions at coating yard
Method and equipment for surface preparation
Ranges of temperature and relative humidity
Application methods
Time between surface preparation and first coat
Minimum and maximum dry film thickness of a single coat
Number of coats and minimum total dry film thickness
Relevant drying characteristics
Procedure for repair of damaged coating
Methods of inspection—for example, adhesion testing and holiday detection

Surface preparation and application of coating should be carried out when the surface temperature is more than 3°C above the dew point or when the relative humidity of the air is below the limits recommended by the coating manufacturer.

Coatings are usually applied to steel in the splash and atmospheric zones and to internal spaces which are exposed to seawater. In the case of sealed internal spaces permanently filled with seawater, corrosion inhibitors may be added to the water prior to sealing.

Sacrificial anodes are normally used to protect steel below water. They must be carefully installed in accordance with the specifications to ensure that they cannot become dislodged during transport, launching, installation, pile driving, and service. An adequate electrical connection between sacrificial anodes and the steel structure is essential.

Impressed current cathodic systems are sometimes installed in lieu of sacrificial anodes. They are prohibited in closed spaces or where water flow is restricted because of the possibility of generation of hydrogen.

Coatings may be applied to members which will be underwater in service in order to minimize the requirements for protection, provided the coating has adequate resistance to cathodic disbondment.

In the splash zone, additional protection may be provided by means of monel wrap, copper nickel, austenitic stainless steel, or carbon steel plate wrap or simply by allowing an added steel thickness in order to provide for some corrosion. Allowances of 0.1–0.3 mm/yr are made, with the higher values being used in locations where silt or ice in the water can remove the protective corrosion products, exposing new surfaces to attack, and in aggressive areas such as the Arabian Gulf.

Recently, precoated steel tubular members have become available from Japan. Polyethylene coatings are applied in the plant. Polyurethane coatings are applied both in the plant and in the field.

Spray-applied dense polyurethane coatings,

dense epoxies, and zinc-enriched epoxies have been developed for application to tubulars and structural steel. These not only give good corrosion protection but also possess good resistance to abrasion.

4.2 Structural Concrete

4.2.1 General

Reinforced and prestressed concrete has been used for some 20 large offshore platforms, most of them in the North Sea. This material lends itself to the gravity-based caisson type of structure and has often been used when offshore storage has been required.

Structural concrete has also been used for Arctic offshore platforms. Concrete is also used in conjunction with structural steel in hybrid and composite designs. Cement grout is used in steel offshore platforms, principally to bond the piles to the skirts and/or jacket legs.

Structural concrete is itself a composite material, consisting of aggregate, cement mortar matrix, reinforcing steel, and prestressing tendons.

Structural concrete should conform to the best practices of concrete construction and codes as set forth in building codes and recommended practices for bridges, nuclear reactor containments, and so on, as applicable. As in the case of steel offshore structures, the harsh environment and the special loading combinations and operating requirements make it necessary to supplement such general documents with recommended practices and rules for marine and offshore concrete structures. The principal rules used are listed in the Introduction and in the Bibliography. Excerpts relating to construction are presented in the appendices.

The concrete portion of structural reinforced and prestressed concrete has traditionally employed natural sand and gravel aggregates, giving a unit weight of about 2300 kg/m^3 (145 lb/ft^3). More recently, especially for Arctic platforms, where shallow water limits the draft, structural lightweight concrete has been employed, having a unit weight of about 1830 kg/m^3 (115 lb/ft^3). Both these unit weights are wet weights, without including the reinforcing steel.

Structural concrete as a whole and its individual components must be designed to work together in effective composite action and must be durable under exposure to the sea and the air. Emphasis is placed in design and construction on quality control so as to assure a long life with minimal maintenance.

4.2.2 Concrete Mixes and Properties

For modern offshore construction, the desired properties are often complex, demanding, and occasionally conflicting to some degree, thus requiring development of an optimal solution.

Compressive strength has historically been the controlling parameter by which concrete quality has been measured. We now know that while it does have importance in determining the resistance to external forces, it is not necessarily an accurate nor adequate measure of other qualities. Recent advances in concrete technology have led to significant increases in concrete compressive strength, and the trend continues.

Tensile strength determines the onset of cracking and shear strength and influences fatigue endurance.

Reinforced and especially prestressed concrete shows excellent *fatigue endurance* in the air, as long as the concrete is not cycled into the tensile range to a greater level than half its static tensile strength nor into its compressive range more than half its compressive strength. These limits are normally readily met in practical design.

When submerged, conventional concrete shows a reduction of fatigue endurance, apparently due to high pore pressures generated within the microcracks. Interestingly, structural lightweight concrete using modern high-strength, lightweight aggregates shows no such reduction.

Fortunately, even the reduced S–N curve or equivalent for submerged prestressed concrete structures of normal-weight concrete is adequate for water depths of current interest. For the greater depths envisaged in the future, use of either lightweight aggregates or silica fumes or both may be indicated.

Permeability is an extremely important property. Low permeability to seawater and chlorides is desirable to minimize the occurrence of corrosion. Low permeability to water and especially water vapor is important to the prevention of freeze–thaw damage in cold environments. The use of cement having a moderate tricalcium aluminate (C3A) content is beneficial in that it combines with the chloride ion to form an insoluble compound that blocks the pores. Permeability in concrete occurs primarily along the interfaces between the aggregate and the cement paste matrix. It can be minimized by selection of a mix with minimum bleed, by the use of aggregates having surface characteristics that promote physical or chemical bond, and by adopting a low water-amount ratio.

Both the *quantity and the quality of entrained air* (its pore or bubble size and its spacing) are important to ensuring durability in low-temperature areas. Because the concrete is being used in the marine environment, it has the potential for water absorption and saturation; when such a condition exists, the number of freezing and thawing cycles to cause freeze–thaw disruption is significantly reduced. The sea itself, rising with the tide or the waves splashing over the cold concrete surfaces, thaws the concrete, thus drastically increasing the number of cycles when the air temperature is below freezing.

Abrasion resistance formerly was felt to be determined by the hardness of the aggregate. Now it is recognized that the strength of the cement paste and the bond with the aggregate are prime factors. The use of condensed silica fume is especially advantageous in critical applications.

Sulfate resistance is almost a non-problem in seawater when rich and impermeable concrete mixes are employed, except in areas having extremely high sulfate contents (the Arabian Gulf, for example). The addition of pozzolans is especially useful in both reducing permeability and in combining with the free lime so as to reduce the chemical attack of the sulfate ion.

Resistance to petroleum compounds and crude oil is provided by normal high-quality concrete. Resistance can be enhanced by the inclusion of pozzolan in the mix. This also enhances the resistance to the sulfate/sulfide-producing bacteria such as *Theobacillus concretivorus* which blossoms from anaerobic bacteria released with the oil.

A high *modulus of elasticity* increases stiffness; a low modulus enhances energy absorption and ductility.

Creep has long been considered an undesirable property because it reduces the effective prestress and produces permanent deformations such as sag. However, it also is beneficial in enabling concrete to adjust to sustained concentrated loads, thermal strains, and differential settlement. Thus it often reduces the onset of cracking.

Fire resistance is important in the structural portions embodying operating facilities such as utility or riser shafts or where hydrocarbons may be accidentally released. On the typical offshore structure, the principal elements of fire resistance which are of interest are spalling, thermal conductivity, and creep at elevated temperatures.

Bond properties are of special importance when mortar or grout are used to transfer shear from piling to skirt or jacket sleeves. It is also of importance in the anchoring of reinforcing steel and of prestressing tendons.

Heat of hydration must often be limited in order to reduce the temperature gradients which later arise when the outer surface cools or when the element as a whole cools but is restrained. Too high a heat of hydration can lead to thermal cracking. Heat of hydration may be reduced by using coarser-grind cement, by controlling the cement chemistry (e.g., using blast furnace slag cement), and by cooling the mix. The mix is usually cooled by one or more of the following methods:

1. Water soaking of aggregate piles, so as to cool by evaporation
2. Shielding of aggregate piles from the sun
3. Shielding of batch plant, cement silos, delivery trucks, and so on from the sun or using reflective surfaces
4. Mixing ice instead of water
5. Introduction of liquid nitrogen into the water or concrete mix.

Temporary insulation may be used to reduce thermal gradients. Other properties may become of importance for specific applications offshore, such as floating or submerged cryogenic storage.

The concrete mix will typically consist of the following ingredients:

1. *Cement.* This should be similar to ASTM Type II, except that a C3A content of 8–10 percent seems optimum to minimize chloride attack. With very thick sections or large masses, blast furnace slag–portland cement in the ratio 70:30 may be employed. Alkali content should be limited to .65 percent (Na_2O + .65 K_2O).
2. *Coarse Aggregate.* Natural or crushed limestone or siliceous rock (gravel), maximum size 20–25 mm for normal sections, but may be as small as 10 mm for congested and thin sections. For lightweight concrete, use structural lightweight aggregate, normally of sealed-surface type, having minimum water absorption characteristics.

3. *Fine Aggregate.* Natural or manufactured sand conforming to standard grading curves but emphasizing minimum fines.
4. *Pozzolanic Additions or Replacements.* Use pozzolans, ASTM Class F (fly ash) or N (natural) with limitations on free carbon, sulfur, and CaO. In special cases, use condensed silica fumes in order to achieve very high strengths and impermeability
5. *Water.* For all reinforced and prestressed concrete, use only fresh water, with appropriate limits on chloride ion and sulfate ion.
6. *Water/Cement Ratio.* Normal maximum, 0.42.
7. *Water-Reducing Admixtures.* Consider use of high-range water-reducing agents (superplasticizers) where reinforcement is very congested or where high strength is desired.
8. *Slump.* With conventional water-reducing agents, 50–100 mm; with high-range water-reducing agents, 150 mm.
9. *Retarding and/or Accelerating Admixtures* as required. Do not use $CaCl_2$ as accelerator in reinforced or prestressed concrete.
10. *Air Entrainment Agent.* To give proper amount of entrained air and proper pore size and spacing in hardened concrete.

In recent years there has been a revolution in concrete technology, resulting in the ability to specially design concrete mixes for specific performance characteristics and environments. Thus the state of art concrete mix requirements become a recipe, including not only mix proportions but sequences of addition.

Fresh water should be used for all structural concrete which is reinforced or prestressed. The chloride content of the mix is an important factor in ensuring protection against corrosion of the steel.

However, salt water can be used in unreinforced concrete, such as breakwater armor units (Dolos, Tribar, Tetrapod, etc.). The concrete mix should be verified by trial, since the salt water tends to accelerate set and since it may be incompatible with certain admixtures. Added dosage of retarding admixture may be necessary.

Salt water can also be used with the unreinforced underbase concrete for gravity-based structures such as the concrete offshore platforms of the North Sea. There, salt water has been used with heavy doses of retarder.

Salt water may be suitable also as the mixing water for the concrete of belled footings and the grout for cementing deep piling where located below the seafloor or completely encased by steel, so that no oxygen is available.

With so many components, compatibility is essential. Many of the problems which have developed can be traced to incompatible trace chemistry in the various components. For this reason, trial batches are always recommended, duplicating insofar as practicable the placing and curing conditions and temperatures.

Because of the above considerations and because of the worldwide spread of offshore construction, no attempt has been made to tie the components to specific national specifications nor to specific quantitative values. These can be obtained in general terms from handbooks on concrete technology and in specific terms from consultants specializing in this field and knowledgeable of the construction area and the service environment.

Special warning is given concerning application to new environments or requirements. The concrete mixes which are giving highly satisfactory service in the North Sea will probably require important modifications for use either in the Arctic or the Middle East.

4.2.3 Conveyance and Placing of Concrete

Concrete, properly mixed, may be conveyed by a wide variety of means to the site for placement. See Figures 4.2.3.1 and 4.2.3.2. It is important that its properties not be significantly altered during conveyance.

If conveyed in trucks, continuous agitation of the

Figure 4.2.3.1 Floating concrete plant produces concrete for the Ninian central platform.

Figure 4.2.3.2 Distributing concrete by pumping, Ninian central platform.

entire batch is normally required. This means that blades must not be excessively worn and that there not be significant buildup of hardened concrete inside the mixer. If conveyed by conveyor belt, covering may be required in intense heat to prevent premature stiffening or flash set or, in rainy weather or extreme cold, to prevent excessive bleed and slump loss. In all cases segregation must be minimal. If conveyed by pumping, the pressure may be so great as to cause water absorption into the aggregates, causing a loss of slump. Entrained air content may be seriously reduced, although experience shows that spacing factors may usually remain satisfactory.

All the above means have been used satisfactorily, but only when recognition has been given to the various factors and appropriate steps taken.

Concrete when placed must be properly consolidated. In general, internal vibration is required, even when high-range water-reducing admixtures are used, in order to ensure complete consolidation and filling of all spaces and interstices between reinforcement.

When placing in hot weather (above 30°C, 90°F) or in cold weather (below 5°C, 40°F) adequate procedures must be implemented to ensure not only a proper mix temperature when placed but protection until it has gained adequate maturity.

4.2.4 Curing

Current research has greatly revised requirements for curing. Because most of the mixes used for sophisticated offshore structures are highly impermeable, emphasis today is primarily on sealing the surface against loss of moisture and heat rather than supplying additional water, which in most cases was never carried out thoroughly anyway.

Membrane-curing compounds represent one form of sealant, the white pigmented variety being especially suitable for reflecting heat in hot climates. However, heat, whether from the sun, the internal heat of hydration, or steam curing, degrades the curing compound, and so one or more additional applications may be necessary during the first day.

Where coatings (such as epoxies) are otherwise required, these may often be best applied to the concrete as curing membranes as well, provided they will properly adhere to damp concrete.

4.2.5 Reinforcing Steel

Conventional reinforcing steel is also called passive steel because it is nominally under a state of no stress when enclosed in the fresh concrete. Actually it usually ends up being in mild compression due to shrinkage and differential cooling of the concrete after hardening.

The typical offshore structure uses very heavy concentrations of reinforcing steel, far more than the usual structure. To provide adequate space for concrete placement, the bars may be bundled in groups up to four. Special consideration should be given to the design of the concrete mix to ensure that adequate "paste" (cement–sand mortar) is available and that thorough vibration, preferably internal vibration, is employed, so as to completely fill the interstices between bars.

Cover over the reinforcement is important for durability in the marine environment. Too much cover increases the width of cracks, while too little leads to easier access by chloride and oxygen and in extreme cases to loss of mechanical bond.

Reinforcing steel requires an adequate development or bond length at the end of each bar, in order to be able to develop the full strength of the bar without pullout. Because of the dynamic and cyclic nature of loadings on offshore structures, bond lengths are often set to be greater than those specified for static loads, up to as much as double. The positioning of the ends thus becomes of importance, so as to ensure that the anchorage is in a zone that is able to develop bond, that is, within a compressive zone. Tie wire for steel reinforcement should be black (uncoated), soft, iron wire, never copper or aluminum; otherwise local pitting may ensue. Similarly, welded fixing of bars may lead to pitting corrosion and loss of fatigue endurance.

Stirrups are extensively used in offshore structures. It is desirable that the tails of all stirrups be anchored back within the confined core of shells, slabs and beams, but this is often impracticable. Tying of the free end helps. Because of this problem and the fact that at high loads the concrete under the bend of stirrups crushes, stirrups rarely develop their full yield strength. Because of the bends in stirrups, their fatigue strength is limited. Therefore, headed bars, such as a forged or fabricated T-bar, anchored behind the in-plane reinforcement have been developed so as to provide a more efficent tie and shear-resistant element. See Figure 4.2.5.1.

Splices of reinforcing bars may be made by lapping; however, the long splice lengths and the congestion when trying to lap-splice bundled bars raises many difficulties. Mechanical splices and possibly welded splices appear to be desirable for the larger bar sizes. Welding of reinforcement must employ low-hydrogen electrodes.

Working drawings must show splice details so that congestion may be evaluated and the weights of steel accurately calculated, the latter for purposes of draft and stability control.

Splices of bars destined to work only in compression may, according to the code, be end-bearing. However, as a practical matter, in most cases the reinforcing steel in offshore structures must work in both tension and compression under some loading conditions. Therefore, it is considered good practice to design splices to transmit both. If lap splices are used, the bar size should be limited to 32 mm (1¼ in.) diameter and the laps should be well tied at both ends. Splices in adjacent bars must be staggered. Mechanical or welded splices are preferred for the larger-diameter bars.

Corrosion protection for the reinforcement has traditionally been provided by selection of the proper concrete mix for low permeability and cement chemistry, along with an adequate cover and good consolidation and curing. The constructor is responsible for ensuring that the specified cover is achieved by use of appropriate chairs and spacers.

In recent years, the use of epoxy-coated reinforcement for bars located in the splash zone and decks of bridges has grown extensively, especially in the United States. This epoxy is normally applied as a powder, electrostatically, and hence bonds tightly, so that bars can be field-bent to some degree without loss of coating integrity. To prevent abrasion damage, bars should be handled with fiber slings. Abraded areas should be touched up in the field.

A practical problem in construction arises when bars of different steel grades or both coated and uncoated bars are to be placed in the same general regions. Clearly identifiable coding must be employed or else all the steel should be of the highest grade and coating. Clearly identifiable color coding is preferable to the marks stamped on the bars' deformations; the latter are of use in the warehouse but not in the field for installation and inspection.

The supports for reinforcement should be concrete "dobe" blocks or plastic chairs. Stainless steel chairs may lead to local pitting of the bar in a salt-water environment.

Recently, practicable methods of cathodic protection of embedded reinforcing steel have been developed and applied where serious corrosion has appeared.

Bending in the field must follow carefully detailed procedures. Reference is made to the ASME–ACI *Code for Nuclear Containment Structures*, where provisions for field bending are carefully prescribed.

Small tie bars, 12 mm and less, of weldable steel may be field-bent as required.

4.2.6 Prestressing Tendons and Accessories

Prestressing tendons usually consist of seven-wire strand for long lengths and threaded bars for short lengths or where the tendons must be successively extended.

The tendons are placed within ducts, stressed and anchored, and then grouted.

Ducts should have thick enough walls to prevent local sag, be tight against the entry of mortar from concrete mix when it is vibrated, and have smooth

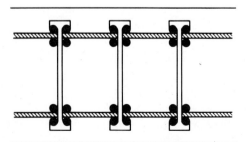

Figure 4.2.5.1 Tee-headed stirrup bars for peripheral wall of structure subjected to impact or "high flexure plus shear."

interiors so as not to snag the strands when they are pushed or pulled in.

The ducts should be tied to the reinforcing steel so as to maintain a smooth profile in accordance with the design and within the limits of the prescribed tolerances. Ducts must be protected against the accidental entry of debris and concrete materials. Vertical ducts should be delivered with plastic caps attached. Prior to the general use of plastic caps, ducts have been blocked by such miscellaneous waste as screwdrivers, concrete aggregate, rags, and even soda bottles! Duct ends must be protected from burrs. Ducts should be sawed, not burnt, and a cap attached.

Vents should be provided at all high points in the tendon profile.

The seven-wire strand is high-tensile steel and must be treated with care. It should be stored in a dry, weatherproof warehouse: German specifications require that the warehouse be heated so as to reduce the relative humidity well below the dew point. When inserted in the ducts, it should be entered through a smooth, abrasion-free trumpet or funnel.

The present practice is to procure strand that is coated with water-soluble oil. During insertion, more water-soluble oil may be daubed on. The ends of the duct are then sealed against moisture entry until stressing.

As an alternative to the use of water-soluble oil, a vapor-phase inhibitor (VPI) powder may be dusted on.

Stressing is carried out normally from one end for straight tendons, from two ends for curved tendons. For curved tendons, friction losses may be minimized by one or two cycles of pulling, first from one end, then from the other.

The grouting of ducts can be carried out in accordance with standard practice. The grout mix should be selected for minimum bleed. The ducts should not be flushed with water. Where water-soluble oil has been employed, the grout should be pumped through until all oil-contaminated grout is expelled from the far end. Vents should be closed and the grout forced out through the strand ends.

Vertical ducts and those having significant vertical components require special consideration, especially for offshore structures. What happens is that after the duct has been pumped full, the head of grout will force the water out into the strand interstices which will act like a wick, expelling water through the strand ends at the top. This allows the surface of the grout to settle, leaving a hidden void at the top, which may be several meters in length.

Several means are available to help prevent this:

1. Minimize bleed in the grout by selection of mix and admixtures.
2. Use a thixotropic admixture which causes the grout to gel as soon as pumping stops.
3. Use a standpipe so as to maintain a reservoir of grout above the top of the upper anchorage.
4. Top up the grout a few hours after initial grouting.

Steps 3 and 4 require an extra hole in the anchorage plate.

Usually two or more of these steps will be found necessary.

In cold weather, the grout must be protected from freezing. Prior to grouting, the temperature of the concrete structure is to be at least 5°C (40°F), and this should be maintained for two days after grouting.

Anchorages should normally be of the recessed type. For offshore structures, these are the zones most likely to show corrosion. The provision of proper anchorage patches requires a high degree of care in detailing and workmanship. See Figure 4.2.6.1.

In temperate climates, coating the pocket with epoxy bonding compound or latex, pouring a high-quality grout, using the window box technique to ensure complete filling, and provision of two or more steel ties, bent down from the structural concrete in the pocket patch, represent the best practice. In some cases a strand extension may be used for this purpose.

For structures which will be subject to a freeze–thaw environment, the epoxy should be omitted. Instead, the surface of the pocket should be carefully cleaned by wire brushing and latex applied.

4.2.7 Embedments

Embedment plates should be accurately installed prior to concreting: They must be sufficiently fixed so as to avoid dislocation during slip forming and vibration. In some current practice, it is required that embedment anchors be electrically isolated from the reinforcing steel. This is usually done by tying on of concrete dobe blocks, using fiber or plastic cord.

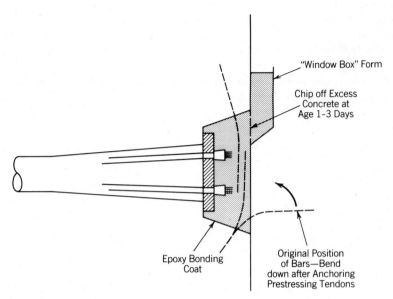

Figure 4.2.6.1 Recommended anchorage pocket details to provide protection to prestressing tendons.

After concreting, the embedment plates should be sealed at the edges to prevent crevice corrosion behind them. This is especially necessary if padeyes or similar attachments are later welded to the plates, as this causes heat distortion. Sealing is facilitated if a rubber or wood strip has been attached to the plate all around; this then forms a recess to receive the epoxy sealant.

Dissimilar metals must be avoided: Stainless steel of many grades is highly cathodic to steel. A very serious and expensive error was made in which stainless steel lugs were used to position cast steel embedments: the latter corroded seriously in a matter of months. Copper and aluminum can also lead to corrosion of the steel.

4.2.8 Coatings

If coatings are specified, it is important that the surface condition of the concrete be properly prepared. This may involve filling of air or water "bugholes" or honeycomb, and light (not heavy) sandblasting of the overall surface.

Primers and coats should be applied and cured in accordance with the manufacturer's recommendations.

Because of the prevalent damp conditions around offshore structures, use of a hydrophobic epoxy which permits placement on damp (not wet) surfaces will be found desirable.

Many otherwise satisfactory coatings develop "blowouts" and "pinholes" a week or two after application due to water vapor pressure from underneath the coating. Procedures and materials should be selected to minimize these and, where they occur, to fill them.

4.2.9 Construction Joints

Waterstops are no longer considered necessary in most cases where reinforcing steel extends across the joint.

For horizontal joints, rough screeding and cobbling of the top surface, followed after initial hardening by a waterjet to clean off all laitance, can produce a sound, watertight joint. The indentations should be sufficient to engage the coarse aggregate, normally 6 to 10 mm deep.

The subsequent concreting should start with a lift of the regular mix, minus the coarse aggregate, then be followed by the regular mix. The two should be vibrated internally so that the first lift is well engaged by the second.

Vertical construction joints can be prepared after hardening by wet sandblasting or high-pressure waterjet, so as to expose the coarse aggregate. Reasonably good success has also been experienced by placing metal wire mesh against the form of the first pour and then stripping it when the form is removed.

4.2.10 Forming and Support

Vertical elements of offshore structures—slabs, shells, walls, and so on—usually constitute the major portion of offshore concrete structures. The basic forming systems employed are panel forms, slip forms and flying forms.

Slip forms enable the steel to be installed as the forms are raised; thus the concrete is placed at deck level, where it is not dropped, that is, segregation is not a concern, and where it is accessible for internal vibration. The rate of rise of the forms is controlled by the rate of production of the concrete and the rate of placement of the reinforcement, prestressing tendons, and embedments. The time of initial set of the concrete is then controlled by the use of admixtures, so that the concrete emerging from the lower end of the slip forms will not slump or fall out. The temperature of the mix and of the air will also affect time of set.

When reaching the top of a slip-forming operation, there is no longer the weight of fresh concrete to hold the last lift down: jacking up of the forms after initial set may cause horizontal cracking. One solution is to stop them until after final set and then loosen them and jack up a short distance.

During concreting of slip-formed structures, the vibrator should be marked so that it will always be inserted through the present lift and extend into the previous lift; this to prevent horizontal cracking.

The use of pozzolans and especially silica fumes makes the concrete adhere to the slip forms. Methods being currently tried as a means of overcoming this include special coatings on the forms and possibly the incorporation of a pumping-aid admixture in the concrete to reduce friction. In any event, steel slip forms should be coated so as to prevent rusting, since rusting greatly increases surface friction. Stainless steel forms, coated with dense polyurethane, have recently been adopted by a leading offshore concrete contractor.

Slip forms have been used on offshore structures to construct tapered sections and to construct inclined sections. See Figure 4.2.10.1.

For the casting of the cell walls of the Oseberg A platform, Norwegian Contractors constructed a giant slipforming yoke, giving more room for installation of reinforcing bars. The side panels of the slipforms are galvanized to prevent corrosion and some are coated with polyester to reduce friction.

"Flying forms," that is, panel forms which are

Figure 4.2.10.1 Norwegian Contractors demonstrate their skill in slip-forming with this remarkable inclined and tapered shaft constructed on land at Stavanger. This was a half-scale model of a leg on the NC-T300 platform concept, proposed by Norwegian Contractors for the Troll Fields in 300 m of water.

progressively raised after a lift has gained sufficient strength, usually at age 1 or 2 days, are employed where the congestion of reinforcement steel, prestressing ducts, and embedments is so great that an impracticable number of workers would be required to maintain the minimum rate of slip. Typical flying-form panels are 3 to 6 m high; they may be equipped with window boxes to minimize the height of fall of the concrete (hence, reduce potential for segregation) and to facilitate vibration. Elephant trunk tremie pipes may be used to prevent segregation when pouring in deep sections of thick walls.

Typically, inserts are installed at the top of each lift to support the lower edge of the form in its next lift. The forms are sealed by gaskets along the bottom edge so as to prevent mortar leakage.

Horizontal slabs must of course be supported.

Typical scaffolding may be used in some cases, but in many cases the slab will be high above any supporting structure. Then trusses may be used, supported on and spanning between walls.

Stay-in-place forms may be employed to eliminate the need for stripping; this may also be applied to the trusses.

The reinforcing itself may be prefabricated so as to serve as an internal truss, with stay-in-place forms hung from it.

When the walls are slip-formed, blockouts and bent-up dowels may be left, so that the wall construction may continue past and the horizontal slab can be constructed later.

Horizontal and sloping surfaces may be formed using precast segments set in place, aligned, and joined with reinforcing and cast-in place joints or with prestressing and epoxy joints. Where heavy craneage is available, this enables very rapid construction of complex elements. See Figure 4.2.10.2.

Match casting techniques may be effectively employed, so as to ensure accurate fit of mating surfaces and tendon ducts. See Fig. 4.2.10.3.

Precast elements may also be cast as half-depth segments, to be completed with a top pour of cast-in-place concrete or as merely stay-in-place forms.

If such half-depth segments are designed to work in full monolithic action, then their surfaces must be roughened and reinforcing ties provided at relatively close spacing so as to prevent laminar cracking.

4.2.11 Tolerances

Construction tolerances will generally be given for at least the following:

1. Geometry of cross-section
   a. Deviation from true position along vertical axis
   b. Deviation from true circle or polygon on a horizontal plane
   c. Deviation from best fit circle or polygon
2. Verticality: deviation from vertical and horizontal axis and planes
3. Distances and bearings between shafts
4. Thickness of members
5. Positioning of reinforcement (through thickness and along wall)

Figure 4.2.10.2 Erecting precast concrete shell elements on Ninian Central Platform.

6. Cover over reinforcement and prestressing ducts
7. Positioning of embedments
8. Deviation of prestressing ducts from design profile
9. Deviation in fresh unit weight of concrete

4.3 Hybrid and Composite Steel–Concrete Structures

The combination of structural steel and concrete in offshore structures is an obvious development that has finally emerged as the synergistic benefits of combining the two materials have been recognized. Two forms of steel–concrete construction offer potential advantages.

Figure 4.2.10.3 Match-casting techniques were used to construct the breakwater segments for the Ninian central platform.

4.3 Hybrid and Composite Steel-Concrete Structures

4.3.1 Hybrid Structures

The first form is where elements of structural steel are jointed to elements of concrete. Each works in its own way, but the joint must transmit structural forces in such a way as to preserve the integrity of the joint. Examples are:

1. Structural steel superstructures jointed to concrete substructures
2. Structural steel frames supporting exterior concrete walls
3. Steel "cardan" (articulated hinge) providing the articulated joint between a concrete base and a steel or concrete column of an articulated loading buoy.

These will henceforth be referred to as *hybrid structures*.

For these hybrid structures, the principal concern is the working of the joint under cyclic–dynamic loads. This can properly be taken care of by prestressing the steel element to the concrete element. See Figure 4.3.1.1. To ensure successful performance of such a prestressed joint, several aspects have to be given special consideration.

1. Because the prestress tendons are usually relatively short, the effect of takeup and seating losses will be significant. For this reason, prestressing hardware has been developed in which the final securing is done by threaded nuts or couplings rather than by wedges.
2. The anchorage zone of the prestressing tendons within the concrete must be carefully detailed to ensure that cracking and progressive degradation cannot occur around and behind the anchorage. The concentrated bearing of the anchorage slate produces high strains which in turn lead to radial tension: the well-known bursting forces associated with a high-capacity anchorage. For this phenomenon, transverse reinforcement is required.
3. Full bearing is achieved between the steel bearing plate and the concrete.

Less well recognized is the fact that where the tendon is only stressing a portion of the concrete, that portion tends to be pulled out from the remaining concrete. Hence anchorages in the concrete should be staggered and fully adequate passive reinforcing steel provided to distribute the tension back into the concrete structure.

The steel plate (P) of Figure 4.3.1.1 must be sufficiently thick to prevent local warping under the high concentrated forces imposed on it by the tendon anchorages. Obviously the weld (W) must have had adequate procedures, such as pre- and post-

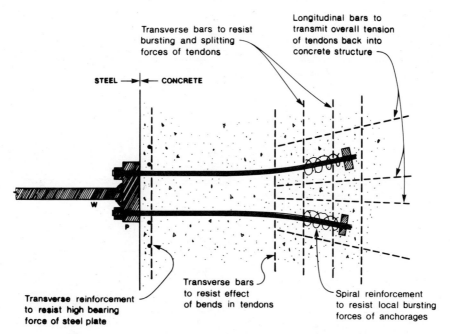

Figure 4.3.1.1 Connection of steel elements to concrete structure.

heating, to ensure full development of the connection. The plate P must have proper through-thickness to prevent laminar tearing.

As noted, the bearing of plate P on the concrete must be uniform. Two methods are used. In the first, the gap between plate P and the concrete is filled with cement mortar or a special bearing compound (noncorrosive) or is injected with epoxy. In the second, the concrete is ground and the plate milled, both to a fine tolerance such as 0.1 or 0.2 mm., the allowable tolerance depending on the size of the plate and its stiffness (thickness).

4.3.2 Composite Construction

The second form of steel–concrete construction is where steel plates are joined with concrete so that the two form one structural element in themselves. The two will normally have mechanical means for developing shear between the steel and concrete components.

Typical arrangments are:

1. A steel beam is joined to the concrete by welded studs. The concrete takes compression; the steel plate takes tension and transverse shear.

2. Two steel plates, spaced apart, are filled with concrete. The two plates are tied together with steel diaphragms, bolts, or the like or perhaps even short stressed bars.

The concrete ensures plate and shell action to distribute local loads and transmits horizontal shear.

The external steel plates provide tensile capacity: The internal concrete provides compressive capacity, and the diaphragms provide shear capacity. With such sandwich construction, the steel plates will of course be forced to carry any compressive strain imparted to the concrete. Buckling inward is prevented by the concrete fill; buckling outward is thus the typical mode of failure and must be restrained by the through-thickness ties. Thus not only their size but their spacing and welding become important. Figure 4.3.2.1 indicates some forms of composite construction.

Where welded studs are used, the amperage and procedures must be in accordance with the manufacturer's recommendations. These welds are heavily stressed in shear and bending at points of high beam shear. Friction welds are preferable.

The surface of the steel should be clean so as to develop good bond with the concrete.

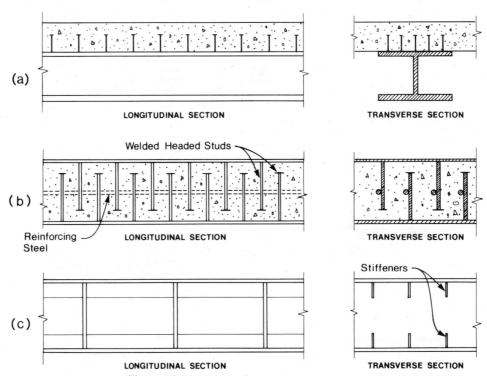

Figure 4.3.2.1 Forms of composite construction.

Placement of concrete into sandwich composite systems requires careful planning and procedures so as to ensure complete filling. Small holes may be necessary in internal stiffeners and diaphragms to prevent trapping of air or bleed water.

It will frequently be found practicable to leave temporary (or sometimes permanent) holes in one of the plates for access for placing the concrete by pump.

The hydrostatic head of fresh (unset) concrete, especially when placed under pressure by pump, may deflect the plates. This needs consideration and control.

Composite construction appears especially attractive where offshore structures must resist high local impact forces, such as those due to ice or boat/barge collision.

4.4 Plastics and Synthetic Materials

Increasingly, plastic and similar synthetic materials are being utilized in the marine environment. Uses range from fiber-reinforced glass for pipelines to neoprene and natural rubber fenders and bearings, to polyethylene bags for slope protection and polyurethane foams for buoyancy. Porous filter fabrics are extensively employed under riprap to prevent leaching of sand. Epoxies are injected for repair or applied as jointing and coating compounds. Polyethylene pipes have been used for cold water pipelines in depths up to 2000 feet off the island of Hawaii, and Kevlar and nylon mooring lines are in common use in floating offshore operations.

From the constructor's point of view, certain aspects must be considered.

1. Where the plastic is to be applied in the field, the joint surfaces must be clean, of proper texture (roughness), dry, and at the proper temperature to permit curing. Epoxies are especially sensitive to moisture and dilution by water, unless they are especially designed with a hydrophobic component to be used underwater or on damp surfaces.
2. Many jointing materials require a closely controlled thickness over which they are effective, with a narrow tolerance range.
3. Extreme temperature changes may lead to delamination of steel–neoprene bearings and fender units, due to change in properties and differential thermal contraction–expansion.
4. Many plastics, especially polyethylene, are subject to ultraviolet (UV) degradation, unless they are formulated with a pigment to give them added UV protection.
5. Most plastics are positively buoyant in water; this makes them difficult to place underwater. Some filter fabrics are now purposely manufactured so as to have negative buoyancy and thus facilitate placement. Fiber-reinforced glass pipe may require supplemental weighting to ensure stability during placement.
6. Membranes and fabrics are generally supplied in rolls and sheets, which are to be overlapped in installation. The laps are a cause of many difficulties both in placement and service. Adequate, even excessive overlap is usually a wise precaution for the constructor in order to accommodate irregularities in the seafloor and the tolerances associated with work underwater and under wave action.
7. Rigid plastics are susceptible to damage from impact. They generally require special softeners for handling.
8. Plastic pipe such as polyethylene is subject to internal fatigue at locations of concentrated stress. Therefore all attachment points must be reinforced and care taken to ensure a distributed rather than a concentrated bearing force. A polyethylene pipe for an OTEC project was under tow near Hawaii. Its weighted end was being held up by wire rope, which in turn was attached to the barge. Under sustained load, with cyclic stresses due to waves, the pipe ruptured, allowing the line to break loose. A major salvage effort was required for recovery.
9. Polysulfides can be rapidly disintegrated by bacteria under some conditions. In one case in the Arabian Gulf, the disintegration took place within a few months, before the construction contract was completed and accepted.
10. Flexibility properties change rather dramatically with temperature, so that a material which is very pliable at 15°C may be stiff and brittle at −10°C.

Having cited some of the problem areas, it is important to look at the many advantages offered by the use of such materials. They are free from corrosion. They generally have low friction factors:

Dense polyurethane and dense epoxies are used on icebreaker hulls to provide corrosion protection and reduce ice friction. Teflon pads are used to reduce friction when skidding heavy jackets onto barges and again during launching.

The light weight of these materials eliminates many handling problems. Many elements can be floated.

Polyethylene pipes can be "buckled" transversely during installation and later regain shape, all without damage. Their positive buoyancy permits them to be installed above the seafloor and anchored down at intermittent points.

Coflexip pipelines, comprised of specially bonded steel-reinforced neoprene, are extensively used for flexible risers and connections for offshore oil production. See Section 13.3.

Kevlar is being widely adopted for mooring lines and light-duty hoisting lines, especially in deep water, where its almost neutral buoyancy is highly beneficial. Kevlar can be obtained in either highly flexible or relatively stiff form. Nylon is also used for such lines but has a very low modulus.

Filter fabrics can be laid as mats underwater, ensuring far more complete, economical, and effective protection against sand migration than can be attained by graded rock filters.

In some cases, polyethylene liners are bonded to steel or concrete pipe in order to provide corrosion protection, reduce ion transfer into the fluid, and reduce friction.

Polyethylene is being increasingly utilized to provide corrosion protection to steel pipe piles and to wire rope cables. The guys for the guyed-tower structure Lena are encased in polyethylene for long-term corrosion protection.

It appears certain that the use of plastics in the marine environment will continue to grow. Because of the wide variety of properties of the various materials, the constructor must take special care to verify the special requirements for installation under the ambient conditions involved.

4.5 Titanium

Titanium is used in sophisticated marine vessels (e.g., submarines) where its special characteristics justify the increased cost.

Titanium structural elements can be rolled with the following properties:

Yield Strength	120,000–160,000 psi (800–1200 $\frac{N}{mm^2}$)
Endurance (fatigue)	60,000–70,000 psi (400–500 $\frac{N}{mm^2}$)
Weight	300 lb/ft^3 (4.8 tons/m^3 = 48 $\frac{kN}{m^3}$)
Cost	5 times that of steel

Future developments in metallurgy may make titanium more available at lower costs.

4.6 Rock, Sand, and Asphaltic–Bituminous Material

Rock is, of course, extensively used in coastal construction for seawalls, breakwaters, and revetments. For offshore construction, it is used to protect the foundations of structures from scour, to protect submarine pipelines from current-induced vibration, trawler boards, and impact, and to protect the slopes of embankments from wave and current erosion.

The three principal properties specified in the design relate to specific gravity, size, and durability. Increased density of the rock permits use of smaller-sized rock while still maintaining stability against erosion. The stability of rock under wave and current action is approximately proportional to the cube of the underwater density; that is, $S \cong$ (sp. gr. − 1)^3. Thus the contractor may be able to obtain the required results with material that is easier to handle and place.

Size of rock fragments is usually specified as a nominal maximum dimension or weight, with a gradation of smaller rock fragments permitted. In most cases, the most difficult and expensive rock for the constructor will be the larger size. In quarrying, therefore, he will choose methods designed to maximize the production of the larger rock. Usually he will still end up with an excess of finer material, some of which may be used elsewhere on the project, some of which may be used for his temporary roads, and so forth.

Conversely, to prevent sand migration through embankments, filter rock must be well-graded. This usually requires crushing and screeding.

A significant problem for the constructor is to

ensure that when a range of sizes is specified (a gradation curve), each area and zone in the completed structure ends up with a reasonable approximation to that gradation. Although he may have complied with the specified gradation in total, it requires experience and skill to ensure that each load is properly blended.

When rock is transported, there is always some abrasion and breakage. Thus there will be a layer of fines which will accumulate on the deck of a barge. In many cases this can be placed or wasted without harm; in other cases where porosity and permeability are important, precautions must be taken to ensure that this fine material is not allowed to contaminate the specified rock.

Typically the problem occurs at night, when overzealous operators "clean up the barge," often with serious results.

Durability of rock is normally specified to ensure against disintegration in seawater, primarily due to sulfate expansion. Rock produced from arid and desert regions is particularly suspect, as it has not been subjected to normal weathering.

Sand is generally specified by internal friction angle and gradation, the two of course being related. Sand is often produced hydraulically, which allows washing overboard of fines in order to achieve the prescribed gradation.

The principal problem in sand production is that of excessive fines, which prevents proper densification and makes the embankment subject to liquefaction and, above water, to frost heave.

Asphalt binders are used to bind rock and sand into flexible yet scour-resistant mattresses. Great technical advances have been made in this regard by Dutch engineers, who have perfected the ability to place asphaltic and bituminous materials underwater, including placement while hot. They have also developed various gradations and percentages of binder material so as to permit or restrict permeability as desired for the particular application.

Rubber asphalt, for which old tires have been blended into asphaltic mixes, has a number of desirable properties for marine application, since it has greater extensibility and flexibility over a range of temperatures.

A rubber asphalt layer was used on Global Marine's Super CIDS platform between the concrete and steel elements so as to provide uniform bearing while later providing high lateral shear resistance in the cold Arctic waters.

Roll on, thou deep and dark blue ocean, roll!
Ten thousand fleets sweep over thee in vain.

Lord Byron, "Childe Harold's Pilgrimage"

5

Offshore Construction Equipment

The demands of the offshore working environment coupled with the demand for large-scale sophisticated and very expensive structures have led to the development of a great many types of specialized and advanced construction equipment. Indeed, the response of constructors and equipment manufacturers to these needs has been rapid and effective; the availability of construction equipment of greater capability has in turn played a major role in altering methods and systems employed. These developments will continue as long as the offshore petroleum and other industries continue to push into new environments and greater depths of water.

The major equipment is designed to work in or under the sea and hence has drawn heavily on naval architecture to ensure stability, with appropriately limited motion response under the prevalent sea conditions. This extension from conventional barges and ships, destined primarily to transport cargo, has in turn forced the naval architectural discipline to develop a methodology adaptable to a wide variety of configurations, attitudes, and dynamic actions. Most transport vessels are designed to alter course and speed to avoid excessive dynamic response, and even to run from severe storms such as hurricanes. Icebreakers have been designed to follow open leads and avoid deep multiyear ridges. For the offshore industry, these tactics of avoidance and mitigation may no longer be practicable.

Life safety must be paramount in offshore operations. The nature of the work is inherently dangerous and demanding. Similarly, the cost of operations per hour or per day is extremely expensive; the equipment must be designed with reliability and redundancy to enable operations to proceed despite unfavorable weather.

Construction engineers must understand the capabilities and limitations of the equipment they use. They must be alert to detect early signs of problems before they develop to catastrophic proportions. Thus a full understanding of equipment performance is essential.

In subsequent subsections of this chapter, principal generic types of offshore construction equipment will be discussed.

There are a number of basic considerations applicable to all offshore construction equipment. These are motion response, buoyancy, draft, freeboard, stability, and damage control.

5.1 Basic Motions in a Seaway

A typical floating structure has six degrees of freedom and hence six basic response motions to the waves: roll, pitch, heave, surge, sway and yaw. See Figure 5.1.1.

Wave action on vessels is exhibited by two effects. The first-order effect is the oscillatory force at the period of the waves. The second-order force, often called the wave drift force, is a relatively steady force in the direction toward which the waves are propagating. In irregular seas, it does vary slowly, with a period of one minute or more.

In a real sea, the vessel will be responding to a complex set of excitations, differing in direction, frequency, phasing, and magnitude. The responses are therefore a combination of all the above.

The typical construction vessel is moored with mooring lines of rope or chain, which give nonlinear restraint to the movements of the vessel. Energy is stored in these lines, to be subsequently released to the vessel as the restoring forces return the vessel to its mean position. Although mooring lines are normally designed to prevent low-frequency dis-

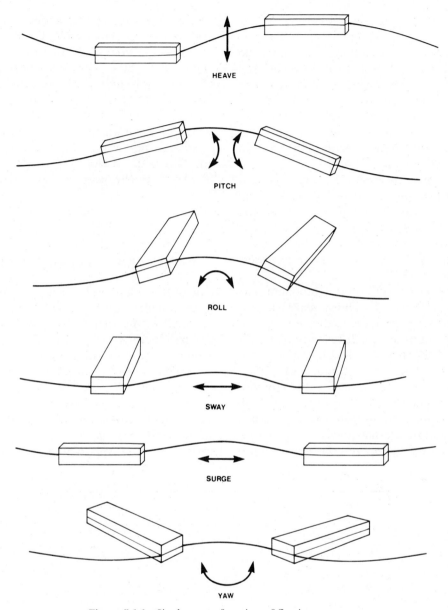

Figure 5.1.1 Six degrees of motion of floating structure.

placements in surge, sway, and yaw, they do incidentally act to force both low- and high-frequency excitations into the vessel.

The various types of motion interact so as to reduce or amplify the motion of any individual point on the vessel. A point of special interest in derrick barges, for example, is the boom tip. Due to the interaction of the six response motions, the boom tip may describe a complex three-dimensional orbit in space.

Vessel motions are affected by hydrodynamic forces when the vessel is in close proximity to a boundary—for example, when in shallow water, so that there is little water between the hull and the bottom.

Vessel response is highly frequency-dependent. Each vessel has its natural period of response in each of the six degrees of freedom. Note that in addition to a resonant period of the vessel itself at high frequency, there may in addition be a natural period at low frequency for the full vessel–mooring line system.

80 Offshore Construction Equipment

Maximum input excitation in pitch will occur when the effective wave length parallel to the vessel's longitudinal axis is about four to five times the vessel's length: Then the vessel will be riding the slope of the wave. Under such a condition it will also have maximum surge; in effect, it will be trying to surf. This is why the trend in design of offshore construction vessels is to make their length equal to 50 percent or more of the maximum wave length in which they are expected to work.

This is also why such great difficulty is experienced when using conventional offshore floating equipment in areas such as off West Africa or the southwest coasts of Africa and Australia, where very-long-period waves from the southern ocean, even though of moderate height, cause excessive response of the vessel.

The effective wave length is the component parallel to the longitudinal axis. Thus a sea of shorter wave length, acting at an angle to the vessel's axis, can produce a significant pitch response. The directional spread of waves can also cause a response in pitch even in a beam sea; see Figure 5.1.2. Similarly, even when the barge is headed directly into the sea, there can be a significant response in roll.

In shallow water, the long-period swells are shortened in length, so that they may approach critical length in relation to response of a typical barge. For example, with a barge of 120 m and a swell period of 20 seconds, the deep-water wave length will be $5T^2 = 2000$ ft or 620 m. In water depths of 15–20 m, these will shorten to 500 m or so, resulting in high pitch and surge.

In roll, most barge-type construction vessels have a natural period of 5 to 6 seconds. The average wind waves in a workable sea have a 5- to 7-second period, giving a wave length of 125–240 ft (40–70 m). This is the reason why offshore construction vessels are usually designed with beams greater than 25 m.

5.2 Buoyancy, Draft, and Freeboard

One of the oldest engineering principles of history is Archimedes' principle that the displacement will be equal to the weight. The same results can be reached by integrating the hydrostatic pressures acting on the vessel under still-water conditions.

Weight control is always a concern during the fabrication of structures. A check on weight can of course be kept by measuring the displacement. There are, however, a number of factors which may act to reduce the accuracy of simple draft measurements and displacement calculations:

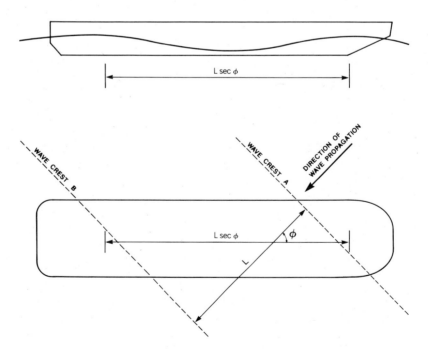

Figure 5.1.2 Effect of quartering waves on long offshore construction barge in producing both bending and torsion.

1. Variations in the density of the water
2. Deflections and deformations of the structure
3. Tolerances in the underwater dimensions, and hence in the displaced volume
4. Inaccuracies in calculation of ballast water and inadvertent drainage water
5. Absorption of water, for example, into concrete

Draft is determined by geometry and displacement; it is the depth below sea level of the lowest point of the structure, as measured in still water. Phenomena acting to increase draft are "squat," the hydrodynamic pulldown force acting on the hull when the vessel is moving through shallow water, "heel," which is the list of the vessel under wind and eccentric loading, and variations in trim. Pitch, roll, and heave can also of course reduce the bottom clearance.

Reduced bottom clearance can adversely affect the vessel's movement through the sea. Water pressures build up underneath the vessel, tending to increase the added mass and reduce speed: more water is pulled through the surrounding sea. The vessel loses its directional stability and starts to make rather extreme excursions in yaw and sway.

Freeboard is the deck height above the still-water level; like draft, it is influenced by squat and heel in shallow water, by list and trim, and temporarily by pitch and heave.

Freeboard is usually designed to minimize the wave which can build up and overtop the deck. The wave height adjacent to the vessel's side is built up above normal levels by refraction and mach-stem effects, the latter being the result of excess energy spilling lengthwise along the side.

Freeboard may also be selected to reduce the spray on deck; this is often due to the high winds acting on the top of the standing wave (clapotis), which upon reflection from the side can reach a height twice that of the incoming wave.

5.3 Stability

There are three major parameters controlling stability: the center of gravity, the center of buoyancy, and the water plane inertia. See Figure 5.3.1. Submerged vessels, with no water plane, depend on the center of gravity remaining below the center of buoyancy.

The formula for stability of a surface floating vessel is

$$\overline{GM} = \overline{KB} - \overline{KG} + \overline{BM}$$

where $\overline{BM} = I/V$, K is the geometrical centerline at the hull bottom, G is the center of gravity, B is the center of buoyancy, M is the metacenter, I is the transverse moment of inertia of the waterplane area, and V is the displaced volume.

The vessel will have inherent stability at small angles of roll as long as $\overline{GM}$ is positive. See Figure 5.3.2.

The transverse moment of inertia of the waterplane of a typical barge or other rectangular vessel is given by $I = b^3 l/12$, where b = beam and l = barge length. Since V is $b \cdot l \cdot d$, where d is the draft, I/V reduces to $b^2/12d$.

When the vessel rolls, B translates towards the downside. The easiest practical way to find the approximate location of B is to lay the cross-section out on graph paper and count the squares. This is especially useful for a vessel with a complex below-water configuration, for example, a semisubmersible.

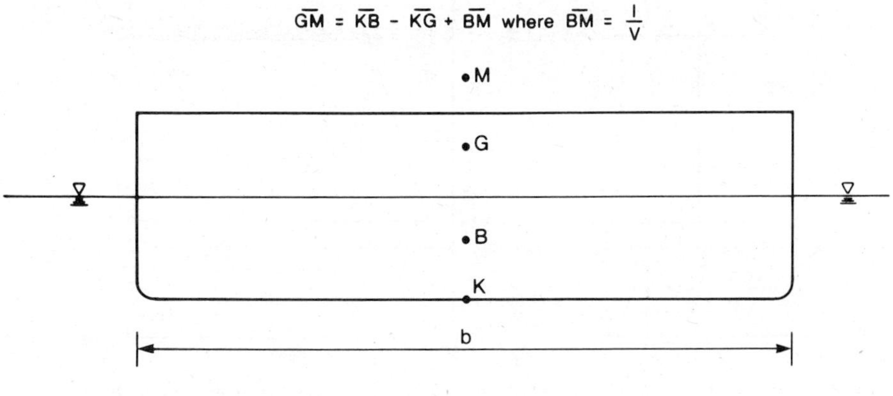

Figure 5.3.1 Relationship between centers of buoyancy and gravity and metacenter.

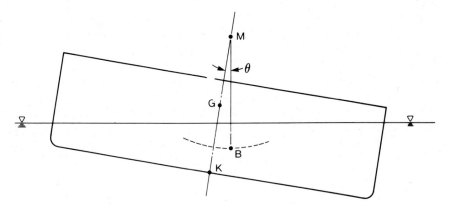

Figure 5.3.2 Effect of metacentric height.

The righting force acts vertically upward through B. The gravity force acts vertically downward through G. The righting moment is equal to the displacement times the righting arm, $\overline{GM} \sin \theta$. For small angles of roll, $\sin \theta$ can be assumed equal to θ, where θ is in radians.

The moment of inertia of the vessel's waterplane area is diminished by the moment of inertia of the waterplanes of any partially flooded spaces. The reduction in metacentric height and stability is known as the *free-surface effect*.

For structures having columns or shafts which extend through the waterplane, the moment of inertia is approximately proportional to $\Sigma A r^2$, where A is the area of each shaft and r is the distance from each shaft's centerline to the vertical axis of the structure. Thus the most efficient columns, insofar as stability is concerned, are those located furthest from the axis.

The above criteria are of value only for small angles of roll. At large angles of roll the geometry may change rapidly. The following are typical areas of sensitivity:

1. When the deck goes awash: The waterline plane diminishes rapidly, and the moment of inertia diminishes as the second or third power.
2. When the center of gravity is very high, as in a jack-up with legs raised. Then the stability becomes excessively sensitive to the instantaneous transverse moment of inertia.

Special care must be taken when the waterline area reduces materially with an increase in draft. See Fig. 5.3.3. Such a stability problem occurs with semisubmersible vessels, with conical structures such as those used in some Arctic offshore platforms, and with gravity-base structures which have a large base but only shafts extending through the waterline.

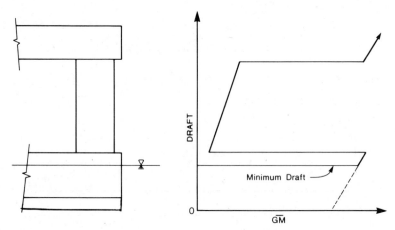

Figure 5.3.3 Typical variation of metacentric height with draft for semisubmersible configuration.

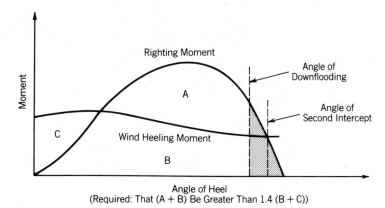

Figure 5.3.4 U.S. Coast Guard and American Bureau of Ships intact stability criterion.

When computing stability of a crane barge or derrick barge, the load picked is computed as though it were placed at the boom tip.

A more meaningful guide to stability is provided by the curves of righting moment versus wind heel. A typical curve for a relatively stable vessel is shown in Figure 5.3.4.

Many offshore organizations have adopted the U.S. Coast Guard and American Bureau of Shipping (ABS) rule, which is essentially as follows:

The area under the righting moment curve to the second intercept or downflooding angle (or angle which would cause any part of the vessel or its load to exceed allowable stresses), whichever is less, is to be not less than 40 percent in excess of the area under the wind heeling moment curve to the same limiting angle.

See Figure 5.3.4.

The above would apply, for example, to a barge carrying a large module, and the limit on allowable stresses would apply to the tie-downs as well as the barge and module.

Often a minimum $\overline{GM}$ for example, +1.5 m, is also required. When moored, the effect of mooring and anchor lines should also be considered.

A relatively critical situation exists for a vessel with a righting moment curve like that shown in Figure 5.3.5, which may be found with some jack-up barges in transit with legs raised because of shallow water.

A catastrophic accident occurred with a construction barge carrying concrete materials. As the materials from below deck were depleted, the $\overline{CG}$ increased, the draft lessened and the $\overline{CB}$ decreased. The $\overline{GM}$ became negative and the barge capsized, with tragic loss of life.

Serious problems of stability as well as potential collapse of the boom may occur if the load is allowed to swing with the roll. Use of tag lines, properly sized and secured, can reduce this tendency. In the event that a load gets away—that is, swings wildly free—the only solution may be to lower the load into the water in order to reduce its weight and dampen its motion. The act of lowering may also lengthen the period of swing, bringing it out of resonance with the natural period of roll of the barge.

5.4 Damage Control

Offshore construction vessels are subjected to collision from barges and boats to a far greater degree than normal vessels engaged in transport. The latter avoid close proximity to other vessels, whereas an offshore construction vessel must work with these other craft alongside. The offshore construc-

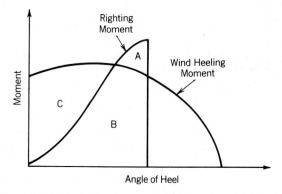

Figure 5.3.5 Example of an unacceptable response despite initial high GM (typical of some jack-up rigs with legs raised).

tion vessel is frequently picking and setting anchors; it is not unusual for the fluke to rip into the side. Finally, this rig must also work adjacent to platforms and other structures. Every precaution is taken to avoid collision with the structure because of the danger to the equipment, facilities and wells: for an operating platform there is also the danger of fire from hydrocarbon release.

Damage control considerations require that vulnerable areas be subdivided into smaller compartments, that all manholes and most doors be equipped with watertight gaskets and dogs, so that they may be kept closed except when actually in use, and that areas where anchors will rub or boats lie alongside be armored or fendered as appropriate.

During tow or when moored in heavy weather, green water will come over the decks of most barge-type vessels. An inadequately closed manhole will let in a large amount of water within a short time. See Figure 5.4.1.

Temporary attachments and supports are frequently welded to the deck. If welded only to the deck plating, they may pull free; the welds are in tension normal to the deck plate, and the deck plate may be unsupported from below at that point. Therefore holes are frequently cut in the deck plating so that attachments may be welded to the bulkheads below. These must be seal-welded to the deck plate to prevent water entry.

These temporary attachments, especially padeyes, winch foundations, and the like, are often subject to extreme lateral impact loading. The connection should be detailed so that failure occurs in the connection, not in the vessel's structure.

Figure 5.4.1 Offshore vessels are subjected during storms to heavy bending moments as well as green water on deck. (Courtesy Ocean Beach Outfall Constructors.)

Construction vessels often take a sudden list due to a shift in the load, for example, as a crane swings or a heavy deck module is moved to one side or lifted off. These sudden lists may coincide with a roll and temporarily submerge an above-deck door, vent, or other opening below the oncoming wave.

Other flooding accidents have occurred due to broken portholes (or ports left open).

Boats and small barges are often moving with heavy mooring lines, whose weight may cause a temporary trim down by the stern or bow, resulting in wave overtopping. Boats pulled or running astern are especially subject to taking water over the stern sheets.

Flooding into the stern well may have several severe consequences. It may enter the engine room or control room spaces and short out the power. Even a small amount of water in a compartment gives a free surface which reduces the metacentric height; that is, the I available at the waterline of the vessel is reduced by the waterplane I of the partially flooded compartments.

Watertight closures designated to be closed during operations must be kept closed and dogged. Two serious accidents occurred to workboats during the 1970s when, because of the calm seas and warm weather, engine room doors were left open for ventilation. An operational event caused water to be shipped over the deck; this in turn flooded the engine room spaces. Both boats sank, in each case with serious loss of life.

In a somewhat similar case, in dead calm weather, a semisubmersible derrick barge was making a heavy lift. The load swung to one side, the vessel heeled, and the swing engines were unable to hold the crane, which rotated to the beam, causing a very extreme heel, so that the upper deck was awash. Doors on the upper deck, supposed to be closed at all times, were wide open. In this case the crane operator prevented a catastrophe by lowering the load into the sea.

Construction vessels are usually equipped with ballast tanks so as to enable the list (heel) or trim to be controlled. These ballast tanks typically have vents which extend in a gooseneck above deck and are equipped with a flame-arresting screen and a flap valve. One purpose of these vents is to prevent accidental overpressurization of the tank, which might rupture a bulkhead and flood an adjacent space. However, these vents may become plugged or may be intentionally blocked off, for example, in order to store cargo on the deck space in question. The result may be overpressurization and a rupture of an internal bulkhead.

The more sophisticated offshore equipment of today—derrick barges, pipelaying barges, launch barges, and semisubmersibles—have complex ballast systems to enable their list and trim to be rapidly controlled, even as operations are being carried out. Accidents, even capsizing, have occurred when the controls were shorted. Therefore, emergency manual controls are also provided. The crew must be trained as to procedures after a malfunction.

Valve stems sometimes break loose, so that they appear "closed" when actually the valve gate itself is still partially open. Critical valves should be equipped with remote indicators.

Critical valves have been opened for testing and then inadvertently failed to be closed afterward. These should be equipped with locks or tags as appropriate.

Steel working under cyclic or impact loads can be subject to fracture, especially at low temperatures when it drops below the transition value. Usually these cracks start and propagate with repeated cycles. Careful inspection of critical areas can locate these incipient cracks before they have propagated to a dangerous degree. They can then be repaired or a crack-arresting hole drilled or strap installed. Decks of barges are especially vulnerable since they are exposed to low temperatures and high stresses.

Closed compartments inside steel vessels can be very dangerous to human life. The steel corrodes slowly but continuously, using up the oxygen in the compartment. In other cases, heavier-than-air gasses such as carbon monoxide may accumulate in the lower areas and bilges. Therefore, all compartments which have been closed and all tanks should be thoroughly ventilated before entering.

Fire aboard a vessel at sea is one of the traditional worries. For many fires, the most effective way of fighting it is to close off all air supply and cool down the adjoining bulkheads and decks by water spray. Fires can jump across steel bulkheads by igniting the paint on the other side. Electrical fires and hydrocarbon fires should not be fought with water. Acetylene tanks must be chained or strapped tight to prevent "falling, fracture, and ignition."

Materials, such as casing and all separate units stored on the deck of barges must be secured against displacement in the event of a sudden list. Casing and pipeline pipe are especially dangerous because of their ability to roll and the large tonnages involved. Shifting loads can cause the vessel to cap-

size. Braced steel structural posts well secured to the internal bulkheads of the barge are perhaps the best method of containment.

Often a large module is transferred from a cargo barge or a shore base to sit on the deck of a derrick barge. Even though the duration of exposure is relatively short, the module should be adequately secured with chains or wire lines so that it cannot shift even if the derrick barge rolls.

All life-saving equipment should be maintained in full operating condition at all times. When a capsule or raft or fire-fighting gear must be removed in order to carry on operations, it must be relocated or reinstalled immediately afterward. Emergencies can occur at any time and, according to "Murphy's second law," will occur at the worst possible time.

5.5 Barges

An offshore construction barge must be long enough to have minimal pitch and surge response to the waves in which it normally works, wide enough in beam so as to have minimum roll, and deep enough so as to have adequate bending strength against hog, sag, and torsion. The deck plating must be sufficiently continuous to enable it to resist the membrane compression, tension, and torsion introduced by wave loading.

Impact loadings can come from wave slam on the bow, from ice, and from boats and other barges hitting against the sides. Unequal loads may be incurred in bending of the bottom hull plates during intentional or accidental grounding and of the deck plates due to cargo loads. Corrosion may reduce the thickness of hull plates.

The internal structure of a barge is subdivided by longitudinal and transverse bulkheads. Because of the relatively high possibility of rupturing of a side plate, with consequent flooding of the adjacent compartment, the longitudinal bulkheads are usually spaced at the middle third of the beam. A single centerline bulkhead could allow flooding of one entire side, causing excessive heel and possible capsizing.

Longitudinal bulkheads plus the two sides provide the longitudinal shear strength of the barge.

The transverse bulkheads are usually spaced with one just aft the bow (the collison bulkhead), one forward of the stern, and one or more in the midships region. These provide the transverse shear strength.

86 Offshore Construction Equipment

Quartering waves produce torsion as well as bending in both planes. The torsional shear runs around the girth of the vessel: sides, deck, and bottom.

Typical offshore barges run from 272 to 400 ft (83 to 122 m) in length. Width should be about one-fourth the length. Thus a barge may be 272 × 70 ft (83 × 20 or 21 m) to 400 × 90 or 100 ft (122 × 27–30 m).

Depth will typically run from 1/12 to 1/15 the length: Such ratios have been found to give a reasonable balanced structural performance under wave loadings.

Shallower-depth barges are often used in rivers and lakes; these can be very hazardous offshore where not only high quasi-static bending stresses can develop but also dynamic amplification and resonance.

Offshore barges typically have natural periods in roll of 5 to 7 seconds. This is unfortunately the typical period of wind waves; hence resonant response does occur. Fortunately, damping is very high, so that while motion in a beam sea will be significant, it does reach a situation of dynamic stability.

The corners of barges are subject to heavy impacts during operations; thus they must be heavily reinforced. Fenders should be provided on the corners to minimize impact damage to other craft and structures. Fenders should be provided along the sides to minimize damage to the barge itself from other boats and barges as they are docked. These may be a combination of integral fender strakes plus renewable fenders.

Bitts are provided at the corners and at intervals along the sides to enable the securing of the barge and any other craft which come alongside. Towing bitts are provided on both bow and stern.

Consideration must be given to the need to temporarily weld padeyes to the deck in order to secure cargo for sea. These padeyes must distribute their load into the hull; they cannot develop proper strength by just welding to the deck plate. They will be subjected to impact loads in both tension and shear. In modern offshore barge design, special doubler plates are often affixed over the internal bulkheads so that padeyes may be attached along them. Low-hydrogen electrodes should be used.

Alternatively, posts may be installed, running through the deck to be welded in shear to the internal bulkheads. The slot through the deck should be sealed.

The deck is often protected by timbers to absorb the local impact and abrasion of the load. This is especially needed for barges which will carry rock, which will be removed by clamshell or dragline bucket, or upon which a tracked crane or loader will operate.

Manholes are provided in the deck for access to the inner compartments. These must be watertight. There should be a heavy coaming to protect the dogs or bolts which secure the manhole.

Once again the warning must be made about entering inner compartments which have been closed for a long period. They are probably devoid of oxygen and must be thoroughly aerated before entry.

Offshore barges are often intentionally grounded (beached) in order to load or unload cargo. The beaching areas must be well leveled and all boulders and even large cobbles removed in order to avoid holing or severely denting the barge. Once beached, the barge should be ballasted down so that it will not be subjected to repeated raising and banging down under wave action at high tide.

When heavy loads are skidded on or off a barge, they punish the deck edge and side because of the concentration. Skid beams are often arranged to partially distribute the load to interior bulkheads. A timber "softener" may be temporarily bolted to the deck edge. The barge must be analyzed structurally for each stage of loading to ensure that a side or bulkhead will not buckle under temporary overload.

Cargo must be secured against movement under the action of the sea. See Figure 5.5.1. Thus sea

Figure 5.5.1 Large-diameter steel cylinder piles being transported by barge from California to Alaska. Note that both chain and structural fastenings are being employed. .

fastenings are designed to resist the static and dynamic forces developed under any combination of the six fundamental barge motions: roll, pitch, heave, yaw, sway, or surge. The dynamic component is due to the inertial forces which develop due to acceleration as the direction of motion changes. Roll accelerations are directly proportional to the transverse stiffness of the barge, which is measured by its metacentric height. Since barges typically have large metacentric heights, accelerations are severe. Conversely, if due to high cargo, the metacentric height is low, the amplitude of roll and the quasistatic force imparted by the load are greater.

These loads are cyclic. Seafastenings tend to work loose. Wire rope stretches; wedges and blocking fall out. Under repeated loads, fatigue may occur, especially at welds. Welds made at sea may be especially vulnerable because the surfaces may be wet or cold. Low-hydrogen electrodes will help. Chains are a preferred method for securing cargo for sea, since chain does not stretch. If structural posts are used, they should be run through the deck to be welded in shear to the internal bulkheads. The slot through the deck should then be seal-welded to prevent water in-leakage.

The effect of the accelerations is to increase the local loading exerted by the module, jacket, or the like on its supports, up to a factor of two or more. Flexing of the barge can also have a major effect on support forces and the seafastenings. Thus deeper and hence stiffer barges will experience a smaller range of loads than shallow, less stiff barges.

With important and valuable loads such as modules or jackets, sufficient freeboard should be provided to ensure stability even if one side compartment or end compartment is flooded.

Barges are normally designed only to the standard loading criteria of the classification society. These criteria are usually based on submergence of the hull to the deck line, plus an arbitrary load of 3 m of water on deck. Thus the barge is usually not designed for complete submergence.

Proposals are often made to build a structure on a barge, then submerge the barge by ballasting, and float the new structure off. Actually this has been successfully carried out in a number of cases: the construction of the pontoons of a drydock in northern Spain, the manufacture of several hundred shallow draft concrete hulls and posted barges near New Orleans, and the construction of Arctic offshore caissons in Japan. See Figure 5.5.2.

However, there are three key items that have to be taken care of:

1. When a barge is submerged by partially filling compartments with ballast, the external pressures are essentially the same as if the barge were empty and submerged to that depth. The hull must be designed for the deepest submergence.

2. Once the deck of the barge is submerged, stability of the barge itself is lost, although the structure on deck may give an effective waterplane and thus provide stability during submergence. Once the structure floats free, while the barge itself may initially have stability due to the low center of gravity of the ballast water, as it is pumped, the barge will rise up in an unstable mode, out of control. Stability and control can be provided by columns at the ends which always extend above the waterline. See Figure 5.5.3a and b.

3. The third problem is that of depth control. The support barge must be neutrally or negatively

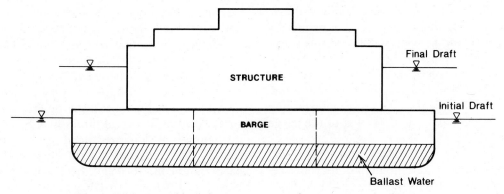

Figure 5.5.2 Launching a buoyant structure from a construction barge.

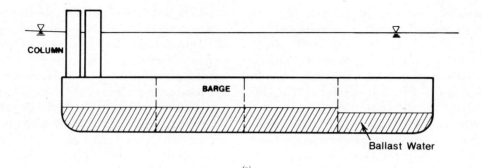

Figure 5.5.3 (a) Maintaining stability and draft control by use of columns at one end. (b) Submersible deck cargo barge for transportation of large structures over long distances.

buoyant in order to launch the structure. Control of depth can then only be achieved by:

a. Carrying the operation out in shallow water of known depth.
b. Use of lowering lines from a surface barge. However, the load distribution may shift, causing lines to part successively.
c. Use of columns penetrating the waterline, as recommended for stability control.

If a barge is to be seated, fully submerged, on an underwater embankment or on the seafloor in relatively shallow water, then it can be tipped down, one end first. Thus the beam of the barge and the inclined waterplane provide stability at this stage. Then the barge end touches bottom. Now the barge may be fully submerged, gaining its stability from the end of the barge reacting against the bottom. See Figure 5.5.4. This practice is normally limited to a water depth about one-third that of the barge length.

Note that as the barge is tipped down, the transverse waterplane area and moment of inertia is reduced to about one-half of normal. Therefore, transverse instability can develop and the barge can roll. This acts to limit the depth of water suitable for such an operation.

To recover the barge from the seafloor, the reverse procedure is followed, raising one end first.

A barge seated on a mud or clay seafloor develops a suction effect, consisting both of adhesion and a true suction due to differential water pressures. To break the barge loose requires that full hydrostatic water pressure be introduced under the bottom and that the adhesion of the clay to the barge be broken.

Extensive experiments by the Naval Civil Engineering Laboratory at Port Hueneme, California, and confirmed by practical experience in the Gulf of Mexico shows that breaking loose can best be accomplished by sustained water-flooding at a low pressure, less than the shear strength of the clay. Higher pressures will just create a piping through to the sea and prevent development of any pressure. Periods of several hours may be required to develop a fully equalized head of water under the structure. Then by deballasting one end first, the barge can be lifted clear. This method was used to free the GBS-1, (Super-CIDS) exploratory drilling

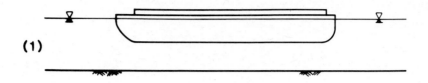

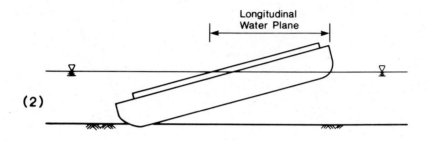

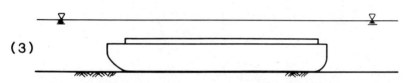

Figure 5.5.4 Tipping a barge to the seafloor in shallow water.

platform, after one year's service, seated on a clay foundation. It proved very successful.

As with ship salvage, a fully submerged barge must be given only limited positive buoyancy; otherwise it may break loose suddenly. If compressed air has been used to displace water in open compartments, the vessel will achieve additional buoyancy with every meter of rise due to the expansion of the air inside and may become uncontrollable.

For all the above reasons, submergence of standard barges must be considered only to shallow depths. For deeper submergence, special construction, interior pressurization, and so on are required; these are described in Chapter 19.

5.6 Crane Barges

The term *crane barge* is used to denote an offshore barge equipped with a sheer-legs crane which can pick loads and luff but not swing. The sheer legs essentially consist of an A-frame made up of two heavy tubulars held back by heavy stays to the bow. See Figure 5.6.1.

The sheer legs are maneuvered by deck engines, tugs, or mounted outboard engine propellers. The crane moves in to the side of the cargo barge, picks the load, then moves as necessary to set the load in exact position. Modern torque-converter deck engines and propellers with variable pitch allow a high degree of accuracy in positioning to be obtained, for example, of the order of 1 in. (2 mm). See Figure 5.6.2.

One of the advantages of a crane barge over a fully revolving derrick barge is that the load is always picked over the stern end, hence preventing list from the swing of the crane.

The sheer-legs crane is also much less costly, both in first cost and in maintenance. Because of the need to move the entire barge to proper position to set the load, its operations are slower than those of a derrick barge. Further, it cannot choose its

90 *Offshore Construction Equipment*

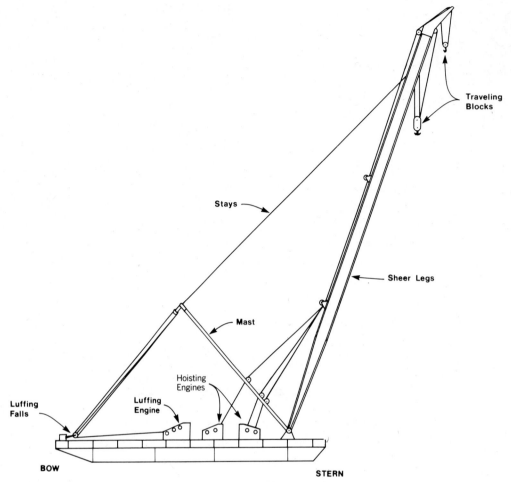

Figure 5.6.1 Crane barge.

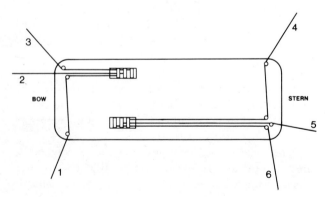

Figure 5.6.2 Deck engines and mooring line arrangement for crane barge.

heading so as to minimize motion response to the sea. See Figure 5.6.3.

A sheer-legs crane barge is normally capable of ballasting down by the bow, so as to offset the trim induced by picking of the load. The barge must of course be designed to resist the hogging moment which occurs when the maximum load is picked at the flattest angle of the sheer legs.

The ability of a sheer-legs barge to lift a module or other large spatial load to a height (for example, in order to set it on a platform deck) is limited by the necessary length of slings and by the interference between load and the sheer legs themselves. The load cannot be allowed to swing into the sheer legs or it may buckle them. Swinging of the load due to pitch will of course increase this danger. To prevent such fore and aft swing, tag lines should be used to suck the load slightly toward the stern;

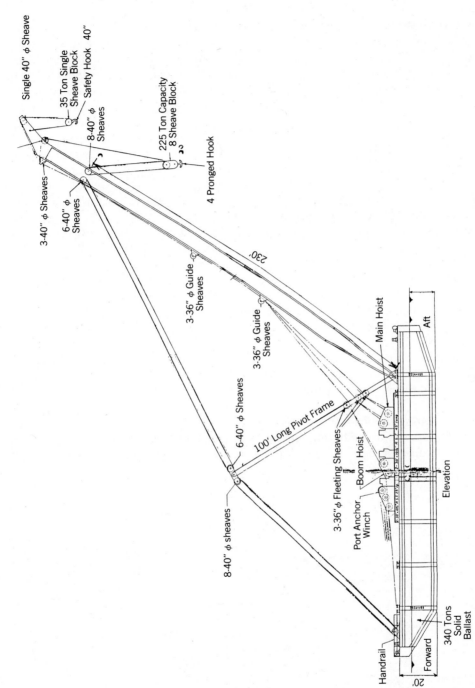

Figure 5.6.3 Crane barge used to construct Hay Point Terminal No. 2, Queensland, Australia. 225-ton capacity at 20 m over stern.

this will prevent it from swinging in this direction. Swinging tranversely can be snubbed by the use of tag lines as well.

Typical tag lines for offshore crane barges are ½- to ⅝-in. 6 × 37 wire lines to ensure flexibility, controlled by air or hydraulic hoists. Care must be taken to prevent their chafing as the load is moved to new positions in three-dimensional space. Softeners should be provided as necessary.

To pick loads from a barge at sea and then set them on a platform, the sheer legs are usually fixed at the appropriate angle to serve both. Luffing of the sheer legs, that is, raising the sheer legs themselves, is awkward and slow and should normally be avoided.

The load should be hoisted from the barge at the top of the heave (of the barge) so that 6 seconds later, on the next cycle of heave, the load will be clear of the barge. The operator (and foreman) will watch and try to catch a relatively higher wave on which to start the pick.

Hoisting speeds depend on the number of parts of line in the blocks and of course on the rated speed of the engine and the amount of wire on the drum.

When it comes to setting the load, the problem is reversed. The load will set on the bottom of the heave cycle; 3 seconds later the crane barge will lift it up again. With any significant sea state and pitch response, the load becomes a battering ram. Therefore, the most modern crane barges are fitted with a free overhaul capability so as to allow the load to remain seated. In any event, the skillful operator will try and set the load during a period of minimum motion and as close to the top of the crane barge's heave cycle as practicable.

The slings used to lift typical modules and the like are very heavy and awkward. A whip line, single-part, is run over a sheave at the boom head so as to help lift the eye of each leg of the slings over the hook.

The deck engines of a sheer-legs crane barge must be adequate to control the barge's motion in yaw, sway, and surge to a very close tolerance despite the state of the sea. This requires an excess of power as well as torque-converter controls or equal. Fairleads must be carefully laid out so as to ensure a proper fleet angle from the winch and to ensure that they will properly follow the changing position of the barge. See Figure 5.6.4.

Sheer-legs crane barges were used to set the 200-ton precast concrete dome and breakwater segments of the NINIAN central platform. Three crane barges, rigidly positioned relative to each other by multiple lines, were used to lift the quarters modules onto Statfjord A platform. This is inher-

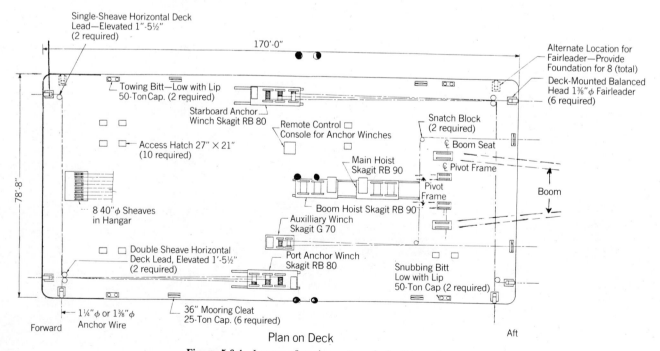

Figure 5.6.4 Layout of equipment on deck of crane barge.

Figure 5.6.5 3000-Ton-capacity sheer-legs crane barge.

5.7 Offshore Derrick Barges

Fully revolving derrick barges are the workhorse of offshore construction. As with crane barges, they are fitted with deck engines and full mooring capability, only here the emphasis is on security against failure rather than close control in positioning, since the derrick barge normally remains stationary during any particular operation. See Figure 5.7.1.

The typical offshore derrick barge has a crane capacity of 500 to 1000 tons. To handle ever larger modules and deck sections, capacities have been rapidly increased in recent years, with the latest derrick barge having two cranes, rated at 6000 metric tons each, or a total of 12,000 T.

The derrick barge represents a compromise (or optimization) of opposing demands. Structural and naval architectural considerations require it to be located forward of the stern a distance 20–25 percent of the length, that is, at the one-quarter or one-fifth point. The barge should be wide enough to minimize list as the crane swings and to provide adequate distribution of the structural load.

On the other hand, the effective reach of the crane and its load capacity is diminished by the distance from the boom seat to the stern or side of the barge.

One way to meet these two contrary demands is by the use of a large swing circle which moves the boom seat closer to the barge edge while maintaining the center of rotation and support well back.

ently a hazardous operation, but was carried out successfully because of the provision of oversize lines and adequate deck engines interconnected to a single control station on the central barge.

A crane barge was successfully used on the Port Latta, Tasmania offshore terminal to set jackets into preinstalled frames with a tolerance of only a few inches. Thus the capability of a crane barge for extensive use offshore should not be eclipsed by the currently more popular but much more expensive fully revolving offshore derrick barges. See Figures 5.6.5 and 5.6.6.

Figure 5.6.6 Heavy-lift sheer-legs crane barge setting 1400-ton prestressed concrete deck sections for the Saudi Arabia–Bahrain bridges. (Courtesy Ballast–Nedam Group).

Figure 5.7.1 600-ton-capacity offshore derrick barge *Cherokee*. Note that main falls and traveling block are housed in boom. (Courtesy Santa Fe International.)

This is one of the essential differences between a land crane and an offshore derrick.

A major consideration is the list of the barge under fully loaded or no-load conditions. The counterweight is usually designed to limit the list under almost full load, hence under no-load the barge may list opposite to the boom. This can be reduced during operations by booming down while swinging under no-load.

The swinging is carried out by swing engines driving the bull wheel. Due to list the crane is often forced to swing "uphill" under full load. Offshore cranes are therefore provided with two and sometimes three swing engines.

The list also places heavy structural loads on the crane tub, which forms the structural connection to the barge. Hence its design must provide proper structural reinforcement for bending and to prevent buckling under inclined compression loads.

The advantages in operations of a fully revolving derrick are many: the ability to pick off a barge or boat alongside or even from the deck of the derrick barge itself, the great control in positioning so as to be able quickly to reach any point in three-dimensional space with one set of controls, the ability to follow the surge motions of a boat or barge alongside in order to pick a load from off it, and the ability to orient the derrick barge in the most favorable direction to minimize boom tip displacements and accelerations.

When setting large and heavy loads, it is the boom tip motions which control. These are affected by motions of the barge in each of the six degrees of freedom. When working far out over the stern, pitch amplitudes will be amplified. When working over the side, it is roll which causes the most difficulty. Computer programs have been developed to assist in selecting the proper heading, which treat the barge and load as a coupled system.

A skillful barge superintendent and crane operator will take advantage of the "groupiness" of waves so as to perform a critical pick or setting operation during a succession of low waves.

As with a crane barge, tag lines must be used to control the swing of the load. As contrasted with the sheer-legs crane barge, the position of the load relative to the barge is constantly changing; hence the tag line engines are fitted to the crane.

A load suspended from a boom tip is a pendulum. While the load line length is always too long for direct resonance, the load may tend to get dynamic amplification from lower-frequency energy. The practical solution is to raise or lower the load quickly through those positions which develop amplified response.

Marine cranes are usually designed to work under their rated loads up to a 3° list. The load capacity ratings for marine cranes are based on 2° roll at a period of 10 to 12 seconds, which equates to an acceleration of 0.07 g.

The swinging of the load develops lateral forces on the boom. Hence offshore crane booms are designed with a wide spread at the heel (usually $1/15$ of boom length or more). This in turn means that the boom lacing (bracing) members will be more subject to buckling; they must be properly designed to prevent this mode of failure.

Booms today are made of high-strength steel, usually round or square tubulars. This makes them lighter and hence increases the effective load capacity of the crane and reduces the inertia in swing. However, it means that welds are more critical and that buckling becomes a common mode of failure. Good design and fabrication will take care of these. It also means that the boom is much more sensitive to impact from the load itself or to failure under an accidental loading from a line. It means that attachments such as padeyes for snatch blocks and so forth must be affixed to the boom only after careful engineering and fully controlled welding procedures suitable to the grades of steel involved.

One of the potential hazards with offshore derrick barge operation is that, although the lifts have been carefully engineered for load and reach, in the actual situation the derrick barge surges further away from the platform and moves laterally. The operator, intent on the load and the landing site, booms out and swings beyond the crane's capacity. The resultant failure may be a direct failure of the boom or may result in a loss of swing control which accelerates as the barge lists.

Offshore derrick barge cranes are fitted with automatic warnings to alert the operator when allowable load–radius combinations are being exceeded, but swing control is normally a matter of judgment.

To snatch a light load out of a supply boat, a single line, the whip, is preferred. It can raise the load fast enough to prevent an impact on the subsequent heave cycle. Raising a heavy load from a barge is more difficult since there may be 24 or more parts in the line and the barge will rise as the load is lifted.

A similar problem occurs when setting a heavy

load. When setting on a platform, the deck will usually be above the sightlines of the crane operator: He is working blind, dependent on signals. Hence one or more guiding devices may be usefully employed.

Tag lines from the crane barge may bend over the edge of the platform deck; if they chafe, they may part at the worst possible time. Softeners should be provided.

Structural guides may be preinstalled on the platform so that the load, once set within 0.5 m or so of position, automatically guides down to the correct location. These guides must have sufficient height so that the load does not ride up out of them on the next pitch–heave cycle. If that were to happen, they could puncture the load rather than guide it.

Taut guide lines can be employed to help pull the load to correct position. A system of guides that often works well is to use four columnar guide posts. Suspended loosely from the load are four pipe sleeves of larger diameter. These can be hand-fitted over the posts; when the load is lowered, the sleeves will guide the load into place; see Figure 5.7.2. Alternatively, loosely hanging pins (smaller-diameter pipe) may be entered into the tubular posts. Tag lines and winches may be installed on the platform to assist in guiding the load into place.

Another solution is to set the load only to an approximate location, landing it on softeners such as timber or rubber fender units or used earthmover tires. After it has been landed in approximate position, it can be skidded to final exact position by hydraulic-jacking equipment of the type commonly employed on oil-drilling rigs.

There is an arbitrary functional division that exists between offshore construction crews and offshore drilling crews. Neither seems to fully appreciate the problems of the other. This has resulted in much needless work and not a few accidents. Close coordination and communication are essential.

The lower block and hook of a large offshore derrick can weigh 20–30 tons or more. As it is brought up close to the boom housing, it may get into resonance with the roll of the barge. A special hook control tag line is required.

The traveling block–hook combination should never be left hanging at short scope. A sea may come up that excites the hook and makes it impossible to secure. Thus, except when the crane is being used, the block should always be fully stowed

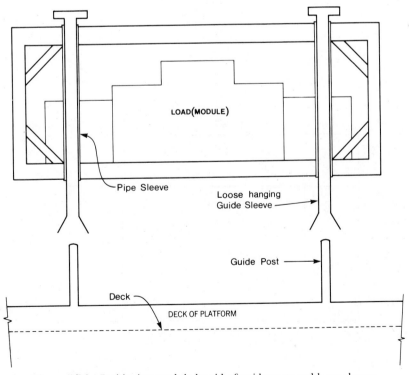

Figure 5.7.2 Positioning module by aid of guide posts and loose sleeves.

and the boom lowered into the boom cradle and secured. This will also reduce fatigue wear on the swing gear.

When a derrick barge is working alongside a platform, the moorings are laid out in a pattern which allows the barge to reorient and relocate as necessary to reach as many parts of the platform as possible. Care must be taken that during a reorientation the mooring lines are not allowed to cross one another. Although there are exceptions, as a general rule, mooring lines should never cross; it prevents retrieval of the underneath line, and it may lead to erratic reactions from the lines as the load in one changes its catenary and affects the other. Worst of all, it may snag the anchor of the other line.

5.8 Semisubmersible Barges

While the standard barge, whether serving for cargo transport or as support for a crane or other operational equipment, has good stability and load displacement characteristics, unfortunately it has excessive response to the wind-driven waves. These then limit the workability of the vessel.

In areas such as the Bass Strait and the northern North Sea, where the persistence of low-sea states is short, a conventional barge may encounter excessive weather downtime, which may extend the construction schedule beyond the summer "weather window" and thus require an extra year for completion.

The semisubmersible concept was first developed for offshore exploratory drilling but has since been extended to both derrick barges and pipelaying barges. It is a simple concept: a large-base pontoon or pontoons which are fully submerged during operations, supporting four to eight columns which extend through the waterplane and in turn support the deck.

Thus there is a large submerged mass and large displacement combined with minimum waterplane. The vessel is therefore subject to minimum exciting and righting moments. Some have referred to the concept as "transparent" because the waves sweep right through between the columns or shafts, with little effect on the barge motion. See Figure 5.8.1.

This lack of response to the typical wind-driven seas is due both to the relatively small change in gross displacement and to the much longer natural period of the vessel, especially in roll, pitch, and heave. Whereas the standard barge has a natural period of 5–6 seconds, the typical semisubmersible has a natural period of 17–22 seconds. See Figure 5.8.2.

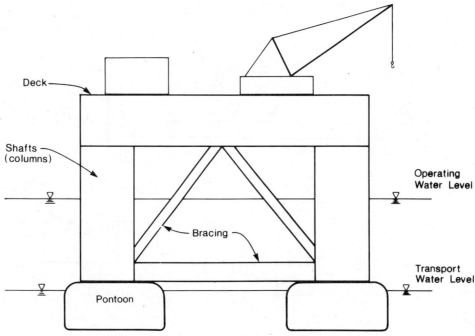

Figure 5.8.1 Semisubmersible concept.

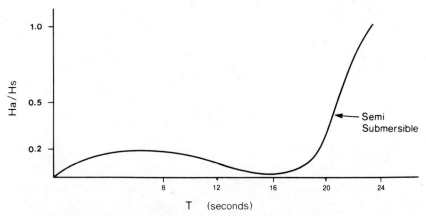
Figure 5.8.2 Typical response of semisubmersible to heave.

There are three penalties to pay for this favorable performance.

1. The semisubmersible has much greater response to externally applied loads such as weights, loads, and ballast. Another way of stating this is to say that its righting moment and metacentric height are much lower than those of a standard barge.
2. The semisubmersible has much reduced topside cargo capacity. It relies on a low center of gravity to maintain stability.
3. The semisubmersible costs more to build and to operate. Ballast controls are similar to those for a submarine.

However, semisubmersibles are increasingly used in drilling construction, for floatels, (floating quarters vessels), and even for floating production because of their ability to carry out their operations over extended periods without interruption due to weather downtime.

As indicated above, the semisubmersible must have a very complete and effective ballast and drainage system, with high-capacity pumps and quick-acting controls. The system must have a high degree of reliability; failure of a valve to close can cause a catastrophe. There must be positive valve position indicators, tank level indicators, and sensitive list–trim indicators in the control room. Redundant venting systems must be provided to prevent accidental overpressurization.

The semisubmersible normally rides upon its base pontoons for transit, going into the semisubmersible mode only after arrival on station. See Figure 5.8.3. As with all vessels having a drastic change in section, at the draft when the waterplane crosses over the base pontoons there is a sudden loss in metacentric height to almost zero. This is further compounded by the action of the waves breaking across the tops of the pontoons and impacting the shafts. Thus this stage is one of unpredictable response and instability. Therefore, no other operations should be attempted during submergence until there are 2–3 m of water depth over the pontoons.

The effect of accidental holing of a shaft, for example, can be much more serious than for a similar holing of the side of a standard barge. Therefore, the shafts of many modern semisubmersibles are double-hulled and protected by heavy timber and rubber fenders.

Because of low topside capacity, deck winches are usually mounted low in the shafts. Mooring lines leave the barge through swivel fairleads located on the base pontoons; this keeps them well below the keel of attendant boats and barges.

The safety of a semisubmersible against capsizing can be immeasurably improved if appropriate damage control systems are built into the vessel and enforced in operations. For most semisubmersibles, the deck is watertight and has the structural strength to act as an upper barge hull. Then if the vessel should heel over so that the deck enters the water, the righting moment jumps back up significantly.

However, operational carelessness often negates this in practice. Watertight access doors and portholes and vents are left open, especially in warm climates. Internal subdivisions are modified so as to lose their watertight bulkhead capabilities. Gear is left loose on deck to shift with the list. These have

Figure 5.8.3 Semisubmersible derrick barge *Choctaw I* during transit, floating on lower hull. (Courtesy Santa Fe International.)

contributed to the loss of two semisubmersible vessels, the *Alexander Kjelland* in the North Sea and the *Ocean Ranger* off Newfoundland. Operational mishaps and structural defects have apparently also been involved.

The semisubmersible is subjected to high stress concentrations and cyclic loadings at the connection between shafts and pontoons and bracing. When sea and operating conditions permit, the semisubmersible should be ballasted up so that these can be visually checked to make early detection of any cracks.

There is another advantage to the semisubmersible, and that is the high elevation of its deck, especially when ballasted up onto the pontoons. From such an elevated situation, a crane can reach further out over the deck of a platform, and thus interior modules can be more easily placed.

Because of its low heave response, the semisubmersible can be moored as a tension leg platform, that is, with vertical mooring lines to clump weights or anchor piles on the seafloor. Pulled down against these reactions, this temporary tension leg platform (TLP) can hold itself accurately in the vertical direction, thus enabling it to carry out heave-sensitive operations such as screeding, setting, and fitting large individual pipes or underwater vehicular tube sections.

These favorable properties of the semisubmersible concept have been adopted by offshore contractors and operators for a variety of small special-purpose rigs. These may in turn be tended by large offshore derricks, since the small semisubmersible has limited versatility, usually being intended for one specific operation. Moored, for example, as a TLP and working in conjunction with a derrick barge, it can carry out operations on the seafloor which require minimum or no heave.

While the semisubmersible itself is stable in moderately high sea states, it has problems in transferring piles, pipe lengths, deck equipment, and so forth from a standard cargo barge because of the latter's motions. The semisubmersible provides little lee: the waves sweep right through its columns.

Large supply boats are therefore often used to deliver such items to the extent practicable. The supply boat can run a stern line to the semisubmersible, then run ahead slowly, with its bow headed outward. Thus it can lay alongside but still be free from direct contact.

For deck modules and the like, it may occasionally be most practicable to tow the semisubmersible into protected waters, load the module, and tow back to the offshore site. If the load is hanging on the boom(s), over the stern, then the load must be blocked to the barge and held in with taut tag lines of adequate size so as to prevent swing and heave.

Finally, long piles of large diameter may be delivered to the site afloat, to be picked from the water by the derrick barge.

5.9 Jack-Up Construction Barges

The jack-up barge has proven itself to be a very useful construction "tool," especially when working in turbulent sea areas, such as shoal or coastal waters, and in swift currents. Where a great many operations must be carried out at one location—for example, at an offshore terminal or bridge pier—the jack-up construction barge is especially valuable. See Figure 5.9.1.

The barge is outfitted with four to eight large jacks and legs, built either of tubulars or latticed steel. The barge is towed to its work position and jacked up free of the waves to perform its work.

The typical sequence starts with the barge moving to the site with its legs raised. Upon arrival at the site, it is moored with a spread mooring. Jack-ups can operate only in relatively shallow water, 30–60 m, with 100 m as an extreme, so the use of a taut mooring is practicable.

With the sea state being calm (waves and swells must usually be less than 1 m), the legs are lowered to the seafloor and allowed to penetrate under their own weight. In some soils, penetration can be aided by jetting and vibration. Using the jacks on one leg at a time, the barge acting as the reaction, the legs are forced into the soil.

With all legs well embedded, the barge is jacked up clear of the water. This is the most critical phase, since wave slap on the underside of the barge may cause impact loads on the jacks and may shift the barge laterally, bending the legs. To cushion the impact, special hydraulic cushioning may be connected to nitrogen-filled cylinders; alternatively, neoprene cushioning may be employed. Once well clear, the barge is raised up to its working height.

Since uneven settlements may take place as a result of time, operations, and wave energy input into the legs, the jacks have to be periodically reactivated so as to equalize the load at each. This is especially necessary during the first few days at a site.

To leave a site, the sea must again be calm, with waves and swells usually less than 1 m. The mooring lines are reattached, slack. The barge is then jacked down until it is afloat. Once again the critical period is when the waves are hitting the underside. The

Figure 5.9.1 Large jack-up construction barge used in offshore bridge and terminal construction.

mooring lines are tightened. Then the legs are jacked free, one at a time.

If legs do not pull out easily, several techniques can be applied. The fastest is jetting. In clays, a sustained load may eventually free the leg. Also in clays, water injection at low pressure to break the suction may be more useful than high-pressure jetting, which leads to the formation of escape channels.

In no event should an attempt be made to free the legs by lateral working of the barge. A bent or jammed leg may result, with very serious consequences. In one case in Cook Inlet, Alaska, the leg was jammed by the high current working on the barge side. Then the tide rose some 6 m, flooding out the jack-up barge itself.

When the seafloor is sand, the current may create local eddies around the legs, leading to scour and loss of lateral capacity. A similar problem may arise when working in strong currents. See Figure 5.9.2.

A relatively new development is to build mats onto the bottom of the legs, so that when the legs are jacked down they take their support from the seafloor. A short stub leg may penetrate below the mat to provide shear resistance against sliding.

Since jack-up performance is so highly dependent on the seafloor soils, it is essential that a thorough geotechnical evaluation, including at least one boring, be made at each site. Of particular concern are layered soils, in which a leg may gain temporary support but then suddenly break through.

In clay soils, where jack-ups have previously

Figure 5.9.2 Jack-up construction derrick foundering due to erosion around leg during storm on California coast.

worked around the site, holes will have been left which now may be partially empty or filled with loose sediments. If a leg is now seated adjacent to such a hole, it may kick over into it, losing both vertical and lateral support and bending the leg. A general rule of thumb is to plot the previous leg positions (if known) and to space the new leg locations 4 to 5 diameters away. This of course is another advantage of the mat-supported jack-up legs: The mats can span local anomalies.

Walking jack-ups have been built, varying in size from a small test-boring rig capable of walking through the surf to a monstrous dredge hull on jack-up legs. These rigs are equipped with two sets of legs (six or eight in all) supporting a double-framed hull (or segments) so that it can successively launch forward, lower the legs, take its full support forward, pick up the legs behind, and retract the rear into the forward section. The rear set of legs are now lowered so as to give added support during operations and then to enable the forward set to be picked up once again. Such walking jack-ups eliminate the need for the hull to be lowered into the sea in order to move.

Large jack-up construction rigs are most applicable where the sea conditions are highly variable, with frequent periods of calm, so that the rig may find convenient times to move. On the other hand, if numerous moves are required—as, for example, in laying outfall sewer pipes—then persistently rough seas may delay moves so long as to render the jack-up uneconomical.

One disadvantage of the jack-up occurs during the transfer of loads from barges or supply boats. Here the jack-up again becomes weather-sensitive, for the barges must not be allowed to contact the legs or they may damage them.

Jack-ups provide a fixed platform, free from motion response to the seas. See Figure 5.9.3. Hence, they are ideal for carrying out operations such as grinding a rock foundation in order to seat a caisson, as was done on the Honshu–Shikoku bridge (Koyama–Sakaide Route, Pier 7A). They are also ideal for screeding the foundation site. Jack-up barges have been used as dredges in the Netherlands.

Statistical studies covering both jack-up drilling rigs and jack-up construction rigs show that they are six times more likely to suffer serious damage or loss during relocation and transit than they are when on location.

As with the semisubmersible concept, the jack-

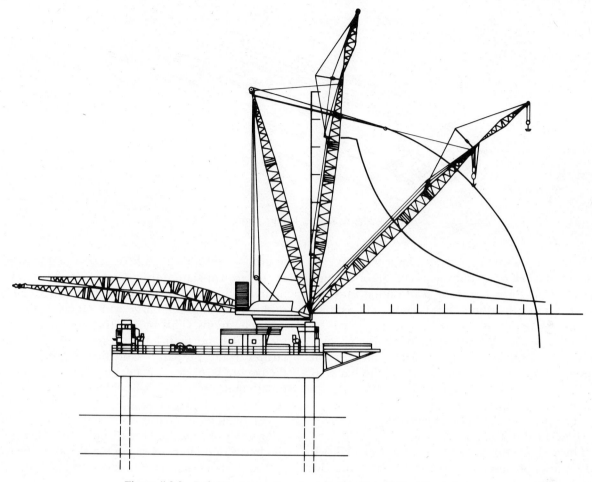

Figure 5.9.3 Jack-Up construction derrick. (Courtesy Volker–Stevin.)

up principle has been applied to smaller, special-purpose construction rigs tended by an offshore derrick. The derrick barge with its large mooring system can be used to position the jack-up and if necessary to help in penetrating its legs and later to help in retracting them. Meanwhile the jack-up rig forms a vertically stable work platform for such sensitive operations as coring and sampling operations or for submarine pipeline repairs.

The jack-up can be used to set heavy loads. In this case, a barge carrying the load is floated in between the legs. The load is then lifted by direct hoisting from the jack-up deck above the barge, removed, and the load lowered to the seafloor. This operation, with a barge between the legs, is obviously highly hazardous and should only be attempted under ideal sea conditions, with adequate lateral controls to ensure that the barge cannot hit the legs.

5.10 Launch Barges

One of the most dramatic developments in offshore construction practice is the use of the launch barges for the transport and launching of jackets. They have also been utilized to deliver and launch subsea templates. See Figure 5.10.1.

The typical launch barge is a very large and strongly built barge, long and wide, subdivided internally into numerous ballast compartments. Since it must support a jacket weighing thousands of tons, it must have strong longitudinal and transverse bulkheads.

Heavy runner beams or skid beams extend the length of the barge. See Figure 5.10.2. These girders distribute the jacket's load to the barge structure. The stern end of the barge, over which the jacket will rotate and slide into the water, requires special construction.

102 *Offshore Construction Equipment*

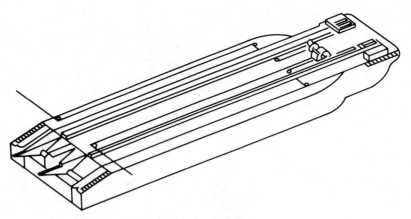

Figure 5.10.1 Launch barge.

First, for a short period of time it will have to support the full weight of the jacket. Second, since this reaction force has to be transmitted into the jacket, it must distribute the reaction over as long a length as feasible to avoid a point reaction. The jacket will be sliding on its specially reinforced runners; even so, they need a distributed rather than a point reaction. Hence the stern of the barge is fitted with a rocker section that rotates with the jacket as it slides off. See Figure 5.10.3.

The operational aspects of launching of a jacket are described in Chapter 9.

There is one other structural aspect to a launch barge: intentional grounding. For loading out the jacket at the fabrication yard, the usual method is to ground the launch barge on a screeded sand pad at the appropriate depth so that the barge deck matches the yard level. Then the jacket can be skidded out onto the barge with no change in relative elevations. This means that the hull bottom must withstand high local pressures from irregularities in the prepared sand bed. Not only must the bottom be of heavy plate, but the stiffeners must be adequate to prevent buckling as the jacket is moved onto the barge.

A launch barge is also fitted with heavy winches or linear jacks on the bow to pull the jacket onto the barge and later, by rerigging through sheaves on the stern, to pull the jacket off the barge during launching.

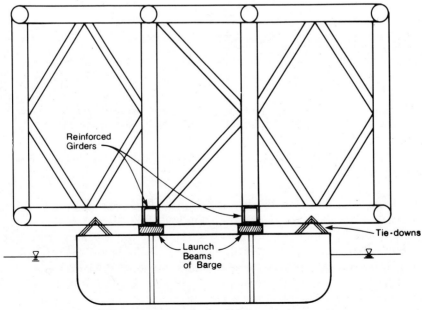

Figure 5.10.2 Jacket loaded out onto barge.

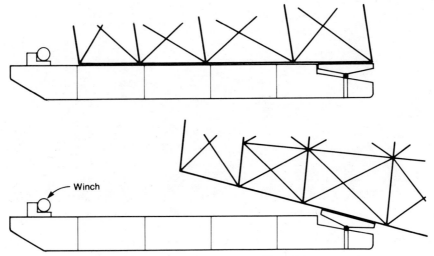

Figure 5.10.3 Rocker arm stern of launch barge.

The typical launch barge has a beam of 90–106 ft, the latter limit being set to enable passage through the Panama Canal. The base of a deep-water jacket may be 160–200 ft wide, overhanging the sides significantly. The largest launch barges constructed to date are 196 × 52 m and can carry and launch a jacket of 40,000 tons. A 1000 ft. (300m) long launch barge is now being fabricated in Japan to transport the 55,000 ton, 1365 ft. long, Bullwinkle jacket.

During transport the barge must have enough freeboard to prevent the outside legs of the jacket from dipping into the waves as the barge rolls. The beam width of the barge and its freeboard are designed to give stability transversely during launching. This is often the critical condition during launch; if the barge lists and the jacket rolls sidewise, it may buckle a jacket leg.

5.11 Offshore Dredges

Dredging offshore is rendered extremely difficult because of swells, depth, and position control. Quantities are usually substantial. Disposal, while usually not restricted, must be sufficiently distant from the excavation to prevent its running back in. There are the usual problems of trying to work with auxiliary craft, each responding in its own way to the waves. Nevertheless, efficient and effective dredging systems have been developed for use offshore.

Dredging operations themselves will be discussed in Chapter 7, as will the minor equipment often built into structures such as air lift and eductor devices. Ocean mining will be described in Chapter 19. Use of jet sleds and plows to bury pipelines will be discussed in Chapter 12.

The trailer suction hopper dredge is perhaps the workhorse of ocean dredging. This is a self-propelled vessel of standard ship hull configuration. It supports a long "ladder," a pipe with structural framing capable of reaching to the seafloor at an angle of about 30°. At the end of the ladder is the entry port, equipped with jets to break up sand so that it can flow freely into the ladder or with ripper teeth to break up cemented or hard material.

On board, a suction pump moves a column of water and dredged material up the ladder at a velocity sufficient to keep the material in suspension.

The pump discharges the material into hoppers of the vessel; the solids settle out and the surplus water overflows. By controlling the flow into the hoppers, the material can be roughly graded, with silts and other fines being washed overboard.

The trailer suction hopper dredge can now transport the material as far as desired, discharging it by dumping from the hoppers by opening the gates. Alternatively, the trailer suction hopper dredge may agitate and slurry the material in the hoppers and pump it ashore. See Figure 5.11.1.

When dredging harder material, the dredge may use the power of its propellers to give added thrust to the rippers attached to the lower end of the ladder (the dredge head) or even use the momentum of the vessel. For example, when dredging soft and fractured rock, it may circle around to gain speed and then make a run at the dredged area. As it

Figure 5.11.1 Trailer suction hopper dredge under transit in Beaufort Sea, Canada.

reaches the area, it lowers its ladder and rips and sucks up the rock.

A trailer suction hopper dredge performs its work in successive long runs. Thus it is ideally suited to excavation of a trench.

One of the problems facing the operation of a trailer suction hopper dredge in the open ocean is that of swells, which raise and lower the dredge head and may damage it, as well as giving an uneven dredge depth. Therefore, the suspension system, by which the dredge ladder is suspended from the hull, may have a heave compensator installed in it. In its most sophisticated installation, the heave compensator may be activated by a sonic depth indicator so as to maintain an appropriate elevation relative to the seafloor regardless of the heave or pitch of the vessel.

The hydraulic suction dredge is an extremely efficient piece of machinery for moving large masses of material for moderate distances. In the Beaufort Sea, for example, a hydraulic dredge has moved up to 100,000 m^3 of sand per day to deposit it in an underwater embankment.

The standard hydraulic dredge suspends its ladder from an A-frame over the stern. This places high bending moments in the hull. Various other means of supporting the ladder are shown in Figure 5.11.2.

The lower end of the ladder is fitted with a cutter head to cut the soil so that it can be swept up into the flow of water up the ladder. This cutter head has traditionally been a set of blades rotating around the axis of the ladder. More recently, wheel excavators have been employed, rotating around the axis normal to the ladder. These latter are driven by a submersible motor, either electric or hydraulic. As with the trailer suction hopper dredge, swells cause the dredge head to raise and lower. This may severely damage the cutter head. A heave compensator is therefore necessary when working in the open sea. One form of heave compensation may be obtained by the use of an articulated arm and spar buoy to support the end of the ladder; this eliminates the effect of pitch of the hull and minimizes the heave response due to the swell.

Powerful cutterheads can dredge material which would be classified as soft rock. In very soft or loose material, on the other hand, the cutter head may be replaced by a system of jets.

In deep dredging the pump on board the vessel loses efficiency. For depths over about 25–30 m, therefore, additional means are employed to keep the water column moving at high velocity.

Eductor jets ("booster jets") may shoot water at

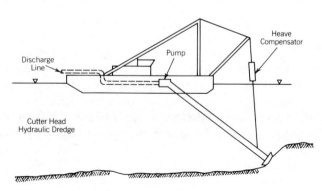

Figure 5.11.2 Offshore dredges. Note: Mooring systems not shown.

5.11 Offshore Dredges

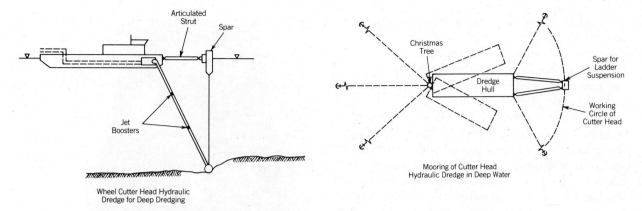

Figure 5.11.3 Adaptation of hydraulic dredge to deep water.

high velocity into ports located at several elevations along the column in the ladder. This overcomes the frictional and entrance losses in the suction pipe and accelerates the water to working velocity so that the dredge may function at full efficiency despite the greater depth. Alternatively, the dredge pump may be submerged or even located down on the ladder so as to minimize the suction length and utilize the discharge head instead.

In the depths and sea conditions found offshore, spuds cannot usually be used. The hydraulic dredge must operate off wire lines to anchors. To allow the dredge to sweep laterally without undesired translation, the three aft-leading lines are brought in through a "Christmas tree" fairlead system, with the three fairleads arranged on one vertical axis. See Figure 5.11.3.

Special concepts for cutterhead dredges have been developed to enable mining and heavy-duty dredging at depths of 100 m and even more. See Figure 5.11.4.

The clamshell dredge is another tool that can be used effectively offshore, especially when digging rock or other hard materials or when excavating within limited areas. It is not affected by pitch, heave, or roll. Surge, sway, and yaw will of course make the control more difficult but do not affect the ability to dig. The clamshell bucket is designed to penetrate by its own weight, acting on the lips of the bucket or on the teeth. It is closed by raising on the closing line or by hydraulic activators.

For practical reasons, offshore clamshell buckets are large and heavy, weighing up to 100 tons, as in the case of the deep dredging for the piers of the Honshu–Shikoku bridges. See Figures 5.11.5 and 5.11.6.

After closure, the bucket is lifted to the surface; in deep water, this is the slowest part of the cycle. Then, conventionally, the boom swings to starboard for disposal into alongside hopper barges. With a long boom and favorable currents, the material may just be sidecast when the boom is abeam. Then the boom must swing back. The bucket is allowed to fall, usually freely if velocity does not get too high, since it is restrained by friction of the wire lines over the multiple sheaves or else braked as necessary. The cycle then repeats. See Figure 5.11.7.

Swinging also is time-consuming, due to the inertial effects in starting, stopping, and reversing the swing. One operator minimizes this by swinging in a 360° circle, without slowing at discharge, that is discharging as the boom and bucket pass abeam, and then continuing on around.

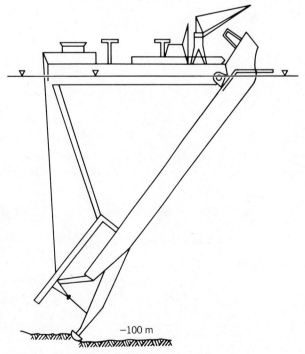

Figure 5.11.4 Deep-sea cutter head dredge. (After John B. Herbich, *Coastal and Deep Ocean Dredging*, Gulf Publishing Co., Houston, Texas 1975.)

Another solution is not to swing at all, but to position the barge transversely at the stern. With the boom well out, the bucket is raised and then pulled in to discharge into the disposal barge. This can reduce the cycle time significantly, but in turn requires a long boom. See Figure 5.11.8.

To help the clamshell bucket teeth penetrate hard material, vibratory exciters have been fitted; these have so far been plagued with maintenance and practical problems. In one case in very stiff coral, the teeth penetrated so far that the bucket was unable to close. In other cases, the vibrators have led to fatigue failures.

The hydraulic backhoe has been adapted to moderate dredging depths, with reaches up to 25 m below sea level. This tool is especially effective in removing ledges or strata of limestone and cap rock, such as are found in the Arabian Gulf. It has excellent power, applied optimally in a leverage fashion. In calm water, it also gives accurate depth control. However, this equipment is not able to work effectively in swells and is of course severely restricted as to depth. Continuous bucket dredges have been developed for tin mining in Southeast Asia; they are able to work to 60 m depth but are restricted to almost flat, calm sea conditions.

For further discussion of dredging operations, see Chapter 7.

Figure 5.11.6 99-Ton clamshell ("grab") bucket used to excavate broken rock for offshore bridge construction in Japan.

Figure 5.11.5 Clamshell dredges at work in Inland Sea of Japan excavate pier foundations for Honshu–Shikoku bridge.

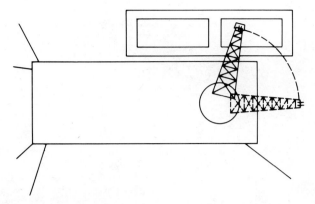

Figure 5.11.7 Clamshell derrick discharging to hopper barge alongside.

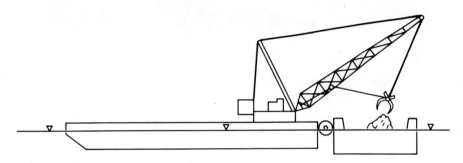

Figure 5.11.8 Clamshell derrick discharging to hopper barge at stern.

5.12 Pipelaying Barges

The pipelaying barge is a highly sophisticated vessel which constitutes the key element in a total submarine pipeline installation system. The system itself and its operation are discussed in detail in Chapter 12. In this section, attention will be devoted to the pipelaying barge proper. See Figure 5.12.1.

The functions of the barge are to receive and store pipe lengths, assemble and weld them into a single length, coat the joints, and lay the pipeline over the stern to the seafloor. Operations involved in accomplishing the above include:

1. Positioning the barge
2. Handling pipe lengths from a barge or supply boat to the barge deck
3. Double-ending (optional)
4. Lining up and completing the initial hot pass weld
5. Completing the welds
6. X-ray
7. Applying tension to the pipeline
8. Coating the joints
9. Laying the line out over the stern, usually by means of a stinger
10. Moving the barge ahead on its anchors
11. Shifting anchors continuously ahead
12. Recording positions of laid pipe accurately
13. Radio communications to boats, shore, and aircraft
14. Helicopter and crew boat personnel transfer
15. When weather conditions dictate, "abandoning" pipeline onto the seafloor in an undamaged unflooded condition
16. "Recovering" an "abandoned" line and recommencing pipelaying operations

108 Offshore Construction Equipment

Figure 5.12.1 "Second-generation" large offshore pipelaying barge at work in Gulf of Mexico.

17. Davits to permit supporting a section of the line uniformly for riser tie-in or repair
18. Diving support for inspection
19. Housing and feeding of up to 300 men

In the above listing, the word *abandonment* is to be construed as a temporary cessation of work and laydown due to the real or threatened onset of a storm.

Such a long list of requirements inevitably re-

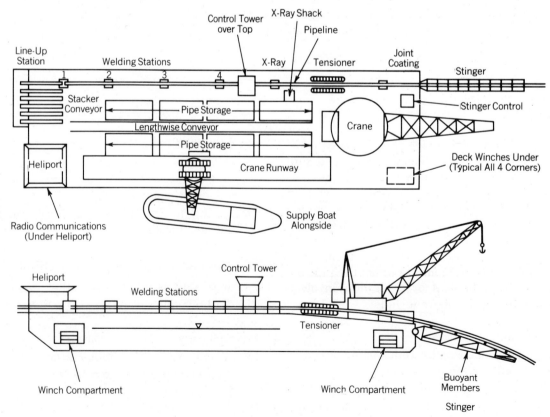

Figure 5.12.2 Second generation pipeline laybarge.

quires a large offshore barge. Both heavy-duty standard offshore barges and semisubmersible hulls have been used. The length of the barge is further dictated by the number of welding stations required in order to maintain the desired rate of progress. Since deep-water pipelines inevitably have thick walls, many passes are needed in order to complete full-penetration welds. The more stations there are, the less time needs to be spent at any one station, and hence the ability to increase the lay rate. See Figure 5.12.2.

To move the barge ahead requires many mooring or anchor lines. The barge must work more or less in alignment with the pipeline route, which may place it at an unfavorable angle with regard to waves, wind, and current. Hence it is usual to have 8 to 12 anchors in use. In order to make progress, two anchors are always in the process of being moved ahead by the anchor-handling boats.

The winches must be constantly monitored and controlled so as to maintain the tension of each line within desired limits and to keep the length of each line within a prescribed range.

The mooring lines lead from the winches over direction-changing sheaves to submerged fairleads and thence to the anchors.

To handle the pipe lengths onto the pipelaying barge, a large crawler crane is used, one which can quickly snatch a 40-ft length from a tossing supply boat or barge at the top of the heave cycle. Thus a single line is usually used, and the pipe is transferred in a single length.

A number of heave-compensating devices have been tried, with a signal line from the hook to the boat, for example; however, the snatch method still seems most effective. Once the pipe is stored on board, the next operation may be double-jointing. This usually does not speed up the overall pipelaying, but does reduce the number of specially skilled welders required.

The pipe length, single or double, is then conveyed end-0 and sideways to the lineup station. The pipe rolls onto the rack, which is hydraulically controlled so as to accurately line up and position the pipe. An internal lineup clamp is applied to join the new section to the previous one so that the first "hot pass" weld may be made. The joint then moves iteratively to the several welding stations, where the weld is chipped and cleaned and new metal deposited.

The weld, once completed, moves to the X-ray station, where pictures are taken, reviewed, and approved. In the case of a reject, a cutout must be made and the removed weld rewelded and reinspected.

Either before or preferably after the X-ray station, the tensioner is installed. See Figures 5.12.3 and 5.12.4. These are usually of the caterpillar track type, using polyurethane tracks pushed tight against the rough coating by multiple hydraulic jacks. The tensioning force is thus applied to the pipeline by friction.

At the next station, the joint is coated with bitumastic. Steel forms now encase the joint into

Figure 5.12.3 Pipeline tensioner maintains constant tension on pipe despite irregularities in coating.

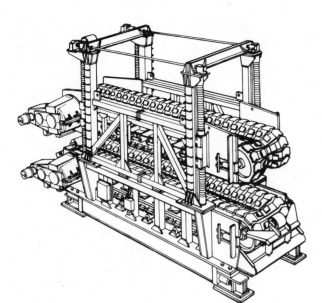

Figure 5.12.4 Pipeline tensioner. (Courtesy Western Gear Corp.)

which concrete is poured. Sheet metal forms may be left in place to protect the fresh coating. The pipeline is now ready to move down an inclined ramp and out over the stringer.

Early (first-generation) stingers were long, hinged ladders, partially buoyant, not unlike a dredge ladder in concept. They in effect formed a ramp down which the pipeline ran to the seafloor, with minimum bending stress. Wheels or rollers were provided to reduce friction and to prevent abrasion of the coating.

Second-generation stingers were articulated so as to accommodate the higher-frequency wave motions so as to reduce the stress in the pipe. These stingers were also buoyant, some even employing the semisubmersible or spar principle so as to minimize heave response to the waves.

With the development of improved tensioners has come the third-generation stinger, a curved extended ramp that guides the overbend of the pipeline down to its point of departure.

Early welding lines, ramps, and stingers were also put on one side of the barge, usually the starboard side, originally as an appendage to an offshore derrick barge. With higher tensioner forces, the tensions in the anchor lines leading forward became critical. The most recent pipelaying barges therefore have the welding line and stinger on the centerline of the vessel.

Control of the pipe on the stinger and consequent control of the tensioner force require the use of load cells or similar devices on the stinger so that the pipe reactions and point of departure may be read out in the control room.

For abandonment and subsequent recovery, a large, constant-tension winch is required, positioned so that it can lead its line down the pipelaying alignment.

Finally, there must be provided all the housing, feeding, and support functions: cabins, mess room, recreation hall, machine shop, power generation, pumps, and winches.

A large crane is on the stern. In part, the original cranes were there to enable the pipelaying barge to also double as a derrick barge. However, a long-boom-crane capacity is also needed for setting risers and for installing and removing the stinger.

5.13 Supply Boats

A supply boat is a boat having a large open bay astern, as wide and as long as feasible, so as to enable the boat to deliver cargo and supplies of all kinds. The "well" or open bay should be long enough to accommodate pipe lengths which, although nominally 40 ft in length, may run 6 ft or more additional. So a 50- to 60-ft well is common. The boats are constantly increasing in displacement and capacity; 1000 tons displacement was a large boat until recent North Sea needs, sea states, and distances led to increases to 1500-, 2500-, and even 3500-ton displacement. See Figure 5.13.1.

Figure 5.13.1 Supply boat tending platform Cognac in Gulf of Mexico.

While the boat is designed primarily to transport cargo, it must have maneuvering ability for close-in work alongside. It also needs reinforced gunwales and heavy fendering to absorb the impact of contact with other vessels.

5.14 Anchor-Handling Boats

The anchor-handling boat is specially designed to pick and move anchors, even in a rough sea. Therefore, it is a short, highly maneuverable vessel. Its stern is open and armored so that wire or buoys can be dragged in over the stern as required. It has a winch at the forward end of the well so that by means of a line, a wire line pendant or buoy can be quickly dragged on board.

5.15 Towboats

Towboats are of several basic types. The large, ocean-going, long-distance towboat is capable of going 20–30 days without refueling. It is designed to move to any part of the world to carry out a major towing job. Such vessels may be up to 250 ft or more in length and carry a crew of 16–20. They can run light at speeds of 12–15 knots.

Towboats are often described in terms of horsepower, but this can be misleading. *Indicated horsepower* (IHP) measures the work done at the cylinders of the engine. *Shaft horsepower* (SHP) is the work actually delivered to the propeller shaft and may be 15–20 percent less than IHP. Long-distance towboats typically have IHP ratings of 4000 to 22,000 HP.

Bollard pull, a much more meaningful measure, is the force exerted by the boat running full ahead while secured by a long line to a stationary bollard; that is, the boat is making no headway through the water. A rough relationship exists between IHP and bollard pull: a 10,000-IHP boat can exert 100–140 tons of static bollard pull. However, the relationship varies with the size of the propeller(s), whether single- or double-screw, and the draft of the towboat. The effective bollard pull falls off as the speed through the water increases. See Figure 5.15.1.

Large oceangoing towboats are fitted with the latest in navigational equipment: Loran C, satellite navigation, radar, and sonar. They can communicate by voice radio anywhere in the world. These boats may be fitted with a towing engine to enable them to maintain a constant tension on the towline

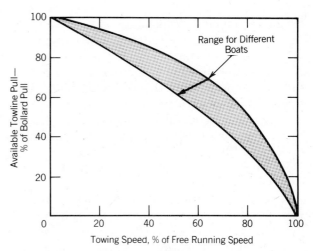

Figure 5.15.1 Effect of speed on tug towline pull.

despite the varying response of the boat to the waves. Other operators prefer to rely on the use of a long catenary, adjusted during tow so as to span a full wave length or more. Boat length should be 11 or more times the expected maximum H_s for safe and efficient operation. In seas greater than this, the boat may have to cut loose.

Shorter in length but still powerful are the boats designed for general operations in a specific theater of operations such as the North Sea. These boats are highly maneuverable, often fitted with a variable-pitch propeller that enables them to keep the engine running at full speed during critical positioning operations. They are usually equipped with bow thrusters to enable them to turn up into the wind while making no headway.

Ocean towboats range from 4000 HP for use in moderate seas to 11,000 HP for all-weather ocean tows to 22,000 HP for towing the largest offshore platforms. Up to six large boats have been used in tandem to tow a platform displacing 600,000 tons.

The smaller harbor and coastal tugs are short-range, meaneuverable boats, designed for short-term jobs near to port. They may have a heavily reinforced and fendered bow to enable them to push as well as tow on a line.

A few towboats have been ice-strengthened so as to enable them to tow through broken ice.

5.16 Drilling Vessel

Normally one does not think of the drill ship or semisubmersible vessel as construction equipment. However, such vessels are often available at the site,

112 Offshore Construction Equipment

having been used for exploratory drilling. They are large offshore vessels, fully equipped, including appropriate mooring gear. They have heavy lift equipment for direct lifts, and they have a central moonpool (open well) which provides direct and often partially protected access to the sea below with minimal wave action at the interface. They have the ability to work at great depths. Thus they have been used for many offshore construction tasks, from setting subsea templates to pipeline repair and seafloor modifications.

The offshore drilling vessel may be a semisubmersible, with response characteristics as previously described, or may be a large ship hull, especially configured to minimize roll. Nevertheless, such a ship-shaped vessel does have inherently more roll response than a semisubmersible.

The drilling derrick is equipped with a large hoist with perhaps 1 million lb (500 tons) direct lift and is often equipped with a heave compensator.

5.17 Crew Boats

Crew boats are used to transfer personnel from shore to offshore operations wherever sea conditions permit this to be carried out in a reasonable and practical manner. Crew boats are seldom used in the North Sea: distances are too great and weather conditions unpredictable. Crew boats are used in the Gulf of Mexico and offshore Southern California.

A rough rule of thumb is that the length of the crew boat in feet should equal ⅔(number of passengers − 15) + 40. The most economical speed in knots will be 1.3 $\sqrt{\text{waterline length in feet}}$.

Economics dictate that the boat should have as high speed as practicable. For nonplaning boats, the required horsepower is proportional to the square of the velocity.

Consideration has to be given to the boat's motions en route: one does not want the entire crew change to arrive seasick. Generally speaking, accelerations should be minimized by adopting a boat with as low a metacentric height as is consistent with safety. A high GM means a quick roll response and physical discomfort to the passengers.

A boat may get into pitch resonance with head-on or nearly head-on seas. This may be modified by changing the speed or heading or both. If the boat's length exceeds the wave length, pitch response is reduced; however, this is usually only practicable in the Gulf of Mexico, not in the Pacific with its longer waves.

Discharge and transfer of personnel at sea will be discussed in Section 6.4. In relatively low-sea states, direct transfer can be made to a large derrick barge or pipelaying barge by coming alongside the lee side or stern while heading into the sea, thus using the derrick barge as a floating breakwater.

5.18 Special-Purpose Equipment

Special-purpose equipment has been developed for offshore use, usually to carry out specialized functions associated with a very large project such as the Oosterschelde Storm Surge Barrier in the Netherlands. (See Section 11.12 and especially Figure 11.12.2.)

Other special-purpose rigs include the reel barge for laying small-diameter flow lines, pipeline "bury barge" and jet sleds for burying submarine pipelines, and special rock placement ships for placing rock around structures and over pipelines and deep-sea nodule-mining equipment. Seafloor crawler tractors to support dredging, trenching, surveying, and sampling operations have been developed and used in special cases where the characteristics of the seafloor are appropriate. As applicable, these will be described in following chapters where particular offshore operations are discussed.

The waves tell of ocean spaces,
Of hearts that are wild and brave.

ROBERT SERVICE, "THE THREE VOICES"

6
Marine Operations

In this chapter, the principal offshore operations common to all types of construction will be described. These include towing, mooring, ballasting, handling heavy loads at sea, personnel transfer, and surveying and diving.

6.1 Towing

Certain basic principles apply to towing. One is that the attachments to the structure or barge must always be sufficiently strong so as not to fail nor damage the structure under the force that parts (breaks) the towline. The actual breaking strength of wire rope is typically 10–15 percent greater than the guaranteed minimum breaking strength. Actual breakage will usually occur under a dynamic load rather than a static load. It is important that under such overload, the structure or vessel being towed remain undamaged.

A usual requirement is that the ultimate capacity of any towline attachment to the unit be at least 4 times the static bollard pull and at least 1.25 times the breaking strength of the towline from the largest tug to be used on that attachment.

At least one spare attachment point, with pennant, should be fitted for towing ahead, to be used in case of emergencies.

A second principle is that the towing force can be exerted through a significant range of horizontal and vertical angles, thus imparting shear and bending, as well as tension, on the fitting.

If a towline does break at sea, it is desirable that it fail at a known "weak link" so that it may readily be reconnected, even in severe sea states. A typical arrangement when a single boat is towing with a bridle is shown in Figure 6.1.1. If the towline is subjected to a high-impact overload, the short pendant between B and C breaks, the shackle at B is pulled back on deck by means of a fiber rope pendant, a new pendant fitted (BC), and the towline reconnected. To reduce shock loads in the towline, either a highly elastic pendant, such as nylon, or a length of chain may be used. See Fig 6.1.2.

When towing an unmanned vessel or structure, trailing lines of fiber rope should be fitted to facilitate picking up the tow if it should break loose in heavy seas.

During maneuvers and positioning, the towline may be shortened in scope so as to permit better control. If it is too short, however, the thrust of the propeller's wash will react against the towed vessel.

When one of the large GBS caisson structures was being moved in Stavanger Fjord, Norway, the lines had very short scope in order to control movement between rock islands. The thrust of the propeller wash against the 120-m-wide and 50-m-deep projected area of the caisson resulted in inability to get the structure to move. The solution was to place the primary tugs at the rear of the caisson, pushing in notches fabricated of steel and timber. Thus the full efficiency of the propeller's thrust could be developed. See Figure 6.1.3.

The inertia of a towed structure, especially a large one such as an offshore caisson, is tremendous. Thus it tends to keep moving ahead long after pull has ceased. A constant concern of boats towing in congested traffic conditions or in ice is that if the boat is stopped, the towed vessel or structure may overrun it.

Further, due to the inertia of the towed structure, it is difficult to slow it or change direction. In a narrow channel therefore, additional boats may be used alongside and also astern. The boats located astern are being dragged backward; when needed, they can go ahead on their screws and thus slow the towed structure. However, being dragged astern, there is a tendency for them to be pulled down

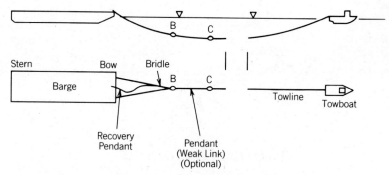

Figure 6.1.1 Typical towing arrangement for barge on ocean tow.

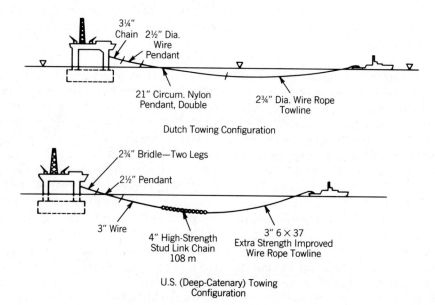

Figure 6.1.2 Ocean towing configurations.

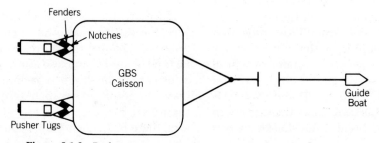

Figure 6.1.3 Pusher tugs are effective in moving a massive caisson.

and swamped. Thus special stern sheets are usually fitted and special attention paid to watertight closures on the boat, since otherwise the engine room door may be left open, regardless of the published instructions.

When towing out in the open sea, the boats lengthen out their towlines so as to offset the wide range of loads in the lines due to the waves and swells.

When towing a very large offshore structure, a single lead boat may run ahead to verify route, confirm depths by forward-looking sonar, and pick its way through underwater obstructions, ice, and so forth. Such a lead boat can also warn other shipping, even though there may have been published Notices to Mariners that should have cleared the route.

If the towed structure is a deep-draft vessel (some of the offshore platforms in the North Sea have drawn 110–120 m) then the towline, if attached to the structure below water, near to the center of rotation, may have a steep inclination. This will tend to pull the stern of the boat down into the water. Therefore, the towline may be led up to a buoyant line or buoys, which will resist the vertical component of the towing force. Such a buoy should be foam-filled to prevent flooding in event of a leak or hole; see Figure 6.1.4. Such a system may also be useful when towing through broken ice, so as to minimize the shock loads in the towline itself.

In the case of the Andoc Dunlin platform and the Doris Ninian Central Platform, flotation units shaped like sausages were clamped to the towlines. These floaters were filled with polyurethane, each giving approximately 5 tons net buoyancy. In Figure 6.1.4, note the pendant leading up to the deck, to permit reconnection in case of a tow line breakage.

Most channels, harbors, and shallow offshore coastal areas have been extensively surveyed, with the results published on hydrographic charts. Unfortunately, the depths of interest were those which pertained to ships having drafts of 10–20 m. Hence when the survey ship got into deeper waters, they usually only recorded depths on a grid, with no interest in determining possible rises, shoals, or pinnacles as long as they did not present a hazard to shipping.

A similar problem arises when towing vessels or structures in areas not normally used by shipping. Hence careful surveys need to be made, using sonar, side-scan, and profiler acoustic equipment so that both the route and its full swath, including sway excursions, are thoroughly scanned.

Required channel widths in sheltered areas are usually twice the beam, but this must be considered in relation to the environmental conditions and navigational accuracy. For exposed areas, the required width will depend on currents, and navigational accuracy and thus may vary from about 600 to 1500 m for relatively short distances of 1–2 km.

During the tow between islands or shoals, accurate electronic position-finding systems need to be set up. Unlike the case of a towed ship or barge having a draft of perhaps 8–10 m and a width of 30–40 m, an offshore structure such as a deepwater caisson may have a draft of over 100 m and a width of 100–150 m. Therefore, it is not enough to only plot the position of the "bridge"; the extremities must also be charted.

Detailed current surveys, both surface and at depth, must be made in restricted areas.

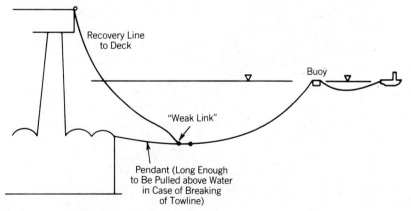

Figure 6.1.4 Towline arrangement for deep draft tow.

A structure under tow will experience sway and wander somewhat on its course. In confined waters, a band may be plotted, shaded in color, within which the structure is safe. Then as the edge of the structure approaches the band edge, corrective action can be taken. This will eliminate excessive "hunt" back and forth trying to stay exactly on a course line. See Figure 6.1.5.

Towed vessels and shallow-draft structures may have an actual draft greater than their mean draft. This may be due to trim—for example, trimmed down by the stern—squat, list, or wind heel. It may be due to the lower density of fresh water discharging from a river into the adjacent sea: Fresh water reaches long distances from the mouths of such rivers as the Orinoco, Amazon, and Congo. In some cases, especially if crossing a bar, heave response may need to be considered.

The usual requirement for underkeel clearance is that the distance between maximum static draft and minimum water depth should not be less than 2 m or 10 percent of the maximum static draft, whichever is lesser, plus an allowance for motion. The maximum static draft should be the actual measured draft at the deepest point with allowance for errors in measurement, initial trim, and water density change.

The motion allowance should include the maximum increase in draft due to towline pull, wind heel, roll and pitch, heave and squat. These values can best be determined by model tests.

Air cushions may be used to reduce draft when crossing local areas of limited water depth. In general, the use of air cushions should be employed only to increase underkeel clearance above the minimum values determined as above so as to ensure the structure will still not hit in event of loss of air.

It is important that the reduction in metacentric height and stability due to an air cushion be considered. After the crossing of the shoal area, the air cushion should be completely vented.

When using an air cushion, an adequate water seal height must be left so as to prevent loss of air. The height of the seal will depend on the speed, since this will cause some air to be sucked out. Typical water seals vary from 0.5 to 2.0 m in height.

To optimize the reduction in draft, large air bags, each 11 × 11 × 4 m, of PVC-coated polyester fabric, were inserted under the base of the Andoc Dunlin platform. This enabled the full depth of the skirts to be utilized for buoyancy, without need for a water seal. Free air was also used in the compartments (in the small spaces around the bags) so as to minimize the changes in external pressure on the bag due to heave.

Communications between multiple boats during a critical positioning operation is all-important. Voice communication is used exclusively; however, it must be remembered that the tug skippers are of all nationalities. While English is usually the common language, lack of full comprehension and misunderstandings have led to serious mistakes. To obviate these, a carefully agreed set of common procedures should be adopted and reviewed so that there will be a clear understanding of all commands.

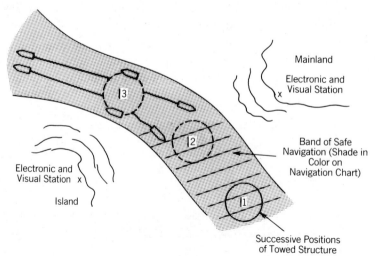

Figure 6.1.5 Laying out a safe swath or band for navigating in constricted channels.

If there are one or more captains who are not fluent in English, it may be desirable to have an interpreter available at the command post.

Procedures should be adopted to handle the case of a broken towline. The boat in question must take in the line and circle back. The towed vessel or structure should recover the bridle or pendant. As the towboat returns alongside, a messenger can be passed and the towline brought on board and made fast. All this is simple with one boat in a calm sea. It is very complex when it occurs at night in heavy seas and wind and the boat is one of three or four.

For very long tows, provision must be made for refueling enroute. One boat at a time can be fueled from a spare boat.

The dynamic accelerations of the towed structure should generally be limited to 0.2 g, so as to minimize forces acting on the tie-downs and to minimize adverse effects on personnel.

Emergencies which must be included in the planning for a tow are fire, flooding, and man overboard. While in congested areas near the exit port, a special fast boat "guardship" should be employed for the dual purpose of picking up a man overboard and warning away sightseeing boats.

When the Condeep Statfjord A sailed from Stavanger, private cruise boats advertised an excursion alongside. This involved hundreds of people whose safety must be paramount. News photographers fall into a similar category. Imagine what would happen if a tour boat gets overturned by running up on a submerged towline! Harbor police can often be engaged to keep the route clear.

In the case of the Statfjord A Condeep, the ceremonies were held on the advertised day, the flags flown, the cruise boats ran around to take pictures, and the towboats blew their whistles. Meanwhile the caisson was securely moored. Two days later, with no fanfare, the flotilla actually got underway, with no unwanted interference.

When large and valuable structures are towed (Statfjord B and C each had a value of approximately $2 billion), the insurance surveyors require a full manning, with adequate pumping capacity, power generation, and firefighting capabilities.

Manning of large and important structures under tow may require a crew up to 30 in number. This crew will probably contain not only marine and ballasting crews, but also a meteorologist, sonar specialist, and navigating surveyors.

When towing in thin or broken ice, an icebreaking vessel will usually open a clear channel. Crowley Maritime has utilized an ice-strengthened barge, with a pusher tug, to open a channel around Point Barrow into the Beaufort Sea. Other similar tows have been preceded by several icebreakers.

Even when a channel has been made through the ice field, there remains the problem of clearance. An offshore platform for the Arctic may have a beam width of 100–200 m. Underkeel clearance will usually be minimal. Somehow the broken ice must be forced to clear around the sides so as not to jam the tow. Boats at the sides can clear ice.

Towing in ice has the further problem of fog obscuring visibility. Radar is fine for locating other vessels but usually cannot distinguish floating ice growlers, bergy bits, and other broken ice floating half-submerged from wave-reflected echoes.

If a lead boat, towing ahead, is on short scope to facilitate maneuverability, it is in danger of being overrun by the tow in the event it is stopped by ice. For these reasons, the use of stern pusher tugs may be an appropriate method of moving an offshore structure through heavy ice. See Figure 6.1.6.

Some of the structures proposed for the Arctic have a conical shape. In the open sea, waves can run up over the lower sides, leading to erratic response. The waterline diminishes rapidly with immersion; hence stability can be significantly reduced. Tows of such structures may require trim

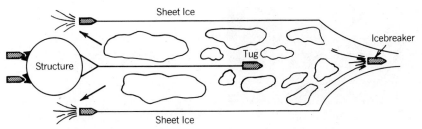

Figure 6.1.6 Towing a large structure through heavy ice.

down by the stern, thus increasing draft. In broken ice, masses may ride up over the sides. The effect of all these can best be evaluated by model tests.

Tows of lesser value may be manned or unmanned. If manned, the Coast Guard will require adequate life rafts and communications to ensure the safety of those on board. If unmanned, trailer lines of fiber should extend from the four corners or quarters to enable a boat to pick up the tow and put personnel on board.

When towing a deep-draft structure through shoal water, the tidal conditions must be carefully plotted. Delays of a few hours, common to marine operations, must be anticipated; otherwise the structure may arrive at the critical zone at low tide.

On the other hand, an advantage of traversing a shoal at midtide is that, if the structure does ground, it will probably raise off at the next high tide. Of course, shifting ballast or deballasting may also be attempted, but not so much as to endanger stability once the structure floats free.

Currents affecting the tow of large structures are principally tidal in nature. When towing the Ninian central platform through the Minch, northeast of the Isle of Skye, the necessary height of high tide was unfortunately always preceded by an adverse flood tide. The current was such as to permit almost no headway to be made against the flood. Therefore, the boats had to catch the slack water at the top of the tide and move over the shoal at maximum speed. See Figure 6.1.7.

Summer storms may arise despite the best long-range and short-range forecasting. They may turn out to be so severe as to make it necessary to cut loose the tow.

Lay-by and standby areas along the route should be identified and marked on the chart. The tow can proceed to point A; then, based on the current sea conditions and the short-range forecast, continue to B; and so forth. At each such station, the alternative of standing by can be considered. These areas are selected for having adequate sea room to lee.

When positioning a structure at an offshore site, it is customary for the tugs to fan out in star fashion. Then the positioning is controlled by going ahead on some tugs more than others; that is, all lines are kept taut. Such an arrangement has been used on the North Sea offshore concrete platforms; see Figure 6.1.8. Note the use of buoys in the lines so as to prevent pulling the sterns under. Bow thrusters are very desirable in enabling a boat to turn into the wind without exerting an increased pull on its towline.

As noted earlier, towboats are usually rated by their indicated horsepower (IHP), whereas a more meaningful figure is the bollard pull which they can exert. The ratio between IHP and static bollard pull varies according to the boat's lines, draft, and propeller, but a ratio of 140 IHP = 1 (metric) ton of force (10 kN) is often adopted. A tug will not, of course, be able to maintain its static bollard pull under continuous running conditions at sea, with the bollard pull decreasing with speed.

The towing horsepower selected should be sufficient to hold the towed structure against waves of $H_s = 5$ m, 40-knot sustained wind, and 1-knot current. Obviously these arbitrary parameters have to be adjusted to the region involved.

Limitations and requirements are placed on stability under tow by the marine surveyor. Typical requirements are the following:

Figure 6.1.7 Towing the Ninian central platform through "The Minch" of Northwest Scotland, enroute to the North Sea.

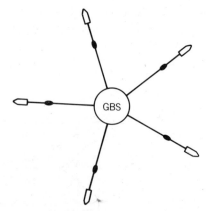

Figure 6.1.8 Positioning of tugs to enable accurate installation of a large gravity-based structure.

1. The metacentric height should have a positive value, typically 1–2 m for a large offshore structure.
2. The maximum inclination of the towed structure under conditions of $H_s = 5$ m, wind 60 km/hr, and full towline pull is not to exceed 5°.
3. The maximum inclination of the towed structure under the 10-year storm for the season involved, with no towline pull, does not exceed 5°.
4. The static inclination under half the total towline pull, in still water, does not exceed 2°.
5. The static range of stability should not be less than 15° at the draft during tow or installation.
6. To insure dynamic stability, the area under the righting moment curve to the second intercept or downflooding angle, whichever is less, is to be not less than 40 percent in excess of the area under the wind-heeling moment curve to the same limiting angle. The wind velocity is that associated with a 10-year seasonal return period, sustained for 1 minute.
7. Model tests are usually required to verify the motion of the towed structure in both regular and irregular (random) seas. These can be used to determine directional stability and any tendency for excessive yaw. Model tests can also be used to investigate the towed structure's behavior under damaged (flooded) conditions, and to determine towing resistance.
8. Inclining tests to verify the $\overline{GM}$ (metacentric height) must be carried out after all superstructure, consumables, and so forth are loaded, immediately before the tow.

In May 1978, the Ninian central platform of prestressed concrete was towed from the Inner Sound of Raasay on the west coast of Scotland to the Ninian Field in the United Kingdom sector of the North Sea. The structure had a base diameter of 140 m, a draft of 84.2 m, and a displacement during tow of 601,220 tons. The draft was severely constrained by the water depth in the Minch at the exit from the Sound of Raasay. The towing distance was 499 nautical miles (925 km) and took 12 days. A fleet of five tugs, totaling 8600 IHP with a combined bollard pull rating of 585 tons (5850 kN) was employed.

In August 1981, the Statfjord B concrete platform was towed from Vatsfjord, near Bergen, Norway, to the Statfjord Field, a distance of 234 nautical miles (433 km). The displacement during tow was 825,000 tons and the draft 130 m. Beam was 135 m. Five tugs, totalling 86,000 IHP, with a combined bollard pull rating of 715 tons (7150 kN) were used. The tow required 6 days.

Detailed data on the tows of concrete platforms in the North Sea are given in Chapter 10 (see Figure 10.1.13.4 and Table 10.1.13.1).

The Thistle platform was one of the largest self-floating steel jackets ever installed. It was built at Graythorp on the River Tees and towed 420 nautical miles (773 km) to the Thistle Field in the United Kingdom sector of the North Sea. Its displacement was 31,000 tons, its dimensions 110 m wide × 184 m long. Two tugs, having a combined bollard pull of 195 tons were used to tow at a mean speed of 3.8 knots.

In September 1975, a slightly smaller self-floating tower was towed from Tsu, near Nagoya, Japan, to the Maui field off New Zealand. The two tugs had a combined horsepower of 15,200 IHP. The structure survived a typhoon in the Pacific with only minor damage.

The steel caisson Molikpaq and the composite steel and concrete caisson GBS-1 were both towed from Japan to Point Barrow, Alaska, and thence to sites in the Beaufort Sea. For the Molikpaq, three tugs with a total of 48,000 IHP were used; for the GBS-1, two tugs totalling 44,000 IHP. Displacements were 33,000 and 59,000 tons, respectively, and the beams were 100 and 110 m. Drafts were about 10 m. Each tow took approximately 50 days.

Four steel gravity platforms were towed 4250 nautical miles from Cherbourg, France to the Loango field offshore Congo. Each structure resembled a jacket, to which three stabilizing bottle-shaped towers had been affixed in order to give stability. Draft of each was 16–19 m; displacement was between 7000 and 8000 tons. Towing speed averaged 3.2 knots, using two boats for each platform, developing 30,000 IHP. Several 5-ton polyurethane foam-filled "floaters" were inserted in each towline.

Further specific information on towing is presented in Sections 9.3 and 10.2.13 (see especially Tables 10.2.13.1 and 10.2.13.2).

6.2 Moorings and Anchors

Vessels working at an offshore site must be held in position despite the effects of wind, waves, and current. The current forces are relatively constant in direction in the offshore zones; in closer-in areas and opposite the mouths of great estuaries, they may vary with the tidal cycle.

The wave forces can be considered as comprised of an oscillatory motion plus a steady, slow drift force. There are both mean forces of a quasi-static nature and dynamic forces to resist.

The standard means of mooring is by means of a mooring system that connects the vessel (or structure) to the seafloor by means of laterally leading lines to anchors. Moorings must be thought of as a system which includes the vessel, the anchor engines, fairleads, mooring lines, buoys, and anchor. In deep water they can be of the catenary type, extending from the vessel in a catenary to the seafloor and thence laterally to the anchor. In shallow water, taut moors may be employed in which the mooring lines are tensioned so as to run relatively straight from the vessel to the anchor or fixed structure.

The dynamic portions of the mooring force must be absorbed. The most often used method is by means of the catenary: the dynamic surge raises the line, using up the kinetic energy in geometric displacement. This concept can be further exploited by including extra weight in the belly of the catenary—for example, a shot or two of chain or a clump anchor weight.

Another means of absorbing energy is by the elastic stretch of the line itself. Wire rope has an initial modulus which increases with use. As the tension force increases due to surge of the moored vessel, the line stretches. As the vessel returns, the line contracts. Use of this system requires either a very-low-modulus material for the line or a very long line.

In practical moorings, both the catenary and the elastic stretch participate.

Very-low-modulus materials are available in the form of nylon and polypropylene. Nylon is widely used for very short lines; unfortunately, it is so elastic that it stores great amounts of energy. If a nylon line breaks, it may not only develop a sudden shock loading but whip back dangerously. Higher-modulus fiber lines are available such as Kevlar.

Steel wire rope is the standard material for mooring lines for construction. Wire lines made with a fiber core and close pitch have somewhat lower moduli than those with a wire core. This is why most mooring lines are wound around a fiber core.

Some large semisubmersible drilling vessels and drill ships use chain mooring lines, with winches specially fitted for this use.

The third method of absorbing energy is by the use of some form of compensator in the system. Hydraulic and steam constant-tension winches are available. Hydraulic compensators may be placed just ahead of the winch. These can be procured with the desired amount of surge accommodation and force displacement characteristics. Even job-fabricated systems with rubber fender units have been used successfully. These latter are effective for very short mooring lines of a temporary or single-time use.

The sheave diameter of fairleads should be at least 20 times the diameter of the wire rope. When mooring lines break, they usually do so at the fairlead, for this is where bending stresses are added to direct tension.

Anchors are of a number of basic types; see Table 6.2.1. First are the reusable drag anchors, which have evolved from ships' anchors; they include the navy stockless type, the Danforth, and the newer Bruce and Stevin anchors. These anchors are designed so that as a horizontal force is applied, they dig down into the soil and mobilize it as resisting force. They are often rated on the multiple of their holding power to their air weight. This is an oversimplification, since it is the soil which they must mobilize, and the resistance varies with the character-

TABLE 6.2.1 Drag Embedment Anchor Efficiencies

Anchor Type	Efficiency[a]	
	Sand	Mud/Silt
Navy stockless	8:1	3:1
Stato	18:1	15:1
Stevfix	31:1	15:1
Stevdig	29:1	—
Stevmud	—	20:1
Hook	12:1	18:1
Bruce	25:1	—
Bruce twin-shank	—	12:1
Doris mud anchor	—	20:1[b]
Danforth	15:1[c]	15:1[c]

[a] Ratio of horizontal holding power when fully "set" to weight.
[b] Exact value unknown but believed to be about 20:1.
[c] Exact values unknown but believed to be about 15:1.

Source: Based on reports by the Naval Civil Engineering Lab, Port Hueneme, Calif.

istics of the soil and the configuration of the anchor. Some anchors are specially designed for soft clays, others for sands.

All these anchors require that the pull be horizontal. In fact, they are purposefully designed so that a vertical pull breaks them loose with little more force than their weight. This means that the portion of the mooring line immediately ahead of the anchor must be heavy enough to stay seated on the seafloor even when the line is under full tension. One or one-and-one-half shots of chain are usually placed in this segment of the line. Sometimes two anchors are used ("piggybacked") by joining them with one-half shot of chain. Tandem anchor arrangements can frequently develop more than twice the capacity of an individual anchor.

There is an exception, however, for the cases where frequent moves are required. Chain cannot usually be accommodated through the fairleads and onto the winches. Therefore, wire lines may be used to within a few meters of the anchor, with extra length provided to ensure a horizontal pull.

At sea, drag embedment anchors of the Danforth, Bruce, or Stevin type typically weigh 10,000–30,000 and even 40,000 lb (4–13 and even 17 metric tons). They are usually placed on the seafloor by lowering directly from the mooring line or from a pendant. The pendant line leads more or less vertically upward to an anchor buoy. This enables an anchor-handling boat to pick up the anchor and move it to a new location. The anchor-handling boat typically raises the anchor only a few meters clear of the seafloor, carries it to the new position, and lowers it back.

For long-distance moves, the vessel takes in on the mooring line until, as it approaches a vertical pull, the anchor breaks loose and swings in under the fairlead. The anchor is now raised until it is housed between plate housings, or else is picked by a crane or davit and placed on deck.

The anchor cannot be left dangling just below the hull; it might swing into the hull under wave action and hole the plates.

Drag embedment anchors are ineffective on rock and erratic on layered (stratified) seafloors.

Another type of anchor is the clump or gravity anchor. This develops its resistance primarily from weight times a friction factor. Such a clump anchor is best used on hard soils (boulder clay or rock, etc.) where the friction factor will approach 1.0 and the dead weight therefore is approximately equal to the holding force. This type of anchor is used on hard bottoms such as rock, boulders, and conglomerate. The navy stockless anchor can be used as a deadweight or clump anchor, although the larger deadweight anchors are usually concrete. Semipermanent deadweight anchors may be open boxes filled after placement with rock or concrete.

Deadweight anchors can also be used for permanent moorings where the long-term characteristics of the soil are little known. They can be used where the direction of pull changes radically from time to time since they are omnidirectional, whereas the Danforth-type anchor will just pull out when the pull reverses direction.

Pile anchors are very effective in many soils. The pile can either be drilled in and grouted, using an offshore mobile drilling rig, or driven in with an underwater hammer or a follower. For practical reasons, the drilled and grouted anchor is most often used in offshore construction practice. The anchor line, usually a shot of chain at this location, can lead from the top or from a distance down the pile.

The anchor pile resists pullout by a combination of bending plus passive resistance (the P/y effect) and tension. In some cases, in rock, a chain has been grouted into a drilled hole, connecting directly to the mooring line. This system was successfully installed off Tasmania to serve as permanent moorings at an offshore iron ore shipping terminal.

Of special concern are soils which have unsuitable characteristics. One of these is calcareous soil, for which little skin friction is developed. Any vertical force applied will lift the pile. Even a straight horizontal force may lead to crushing of the calcareous grains and a degradation of holding power. Extensive grouting of an anchor pile in such soils has greatly improved its capacity as compared to a driven anchor pile. Gravity anchors can also be used.

The most difficult anchoring soil of all is a soft mud, silt, or loose sand overlying a hard material such as conglomerate (off Taiwan) or very dense sand and silt (in the Canadian Beaufort Sea). For these soils the conventional drag embedment anchors tend to skid on top of the hard stratum. Drilled-in anchor piles are not practical if many moves are involved. Deadweight anchors may be used if placed by jetting so as to seat them firmly on the hard material.

Conventional (navy stockless) anchors may be placed in holes excavated by clamshell bucket dredging and then backfilled with dumped rock. These were effectively used in northern Queensland, Australia, where soft muds overlay hard volcanic tuff.

The U.S. Navy has been active in the development of propellant-type anchors, in which the anchor shaft is driven into the soil by either free fall or explosive force. Once penetrated, its flukes resist pullout. These propellant anchors have been extensively used to hold the mooring buoys at the naval base at Diego Garcia, in the Indian Ocean, where hard limestone layers overlain by sand make conventional anchors ineffective.

Suction anchors, which are jetted and airlifted into a sandy seafloor, then continuously pumped so as to maintain a significant pressure differential, have been successfully employed in the North Sea. For the Gorm Field, offshore Denmark, 12 suction piles, each 3.5 m in diameter and approximately 9 m in length, capped at the top, were lowered to the seafloor. By pumping out the water inside, the pile was forced into the seabed. Jets were used to fluidize the sand plugs inside, for removal by the pumps. These piles were designed for a force of 200 tons in any direction.

Propellant embedment anchors are rated up to 150 tons long-term capacity in soft mud and clay soils. Actual values are higher in sand and coral, ranging up to 300 tons. These anchors are multi-directional, are installed rapidly, and function best where drag anchors are least effective.

Once the anchor has been installed, with its adjacent shot or shot-and-one-half of chain, the wire rope line leads to the vessel. If the water is deep, the line may lead directly. However, if the vessels or structure are to be moved on and off many times or if rapid hookup is essential, then mooring buoys will be installed.

The mooring buoy should be designed so that the force is directly transferred from the anchor leg to the barge leg, for example, by running one line through a pipe sleeve. Otherwise, the maximum forces may damage the buoy. The buoy should be designed to resist the maximum hydrostatic head if it is pulled underwater. It should be foam-filled. Many buoys have been sunk by rifle shots from fishermen. The Mini-OTEC riser buoy was not filled: it was sunk by a rifle shot and the entire riser was lost.

To hold a mooring buoy in position requires three anchors. For offshore work, the effective pull which the buoy must resist is usually directional, with a spread of 30–45°. Therefore, two legs of mooring lines are led out to anchors. The third leg is a short leg, either attached to a clump anchor more or less directly below the mooring buoy (it can never be directly below) or leading with short scope toward the mooring position.

Standard API-RP2a, sec. 5.4 2c, warns that if a fully safe anchorage and mooring arrangement cannot be implemented at the site, due for example to constraints imposed by adjacent structures or operations, the orientation should be such that if the anchors slip, the derrick and supply barges will move away from the platform.

Mooring with a single mooring line, thus allowing the vessel or floating structure to weathervane around it in response to the currents and wind, requires that the mooring system have omnidirectional holding capability. Thus a pile embedment anchor or a large clump anchor, both using swivels, are suitable, provided the swivel does not foul. Better yet is a mooring buoy with three or four legs to preset embedment anchors.

When the very large steel jacket for platform Eureka was being towed on its launching barge to the site, it was temporarily moored with a single drag embedment anchor in San Francisco Bay. When the tide changed, the barge drifted back over the anchor and the line snagged it, breaking it loose, so that the barge was adrift, dragging its anchor. The barge eventually went aground, but fortunately there was no serious damage to either barge or jacket.

Moorings designed to hold structures for relatively long periods during construction should be designed on the following basis:

Total Exposure	Return Period
Less than 2 weeks	10 years for season involved
2–8 weeks	20 years for season involved
Greater than 8 weeks	100 years for seasons involved

For very large structures such as gravity-based platforms having long response periods and where the moorings have resilience to absorb shock loads, the design should be based on the significant wave plus a 1-minute sustained wind at the relevant heights and maximum current. Allowance may be made at sheltered inshore locations for reduced wind and waves and the changed geometry of the mooring due to excursion under load.

6.2 Moorings and Anchors

The maximum load in a new or used mooring chain with its associated shackles and fittings, including residual pretension, should not exceed 70 percent of the minimum breaking load, after allowance for corrosion and wear. Note that many manufacturers and classification societies quote an average breaking load rather than the minimum.

Where wire line moorings are used, the maximum load should not exceed 60 percent of the guaranteed minimum breaking load.

The design holding capacity of any anchor, winch, or connection, multiplied by appropriate safety factors to account for gusts, dynamic amplification, and so on, should exceed the extreme storm loading on it. Note that the effect of long-period response motions in sway, surge, and yaw may impose significant dynamic increases in force.

The attachment points to the structure should be designed to resist at least 1.25 times the nominal breaking load of the mooring, without damage to the structure.

For positioning a large offshore structure in the open sea, either several boats or a mooring system may be used. Experience has shown that the use of four or five boats equipped with variable-pitch propellers and bow thrusters can position an offshore structure in calm seas and wind conditions within 5 m or perhaps even somewhat less. However, in many cases the use of such a boat system will not be adequate nor fully suitable.

An example when an offshore mooring system is required is deck mating. Other examples include positioning a bottom-founded structure over a predrilled template, operations in shallow water, and those subjected to significant currents. An offshore mooring system is also required when the structure or vessel must remain on location for a significant period of time, subject to changes in the sea state and wind conditions.

For example, at the site of the installation of the platform Cognac, in 300 m of water depth, south of New Orleans, a group of eight mooring buoys was set out in a pattern around the site. These enabled the derrick barges to be moored in exact location; these in turn controlled the three jackets that were successively placed to make up this record-setting structure.

In shallower waters, off the coast of Queensland, Australia, 10 gravity-based caissons were required to be positioned within 0.5-m tolerance. This was achieved by use of preset mooring buoys. See Figure 6.2.1.

The arrangement adopted, using a spring mooring buoy, gave upward spring action, with the buoy serving as a spring buoy during final setting, thus absorbing shock and dynamic surge loads. This system is also called an "inverted catenary."

For mooring of an offshore derrick barge working in coastal waters, two systems have been found necessary. On the San Francisco Southwest Ocean Outfall Project, a 4-m-diameter buried concrete pipe is being installed 7000 m into the sea in water up to only 30 m deep. An offshore derrick barge was initially moored on a taut mooring in order to enable it to carry out operations requiring accuracy and control. An intense storm of unpredicted long-period waves created enormous surge forces that broke the taut mooring lines and drove the vessel onto the beach, severely damaging it. See Figure 6.2.2.

As noted earlier, long-period waves in shallow water develop elliptical particle orbits and create shorter and steeper waves with increased surge accelerations. When the offshore barge was returned to the site, the taut mooring system (Fig. 6.2.3) was again used for operations, but in addition a survival mooring system was employed.

If the water had been deeper, the survival

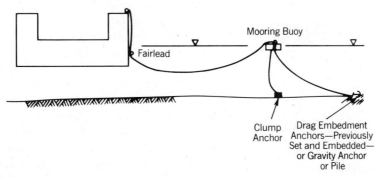

Figure 6.2.1 Mooring line arrangement.

124 *Marine Operations*

Figure 6.2.2 Offshore derrick barge on taut moorings during heavy swells. Shortly after this picture was taken, the taut lines broke and the barge was driven aground. (Courtesy Morrison-Knudsen Co.)

mooring would have been configured as a catenary, perhaps with a clump or chain in the bight to provide greater spring action. See Figure 6.2.4. However, in shallow water the stretching out of the catenary permits very little movement in surge. Hence other means have to be taken. In this case both a spring buoy, giving geometric travel, and a very long wire line, giving elastic stretch, were employed; see Figure 6.2.5. The dynamic surge to be encountered during a severe storm with 18-second period, similar to that experienced in the previous catastrophe, was 27 ft (8 m). This is single amplitude displacement from the mean position, which means that the line will slack at the other end of the cycle. A force of 65 percent of the minimum guaranteed breaking strength was allowed as the maximum load under the design surge force imparted by the highest wave group in a 6-hour storm.

The surge force is of course directly proportional to the stiffness. The lower the stiffness, the less the force. On the other hand, the surge excursion is little affected by the stiffness; see Figure 6.2.6. The stiffness of the system is the sum of the elastic and geometric stretch. Any attempt to restrain the dynamic surge excursion to less than its full value would lead to extreme forces, far beyond the breaking strength of the line.

Consideration was given to the use of two survival lines. However, this would double the stiffness and hence double the force. Further, it would be extremely awkward to work, especially since another piece of floating equipment (a dredge) would be moored only 1000 m to seaward. Also, since the two lines could never be fully equalized, one line would tend to take more than half the total force.

Studies of the dynamic response showed that for the water depth (15–20 m) and barge length (130 m), there was little dynamic response for waves with less than 14-second period. As soon as the wave period increased, the dynamic surge jumps up dramatically. This is strictly a shallow water effect. See Figure 6.2.7. Thus the vessel must be oriented to ride into the longer-period swells rather than any cross waves generated by local wind-driven seas.

The selection of proper lines and appurtenances for moorings is of great importance. Wire rope lines are generally employed: they are flexible, easily coiled on the drum of a winch, and can be changed in direction by means of fairleads and sheaves. However, wire rope is subject to abrasion from sand and corrosion in the spray zone, this latter being a serious matter when the mooring is semipermanent. Galvanized wire rope is sometimes used, although it lowers the strength by 5 percent. Bending over a sheave or fairlead increases the tension in the outer wires: lines usually break in the fairlead. The ratio of sheave to rope diameter should be in excess of 20. Fairleads must be well-designed and properly lubricated in operation so that they will lead fair and not chafe the wire rope. See Figure 6.2.8.

Wire rope tends to become stiffer with use, and this in turn may raise the dynamic loads in the line. If the line becomes flattened over a sheave or if

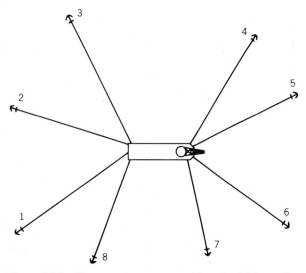

Figure 6.2.3 Taut mooring system positions derrick barge for operations.

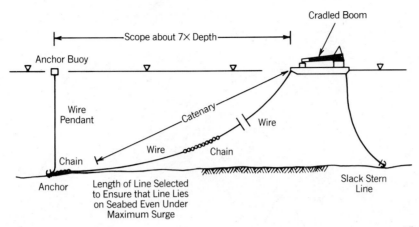

Figure 6.2.4 Survival or storm mooring in relatively deep water, e.g., where depth exceeds 20% of storm wave length. This system accommodates dynamic surge force primarily by changes in geometry of catenary.

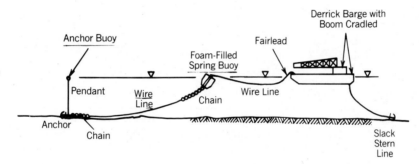

Figure 6.2.5 Survival or storm mooring for relatively shallow water, e.g., water depth less than 20% of storm wave length. This system accommodates dynamic surge primarily by stretch of wire rope.

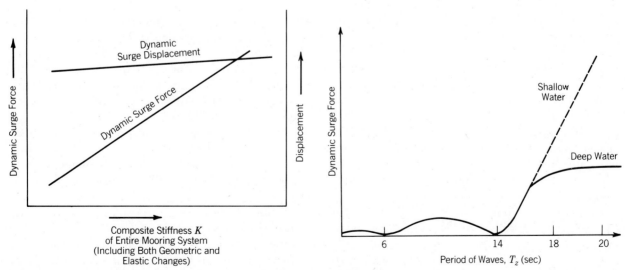

Figure 6.2.6 Effect of stiffness of mooring system on maximum forces and displacements.

Figure 6.2.7 Dynamic surge as a function of wave period, for both deep and shallow water.

125

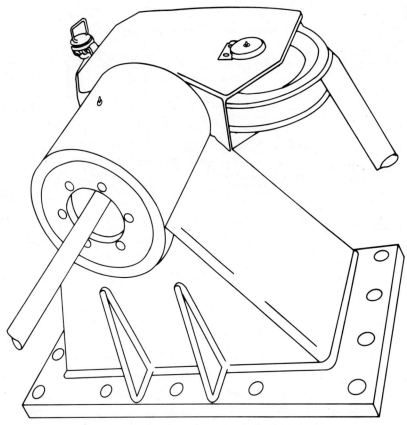

Figure 6.2.8 Fairlead. (Courtesy of A. C. Hoyle Co.)

one or more wires break, its usefulness as a high-capacity mooring line is severely degraded.

Chain is relatively free from abrasion and corrosion problems and does not stretch; thus its extendability to absorb dynamic loads must come from its catenary action, which in deep water is usually adequate.

Synthetic fiber lines made from nylon, polypropylene, polyester, or Kevlar are used extensively where their properties of light weight, flexibility, and low modulus can be utilized to advantage. They are of course very susceptible to abrasion and even to fishbite (including shark attack). Encasing these lines in polyurethane may eliminate these problems. Fiber lines are also subject to a reduction in strength when wet and to fatigue. Polyester ropes have been found superior to other fibers in the marine environment. Fiber lines, especially nylon, tend to store energy elastically, which may give rise to resonant surge.

Composite lines, using chain where weight is required, wire for length, and a short section of synthetic fiber rope for stretch, are becoming more common.

The preset moorings for the construction of the Cognac platform in the Gulf of Mexico in 1050 ft (320 m) of water depth are illustrative of a major installation carefully engineered to ensure that the operations could proceed without difficulty. A 12-point mooring system was adopted. Each of the 12 mooring buoys was built of steel, filled with 2 lb/ft^3 closed-cell urethane foam, and equipped with controllable light signals for guidance of the work boats so they would not run up on a taut line.

The buoys were then anchored by 1300 ft of 2¾-in. chain, which led to pile anchors. The pile anchors were 120 ft long, 30 in. diameter, with 1-in. walls. They were jetted into place by a drilling vessel.

From the buoys, two parts of 2-in. wire rope line led to the barge.

Between the two derrick barges, four 14-in.-circumference nylon lines were run, to absorb shock loads through their elasticity. See Figure 9.9.7.

Dynamic positioning ("thrusters") are being increasingly used to maintain position on some or all of the axes and thus fulfill part of the mooring requirement. These thrusters may be mounted on deck or within the hull. In their most sophisticated arrangements, they are controlled by minicomputers and utilize variable-pitch propellors so that they run at constant speed.

They are frequently employed in conjunction with mooring lines. The latter lead in the directions for which the greatest forces (usually longitudinal) are required, while the thrusters maintain transverse position.

6.3 Handling Heavy Loads at Sea

The installation of offshore structures usually includes the lifting and setting of modules and other heavy loads on the platform. Such lifts may weigh up to 2000–4000 tons and more. The current record is a 5400-ton lift for platform Esmond. New derrick barges under construction will lift twice these loads.

Installation involves motion and hence dynamic loading and impact effects. API RP2A, sec. 2.4, "Installation Forces," recommends specific precautions to ensure safety in the handling of such loads.

The DNV *Rules*, app. H-1, "Lifting," specifies procedures and rules to ensure safe lifting of heavy loads at sea. This document is included in the appendices to this book, by permission of DNV.

When lifting a heavy load, there are both static and dynamic forces to consider. The static forces include the actual load itself, which if not weighed, must be computed to include the design weight, plus adequate allowances for over-tolerance plate thickness, weld material, padeyes, and any supplies stored within. Static lifting loads must also include the slings, spreader beams, and shackles.

The author once investigated a critical lift that was at rated capacity. A careful physical inspection revealed that over 50 tons additional of tools and supplies had been stored aboard by the drilling crews. Worse, many of these, including an acetylene bottle, were loose, that is, not properly secured.

The dynamic forces are those due to acceleration, first as the load line lifts while the load, still resting on the barge, is starting the down heave cycle. Later, both horizontal and vertical accelerations are imposed during swing.

Lifting forces on the padeyes and the structural members of the load to which they are secured have both vertical and horizontal components. Many modules are designed to withstand the vertical quasi-static forces imposed in lifting in the fabrication yard, where bridge cranes or skids may be employed. At sea, however, the slings are usually inclined in two planes. See Figure 6.3.1.

Although the padeyes themselves are usually adequately designed for vertical and horizontal loads, the structure to which the padeye connects must also be able to accept and transmit the total vertical and horizontal forces back into the structure.

Modules fabricated in Houston or Singapore are initially lifted in warm weather. When later lifted at the site, which may be the Bering Sea, cold weather impact properties become important.

Vertical forces on lifting can include the favorable effects of buoyancy where applicable; however, fully or partially submerged structures may pick up an added hydrodynamic mass component. This latter may be an extremely high factor when the submerged surface is horizontal. See Figure 6.3.2.

For example, a proposal was made by one offshore contractor to lower a 50,000-ton boxlike unit to the seafloor of the North Sea. He proposed to utilize the buoyancy of the box to reduce the net

Figure 6.3.1 Lifting 300-ton quarters module onto platform in the Bass Strait, Australia.

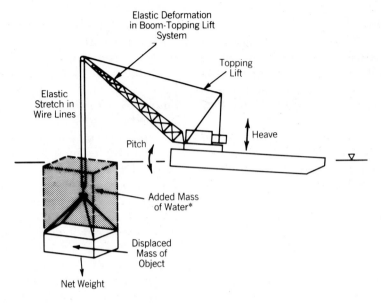

Figure 6.3.2 Schematic representation of dynamics of lowering loads under surface of sea.

weight to a few hundred tons, well within the rated capacity of the derrick barge. What he failed to recognize was that due to added mass effects and inertia, the rectangular box-like unit would react almost not at all to a 6- to 7-second wave, whereas the derrick barge would have significant pitch and heave during that same period. While the stretch of the lines would have accommodated some of the dynamic effects, a detailed analysis showed that the derrick booms would be seriously overloaded.

Instrumentation is now available to enable control of the dynamic aspects of lifting. These consist of sensors on the crane barge, on the crane boom, and on the barge or boat from which the module or other lift is being lifted. Typically, mini- or microcomputers then give readouts of "load on hook," "outreach" (radius), "hook height," "wave height," "wave period," "derating of crane capacity for sea state," "hook speed," "net load on deck of crane barge and effect on stability," "crane hook height," "off-lead" (distance between load and structure), "automatic level luffing," and "warning as to turns left on winch drum." Other programs are available to determine optimum heading of crane barge to minimize boom tip motion and hence the dynamic increment of load during the operation.

The most recent example of an extremely heavy lift, the 5400-ton deck of the Esmond topsides, was carefully engineered for setting with Heerema's derrick barge *Balder,* with its two cranes rated at 2700 and 3000 tons. Positioning was achieved using its conventional taut mooring lines to anchors, along with a computerized ballasting system for the crane barge. The entire operation took only 1 hour.

Lifting eyes are designed to transmit the load to the slings in the plane of the sling. As the structure swings, however, or the barge from which it is being picked sways, a side loading may be imposed. API RP2A recommends that a horizontal force equal to 5 percent of the static swing load be applied simultaneously with the static swing load. It is to be applied perpendicular to the padeye at the center of the pin hole.

When suspended, the lift will assume a position such that the center of gravity of the load and the centroid of all upward-acting forces are in static equilibrium. These relative positions should be taken into account in determining the inclination of the slings. The force in the sling is the resultant of the horizontal and vertical forces at the padeye, as computed for the most severe inclination of the sling. Due to swinging of the load while in the air, the load will not be uniformly distributed on all four slings. This nonuniform distribution must be considered in sizing of the slings and their fittings.

As the load is picked, and again as it is set, the

6.3 Handling Heavy Loads at Sea

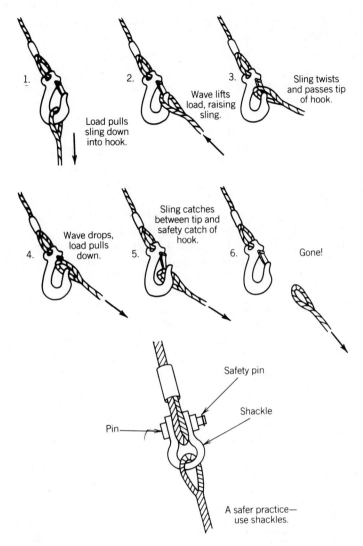

Figure 6.3.3 Potential hazard when lowering load into water. (Adapted from *The Principle of Safe Diving Practice* CIRIA, London, 1984.)

position may vary from the above, due to the horizontal and vertical reactions from the deck of barge or platform, as well as those from tag lines and guides. The change in horizontal and vertical forces so occasioned must be considered in determining the forces and angles of application on padeyes and hooks. Figure 6.3.3 illustrates safe and unsafe use of hooks and shackles. Similarly, Figure 6.3.4 gives right and wrong ways to clip wire rope in the forming of eyes.

API RP2A recommends that for lifts to be made in the open sea, a minimum load factor of 2.0 should be applied to the calculated static loads. This is based on an elastic analysis, so that the actual load factor in ultimate design is significantly greater.

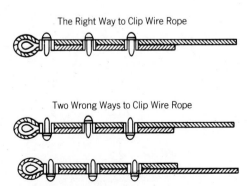

Figure 6.3.4 Proper rigging details such as these are essential for safe operations.

130 *Marine Operations*

Thus it is in general accord with wire rope rigging design, for which a factor of 4 to 5 is normally applied to the minimum guaranteed breaking strength.

The above factors should also be applied to the padeyes and other internal members connecting directly to the padeyes. All other structural members transmitting lifting forces within the structure should be designed using a minimum load factor of 1.35.

During loadout in sheltered harbors, where there are no waves, the load factors can be reduced, but not less than 1.5 and 1.15, respectively. Note that such reduced factors may be technically in violation of local rules and thus require waivers.

As noted above, structural members, padeyes, and so forth are to be designed on the basis of allowable stresses, for example, those specified in the AISC *Specification for the Design, Fabrication, and Erection of Structural Steel for Buildings*, with no increase because the loads are short-term. In addition, all critical structural connections and primary members should be designed to have adequate ductility so as to insure structural integrity during lifting even if temporary or local overloads occur.

Special attention must be given to ensuring weld ductility and prevention of undercutting and adverse heat affects on the surrounding metal. Low-hydrogen electrodes should be used.

The design of padeyes requires special attention and detailing. Given the forces, including dynamic, and range of angles in both planes over which the forces may act, the padeye must transfer the load from the pin of the shackle into the structural frame.

Transverse welds perpendicular to the principal tension, where the member is subjected to impact, are prohibited by some national codes. If they are used, the details, welding procedures, and non-destructive testing (NDT) used to verify them must be such as to ensure full development of ultimate strength and ductility. Fillet welds are especially dangerous under impact tension, whereas properly made full-penetration welds can be safely employed.

Cheek plates have been often used in the past, but they inherently require a transverse fillet weld.

Figure 6.3.5 shows an example of an incorrect (dangerous) design and fabrication of a padeye. In the figure, note the following potential failure areas:

1. The fillet weld at A is unable to develop the full tensile strength of the plate D.

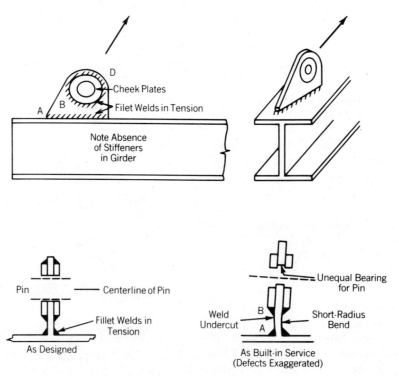

Figure 6.3.5 Example of incorrect and dangerous padeye details.

2. The fillet welds at A and B are both transverse to the principal tensile stress under the impact of lifting.
3. The holes in the cheek plates do not line up with those in the parent plate. Finally, the distance A–B is so short that any swing of the load produces sharp bending in the plate.
4. There is possible weld undercut at A or B.

Unfortunately the details shown in Figure 6.3.5 are often encountered. In practice, it has led to catastrophic and fatal accidents. It should be corrected before a heavy lift is made.

Wherever possible, the load from the padeye to the girder should be transmitted in shear. The use of cheek plates should be avoided. The distance from girder to pin should be at least six plate thicknesses. Where cheek plates are necessary, they should be welded to the main load carrying plates with bevel welds, sufficient for an even transfer of stresses, on the inside and with fillet welds on the outside. The bore of the cheek plates should be flush with the bore of the padeye, so as to achieve uniform bearing on the shackle pin.

The direction of rolling of the parent steel plate should be determined, and this direction should then correspond with the direction of the sling.

If connection between padeyes and structure cannot be transmitted by shear, then full-penetration welds can be employed, using the welding procedures prescribed for the primary structural members, with full NDT inspection. The structural plate to which the padeye is attached must have

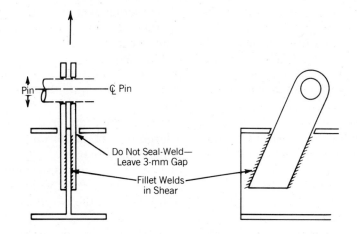

(a) When Lead of Sling Is Parallel to Girder

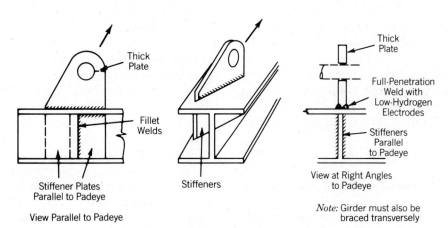

(b) When Lead of Sling Is at an Angle to Girder

Figure 6.3.6 Satisfactory (safe) padeye details.

adequate through-thickness toughness so as to prevent laminar tearing.

Figure 6.3.6 shows satisfactory and safe details.

API RP2A calls attention to the fact that fabrication tolerances and variations in sling lengths can redistribute the actual forces and cause significantly increased stresses in some members. The variation in sling length should not exceed ±.25 percent of nominal sling length nor 1½ in. (37 mm). The total variation from the longest to the shortest sling should not be greater than 0.5 percent of the sling length or 3 in. (75 mm).

Where unusual deflections or particularly stiff structural systems are involved in the lifted load, a detailed analysis should be made to determine the redistribution of forces of both slings and structural members.

The horizontal force in the structural members connecting the padeyes can produce both high compression and bending due to minor eccentricities. It must be checked to ensure that buckling cannot occur. Similarly, spreader beams, if used, must be checked against buckling under compression on both axes, as well as for combined bending stresses between sling attachment points.

Shackles and pins should be selected so that the manufacturer's rated working load, provided it includes a factor of safety of at least 3, is greater than the static sling load.

Lifted loads must be controlled against swinging by use of tag lines.

When lowering objects below the sea surface, special rigging is used to provide the required length of line and to release after setting.

Where a load is to be temporarily set on a platform deck for later skidding to its final location, the first site must be checked to ensure structural adequacy for the load, including an allowance for impact. Jacking points should be designed into the structural frame of the load to ensure that it can be safely supported at these points, which often are offset from the permanent supports. The seat should be especially checked against local buckling due to a combination of direct load and compression.

6.4 Personnel Transfer at Sea

The transfer of personnel from crew boat to offshore derrick barge or onto a fixed platform is a critical operation from the point of view of both safety and efficiency. In fact, the ability to move men on and off can become the limiting criterion for continued operations in a rough sea. All too often this operation is overlooked in the planning phase.

The boat in which the men are traveling to the offshore rig is responding to the wave action in all modes; heave, pitch, and roll being the most critical for the transfer operation.

The use of fixed inclined "ladders" is safe only in a very calm sea; they have been aptly called "widow makers."

Articulated ladders have been developed which essentially eliminate the roll–pitch motions; they utilize compensating pressure cylinders to maintain a steady attitude, subject only to differential heave. These are used on such major operations as the transfer from semisubmersible floatels to fixed platforms in the North Sea. These articulated ladders or gangways are available for smaller boats and operations; however, they are expensive and take time to rig.

More common, therefore, are other means of transfer. It must be also considered that there is usually a substantial height differential between boat deck and barge or platform, for example, 4–20 m. Properly fendered boats can come alongside a large derrick barge under favorable sea conditions, using the barge as a breakwater. In the case of head seas, they may come up to the stern. A notch should be fabricated so that the boat can push tightly against the barge and thus minimize pitch response.

The cargo net concept has been adopted from the World War II troop transport experiences. It is hung from a boom, so that the lower end is at sea level; when the boat moors, the net can be hauled into the boat. It is a relatively simple and safe operation for a man to catch the top of the heave–pitch cycle and climb up the net. When he reaches the boom, however, he faces a dilemma. Somehow he is expected to scramble onto the boom and walk in to the deck. Even if the net is hung directly over the platform side, it is very difficult to scramble onto the deck unless handholds are provided. Hence lifelines (handhold lines) need to be fitted above the boom.

Transfer back from platform to boat is more difficult. Assuming lifelines make it easy to get onto the net and climb down, below is a boat moving up and down several meters in a 5- to 7-second period. There is an instant when he must step completely off, at the top or just before the top of the boat's heave cycle. If a foot catches or the man tries to hang onto the net, he may be jerked clear of the boat as it descends.

For these reasons, the net should be placed in the well of the boat, about midships, rather than at the bow. Then relative motions will be minimized. See Figure 6.4.1.

For more severe sea states, the Billy Pugh net is employed. This can be lowered from a boom on the platform or derrick barge into the boat well and the line slacked so as to leave the net in the boat well despite heave displacements. The men climb on, placing their tools in the box in the middle. Catching the top of a heave cycle, the net is hoisted swiftly so that the net and men are well clear by the time of the next heave. For transfer from platform to boat, the net is lowered down just clear of the heave peak; the operator waits for a relatively moderate amplitude phase. Then as the boat is at the top of the heave, he lowers rapidly. The net contacts the boat as it comes up on the next heave cycle; the operator slacks the hoist line so it runs free, and the net stays in the boat.

This requires that the net be lowered on the brake, not in "power down." Many archaic safety laws contain a general prohibition against lowering on the brake; these are regulations pertinent to land operations but not offshore. In the open sea, attempt to lower under power may result in the men being landed in the boat, then jerked out on the next descending cycle about 3 seconds later.

The other method of personnel transfer is by helicopter. Strict discipline is applied on approaching and entering the helicopter while the blades are running. It is the tail rotor which is most lethal, although in a high wind, the main rotor can tip. All bystanders, men awaiting boarding, and guests must be kept well clear.

6.5 Diving and Underwater Work Systems

Manned intervention in underwater construction is one of the oldest forms of offshore activities, dating back at least as far as the ancient Romans and

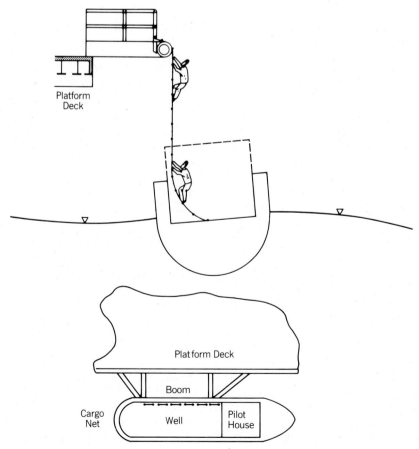

Figure 6.4.1 Arrangement of boom and cargo net for personnel transfer.

probably to even older civilizations. As practiced in the offshore today, diving and the use of highly skilled technicians in the underwater environment is a very advanced technology, supported by extensive research and development in such disciplines as physiology, psychology, communications and control, power systems, and mechanical devices. Underwater tools and electronic–acoustic systems have greatly enhanced the effectiveness of divers.

Among the equipment available to the diver, the *Handbook of Underwater Tools* lists the following. These give an indication of the many tasks which divers may be called upon to perform.

1. Inspection and nondestructive testing (NDT)
   a. Magnetic particle inspection equipment
   b. Ultrasonic equipment
   c. Eddy current/electromagnetic equipment
   d. Radiation monitors, trace leak detectors
   e. Cathodic protection monitoring equipment
   f. Range–level measuring and positioning equipment
   g. Metal detectors
   h. Thermometers
2. Photographic equipment
   a. Still cameras
   b. Cine (movie) cameras
   c. Video systems (TV cameras)
3. Underwater cleaning equipment
   a. Water jetting and grit blasting
   b. Portable brush-cleaning machines
   c. Self-propelled cleaning machines
4. Torquing and tensioning equipment
   a. Manual and hydraulic torque wrenches
   b. Torque multipliers
   c. Stud tensioners
   d. Extensometers
   e. Flange pulling–splitting tools
5. Lifting equipment and holdfasts
   a. Lifting–inflatable bags
   b. Gas generators
   c. Lifting–pulling machines
   d. Magnetic handles and suction pads
6. General underwater equipment
   a. Wet welding and cutting equipment
   b. Underwater machining tools
   c. Chipping hammers
   d. Cutoff saws
   e. Grinders
   f. Drills
   g. Impact wrenches
   h. Hydraulic wire cutters, cable crimpers, spreaders
   i. Breakers
   j. Power-actuated fasteners, cutters
   k. Pressure intensifiers
   l. Grouting and resin injectors and dispensers
   m. Underwater painting machines
   n. Jet pump dredges, air lifts, and ejectors
   o. Subsea marking systems
7. Subsea power packs
8. Diver-held location devices
   a. Cable tracking system
9. Explosive devices
   a. Pipe, chair, casing cutters
   b. Perforators
   c. Shaped charges
   d. Underwater rock drills
10. Underwater lighting
11. Chain blocks
12. Jet burning equipment–thermic lancers
13. Diver-operated geotechnical tools
    a. Impct corer
    b. Miniture standard penetration test tool
    c. Vane shear
    d. Rock classifier
    e. Jet probe
    f. Vacuum corer

The properties of the underwater physical environment which affect a diver's ability to perform work there include the following:

1. *Pressure.* The increase of pressure with depth affects human sensory and reasoning powers and causes gases to be dissolved into the bloodstream.
2. *Temperature.* Low temperatures cause serious loss of body heat. This is especially critical in deep diving and when diving in Arctic or sub-Arctic areas.
3. *Turbidity,* especially near the bottom and around structures, impairs vision. The op-

erations of the diver and his equipment may stir up the sediments and cause a turbidity "cloud."
4. *Currents* tend to sweep the diver away from location and to make his position control more difficult.
5. *Refraction phenomena* of light and acoustic waves are different from those in air.
6. *Waves* endanger the descent and ascent of the diver through the sea–air interface.
7. *Marine growth* shields surfaces and joints from inspection and can rip a divers suit.

Physiological and psychological effects of importance are the following:

8. *Disorientation*—the inability to differentiate the direction of a sound source and the loss of reference planes.
9. *Hearing modes* are changed.
10. *Sight capability* is reduced under the stress environment.
11. *Speaking intelligibility* is greatly reduced; the familiar ducklike sounds of a diver working on a helium–oxygen mixture are well known. Greater depth produces greater distortion.
12. *Physical fatigue.* The work effort at depths is greatly increased.
13. The pressure causes *greater absorption of gases* by the bloodstream.
14. *Miscellaneous.* Diver safety can be affected adversely by noise, electrical shock, debris and fishing lines, and explosive concussion, as well as accidents with lifting and rigging, water jetting at high pressure, and underwater cutting and welding.

If an air mixture, oxygen plus nitrogen, is used, the blood will absorb nitrogen under pressurization. Under subsequent decompression, if carried out too rapidly, the nitrogen will form bubbles in the bloodstream, leading to serious injury and even death. This is the well-known nitrogen narcosis or "bends."

The use of carefully developed gas mixtures such as helium/oxygen has enabled safe diving to be carried out to depths of 150 to even 300 m. Other gas mixtures are being developed and used to permit operations in even deeper waters. One of these being developed by Comex is a 97.5 percent hydrogen, 2.5 percent oxygen mixture called "hydrox," with which they hope to extend safe diving depths to 1000 m.

Decompression rate schedules have therefore been developed to ensure that the blood's natural balance of gases has been restored without the formation of bubbles. Foremost among these and widely adopted by regulatory bodies throughout the world are the U.S. Navy's Standard Tables: "Single and Repetitive Dives, Helium/Oxygen Tables" and "Decompression Tables for Standard, Exceptional, and Extreme Exposure."

Saturation diving systems and "bounce techniques" have been developed which take advantage of the physiological fact that after the blood becomes largely saturated with an inert gas such as helium, then further absorption proceeds very slowly. This enables a person to stay at depths for long periods of time, working short periods, resting without decompression in a habitat or chamber, then returning for another stint of work. Using deck decompression chambers, personnel transfer capsules, and a habitat, the saturation diving system can support several teams of divers for long periods of time. With bounce techniques, the divers periodically descend to work, then return to partial decompression only.

The time required to perform work underwater increases with depth, although not as rapidly as one might anticipate. For example, a number of tests have shown that specific items of work required 20 percent longer when at shallow depths and 50 percent longer at deep depths, as compared to the time required for performance in the air.

The greatest components of time are the times for descent, ascent, and decompression. The rate of descent is usually important only for deeper dives.

Decompression tanks are carried on deck of the diving support vessel to enable saturation diving to be employed and to repressurize a diver who develops symptoms of the "bends."

A major limitation when using divers for underwater work is that of developing a reactive force, "getting a foothold" so that the diver may exert a force. Because the diver is in a state of near-neutral buoyancy, he is like a man in space; if he shoves against a pipe, he is merely moved away. When a diver can plant his feet firmly on the bottom or against a structure, he can exert 100–300 N (22–66 lb) force in push or pull. Various means of attachment have therefore been developed to assist the diver in exerting a force.

There are many forms of diving, based on the

equipment used. These vary from the scuba gear to new diving suits that protect the diver from injury and lightweight helmets with improved vision capability.

Wet suits are most commonly used. For deep diving and diving in the Arctic or sub-Arctic, the suits are heated, usually by warm water circulation.

Free-swimming (scuba) divers can perform inspection tasks; they are limited in communication and of course have no power supply for tools.

Tethered divers can have warm water circulation for their suits, hard wire or fiber optic links for communication, and hydraulic power.

To enhance the diver's capabilities for work, diving chambers or bells may be used. These give the diver more freedom from encumbrances but of course limit his mobility.

Diving bells which operate at atmospheric pressure have been developed; these enable inspection and work by engineers who are not qualified divers.

A complete pressure-resistant diving suit, known as "Jim" has been developed, which enables the diver to stay at atmospheric pressure. A refined version, "WASP," enables a worker to descend to 600 m. depth.

The current trend in diving systems appears to be toward diving bells and similar systems to enable better and safer control. However, there are many tasks requiring entry into congested spaces and so forth which can only be done by a diver.

Diver communications is an area in which there have been major advances in recent years. In addition to hard wire and fiber-optics systems, modulated sonar-frequency carrier systems give ranges of 150–500 m and single-sideband communications can give a 1000- to 1500-m range.

One of the most serious limitations of diver work is inability to determine one's position. This is due to lack of visibility, to disorientation, and to the lack of reference points. Consider an underwater concrete caisson that to the diver presents an endless wall 60 m high. Markings are required. Large orange epoxy numerals have been painted at about 10-m spacings to assist a diver in determining his location.

Wire guide lines have been stretched to serve as guides for a diver and to hold his position in a current. Such wire guide lines are especially important if a diver must enter under or through a structure—for example, into the middle of a braced jacket—or underneath a gravity-based structure while it is still in the floating mode during construction.

To clean off marine growth for inspection, high-pressure water jets as well as hydraulically operated rotary brushes have been developed.

One of a diver's problems is that of marking locations so he can return. Acoustic pinger locators are often used. The diver may use a hand-held sonar to enable a search for a pipeline or dropped object.

Both sonic and tethered guide line systems are employed to guide a diver working under ice back to the entry–exit hole.

A major problem in diver communications is that of transmitting the information to the surface in a form that is fully understood. The distorted "duck-like" talk of a diver on helium–oxygen gas is well known. In addition to transmission of voice communications and data transmission by fiber-optics and so forth, video has become a major means. The ability for an observer on deck to see what the diver sees in real time represents a tremendous advance. Underwater photography can of course be used for recording more clearly specific objects such as a welded joint. Both video and photography require a powerful light source; much recent development has been oriented toward appropriate frequencies to reduce scattering and incident angle refraction distortion.

Many tools and procedures have been developed to enable divers to work effectively underwater. Among these are the following:

1. *Wet welding techniques*, using a high-velocity jet of inert gas to create a water-free zone. Wet welding can be used on low-carbon steels to as deep as 70 m. Although its qualities are strongly influenced by depth, hence pressure, satisfactory welds have recently been completed at 110 m.

2. *Dry welding*, using a habitat and employing gas–metal and gas–tungsten arc techniques. Hyperbaric welding has been carried out to depths of over 1000 m.

3. *Underwater cutting using the electric arc method.* With a skilled diver, steel can be cut almost as rapidly as above water. Arc-flame methods can be used to depths up to 2000 m. At greater depths, potential problems exist as the density of water and gas tend to equalize due to the high pressure. Arcair has developed a sea-jet electrode which will cut both concrete and steel.

4. A wide variety of *hydraulically driven velocity power (explosively-driven) tools* have been devel-

oped in recent years, many of them by the Naval Civil Engineering Laboratory at Port Hueneme, California. These include actuators, impact wrenches, rotary brushes, rock drills, thermic lancers capable of cutting steel and concrete and even rock, explosive (power-actuated) pin-driving tools, grout dispensers (for epoxy injection), and NDT inspection devices. In addition to the use of conventional hydraulic fluids, seawater power supply systems have been developed.

To supplement divers and extend underwater surveying capabilities, a number of underwater vehicles have been developed. These include the work submarine or manned submersible which can survey seafloor sites with remarkable accuracy, using gyrocompasses, acoustic Doppler navigators, inertial guidance, side-scan sonar, and acoustic monitoring by surface vessels.

Work submarines have been used to map boulders on the seafloor of the North Sea. They have been able to locate and identify these far more accurately than has been possible with acoustic means from the surface alone.

Undersea work vehicles are fitted with video, high-intensity light sources, sonar sensors, and manipulators.

Remote-operated vehicles (ROVs) are the newest development. They include "unmanned tethered vehicles" and "unmanned free-swimming vehicles." Combined with fiber-optics and advanced electronic sensors and data collection systems, they can be safely employed at depths up to 2000 m and in congested areas more effectively than human divers. ROVs are now capable of carrying out the following operations: pipeline and bottom surveys, TV imaging, cleaning off marine growth, structure inspection, precision seabed sampling, cable cutting, object retrieval, acoustic imaging, and observation of anchor emplacement and entry of piles into subsea templates.

The ROV *Sprint 101* of Norway has the following equipment and capabilities:

Strobe light
High-resolution TV
Low-level black-and-white photography
35mm still camera
Sonar
Manipulators
Corrosion potential probes
Acoustic navigation
Buoyancy modules
Object retrieval
Stereo photogrammetry
Wire rope–soft line cutter
Pinger dropper

The next step of course is robotics, on which much progress is being made, especially for the performance of tasks where the requirements are of limited variety.

The book, *Undersea Work Systems* by Talkington contains a fascinating description of how undersea work systems might be used to construct a huge array of sensors to detect neutrinos at a depth of 5.5 km (the Dumand project).

Diving and divers are really a transportation system to enable work to be carried out in an otherwise inaccessible environment. Because of the inherent limitations that still exist and of the high costs, experienced constructors make every effort to eliminate or reduce diving requirements. For those still required, extensive planning is devoted to the diver's support, transfer, and work conditions so as to maximize his safety and efficiency. Table 6.5.1 summarizes the manned diving systems employed in offshore construction.

An example of an advanced diving system for use in servicing the subsea well completion template of a floating production system is the Balmoral Field development. This system consists of the following:

1. A diving bell, running on tensioned guide wires
2. Guide rails and traveling cage to support the bell through the sea–air interface
3. Main wire winch for lowering and hoisting
4. Umbilical winch
5. Gas control panels
6. Two decompression chambers
7. Hyperbaric rescue vessel

6.6 Underwater Concreting and Grouting

6.6.1 General

Underwater concrete and grout play an important role in the construction of offshore structures. Un-

TABLE 6.5.1 Manned Diving Systems Employed in Offshore Construction

Type of System	Working Depth (m)	Endurance	Cost Ratio
Scuba (air)	0–40	Very short; interrupted ascent	1
Scuba (air)	40–70	Very short; decompression required	2
Helmet (air)	0–70	Limited by diver's physical endurance	3
Helmet (helium–oxygen gas)	50–100	Limited by diver's physical endurance	4
Bounce (2 divers)	70–100	Few days	10
Bounce (4 divers)	70–100	Over 10 days	13
Saturation (4 divers)	70–300+	Unlimited	16
Manned submersible (without diver lock-out)	600–1500	Moderate	15
Manned submersible (with diver lock-out)	70–200	Unlimited	25
1-Atmosphere gear (e.g., JIM, WASP)	600	Limited by diver's physical endurance	6
1-Atmosphere diving bell	1000	Unlimited	5

Note: Cost ratios do not include support craft.
Source: Based on paper presented at Offshore Goteborg 1983 Conference by Rolf M. Asplund, Sutec A.B., Sweden, as modified by author.

derwater concrete may be placed in forms to serve as the structure itself; more often it becomes the footing block for the structures. It may be used to tie together various elements in composite action—for example, to tie the piling to the footing and thence to the upper portions of the structure. Underwater concrete may also be used to fill pre-excavated holes in the seafloor and to act as a leveling mat. Underwater concrete may also be used as solid ballast to add weight and lower the center of gravity. It may be used to fill under the base of a gravity platform to insure uniform bearing and provide shear transfer. Underwater concrete may be placed in piles or caissons to give added structural strength and to prevent buckling, or it may be placed in belled footings which have been drilled at the tip of the piles in order to increase axial compression and tension capacity.

Underwater grout is also used for many of the above purposes; hence, no rigorous distinctions should be made in classification. Grout is also used to bond piling to jacket legs, to cement well casings, to fill small spaces between elements so as to provide structural continuity, and to fill the voids in pre-placed rock and aggregate.

6.6.2 Underwater Concrete Mixes: Structural Concrete

For many structural purposes, the following mix is suitable:

Coarse Aggregate: Gravel of 20 mm (¾ in.) maximum size. Use 50–55 percent of the total aggregate by weight.
Fine Aggregate: Sand, 45–50 percent of the total aggregate by weight.
Cement: Type II ASTM, 350 kg/m^3 (600 lb/yd^3).
Pozzolan: ASTM 616 Type N or F, 100 lb/yd^3.
Water/Cement Ratio: .42 (.45 maximum).
Water-Reducing Admixture (preferably it is also a plasticizer): *Do not use super plasticizers.*
Air-Entraining Admixture: To give 6 percent total air.
Retarding Admixture: To increase setting time to 4–24 hours, as required.
Slump: 6½ in. ± 1 in.

This mix will develop compressive strengths in the range of 40–50 MPa at 28 days ((5600–7000 psi) (cylinder strength). It will generally flow out on a slope of 6:1 to 8:1 horizontal/vertical and, if properly placed, should give minimal segregation and laitance. It is suitable for placement in voids as small as 30 cm in diameter and can be used for large caissons and bridge piers.

In water depths over 30 m, the entrained air bubbles will be collapsed by the hydrostatic pressure and hence reliance must be placed on the plasticizing admixture to ensure cohesiveness.

For very large pours, larger-size coarse aggregate (40 mm, 1½ in. maximum) has been used but shows little economy because of the need for the large proportion of sand. For smaller pours, the coarse aggregate may be reduced in size, to as small as 10 mm (⅜ in.).

The basic mix recommended above, however, will develop a fairly high heat of hydration (about 35°C above ambient depending on the size of the placement), leading to thermal expansion and possible cracking during the subsequent cooling. Various methods of reducing the temperature rise are available. Their use is justified in special cases. The following is a list of individual means which have been employed (they should not necessarily be combined):

1. Select aggregates with a high thermal coefficient, requiring more heat per degree of temperature rise.
2. Use blast furnace slag/cement in the proportion 70:30 or even 85:15 to reduce heat generation. Slag should be coarse-grind, (<3800 cm^2/gm), with no added gypsum.
3. Increase the percentage of pozzolan. Recent tests indicate the percentage may be increased to as much as 50 percent in fully submerged concrete and unreinforced concrete.
4. Precool the aggregates by water spray–evaporation and use ice as the mixing water.
5. Cool the mix by injection of liquid nitrogen.
6. Subdivide the pour so as to reduce the size of any individual block.

Since various admixtures, pozzolans, and so on behave differently with different cements, trial batches of several cubic meters should be made to ensure that the resultant mix is workable and possesses a high degree of cohesiveness, that is, does not tend to segregate. The addition of 5% of microsilica (condensed silica fumes) by weight of cement will increase the cohesiveness.

6.6.3 Placement of Tremie Concrete

The placement of tremie concrete is carried out through a tube, usually of 10- to 12-in. pipe (250–300 mm). The pipe may be sectional, but joints should be flanged and bolted, with a soft rubber gasket or screwed joints so as to prevent any in-leakage of water. When the mix is poured down the pipe, if there is a gap in the joint, there will be a Venturi effect which will suck in seawater and mix it into the concrete.

The preferred way to start a pour of any depth up to 50 to even 100 m is to install a steel plate on the bottom end with a soft rubber gasket. The plate is tied with twine to the pipe. The tremie pipe must have sufficient wall thickness so that it is negatively buoyant when empty.

The pour is started by placing the sealed pipe on the bottom and then partially filling it with the tremie concrete mix. While there will be some segregation during the fall down the empty pipe, there will normally be adequate remixing at the bottom.

To prevent this segregation on very deep pours, the Norwegians have occasionally used a second, smaller tremie to fill the first pipe. However, a tremie with diameter less than 8 times the nominal maximum size of coarse aggregate will tend to plug.

For the mix suggested in Section 6.6.2, a 10- to 12-in. tremie pipe is recommended.

Another solution that has proven successful is to place 0.5 m^3 of sand–cement grout first and then pour in the regular mix. This first grout mix can be the regular mix, with the coarse aggregate left out. This is the author's preferred method.

When the tremie has been filled to a reasonable distance above the balancing head of fresh concrete versus seawater, the pipe is raised about 150 mm, allowing the concrete to flow out. "Reasonable distance" is that required to overcome the friction head, which may be only a meter or two. The lower end of the pipe is kept embedded in fresh concrete, but no deeper than where the concrete has taken initial set. With a retarder to prevent initial set, the depth of embedment becomes less sensitive. The tip of the tremie pipe should always be immersed about 1 m as a minimum so as to prevent water inflow into the pipe. The flow of concrete should be smooth, consistent with the rate at which concrete can be delivered into the hopper at the top. Similarly, the method of delivery should provide relatively even feed into the hopper rather than large batches being suddenly dumped.

If for any reason the seal is lost, the concrete having flowed completely out of the pipe, then the pipe should be raised and sealed and the pour restarted as in the original program.

When large areas are to be covered, multiple tremie pipes should be used or the tremie pipes reset in new locations within the slope of the fresh pour. The distance tremie concrete can flow without excessive segregation is between 6 and 20 m: the larger distances are obtainable with a very cohesive mix, which prevents excessive segregation and washing of the cement.

The slope of an extensive pour will allow the laitance, silt, and so forth to flow down so that it tends to collect in a far corner, where it might be trapped under good concrete. Therefore, an airlift should be operated in the corner to remove any laitance.

The major problems with tremie concrete have to do with segregation into sand, gravel, and a mixture of cement and water known as *laitance,* from the French word for "milk." Actually laitance may be a very plastic, claylike substance and will eventually harden into a chalklike material. It is very porous and constitutes a weak layer in the structure unless properly removed.

Segregation occurs primarily when concrete is allowed to flow through seawater or where seawater is mixed into the concrete, as by mechanical disturbance, divers walking in it, attempts to vibrate it, or use of a nonplastic mix that tends to build up and then break through in an overflow. Churning of the tremie pipe to promote flow and moving it horizontally through fresh concrete are very bad practices. Leaky joints will act as a Venturi mixer and wash the concrete during its downward flow.

For relatively shallow pours of limited depth, the Dutch have developed the Hydrovalve system, in which the tremie pipe is made of collapsed rubber pipe. A dry concrete mix (80-mm slump) (3 in.) is fed into the hopper in batches; it forces its way down the collapsed pipe in discrete amounts and discharges from the end. Because it is a drier mix, it is less subject to washing from the water, and hence the pipe does not have to be embedded in the fresh concrete, but only rests on it. This permits the pipe to be moved horizontally and thus keeps the surface fairly level. While the Hydrovalve system has been used satisfactorily in shallow water, it does not yet appear to be reliable for deep pours of major extent, due to an inability to maintain "live" concrete on a broad front.

When filling under an upper portion of the structure or where high-quality concrete is needed right at the top, tremie concrete should be overflowed, wasting the laitance until good concrete appears.

A very widespread but generally detrimental practice has been carried forward from earlier years; this is the use of a "go-devil" or traveling plug to start the pour. For example, in the earlier days, a plug of hay or burlap was placed in the pipe, and the concrete poured on top of it, forcing it down and forcing the water out of the pipe ahead of the plug. This practice, while crude, was a moderately effective means for the initial start. It should never be used when restarting a pour after loss of seal or when resetting a pipe into fresh concrete. The water forced down ahead of the plug will be forced to flow through the fresh concrete, washing out the cement. Even in starting a pour, if the bottom is very soft or sandy, this flow of water may cause erosion, mixing, and so on, and so it must be carried out very slowly.

In recent years the plug of hay has been replaced by a ball, usually a volleyball. Such a ball fits nicely in the pipe and is readily forced down by the concrete. A ball is usually inflated to 11 psi (80 kPa). This corresponds to a depth of about 8 m. Beyond that depth, the ball will collapse. If it later comes back to the surface, it may have reinflated, thus disguising the problem. A ball does not provide a suitable answer except in very shallow water.

If a traveling plug or "pig" is used, then it should be of constant dimensions, gasketed so that it loosely wipes the side, and self-buoyant. A wooden cylinder, steel pipe cylinder filled with foam, and polyurethane cylinder (of suitable strength to resist the hydrostatic head) are all acceptable. Rubber wiping gaskets can be installed on the pig. Such a pig may not rise back to the surface—that is, it may become trapped—but in the usual case that is of no concern.

The traveling plug is thus similar to a pipeline pig. It may be necessary to use such a device when very deep pours are undertaken, since over about 100 m depth the use of the plate seal may become less practicable.

Foot valves, mechanically and hydraulically operated, have been tried for many years. They have almost always been a failure due to jamming of the coarse aggregate in the valve or setting of the initial cement grout in the valve mechanism. However, recently-developed hydraulic valves which open to give a smooth bore offer much promise.

6.6.4 Special Admixtures for Concreting Underwater

Recently developed chemical admixtures enable fresh concrete to be dropped a short distance through water without segregation or washing. These proprietary products appear to incorporate condensed silica fumes, a plasticizing admixture, and a thixotropic admixture such as methocel. By its incorporation in the mix, mass concrete can be place on the seafloor, using a bucket which has a closed top. This system appears especially useful for filling voids in riprap, for cyclopean construction, and for underwater repairs. Such a mix can also be placed by direct pumping through a diver-guided hose.

6.6.5 Grout-Intruded Aggregate

In the method using grout-intruded aggregate, coarse aggregate devoid of fines smaller than 15 mm (⅝ in.) is placed within confining walls. Grout pipes are embedded at regular intervals, horizontally and vertically. The exposed surface is covered with a mat or with an extra thickness of rock.

A special grout is then pumped through the pipes so as to fill the interstices between the aggregate. This grout must have excellent flow characteristics and minimal bleed yet retain its general cohesiveness. Various proprietary admixtures have been developed as well as special methods of colloidal mixing.

When the grout from any one level of pipes has reached above the level of the second set, the grout injection points are moved up. Slotted inspection pipes are often employed to verify the level of grout. Electrical resistivity probes may similarly be used.

In selecting the aggregate to be placed, preference should be given to cubical particles as opposed to flat particles, since bleed water tends to be trapped under the flat particles.

The major concern with grout-intruded aggregate concrete is to keep the aggregate clean. Silt, organic growth, and sand particles must be kept out of the placed aggregate. When aggregate is delivered by barge or vessel, fines tend to accumulate on the barge deck at the bottom of the pile due to abrasion and chipping. The lower layer of such rock should not be placed but should be wasted or used elsewhere.

Grout tends to follow the path of least resistance. Fine particles increase the friction head and thus prevent the full flow of grout around all the particles.

Grout-intruded aggregate is an excellent solution for concreting around embedments and instrumentation where these must be kept within very close tolerances.

The fluidity of the grout, an essential quality, also makes it flow out of any gaps in the forms and up through the exposed surface into the seawater. Hence, the pour should be as tightly enclosed as possible.

In very large pours (caissons) the heat of hydration effect may prove serious. Dividing the pour into reasonable-size blocks is perhaps the best answer. In some cases, it may be practicable to embed piping and circulate water to remove the heat. Grout-intruded aggregate has been used successfully on the deep-water piers of the Mackinac Straits Bridge in Michigan and the Honshu–Shikoku bridges in Japan. See Figure 6.6.5.1

6.6.6 Pumped Concrete and Mortar

Concrete made with sand only or with sand and fine aggregate (8 to 10 mm maximum) has been successfully pumped from a mixing barge down the legs of platforms to form bells at depths of 150 m and more for the offshore platforms of the Ekofisk Field in the North Sea and the offshore industrial terminal at Jubail, Saudi Arabia, in the Arabian Gulf. Pipe sizes were selected to maintain a substantial friction head as the concrete was pumped "downhill." A vacuum release valve at the top of

Figure 6.6.5.1 Specialized mortar barge developed for underwater concreting of piers of Honshu–Shikoku bridges.

the pipe prevented a vacuum forming due to too rapid descent of the concrete.

When larger pours are made, using larger coarse aggregate, the placement by direct pumping may lead to serious blockages. It appears more reliable to deliver the concrete by pump to the hoppers at the top of the tremie pipe and then place it by gravity flow through conventional tremie pipes.

It is currently planned to construct concrete bells under the tips of the piles of the North Rankin A platform, at a depth of 250 m, on the Northwest Shelf of Australia. A combination of pump delivery and tremie placement will be used.

6.6.7 Underbase Grout

The development of gravity platforms has led to the need for suitable mixes and methods for filling under a large, usually flat base with grout of special properties. Generally speaking, the desired properties are homogeneity and completeness of filling, low heat development, cohesiveness, low bleed, and long-term stability. The strength and modulus of elasticity required are usually very low; properties equivalent to the natural seafloor soil at a depth equal to the skirt penetration are all that are necessary.

The quantities involved have been up to 10,000 m^3, so the logistics of supply, mixing, and placement in the middle of the North Sea are obviously severe. Therefore, salt-water mixes have been selected, there being no embedded reinforcement for which to be concerned about corrosion. One mix developed used bentonite cement, retarder, and a finely ground filler of inert silica. Other mixes replace 50 percent or more of the cement with bentonite. Of particular interest is a stable foamed mix which uses only cement and seawater plus the foaming agent and a stabilizer.

The materials are usually delivered to the platform by pump and then mixed and fed into a gravity-flow hopper located well down in the utility shaft. From there, the mix flows through pipes to the ejection nozzles. One clever scheme is to suspend a short length of hose under the base slab, held in a horizontal attitude by a chain. The hose then tends to ride on the surface of the soil, ensuring that the grout will be injected at the bottom of the void.

Proper venting must be provided to allow the trapped water to escape. Usually overflow ports are provided at a few meters' height above the base to enable a diver to verify when filling is complete. Electrical resistivity gauges or nuclear methods in the standpipe or overflow pipe can serve a similar function.

Because of the importance of underbase grouting, a model test should be made, using the full thickness and flow length, but of course reducing the width. Measurements should include temperature rise and bleed. Examination should be made for weak layers, trapped inclusions, bleed voids at the top, and laitance.

Excessive pressure can cause piping out under the skirts or even raise one portion of the platform. It could lead to local overpressure damage to the base structure. Measurement of volume is a secondary guide to the completeness of filling and possible losses due to piping under the skirts. Therefore, pressures and volumes should be monitored carefully.

A special case arises when the structure is founded on a preplaced rock fill. The decision has to be made as to whether it is preferred to intrude grout into the rock or to have a grout that will not penetrate into the rock. Too fluid a grout may escape into the sea through passages in the rock. A suitable grout was developed for filling under the bases of the offshore caissons of the Hay Point Terminal, Queensland, Australia, using sand, cement, and a methocel admixture, which gave significant thixotropic properties. Tests indicated little tendency to segregate and little penetration of the rock.

If the underside of the base is essentially flat, it may be desirable to provide inverted channels, which will ensure that the grout will flow under the entire base.

Sandfill under the base is not strictly grout but is used for similar purposes. In the past, with subaqueous vehicular tunnels (tubes), sand has been placed in a semifluid state on one side and allowed to flow down, under, and part way up the other side. The scope of each operation has to be limited in order to prevent the fluid sand from raising the tube.

For wide, flat-bottomed structures, Danish and Dutch engineers have developed sand-flow systems in which fluidized sand is pumped down under low head, to exit at distributed points under the base slab. As the sand spreads out in a circular pattern, the velocity drops and the sand is deposited. Any low spots become channels through which sand continues to flow until they are all filled. Thus the system leads to automatic complete filling.

A somewhat similar flow phenomenon is believed to occur with the thixotropic grout mixes, assuring relatively complete filling under the base.

6.6.8 Grout for Transfer of Forces from Piles to Sleeves and Jacket Legs

Grout is extensively used to "cement" the annulus between pile leg and jacket sleeve. An annular gap of 2–4 in. (50–100 mm) is usually selected. The grout should be pumped from the bottom up. The mix is generally cement plus water. Blast Furnace Slag or pozzolan may be used to replace part of the cement in order to reduce heat of hydration. Silica fume may be added to increase strength and reduce bleed. Admixtures may be used to provide water reduction, retardation, and expansion characteristics. Grout should be overflowed to ensure that the initial mixture of cement and seawater is cleared.

Pressures should be carefully controlled to prevent forcing the grout out from under the sleeve; usually this exit is restricted by a grout retainer, but many times the grout retainer will have been damaged during pile driving. Therefore, a second entry grouting pipe is often provided, to permit the first grout to set and form a plug; then the main grouting is carried out through the upper entry port.

Grouting in connection with piling is addressed in more detail in Chapter 8.

6.6.9 Summary

There are so many variations in the size, shape, conditions, and properties desired of underwater concreting that it is important that recommendations such as those given earlier be treated as a guide only. Test mixes and trial runs are always advised to ensure that the best possible selection of mixes and methods has been made and that the personnel actually carrying out the work are cognizant of the precautions pertinent to that particular project.

6.7 Offshore Surveying and Navigation

Navigation systems used for control of position during tow and emplacement at the site include both satellite fixes and radio navigation positioning systems. When entering a field in which structures already exist and when leaving harbor, theodololite and electronic distance (range) systems may be used to check. Accuracies when underway near shore or structures can usually be kept to ±5 m, with even greater accuracy when stationary.

Many long-range electronic systems suffer from night effects, losing accuracy. They also can be misinterpreted by increments of range steps, giving errors of 50 m.

In the open sea, a system such as Decca Hi-Fix can be utilized for close control; for long-distance tows, Loran C is adequate. Other systems are Sydelis, Artemis, Motorola, Argo, Racal Hyper-fix, and Omega; see Table 6.7.1. Satellite fixes can give accuracies from 5 to 50 m depending on whether one or two satellites are interrogated. The Global Positioning System (GPS), which became commercially available in 1986, now gives positioning accuracy to 5 m worldwide. Short-range positioning systems include Motorola Mini-Ranger, Honeywell Micro-Automatic Station Keeping System, and Simrad.

Underwater acoustic transponders can be preplaced on the seafloor and used to control the final installation of a structure in the open sea. Usually six are preset so as to ensure that at least three will be working when needed.

The use of these for final bathymetric surveys and for borings ensures that when the structure is finally installed, it will be at the correct relative position. They can also be utilized for the underwater assembly of elements and for guidance of an ROV.

Acoustic transponders have proven very satisfactory with steel jackets, but less so with massive concrete structures, which along with the many boats in the area create excessive noise.

A major acoustic transponder system has been developed by Oceano Instruments to enable placing of a jacket near existing pipelines; see Figure 6.7.1. The acoustic positioning system is calibrated with satellite receivers to process Doppler information and thus provide a surface position fix every 1 to 2 hours.

Accuracy in the placement of structures has been steadily improving with the advances in equipment and with experience. In the early and mid-1970s, distances off target averaged 25 m, but by 1980 this tolerance had been halved.

Positioning of important structures should always be backed up by more than one system. Dur-

TABLE 6.7.1 Current Navigation Systems

System	Description	Approximate Accuracy	Availability	Range and Coverage	Comments
Dual-channel transit satellite receiver.	600-MHz high polar orbiting set of satellites; measures doppler shift	30 m	1½ Hours between fixes (average)	Worldwide, 24 hours/day, all weather	Integrate with gyro and log to provide position between fixes; accuracy degrades with velocity, 400 m/knot
Single-channel transit satellite receiver.	Uses only 400-MHz channel	90 m night, 460 day	1½ Hours between fixes (average)	Worldwide, 24 hours/day, all weather	
Omega	Worldwide 10- to 14-KHz phase comparison	2–4 nm	Continuous	Worldwide	0.2-nm differential mode capability; subject to electromagnetic noise.
Loran C	100-kHz long-range precision positioning system; time difference and cycle matching position fix technique	240-m ground wave; −16-m ground wave repeatability; 10- to 17-nm sky wave	Continuous	North Atlantic, North Pacific; 1200-nm ground wave, 2300-nm sky wave	Sky wave susceptible to cycle hopping & electromagnetic noise
Loran A	1.8- to 2.0-MHz time difference technique	0.5- to 1-nm ground wave; 3- to 6-nm sky wave	Continuous	Northern hemisphere; 700-nm ground wave, 1400-nm sky wave	Being phased out
Decca	70- to 130-KHz phase comparison system	90 m at 40 nm; 0.25–1 nm at 150 nm; 2–4 nm at 200 nm	Continuous	500 nm day, 100 nm night; coast of Europe, east coast of Canada	Only local coverage; sky wave contamination at night
Sydelis	450-MHz system		Continuous	100 km	—
Argo	2-MHz system		Continuous	400 km	—
Racal Hyper-Fix	2-MHz system	2–5 m	Continuous	700 km day, 250 km night; Gulf of Mexico, North Sea	Also in Middle East, Malaysia, Australia
Global positioning system	Satellite	5 m	Continuous	Worldwide	Available with greater accuracy in 1989.

6.7 Offshore Surveying and Navigation   145

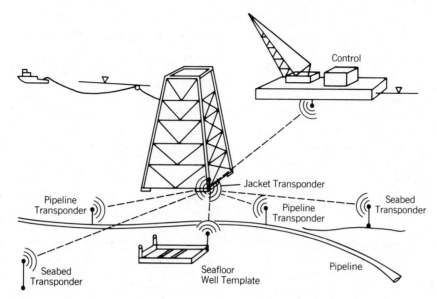

Figure 6.7.1 Acoustic transponder system for installing jacket over template among existing submarine pipelines.

ing the installation of the Tarsiut Island, a survey tower was installed to control the screeding of the seats for the caissons, which was carried out with considerable accuracy. In order to screed the last base, the tower had to be removed and replaced. Due to a combination of human and electronic error, a gross error occurred, resulting in the caissons being installed 20 m off from their intended locations, on unscreeded sand. In hindsight, use of articulated spar buoys, such as the Shelton Buoy or other secondary systems, even taut-moored buoys, would have prevented such gross error.

In many lesser offshore structures, such as offshore terminals and outfall sewers, shore-mounted lasers or even concentrated lights may be set up to provide a range which is directly visible to the barge superintendant and operator. This range, combined with electronic distance systems (EDS) may be used as the primary control, especially if frequent moving is involved.

Bathymetric surveys are carried out by both depth-finder sonic equipment and side-scan sonar. These must be corrected for roll, pitch, and heave and integrated with positioning systems. Such an integrated instrumentation system, called a "profiler," can give a plot of contours within a 200- to 400-m-diameter area.

Depth-finding sonar should be run at two frequencies, high and low, so as to detect the presence of soft, semifluid sediments overlying a firmer bottom. In a deep fjord in Norway, for example, use of the standard low-frequency sonar depth finder gave a depth 25 m greater than actual, since the waves penetrated the very soft soil without reflection.

In areas of strong relief, with steep or near-vertical bluffs and underwater canyon sides, the sonar echoes may come back from the side walls, indicating less depth than the true value. This is because of the conical spread of the beam. Narrow beams can be used to minimize this problem. See Figure 6.7.2.

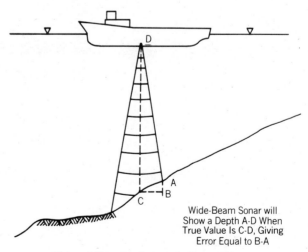

Figure 6.7.2 Conic spread of sonar beam.

The profiler or side-scan sonar should be used to sweep ahead of any proposed route so as to ensure against encountering unplotted anomalies such as pinnacles, coral heads, pingos, and wrecks. Mesotech Systems Ltd. have developed an advanced system of scanning sonar which produces a corrected three-dimensional contour map of an offshore site, embankment, or trench in real time.

Side-scan sonar can produce an excellent two-dimensional portrayal of the seafloor, along with any man-made objects such as pipelines, dropped objects, anchors, even anchor drag marks. Advanced acoustic imaging can now give a map of the seafloor with definition of 1–2 m.

Advanced photogrammetric techniques, using multiple photos, enable a small aarea of seafloor to be mapped to an accuracy of 25 mm in relief. Recent development of extremely sensitive film (ASA 2,000,000), combined with the use of strobe lights, has revolutionized optical seafloor search and survey.

Many new underwater acoustic systems and magnetometers are now available. Many of these can be fitted to an ROV and the data, video, and so on transmitted by telemetry back to the tending vessel. Others are used by divers. They enable the detection of buried cables and pipes, leak detection, high-resolution acoustic imaging of the seafloor, range measurement for short-distance ranging, and guidance for entry of mating cones and piles. See Figure 6.7.3.

Another need for advanced surveying equipment is for detection of sea ice in the Arctic and sub-Arctic. Not only is mapping required but also indications as to size and thickness of floes and the character of the ice, whether first-year or multiyear.

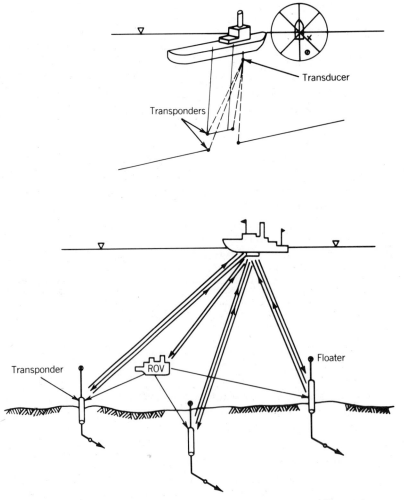

Figure 6.7.3 Use of acoustic transponders and transducers for controlling underwater operations.

These can be given by visual systems or by side-looking airborne radar (SLAR). The most advanced version of this is synthetic aperture radar (SAR), which permits five times better resolution than SLAR.

Present satellite coverage is not able to give all-weather, through-cloud coverage. Planned satellites will be equipped with SAR and by 1989 will enable information to be read out in a timely fashion, identifying ice concentrations, presence of multi-year ice or ice floes, and presence of ice islands. SLAR and SAR flown by aircraft are available today, but coverage is limited in areal extent.

For underwater mating—for example, assembly of underwater structures, positioning of a jacket over a preinstalled template or of an articulated tower over a pre-piled base—a number of systems have been employed. Generally, the control barge, usually a large offshore derrick barge, is preplaced on a taut mooring. Side-scan sonar, electronic positioning, and acoustic transducers are used to verify its position relative to the underwater element. As the structure itself is lowered, short-range, narrow-beam sonars on the structure are used to interrogate acoustic transponders on the submerged element. A video camera with high-intensity lights, mounted on the legs of a jacket, may be used to verify position at close range.

Mechanical devices such as taut wire guide lines and funnels may be employed; the funnels engage a projecting conical dowel or nose and guide it to exact position. Tensioned guide lines (twin wire rope lines connected to a seafloor template and then tensioned to the vessel at the surface) have been used to depths of over 1500 ft to guide subsequent operations. To provide heave compensation, the lines are led over sheaves to hanging clump weights; these provide the constant tension. Alternatively, hydraulic constant tension winches can be used.

Finally, divers may be used to visually verify conformance; their value is in preventing gross error due to some electronic reflections or misunderstanding on topside.

All of these means were employed in mating the three sections of the Cognac jacket in over 300 m of water. (See Chapter 9.)

6.8 Temporary Buoyancy Augmentation

Additional buoyancy may be required at many stages of offshore operations, such as:

1. Reducing draft during floatout from a construction basin and during tow
2. Giving flotation to a pipeline or reducing its net weight in water
3. Reducing weights of structures or elements during installation or salvage
4. Changing the lead of towlines which have been attached below water
5. Providing stability to a structure during deck mating or installation
6. Providing control of draft and attitude during floatout, tow, launching, installation, and/or removal

Temporary buoyancy tanks are usually of steel, although large concrete tanks and hybrid (steel–concrete sandwich) tanks have been proposed for use with gravity structures. They may be pressurized internally for added safety against implosion. Internal subdivision or foam filling may be provided to ensure against in-leakage. See Figure 6.8.1.

The tanks must be designed for the maximum hydrostatic head to which they can be subjected, with an allowance for over-depth submergence in event of credible accident.

The tanks must be connected to the structure in such a way as to ensure against structural overstress globally or locally. The structure being supported must be checked under the upward forces impart-

Figure 6.8.1 Mooring dolphins for offshore terminal utilize temporary buoyancy tanks for control of stability and descent when ballasting down onto seafloor. Hay Point Terminal No. 2, Queensland, Australia. (Courtesy of Rendel and Partners and Utah International.)

ed. Prestressing, welding, and high-strength bolting are acceptable means of attachment, since the tanks will be subjected to cyclic dynamic wave loads, wave slam, and possible vortex shedding. Corrosion-accelerated fatigue may lead to failure at the connections, especially if there are stress concentrations. Buoyancy tanks attached to a jacket may see high-impact and high-acceleration forces during launching. The connections must also be designed so as to permit disconnecting after they have served their purpose.

In removal of a large tank, the procedure must be so planned that it cannot possibly impact the structure, with adequate allowance for wave and current action.

Polyurethane and polystyrene foam blocks have been used to give buoyancy to risers and pipelines. Their attachments must be adequate to withstand abrasion during installation. On one bottom-pull pipeline on the coast of Libya, the temporary floats were broken loose when the steel straps wore through from sand abrasion.

As noted in Section 6.1, inflatable rubber bags were installed under the base of the ANDOC Dunlin A platform to reduce draft during floatout. Such bags might well be used for reducing draft during passage of an Arctic caisson around Point Barrow on its way to the Beaufort Sea.

Smaller inflatable rubber bags have been developed by the Japanese for use in handling moderate weights underwater. Many such balloons were attached to the ill-fated Frigg steel jacket in an attempt to salvage it; unfortunately, their attachment lines fouled and were in some cases torn off by wave action, and so the attempt had to be abandoned.

Collapsible rubber bags are employed in the French "S-curve" system for deep-water pipelaying; they are designed to gradually lose volume and hence buoyancy as they are pulled underwater.

It is important that the same degree of engineering expertise and construction skill be brought to bear on temporary buoyancy tanks as for permanent structures, since failure can have disastrous consequences.

They that go down to the sea in ships, that do business on the great waters, these see the works of the Lord and His wonders of the deep.

Psalms 107:23-24

7

Seafloor Modifications and Improvements

7.1 General

Although in many cases the seafloor has fortuitously been covered and leveled with sediments which have subsequently been consolidated by the action of storm waves, there are many instances, of course, where the site on which the structure is to be placed is not so favorable. There may be soft, unconsolidated, and weak sediments at the seafloor surface. Rock outcroppings may occur, with highly irregular features. Subsurface strata of sands may be capable of liquefaction under prolonged storms or earthquakes. Unstable deposits at or near the site may give rise to slumping, mudslides, or turbidity currents or may be subject to slow and continued creep. Boulders have been deposited by glacial action on many northern seafloors. Weathering of fractured zones during the sea level lowerings of glacial ages may have produced soft layers between hard rock. Calcareous deposits may have formed on windblown sand as it settled through the water. Recent organic silting or volcanic ash deposits may lie almost undetected on top of competent strata.

Any one or several of these or other anomalous situations may exist at a specific site. There are two possible solutions: Either design the structure so as to be stable on the actual seafloor soils as they exist or take various steps to improve or modify the seafloor soil properties.

The first solution has been the one employed to date in most cases of offshore construction. The second solution is frequently employed for major land structures and is increasingly employed for harbor and coastal (shallow water) structures.

This second solution, seafloor preparation, presents some very significant potential advantages for deep water as well. It is being increasingly recognized that there is normally time available in the schedule to do this because of the lead time required for procurement and fabrication of the structure prior to installation.

In some cases, there may also be time after structure installation in which to carry out soil improvement operations; this would be the case, for example, where the structure was installed at the beginning of the good weather season, leaving several months for subsequent soil improvement operations. It is also possible to consider the reduced probability of the design environmental event (e.g., storm) occurring during the first one or two years, thus permitting time, on a stochastic basis, for soil improvement.

It is important always to keep in mind the interactive effects between soil, structure, and the environment. Each acts on and reacts to the others. The environment imposes cyclic loadings on the soil and sometimes physical scour or erosion. The structure imposes forces on the soil, and the soil in turn imposes reactions on the structure. The structure and the waves interact dynamically, as do the soil and the structure, so that as dynamic effects are created in the soils, the soils in turn have a dominant effect on the dynamic behavior of the structure.

The adoption and implementation of seafloor preparation measures have been determined largely on the twin criteria of need and practicality. Large overwater bridge piers, for example, have required a high degree of stability and minimal displacements, tilting, and so forth. At the same time, such bridge piers have until now been almost exclusively located in water depths less than 100 m and in semiprotected waters. Conversely, with the typical large offshore gravity structure, the critical failure modes have been sliding and rocking. Both of these can be significantly improved by seafloor soil strengthening.

With the jacket and pin pile steel platforms, lateral stability (the P/y effect) can be substantially improved, as can the axial capacity of piling. The latter is dealt with under steel pile installation in Section 8.13.

Seafloor foundation modifications are designed to provide a stable base of adequate strength to support the structure and to resist failure and progressive degradation under either a single extreme event or the repetition of cyclic dynamic loads. The foundation must be graded or leveled as necessary to receive the structure and all obstructions removed. In some cases, protective underwater berms will be placed to protect the structure from ice pressure ridges and ice island fragments or from ship collision. Proper controls must be provided to ensure location and grade and to monitor the performance and adequacy of the measures taken.

7.2 Types of Seafloor Modifications

These operations can be arranged in outline form as follows:

1. Seafloor leveling and removal of obstructions (Section 7.4)
2. Dredging and removal of unsuitable soft soils (7.5)
3. Dredging and removal of hard and irregular material (7.6)
4. Removal of boulders from seafloor (7.6)
5. Placement of underwater fills (7.7)
6. Densification, consolidation, and cementation of soils and fills (7.8, 7.9, and 7.10)
7. Scour and erosion protection (7.11)
8. Prevention of liquefaction through control of pore pressures (7.10)

7.3 Controls for Grade and Position: Determination of Existing Conditions

Until recently, one of the more difficult tasks was to properly correlate the relative positions of operations on the seafloor with the structure's final location. Electronic navigation and even satellite positioning have not generally been sufficiently simple and accurate to permit accurate relative positions to be determined and repeated.

In shallow water, relative locations can be adequately marked with buoys. The articulated buoyant staff concept known as the Shelton buoy provides a permanent marker that is little affected by the waves but strongly affected by the currents. In some cases, it may be possible to use inclinometers with appropriate telemetry in conjunction with these articulated buoys.

Acoustic transponders have now been greatly improved in life and reliability. They can be placed on the seafloor surrounding the site; then their true position can be determined by successive iterations of electronic or satellite position-fixing of the surface control ship. For important structures, where operations will continue for a substantial period of time, enough seafloor acoustic transponders are usually placed to assure adequate redundancy in case of malfunction or destruction of one or more transponders.

Bathymetry can be determined by sonar with due consideration to the relief, contour interval, and motion of the vessel. Corrections must be made for change in water level due to tides, barometric pressures, and storm surges. Corrections must also be made for roll, pitch, and heave of the vessel. On the Oosterschelde Storm Surge Barrier, a remote-controlled bottom-crawling tractor tended by a control vessel was able to map seafloor bathymetry at the site of each pier with an accuracy of ±20 mm.

The character of the seafloor can often be determined by video means, using work submarines or a tethered TV camera.

Side-scan sonar is extremely effective in revealing obstructions and sharp breaks in the surface level.

The "profiler," which combines side-scan sonar, automatic compensation, and a directional acoustic beam is very effective in providing a continuous mapping of the seafloor, especially where there are significant changes.

Existing seafloor soil conditions can be determined from grab samples (for surface classification), by cone penetrometer tests, by in-place vane shear tests, by plate-bearing tests, and by bore hole sampling. These bore hole samples can be obtained by core drilling from a vessel or the work platform, by seafloor jacked sample tubes, or by vibratory corers. When deep bore holes are run from drilling vessels, geophysical methods may be used to determine density, resistivity, and permeability.

Experimental work has been done with free-fall or explosively driven penetrometers, which send back their changing rates of penetration (deceleration) by telemetry, enabling a determination of relative density at various depths.

Geophysical seismic and near-surface acoustic surveys are very effective in distinguishing anomalies in subsurface geotechnical properties. When correlated with bore hole sampling, they serve to portray the areal situation much more effectively than linear interpolation between the bore holes by itself.

7.4 Seafloor Leveling and Obstruction Removal

Boulders are scattered over much of the floor of the North Sea as well as many other regions. In general, it has been felt that those less than 0.75 m in diameter were sufficiently small that they would be displaced sidewise or pushed down into the underlying clays by the structure. Boulders larger than this and clusters have been removed.

There are two methods of removal. The most effective one has been to drag the boulders off the site using two tugs and trawl cables and boards, guided by the preset acoustic transponders to the location of the boulder or boulders as previously determined by visual observation from the work submarine. The second method has involved the placement of shaped charges by divers to shatter the boulders; this is obviously limited to those depths and sea conditions in which a diver can effectively work. Thermic lances can be used to cut large boulders into smaller fragments. Ultra-high-pressure water jets with pressures in the range of 15,000 psi can be similarly employed.

Other obstructions can be removed in similar fashion—that is, by dragging or by individual hooking-on by a diver or ROV to an object previously located visually or by side-scan sonar.

Leveling of the seafloor is dependent on having a stable work platform, maintained at a relatively constant grade, from which the drag or screed can be effectively employed. Thus a jack-up rig and a tensioned buoyant platform are especially well suited for such operations.

If the seafloor is generally level but with local ridges and depressions, then dragging of the area with a heavy steel girder can help to smooth out these differences in level. The girder is suspended from a barge by two lines of equal length so that the girder hangs horizontally. As the barge moves across the area, the girder tends to knock down the ridges and fill the valleys between.

The difficulties arise with swell acting on the barge, causing the lines suspending the girder to alternately slacken and then become taut. This can lead to the creation of low and high spots rather than their elimination. If the screed girder is suspended by heave compensators on both lines and if the barge or vessel is always headed normal to the swells, then this method can produce satisfactory results during selected periods of low sea states. Alternatively, spar buoys can be rigged astern of the vessel to act as crude heave compensators.

Another method which has been proposed and engineered is to use a moored drill ship or semisubmersible with heave compensator from which to suspend the drag. The drag is then pulled along the seafloor by two lines which have been run to preset anchors. A major improvement, now practicable with modern technology, will be to equip the drag head with thrusters controlled from the vessel. See Figure 7.4.1.

Screeding frames have been developed for use in preparing a level base on which to seat a caisson or an underwater tunnel segment. Some are bot-

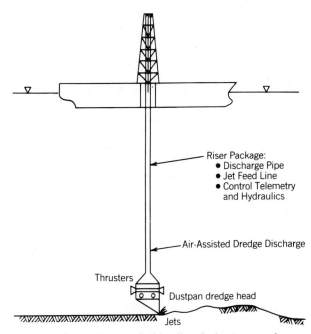

Figure 7.4.1 Deep-seabed leveling device (proposed).

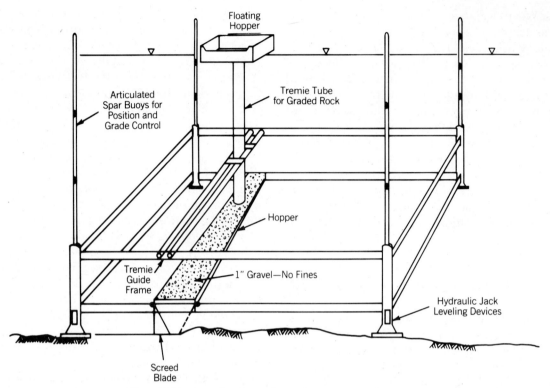

Figure 7.4.2 Screeding frame for underwater fill—schematic.

tom-supported, as shown in Figure 7.4.2. Such a concept was developed by Christiani and Nielsen for use in leveling the base for breakwater caissons in Cape Town, South Africa, and was later modified for use in seating the caissons for an offshore terminal project off Queensland, Australia, located in the open sea at a depth of 25 m.

For underwater tunnel segments ("tubes"), screed barges have been developed, based on a similar screeding arrangement. However, the supporting platform has been a 100-m-long catamaran semisubmersible barge which lowers heavy concrete block weights down to the seafloor and then pulls down against these to stabilize the barge against waves and the effect of tides.

In the Beaufort Sea, a hydraulic dredge has been used to roughly level preplaced sand embankments at a depth of 10–20 m, using a heave compensator to offset the effect of waves. The Dutch similarly used a hydraulic dustpan dredge to level the seafloor for the mattress placement on the Oosterschelde Storm Surge Barrier.

In Japan, for screeding the surface of a trench in Tokyo Bay, a horizontal screw (much like a snow plow) was suspended so that it leveled off high spots by moving the ridge material sideways off the site.

Attempts have been made to screed the seafloor level by diver-manipulated screeds. These have generally been excessively time-consuming and unsatisfactory. A notorious example was the Royal Sovereign Lighthouse in the southern English Channel, where strong tides and storm waves disrupted the work almost as fast as it was done.

For some bridge piers, notably those of the Honshu–Shikoku bridges in Japan, the design has required accurately leveling of bedrock in order to seat a caisson. In such cases, the rock is dredged by the means described in Section 7.6 and then thoroughly cleaned off by jets and airlift.

Grinding has then been employed, using large-diameter grinding bits similar to those used on tunnel-boring machines. Horizontally rotating bits tend to "crawl" over the bottom and hence must be rigidly held by a structural frame just above the seafloor. Counter-rotating bits can be used to offset the net lateral forces.

Grinding with wheels rotating in the vertical plane about a horizontal axis would appear to be perhaps more efficient. These could be extensions of the rock-trenching wheels developed for the English Channel cable crossing and the proposed Straits of Belle Isle crossing between Labrador and

7.5 Dredging and Removal of Seafloor Sediments

Newfoundland. Even as the above was written, an adaptation of such a mining tool was developed and used for grinding shallow underwater concrete on a lock reconstruction project.

7.5 Dredging and Removal of Seafloor Sediments

When weak and unsuitable sediments overlie the seafloor foundation soils, they must be removed or displaced or strengthened. In this section, removal and displacement will be discussed.

For large-scale operations in the ocean, one of the most effective tools is the trailer suction dredge. This vessel, usually self-propelled, uses its speed and momentum, operating through a suspended drag, to excavate the material, which is then sucked up the ladder to the pump and discharged, usually into a hopper for later dumping off-site. The trailer suction dredge can take long runs at a site, lower its ladder as it reaches the near edge, and cut a swath across the site in one run.

The drag may be just a steel plate or it may have ripper teeth or jets or even mechanical screw cutters to help cut the soil. This is an extremely economical means of removing seafloor material. It is only limited in depth to the practicable length of a ladder, in the range 50–60 m. It can remove both soft material and partially cemented materials. Its limitation is that it is difficult to control the depth of excavation. Heave compensators have been installed in some cases to keep the drag at a reasonably constant elevation.

The hydraulic cutter head suction dredge is another tool with a long history of successful large-scale application in inland marine construction. In the open sea, this type of dredge is very sensitive to the swell, which aggravates the movement of the extended ladder and cutter head.

Use of a heave compensator to suspend the ladder is one positive step. Another is to run the ladder from the center of rotation of the hull rather than from its stern.

IHC of the Netherlands has developed an interesting adaptation by mounting a hydraulic cutter head suction dredge arm (or ladder) on a jack-up rig, enabling positive elevation and position control of the cutter head.

With this type of dredge, regardless of whether it is supported from a fixed or floating platform, the lateral thrust must be resisted by either the mooring lines or the legs in order to provide the necessary translation and advance of the cutter head. A monstrous walking jack-up was built in the Netherlands to permit progressive advance of the dredge.

Both of the above schemes are limited in depth to perhaps 50–60 m. By use of jet eductor or airlift devices incorporated in the ladder, the dredge may work to an even deeper level, but it must still be held in position by moorings. See Figure 7.5.1.

Disposal of hydraulically dredged material in the sea may create a turbidity plume which is environmentally objectionable. A cyclone may be used to separate out the coarser sediments for disposal by direct dumping or barge. Other systems have employed coagulants (thickeners) so as to precipitate suspended and colloid materials.

Another approach is to discharge down through a suspended pipe so that the discharge is at the seafloor. In the open sea, this discharge should

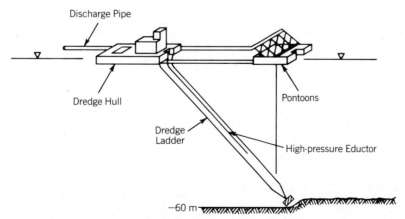

Figure 7.5.1 Deep dredging cutterhead dredge Skookum.

154 *Seafloor Modifications and Improvements*

preferably be located below the thermocline so as to prevent mixing with the surface waters.

Deep excavations may also be performed at sea by the use of large clamshell dredges. On the Honshu–Shikoku bridges in Japan, for example, clamshell (grab) dredges have been used to excavate to depths of 50 m or more. These have very large buckets, up to 99 tons. In deep water, the cycle time for such large and heavy buckets is very long. The hoisting time may be reduced by using especially large winches for maximum line pull so as to reduce the number of parts in the hoisting line and increase the line speed.

The swinging time is again long, due to the inertia of boom and bucket. In some cases, the bucket has been so arranged that it is discharged to a hopper barge moored at the stern of the dredge so that there is no swing of the boom, only a short translation of the bucket along the centerline of the dredge.

Continuous-bucket-ladder dredges have been used to depths of almost 60 m in the calm seas off Thailand digging placer sediments for tin.

The airlift, suspended from a barge or vessel, becomes increasingly efficient with increasing depth. It need not discharge at the surface; discharge above the seafloor may be sufficient. The airlift head may be augmented with jets, so arranged as to feed material to the airlift suction. See Figure 7.5.2. However, this system is effective only in relatively limited areas. The airlift is especially effective when removing material from within enclosures, such as cylinder piles, provided water is continually fed in at the top so as to maintain the external pressure head.

Other systems for dredging of sediments are the jet eductor (see Fig. 7.5.3), the Marconaflo "Dynajet" (see Fig. 7.5.4) and the Pneuma Pump (see Fig. 7.5.5). New submersible pumps equipped with agitators are able to move large quantities of sand

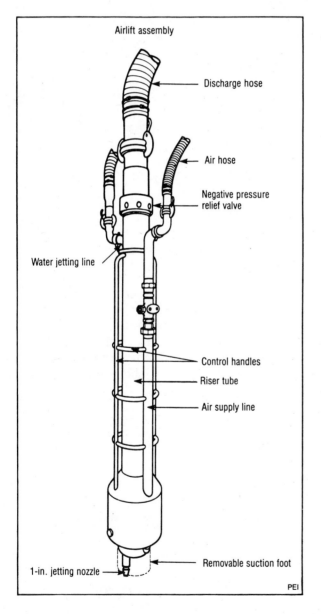

TABLE 1. Typical airlift specifications				
Diam, in.	Water depth, ft	Output solids, tons/hr	Air consumption, cu ft/min	Applications
6	200	33	100	Diver-operated unit for pipeline maintenance, wellhead recovery, salvage, mineral prospecting, etc.
6	500	22	250	Wellhead and pipeline operations
8	625	11	700	Circulation of drilling mud and cuttings on Brent Bravo platform
10	200	55	600	Mineral prospecting, geotechnical investigation, pilot plant mineral production, trenching, and bedrock cleanup.
18	500	110-1,100*	1,975	Platform maintenance, trenching, antiscour, mineral production, mine sludge handling, etc.

*Depending on application

Figure 7.5.2 Airlift excavators and dredges. (Courtesy of Petroleum Engineer International.)

7.5 Dredging and Removal of Seafloor Sediments

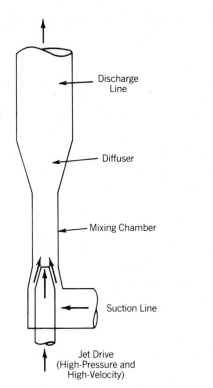

Figure 7.5.3 Schematic of jet drive (eductor) pump.

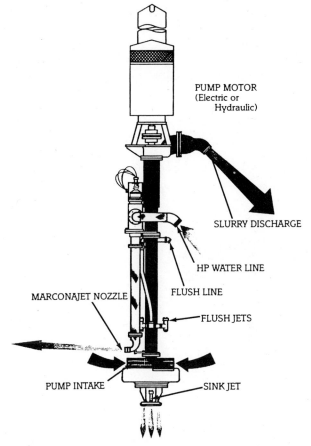

Figure 7.5.4 Dynajet long-shaft eductor–excavator, capacity up to 700 m³ phr. (Courtesy of Marconaflo.)

at shallow and moderate depths and have proven especially effective in excavation within cofferdams and for cleaning underwater trenches. See Figure 7.5.6. All are very efficient in removal of material if it is loose and free-flowing. Hence most such systems include jet systems to break down the soil structure and place the sediments in suspension. In general, these systems are able to do increasingly well the greater the water depth.

The jet burial sled, used in pipeline burial operations, is a very efficient method of soil removal, especially in a trench or limited area. It employs the principles of high-pressure jet cutters combined with either airlift or eductor suction and discharge.

For the deep-sea manganese nodule mining operations, several types of dredge have been developed. One of these employs the suspended drag principle, where a large and heavy base is dragged across the seafloor, cutting by means of jet cutters and sucking up the nodules by airlift or hydraulic transport.

Another system employs the continuous-belt dragline or ladder dredge principle (slack line dredge) whereby a slack line returns the buckets to the seafloor. This system is effective but difficult to control as to depth of cutting. It will, of course, tend to cut trenches.

The Amrod subsea dredge is a remotely-operated seafloor dredge, capable of operating at depths up to 300 m. The operations are controlled by a surface operator using remote electro-hydraulic controls and monitoring operations by video and acoustics. One version, having a 10-in. suction, has a capacity of 500 ton/hr.

With any type of dredge which cuts a swath or trench, there is a tendency for the next cut or run to follow into the previously cut trench, thus producing deep trenches at the site rather than a uniform removal of soils. One means to overcome this at a specific site is to make a number of runs in one direction so as to cover the site and then make a second series at right angles.

156 *Seafloor Modifications and Improvements*

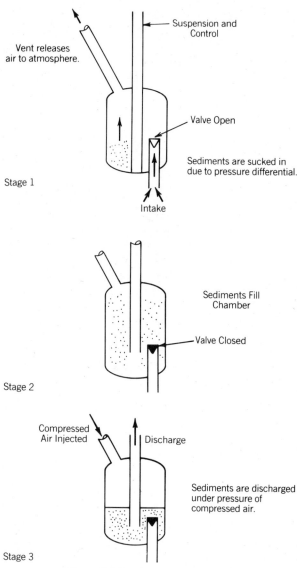

Figure 7.5.5 Pneuma pump cycle.

Figure 7.5.6 Specially designed submersible pump with agitator and counterrotating impellers can excavate up to 500–600 cu. yds./hr against a head of 100 ft. (Toyo Pumps Corporation.)

7.6 Dredging and Removal of Hard Material

The removal of hard material inherently requires the consideration of site-specific data such as stratification, fracture, and bedding planes. Near-surface geological information is necessary to properly plan such work.

Stratified rock, having near-horizontal bedding planes, requires particularly careful evaluation. Any dredging system that works from the top surface may involve an excessive amount of effort, whereas a method which breaks upward from underneath can be very effective. In past years, in channel and harbor dredging work, the dipper dredge was particularly efficient. Recently, large hydraulic backhoes have been adapted for barge mounting and are able to work to depths of 15–20 m. At greater depths, the slack line dragline bucket should be able to break slabs upward.

Drilled-in explosives are generally inefficient for horizontally stratified and layered rock, as most of the explosive energy dissipates through the cracks. In this case, all that may be accomplished is to fracture the rock into very large slabs, which may make them even more difficult to remove.

In some regions, coraline cap rock or limestone layers often overlie extremely soft silt. This condition occurs, for example, in the Arabian–Persian

7.6 Dredging and Removal of Hard Material

Gulf and offshore Hawaii. Explosives placed on top, such as shaped charges, can shatter the hard layer downward.

Shaped charges may be used effectively to fracture boulders and break down ledges. They have also been used to trench through rock. They must be accurately placed and held in the proper attitude so that the blast creates a cavity in the rock. Divers can place a sandbag over the shaped charge to hold it in proper position. This also improves the effectiveness of the blast. Shaped charges can also be assembled on an expendable steel frame, which is then lowered to the seafloor. Shaped charges have proven especially effective in breaking up surface cap rock in order to bury pipelines. Over 300,000 m^3 of calcarenite and limestone were broken up with shaped charges in order to bury the submarine pipeline from North Rankin A on the Northwest Shelf of Australia.

Another type of hard material that must sometimes be removed in preparing a foundation is heavy boulder clay. Very large clamshell buckets with teeth can penetrate the clay and engage the boulders with the bucket.

One of the disadvantages of the conventional clamshell bucket is that the action of the bucket closing line tends to reduce the effective weight on the teeth. Therefore, hydraulic bucket-closing cylinders have been developed capable of operating at significant depths. These enable the full weight of the bucket to work downward and the full force of the hydraulic cylinders to work sideways, that is, to close the bucket.

Another development of use with clamshell buckets in sandstones and conglomerates is the mounting of heavy vibrators on the bowls of the bucket. When the bucket is on the bottom, with teeth pointed downward, the vibrator is turned on, causing the teeth to penetrate. The proper selection of bucket size and weight, line pull hoisting capacity, and vibrator energy and duration is extremely important. There have been cases, for example, where the vibrator was so effective in penetrating that the bucket could not be raised; it had anchored itself into the rock.

Cutter head suction dredges have been developed with tremendous power on the cutters so that they can cut soft rock. However, the action of the typical cutter, rotating around the axis of the ladder, is a grinding action that may involve excessive expenditure of energy. Another new development is the cutter which rotates around a horizontal axis, usually designed to break the material upward, where it can be picked up by the suction.

In many cases the material may be pre-broken so as to facilitate removal. For example, in boulder clay, high-pressure water jets may be used to erode the clay binder, enabling dredges to work more effectively. Surface explosives may break the cementitious bonding of conglomerate formations.

A recent development for quarrying operations on land, which may someday have application at great depth, is the use of hydraulic fracturing techniques. Short bursts of extremely-high-pressure water (up to 15,000 psi) are used to propagate fractures in the rock.

The use of underwater chisels is a method of rock breaking that avoids the use of explosives. In relatively shallow water (15–20 m) the chisel may be a heavy piece of shafting, extending up above water. It may be repeatedly raised and lowered so as to fracture a hard layer; this rather crude but effective process has recently been employed on a large scale in the Arabian Gulf port projects. In some cases, a high-pressure water jet has been incorporated into the chisel so as to wash away loose and broken material.

A more controlled operation is to use an impact or vibratory hammer on top of the chisel, thus driving it into the rock. After penetrating a meter or two, the long chisel is pulled sidewise at the top, breaking off a piece of rock, just like a gigantic clay spade. On some large projects in the Arabian Gulf and the Mediterranean, a bank of such chisels is assembled along the side of the rock-breaker barge so as to methodically break up a hard rock layer for subsequent removal by a hydraulic dredge.

Rock-breaking chisels, driven by hydraulic or vibratory hammers, can also be operated underwater. Their location must be carefully controlled. They use their weight plus the impact of vibration to penetrate. Incorporation of a high-pressure jet may help to dislodge the broken rock and prevent "self-anchoring."

Drilled-in explosive fracturing has a long history in underwater rock dredging. The hole must be cased from above water down into firm material, usually to top of rock. The casing is either driven in through the overburden or drilled in. This latter is known as the Atlas–Copco "OD Method." After the hole is drilled and cleaned, the waterproofed charges of powder are lowered down with either waterproof leads or primacord attached. Sand packing is placed on top of the powder, and the leads or primacord are led all the way out of the casing at the top, with a float attached. The casing is now raised, and the leads are picked up and connected on the barge. After a series of holes have been so charged and connected, the barge pulls away 60–100 m and the round is fired.

The effectiveness in shattering rock is greatly increased by the amount of overburden. Also, the presence of overburden makes it easy to seat and seal the casing, whereas the absence of overburden makes it extremely difficult. Hence, all or some of the overburden should be left in place during the drilling and shooting operation.

Typically, holes are drilled on a 2- to 3-m spacing. The spacing of holes and overdepth (below design final grade) of drilling must be carefully determined to obtain optimum results. From the dredging point of view, aimed at facilitating rock removal, it is generally best to drill below grade a depth somewhat more than half the spacing of the holes so as to ensure against high points (pinnacles) being left above grade. For example, a crude and conservative rule of thumb is "half the spacing plus half a meter." The reason for this is that it is almost prohibitively expensive and difficult to attempt secondary shooting on high pinnacle remnants. Staggered rows of holes appear to give better fragmentation than a rectangular grid.

For the same reason it is considered good practice to use a conservatively high powder factor (e.g., 1.2–1.8 kg/m^3) and a relatively fast powder (e.g., 60 percent). However, such a procedure may cause excessive fracturing of the rock below grade, which may or may not be acceptable from the foundation point of view. If not acceptable, the rock may require subsequent mechanical grinding or pressure grouting, this latter being usually done after the structure is installed.

If it is necessary to minimize the subsurface fracturing, then holes should be drilled on a closer spacing and to a correspondingly lesser depth below grade.

Blasting of rock is always most effective if there is a face toward which the rock can break. Therefore, in some cases it may be desirable to drill and partially excavate a trench, then progressively drill, blast, and excavate, so that there is always an open face.

If there is no open face, then there is a tendency for the rock to fracture, raise, and settle back in the same compact mass. This can be particularly adverse if the powder factor is low and a slow charge is used; the effect will be to fracture large pieces without displacement. Subsequent drilling and shooting may then be very difficult, since the blast will dissipate along the fracture zones without breaking new rock.

When adjoining structures must be protected, line-drilling and cushion-blasting techniques are employed. Pre-splitting along a boundary row prevents extension of the fractures. Use of delays can ensure that the blasted material moves in the desired direction. Air bubbling reduces the water shock effects.

The blasting of solid rock produces fractured material having typically 40–50 percent greater volume, thus raising the level of the seafloor in the blasted area. If the material is to be removed by bucket dredging, the bulk quantity will be increased accordingly. This also applies to the rock which has been blasted below grade. Thus extensive areas of shallow rock removal may involve a 100 percent or more increase of the dredged quantity as compared to the neat solid rock volume.

For deep offshore work, either rotary or percussion drills are employed. Rotary drills are best for deep drilling in competent rock, whereas pneumatic drills are most useful in irregular, erratic material and shallow drilling depths.

Two other methods have been developed, both based on the thermal spalling of rock. These are very rapid when drilling in competent rock but expensive and difficult to control. They are known as "jet piercing" and "thermic lancing." At present, they must be considered in the experimental stage. They appear especially effective in quartz.

Drilling and blasting can also be carried out by divers and/or submersible work vehicles. This has been effective only for relatively small, isolated features and shallow depths of work, such as isolated boulders on the seafloor. However, divers are generally not able to carry out major operations over an extended period of time. Their effectiveness is also limited by their inherent buoyancy. For any major project, working from the surface is currently the most efficient and economical.

Considerable work has been carried out in Japan with acoustic underwater blasting devices so as to eliminate the need of bringing leads up through the water for collection on the barge. These devices appear to be reasonably reliable, provided they do not become silted over.

Subsequent to blasting and excavating rock, it is generally necessary to clean up the foundation by removal of silt, sand, and small rock fragments. This is best done by a straight suction dredge or airlift, aided by high-pressure jets.

7.7 Placement of Underwater Fills

Underwater fills of granular material can be used to provide a reasonably level and uniform support for structures at a practicable and economical level. For example, they can be placed over irregularities and outcrops or used to fill back depressions from which unsuitable materials such as mud have been removed. In deep water they can be used to raise the base of the structure to a more favorable elevation from a standpoint of economy while still staying well below the elevation at which the design wave will have destructive effect.

Such fills can also provide a foundation of known static and dynamic properties from the points of view of stiffness, pore pressure buildup, resistance to liquefaction, and so forth.

Underwater fills may be used to blanket an area so as to contain unstable sands and allow pore pressure relief without sand dispersion. They may also be used to laterally confine unstable materials such as clays, acting as counterbalancing surcharges external to the structure, thus preventing local shear failures.

Underwater rock dikes, placed prior to or during dredging, may be used to stabilize side slopes against shear and erosive failures. In such cases the underwater rock dike migrates downslope and into the sand as the dredging takes place, serving to give steeper and more stable slopes. Clay dikes, using stiff glacial clays, have also been used to retain underwater sand fills.

The use of underwater fills to prevent scour and erosion around structures is dealt with in Section 7.11.

The materials for underwater fills have to be selected with regard to their suitability to the needed objectives, their density and size gradation, and their ability to be placed at the depths and locations desired. Obviously, availability and cost are also factors.

Underwater fills are often placed by discharge from a hydraulic dredge. See Figure 7.7.1. When low-relative-density materials are placed through water, they tend to disperse laterally and to fall through the water at differential speeds. The result is to segregate in layers of different size. In addition, the in-place density of such material is heavily dependent on the permeability and relative gradation of the particles. Relative densities of cohesionless materials (sands and gravels) placed through substantial depth of water may vary from 40 to 70 percent, with 50–60 percent being most common. Lateral spreading is dependent on specific gravity, gradation, particle shape, depth, and currents, but, in general, slopes are very flat. As fine sands impact, they temporarily liquefy, allowing them to flow locally as a dense fluid. Silt content is very critical: Lenses of silt formed during deposition can lead to slope failures at a later stage.

Air content at the time of placement has a very significant effect on segregation, spreading, and density of in-place underwater fills. The air bubbles

Figure 7.7.1 Discharge of sand from hydraulic dredge to form underwater embankment (berm) for offshore island in Canadian Beaufort Sea.

attach to fine particles and give them added buoyancy. The tendency for such segregation can be reduced by thorough saturation of such materials prior to placement.

Sand may be discharged down a tremie pipe whose end is fitted with a special device to force the sand and water to separate and hence enable a steeper slope to be attained. Among the special devices employed are screens of "fabric" mesh, whose openings allow the water to flow out freely while tending to restrict sand passage, and a wide bell-mouth fitting which reduces the exit velocity. At the Tarsiut Island in the Canadian Beaufort Sea, for example, use of such a device resulted in a side slope of 5.5:1 as compared to slopes of 10:1 and even 15:1 when the sand was discharged at the surface.

If the underwater fill is to contain sands and prevent sand dispersion through the fill, then the material should be graded as a filter. In practice, this is extremely difficult to accomplish. One approach is to select a well-graded material that will act over the complete range both as a filter and as a stable rock dike. Another solution is first to cover the area with filter fabric mats with articulated concrete blocks attached. Then rock can be dumped over the mat. See Figures 7.7.2 and 7.11.1. Mats can also be held in place by sandbags or steel pins set by divers. In very deep water, filter fabric mats may be preattached to the structure prior to seating it or may be attached to steel or concrete frames or panels.

One of the best methods of placing underwater rock fills in the ocean is by bottom-dump or side-dump barge. By presaturating the material prior to dumping, segregation is minimized. The mass of the rock hangs together as it falls through the water, thus attaining the terminal speed of the mass, which is usually several times that of the individual particles. The impact of the mass on previously placed material helps to consolidate it.

Tests in the Netherlands by ACZ Marine Contractors have shown that a mixture of stones with a maximum diameter of 0.2 m (200 mm) will reach a maximum terminal velocity of 2.0–2.5 m/s. However, when dumped as a mass from a split hopper or similar barge, the entire mass hangs together so as to develop a fall velocity about twice that of the individual stones, that is, 4–5 m/s. These terminal velocities are reached within a relatively short fall distance (e.g., 2–5 m) regardless of the initial velocity of the rock falling through air or through a pipe.

Other methods which have been employed in-

Figure 7.7.2 Placing a mattress over a sand blanket and then covering with dumped rock. (Courtesy of ACZ Marine Contractors.)

volve placement by clamshell bucket lowered to the bottom before opening (a very tedious process) and placement through tremie pipes. These latter are usually suspended from a barge or from a pontoon-supported hopper laterally restrained by lines to the barge. The pipe must have a large enough diameter to avoid any possibility of plugging: a value of three to five times the diameter of the largest rock pieces is often used. Gravel is less likely to plug than crushed rock. Elongated particles are unsatisfactory.

Another method of placing rock at depths is through an inclined chute such as the modified ladder of a trailer suction hopper dredge. The discharge end is suspended from the vessel by a heave compensator and can be directed to the proper location.

A number of rock-placing vessels have been developed by Dutch dredging contractors which are equipped to place rock either by direct dumping or through a controlled tremie pipe. See Figure 7.7.3.

The selection of proper particle size for long-term stability requires consideration of bottom current velocities due to combined tidal and general currents and storm-wave-induced currents. The effects of the structure itself in generating vertical and eddy currents must be considered. During a storm, the pore pressures in the soil will fluctuate and may make it easier for fill particles to be temporarily placed in suspension.

Another consideration, of course, is the packing factor, which is determined by particle shape, gradation, and degree of consolidation of the fill.

Relative density is extremely important, since we are dealing with the submerged density. Use of a rock consisting of iron ore mineral compounds and having a specific gravity of 4 or higher is much more effective and stable than the typical silica rock with a specific gravity of 2.6, since stability varies approximately as the cube of the underwater net density.

Larger sizes are also more stable and can serve to lock together a fill of varying gradation.

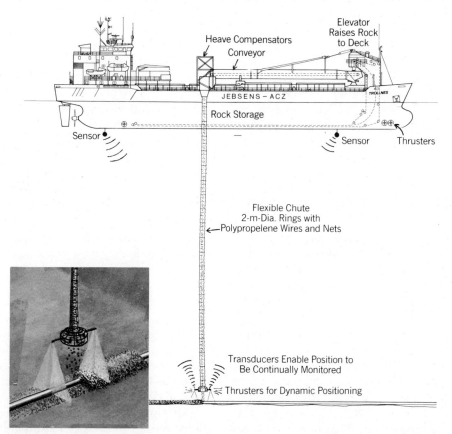

Figure 7.7.3 Flexible chute for placing rock at depths greater than 100 m. See also Figure 19.5.3. (ACZ Marine Contractors.)

162 Seafloor Modifications and Improvements

When fills are placed around a structure—for example, when sand is discharged down a tube as backfill for a pipeline—the fill is temporarily a heavy fluid and has the flow and displacement properties of a fluid. Thus it can run under the pipeline and lift it up or displace it sideways, just as if it were a fluid having a specific gravity of 2. A number of major pipeline installations have been seriously dislocated or even ruptured during backfill in this manner. Of course, the material quickly returns to a solid once the excess pore pressure has dissipated, which usually occurs in a few seconds, but the damage may already have been done.

The proper use of underwater fills would appear to be a major opportunity in the extension of structures and consolidation to deep water, to less competent soils, and to more exposed locations. Properly employed, the fills can enable construction to be carried out at a more favorable elevation or with materials of known and controllable properties.

Underwater rock dikes placed around a structure can be used to prevent collision from ships, to cause icebergs or deep pressure ridges to ground, and to divert mudslides and turbidity currents. They can also, as mentioned earlier, confine soft soils such as clays, in order to lengthen the shear path and to resist local shear failure due to high bearing pressures.

Particularly attractive is the fact that in many cases the underwater rock fills can generally be placed during the period in which the structure itself is being fabricated; thus, the work is not on the critical path.

7.8 Densification and Consolidation of Fills

In-place sands may be so loosely consolidated that they will be subject to liquefaction under prolonged storm or earthquake. Fills such as those described in the preceding section may also require consolidation in order to reduce settlements and to ensure stability.

There are a number of methods and techniques by which soils and fills may be consolidated.

One of these, which applies to granular, cohesionless materials is vibratory consolidation. A large mandrel—up to 1 m in diameter, for example—is inserted into the material by either jetting or vibratory driving. The mandrel has horizontally oriented vibrators mounted in its tip, which are now activated, imparting high-frequency energy into the adjacent soil. This causes the sand particles to reorient themselves. Pore pressures are increased and then relieved by drainage through the adjacent fill.

Several brands of internal vibratory compactors are currently manufactured, some of which are able to work entirely underwater. These are able to penetrate and consolidate material ranging from 75 mm diameter down to fine sands.

It is essential that the pore water be able to drain. Layers of silt or clay will prevent drainage and seriously limit the efficacy of the vibration. If these blanketing layers exist, vertical drain paths such as gravel drains must be provided. These can be installed as drilled wells or even jetted into place. There must also be a horizontal escape path for the water that is expelled; this is usually a sand or sand and gravel layer preplaced on the seafloor, under the base of the structure. Alternatively, drainage may be provided into the structure, from where it is pumped out.

Internal vibration does not compact the near surface layer. For this layer, a vibratory plate compactor must be placed on the surface. A large vibratory plate compactor was employed on the Oosterschelde Storm Surge Barrier where it successfully compacted rock of 350 mm maximum size in layers up to 4 m thick. See Figure 7.8.1a.

Other methods of surface compaction utilize adaptations from above-water landfill practice; these include a roller compactor and a remote-controlled underwater bulldozer. These, however, have so far been limited to relatively shallow water.

Another crude but effective tool consists of a long shaft or pipe with a large plate on the bottom. This is repeatedly raised and dropped to compact the surface. A pile hammer or vibration hammer may be attached to the top, enabling better control than can be achieved by raising and dropping. See Figure 7.8.1b.

Another method, adaptable to a wide range of material sizes and gradations, is dynamic compaction (the "Menard system") which involves the repeated raising and dropping of a heavy weight. Depending on its mass, density, and distance of fall, this dynamic compaction can effectively consolidate up to as much as 10 m of underwater soils or fill. This has been used successfully on both sand and rock underwater embankments.

Air guns, as used in geophysical seismic surveys, may be used to densify loose sands; some experimental work has been carried out. It is important to recognize that this shock causes high pore pres-

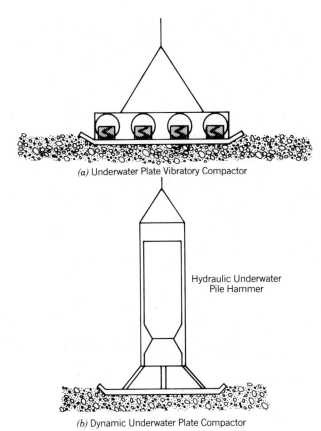

Figure 7.8.1 Compaction equipment for underwater rock.

sures to build up instantly. These are then relieved by drainage through permeable materials or through shear fractures in relatively impermeable materials such as silts. This latter is of course normally undesirable and may result in slope failures.

Similar "shock" or dynamic compaction can be achieved by use of explosives. These are installed by jetting a casing into the fill to about two-thirds of the depth of the stratum to be densified. The size of charge is limited so that craters will not form. Typical spacings of holes are 3–8 m, and delays should be used to separate the times of successive impact.

When using dynamic compaction, strict monitoring and controls must be employed to ensure against the creation of a major slope failure due to buildup of pore pressure and remolding effects. On at least two occasions, dynamic compaction has apparently initiated a major turbidity current flow.

The action of dumping rock fills in large masses so as to dynamically compact previously placed rock has already been mentioned.

7.9 Consolidation and Strengthening of Weak Soils

Sand or gravel "piles" may be driven into clays and silts in order to strengthen them by increasing both bearing and shear resistance. Such piles are usually installed by a driven mandrel into which the granular material is fed. This material is then forced out with air pressure as the mandrel is withdrawn. Since only limited depth of penetration is required—that is, only the weak soils are involved, the process is very rapid. Typically, 1-m-diameter "piles" are placed on 3- or 4-m centers. These are often referred to as "sand columns" and "stone columns."

Vibratory compaction, as described in the previous section, is extremely effective in consolidating loose sand deposits. Perhaps the most extensive use to date was on the Oosterschelde Storm Surge Barrier, where four such compactors which were mounted on a barge compacted the loose sands to a depth of 50 m below sea level.

An extremely effective means of consolidating weak soils is by surcharging, with subsequent removal or redistribution. An underwater fill of sand or rock can be placed by bottom-dump barge and allowed to exert its excess pressure on the soil for a period of a year or more. The consolidation will be even more effective if the pore water has an opportunity to drain through natural or artificial permeable drains.

Silts and clays may be consolidated by means of drainage using vertical sand drains or wicks. This method is especially applicable if the soils are anisotropic with good natural horizontal permeable stratification, as exists in many locations in the Arctic.

Sand drains can be drilled in or jetted in. Wicks are installed by jetting or by attachment to a dropped shaft which penetrates dynamically. Of course, after the structure is in place, drainage wells can be drilled in and even pumped so as to remove excess water during consolidation. The water which is driven out of the soil moves upward through the drains and must then have a means of escaping laterally. Thus a blanket of coarse sand and gravel should first be installed over the seafloor area.

With or without artificial drainage the surcharge effect of a fill over a period of a year or so may serve to stabilize silts and possibly even clays to an acceptable strength to receive and support the structure. Then just prior to the installation of the structure, the surcharge fill can be spread laterally to a peripheral location where it will lengthen the shear path and act as a counterbalancing force against bearing failure.

Cementation of underwater granular soils can be carried out by injection of cementitious material following land-based grouting procedures. The cementing pressures must be regulated so as to displace the pore water yet not cause "fracturing of the formation" through channelization. Cementitious particles must be small enough (fine enough grind) to penetrate the interstices. Addition of a wetting agent to the grout which reduces the viscosity and colloidal mixing both facilitate permeation.

If the cement can be mixed with the soils, even cohesive soils can be stabilized. Various techniques have been developed to accomplish this, usually based on use of an auger-type drill that is jetted and drilled into the clay. The jet water is then turned off and a thin cement slurry injected, to be distributed by the mechanical action of the auger drill. The Japanese have developed such methods and have applied them to shallow seafloor soils. When cement is used, it is termed the *deep cement mixing method*.

A method under development includes the injection of cement slurry under relatively high pressure into clays and silts. The soil is fractured, allowing multiple lenses of cemented material to become interbedded, thus increasing its shear resistance. This is called *jet grouting*.

A similar method of injection uses quicklime (CaO) which is placed in an augered hole in a watertight container so as to avoid premature hydration; the drill is withdrawn, with sand being used as a packer on top of the lime; then the drill ruptures the watertight container. The quicklime draws water from the surrounding soil and is converted to calcium hydroxide, with significant liberation of heat. Stable calcium compounds such as calcium carbonate are formed among the clay particles.

Use of chemicals other than limes and cements is also practicable in permeable and slightly permeable soils. For many years, injection of sodium silicate followed by injections of calcium carbonate has been practiced. In some calcareous soils, if they are sufficiently permeable, a single-stage injection of sodium silicate may be adequate, being fixed by the free lime in the soil.

Organic polymers may also be injected, either as polymers or as monomers which will later convert. These monomers are usually highly toxic, and hence their effect on adjacent marine life must be considered. Some of these have high penetration qualities.

Shell Oil International has developed a penetrating polymer known as Eposand. The soil is first flushed with fresh water, then with an organic solvent, following which the Eposand is injected. This process will be used to provide temporary strengthening of calcareous sands under the tips of the piles of the North Rankin A Platform, located in Australia's Northwest Shelf.

Freezing has been proposed as a method for strengthening the soils and embankments in the Arctic (see Section 20.8) and has more recently been studied in detail for application at the North Rankin A Platform referred to in the preceding paragraph. In seafloor soils, special consideration must be given to the progressive concentration of brine lenses and pockets as the freeze front advances.

The various methods indicated in this section for consolidating and stabilizing seafloor soils have the potential of increasing both the bearing strength and the shear strength by a factor of 2–5. Thus, even very weak soils may show significant increases in their capability to provide support for a structure and to reduce its movements during storms.

Stabilization of weak soils against mudslide and surficial mud mass flows has not yet been carried out to this author's knowledge. Various proposals have been made to construct underwater containment and diversionary dikes by preconstruction of a rock dike. The rock would be dumped in large masses, on a continuous basis, augmented by use of explosives as necessary, so as to displace the weak overlying muds, since mudslide shear planes seldom extend deeper than 20 m.

To date, the provision for mudslide areas has taken an alternative route, that of designing the jacket to penetrate through the overlying mud so that the structure will remain rigid while the mud slides around it.

7.10 Prevention of Liquefaction

Some of the large gravity-based structures in the North Sea are placed on foundations containing strata of fine sands, either at the surface or interbedded between clay layers at shallow depths. To prevent pore pressures building up under the rocking action of the structure due to waves, drainage pads and wells have been installed. The drainage into the structure is controlled so that the ambient pore pressure in these sand strata is less than the hydrostatic head; thus even under storm conditions, excess pore pressure is prevented.

Proposals have also been made to improve the lateral resistance of surface sands around piles by drainage into the jacket legs or mud mats. It appears that provision of short skirts under the mud mats would facilitate such drainage.

A very effective method to prevent liquefaction of soils under the edge of a caisson or other seafloor-supported structure is by the provision of a peripheral apron of graded rock, which allows the pore pressures to dissipate without inducing flow in the sands under the rocking of the structure due to waves. The use of a filter fabric covered by an articulated mat or by rock, as described in the next section, is also highly effective in preventing liquefaction under the edge of a seafloor supported structure.

7.11 Scour Protection

To prevent erosion due to currents, steady and transient, around structures and pipelines, scour protection in one form or another is required. At one end of the spectrum, sacrificial material such as sand is added around the structure or on the berm of an embankment. This is especially applicable to temporary structures such as islands for use in exploratory drilling, designed to last 1 or 2 years, in the shallow waters of the Beaufort Sea.

The most common form of scour protection is by the placement of rock of a size suitable to withstand the currents without dislodgment. This may vary in size from gravel on a seafloor at a depth of 20–30 m to 10-ton rock in the surf zone.

The stability of rock varies as the cube of the buoyant density, that is, (specific gravity -1).[3] Hence trap rock with a unit weight of 190 lb/ft^3 was specified for the breakwaters of the Atlantic Generating Station planned to be built off the coast of New Jersey in lieu of somewhat less expensive silica rock at 165 lb/ft^3. A similarly sized piece of trap rock turns out to have a 50% greater stability factor than that of silica rock.

In general, the larger the individual pieces the greater their stability. However, interlocking of pieces is also very important, with blasted polygonal rock being much more stable as a mass than similarly sized cobbles and boulders. Hence chinking of the crevices and even filling them with concrete can add to the stability as long as there is sufficient permeability (porosity) to allow excess pore water pressures to dissipate.

Wave impact creates hydraulic ram effects that temporarily raise the internal pore pressure. A breaking storm wave can create a hydraulic ram effect that can (and has) hurled a 100-ton block over the breakwater! Breakwater armor typically fails outward, at least initially, due to these effects.

Wave effects can extend to depths as great as 100–150 m, where they can create internal pore pressures of 3–4 tons/m^2 (30–40 kPa).

Sands tend to migrate up into the rock. Conversely, under wave action, the sand immediately under rock fragments liquefies, allowing the rock to work its way down into the sand. To prevent this, filter courses of rock are placed, graded so that the sand will not become interbedded with the rock.

The *Shore Protection Manual* of the U.S. Army Corps of Engineers gives specific guidelines for the sizing and gradation of rock to serve as filters. Placement of several layers of rock of different gradations is difficult enough in the calm water of harbors but much more so in the open sea. It becomes necessary to increase the thickness of each layer so as to compensate for the tolerances in placing. The most practicable way, albeit inefficient in the use of material, is to mix all the materials of different sizes so as to grade from fine to coarse and then place as one combined layer.

Filter fabrics have therefore been widely adopted. These fabrics have specified sizes of pores, which will allow water to bleed through but not sand. To give adequate strength to such fabrics so as to accommodate differential movements, wave and current forces, and the strains induced during installation, the finer fabrics may be backed up by heavier-mesh polypropylene. The heaviest such material even has embedded stainless steel strands.

Fabrics of the two types may be sewn together and laid as a unit.

Most of the filter fabrics are lighter than water and hence hard to place underwater. Concrete "dobe" blocks may be attached with stainless steel staples. The Dutch automated the manufacture of such articulated mats for the Oosterschelde Storm Surge Barrier. In Canada, special filter fabrics are manufactured which are denser than water and hence more easily installed.

These mats may be laid as rolls, unreeled off a large drum, and spread onto the seafloor. See Figure 7.11.1. Alternatively, they may be assembled

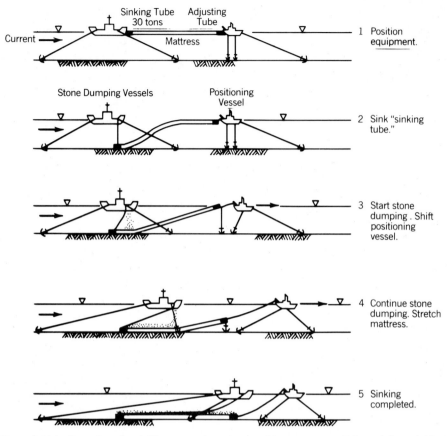

Figure 7.11.1 Procedure for sinking protective mattress with stone cover. (Adopted from ACZ, *Marine Engineering and Construction*, Netherlands, 1979.)

on steel pipe frames, say 30 × 30 m in plan, and set on the seafloor. Mat segments such as these were installed around the offshore terminal caissons at Hay Point, Queensland, Australia, and have successfully withstood the scouring attack generated by cyclones as well as high tidal currents.

On caisson-type structures, the filter fabric may be preattached to the caisson near its base. After seating of the structure, the mats are cut loose and laid out onto the adjacent seafloor. This was successfully carried out on the Ekofisk Oil Storage Caisson.

Filter fabric mats must adequately overlap at joints; this presents some difficulties when trying to lay around the corner of a structure. Rock riprap can subsequently be dumped over the mattress.

Other ingenious forms of scour protection have been developed. Holes left in an external wall of the structure near the seafloor may develop counter currents that tend to cause deposition of sand rather than erosion. Experimentally, steel pipe frames cantilevered out from the structure just above the seafloor and hung with multiple strips of plastic act as artificial seaweed, slowing the local currents so as to cause deposition.

Much experimentation has been carried out on the laying of a mat of concrete or asphalt over the seafloor in order to stabilize it. The Dutch have applied asphalt–sand and asphalt–stone mixes on many of their coastal dikes. These are usually designed to have both flexibility to accommodate local movements and porosity to relieve excess pore pressures. German and Japanese engineers have developed special concrete mixes and admixtures to enable a permeable concrete mat to flow out over the seafloor without segregation.

Where rock riprap of the required size is unavailable or excessively costly, concrete armor units have been used. There are at least 40 different shapes which have been developed, of which the Tetrapod, Tribar, Stabit, and Dolos are among the best known. The Dolos has the best hydraulic char-

TABLE 7.12.1 Seafloor Modification and Preparation

Task	Methods	Task	Methods
A. Survey, Investigation, and Controls	Electronic Satellite Navigation, Spar Buoys, Acoustic Transponders, Coring and Sampling, Grab Samples, Sparker Survey, Side-Scan Sonar, Acoustic Imaging, Foundation Penetrometers, Video, Submersible and Diver Inspection	H. Consolidation and Strengthening of Weak Soils	Sand Piles, Vibration, Freezing, Presurcharging, Surcharging with Membrane and Drainage, Surcharging with Structure and Ballast, Wick and Sand Drains, Drainage Wells, Peripheral Surcharging, Cement Injection, Chemical Grouting, Lime Injection, Deep Cement Mixing, Electro-osmosis
B. Platform	Derrick Barge, Drill Ship, Semisubmersible Jack-up, TLP, Guyed Tower, Heave Compensators	I. Prevention of Liquefaction	Densification as per "H," above, Drainage Wells, Peripheral Apron of Graded Rock
C. Seafloor Obstruction Removal	Drag off with Trawlers. Shaped Charges, ROVs with Manipulators, Underwater burning, Thermic Lancers	J. Leveling of Seafloor or Embankment	Hydraulic "Dustpan" Dredge with Heave Compensator Suspension of Dredge Head, Drags, Bottom-Supported Screed Frame, Screed Frame from TLP or Heave-Compensated Platform, Horizontal Screw Augur
D. Dredging Removal of Sediments	Trailer Suction Hopper Dredge, Cutterhead Hydraulic Dredge, Clamshell (Grab Dredge), Continuous Bucket Ladder Dredge, Slack Line Bucket Dredge, Plow, Jetting, Pipeline Burial Sled, Deep-Sea Mining Drag Excavator, Airlift, Eductors, Remote-Controlled Seafloor Dredge	K. Provision of Uniform Support under Base of Structure	Preleveling as per "J," above, Underbase Grouting, Underbase Sand Injection or Sand Flow, Tremie Concrete, Grout-Intruded Aggregate ("Prepakt"), Mud Jacking
E. Dredging Removal of Hard Material and Rock	Hydraulic Backhoes, Dipper Dredges, Power-Activated Clamshell Buckets, Plows, Shaped Charges, Blasting in Drilled Holes (O.D. Method), Chisels, Hydraulic and Pneumatic Rock Breakers, Driven Spuds, Cutterhead Dredges, High-Pressure (~15,000 psi) Jets	L. Excavation Beneath Structure	Articulated Dredge Arms, Airlift, Jets, Eductors, Drills
		M. Scour and Erosion Protection	Sacrificial Fill, Rock, Filter Rock, Filter Fabric, Articulated Mattresses, Sandbags, Grout-Filled Porous Bags, Skirts on Structures, Aprons and Flow-Control Devices at Base of Structures, Artificial Seaweed, Sand–Asphalt and Rock–Asphalt Blankets, Underwater Concrete Slabs
F. Placement of Underwater Fills	Dikes of Rock or Clay bunds to Contain Sand, Controlled Underwater Deposition, Dump en Masse From Hopper Barges, Tremie, Bucket, Skip, Chute or Ladder	O. Turbidity Suppression	Bentonite–Cement Slurries, Discharge of Fine Sand Blanket
G. Densification, Consolidation, and Strengthening of Fills	Deep Vibration, Surface Vibration, Dynamic Compaction with Dropped Weights, Explosives, or Airgun, Deposition in Mass, Presaturation, Selection of Optimum Grading		

(From B. C. Gerwick, "Preparation of Foundations for Concrete Caisson Sea Structures," OTC 1946, Offshore Technology Conference, Dallas, Tex., 1974.)

acteristics and least material quantities but is weak structurally in the larger sizes. Breakage of the 60-ton Dolos units has been assessed as the principal cause of the failure of the Sines offshore breakwater in Portugal.

Reinforcement in the form of multiple bars has been used for some U.S. armor units, and experimental application of steel fibers in the concrete mix has been tried on the Crescent City breakwater, California, with apparent good results.

For offshore sand and gravel islands in the Arctic, filter fabric has been laid over the slopes near the waterline and then 2- and 4-m^3 sandbags of polypropylene laid in one or two layers. The polypropylene is pigmented to prevent ultraviolet (UV) disintegration. Significant damage has occurred, however, due to sea ice attack on the sandbags, especially when they are still frozen.

Articulated concrete blocks over heavy filter fabric are believed to give more permanent protection. These are currently being produced for use on the Endicott Production Island in the Alaskan Beaufort Sea.

Another type of scour protection mat is available: a membrane with multiple fine holes to relieve pore pressures and containing many tubular sacks, integral with the mat, which can be pumped full of sand or grout after placement.

From the constructor's point of view, the use of the fabric mat with articulated concrete blocks attached seems to be the easiest and most positive method of providing scour protection for offshore structures.

Permanent scour protection is generally the province of the designer. However, it will often be found necessary for the constructor to place temporary scour protection. An example is the landing of a caisson on a sand seafloor where the local currents were quite high; these could then accelerate under the structure's base and cause serious scour just prior to touchdown.

7.12 Concluding Remarks

As structures are being extended into ever more difficult environments, greater depths, and less suitable soils, it is believed that increasing attention will be given to methods of modifying or improving the existing seafloor and the foundation soils.

Application of seafloor modification and preparation concepts and methods for accomplishment are summarized in Table 7.12.1. Further discussion of specialized methods for improvement of the seafloor are discussed in the chapters on gravity-based structures, deep ocean structures, and Arctic structures (Chapters, 10, 19, and 20). Freezing, for example, is discussed in Chapter 20 on Arctic Structures.

There are some days the happy ocean lies
Like an unfingered harp, below the land.
Afternoon gilds all the silent wires
Into a burning music for the eyes.

STEPHEN SPENDER, "SEASCAPE"

8

Installation of Offshore Piles

8.1 General

Piling for offshore structures must be installed so as to develop the required capacities in bearing, uplift, and lateral resistance. For offshore bridge piers, control and minimization of settlements may also be criteria.

Deep water, long, unsupported column lengths, large cyclic bending forces, and large lateral and axial forces all combine to make offshore piles large in diameter and long in length. Piles in most offshore practice are steel pipe piles ranging from 1 m up to 2 m (and even 4 m) in diameter and in lengths from 40 to 300 m. Pile capacities have design ultimate values of 400–6000 tons, far above those of conventional onshore piles. See Figure 8.1.1.

The main piles for platform Lena, the guyed tower in 320-water depth, were 54 in. in diameter and almost 500 m in length.

For resisting axial compression, the pile can transfer its load by skin friction along its outside perimeter and by end bearing on its tip, provided that the tip is either closed or plugged in such a way as not to yield in relation to the pile. Thus for a natural plug of sandy clay, the internal skin friction must be adequate to develop the full end-bearing resistance of the plug.

End bearing and skin friction do not develop their resistances simultaneously and hence are not usually directly additive at serviceability (elastic) levels of load. They may, however, partially augment each other at ultimate load. For this reason, deep pile foundations are usually designed primarily as friction piles.

For resisting axial tension, the deadweight of the pile plus that of the internal plug of soil plus the skin friction are available.

For resisting lateral loads, most offshore structures in deep water (over 30–40 m) depend on bending resistance of the pile interacting with the passive resistance of the soil in the near-surface stratum. Since the soil resistance is a function of its deformation, the analysis is based on the interaction of the lateral load P with the displacement y at each incremental level below the seafloor. Hence this is called the P/y effect. The pile must have sufficient strength to resist the resultant moment at these levels and to prevent biaxial buckling; see Figure 8.1.2. The capacity to resist lateral loads can be improved by increasing the stiffness and moment capacity of the pile in the critical zone near and just below the mudline by grouting in an insert pile and/or by increasing the wall thickness of the steel pile through this zone.

In stiff clay soils and calcareous soils, cyclic lateral loadings may create a gap around the pile, just below the mudline, which increases the lateral deformations of the piles and structure as a whole and increases the moment in the piles. Piling a loose mound of pea gravel, or even high-density rock of small size, around the pile can effectively minimize this effect by filling any gap that does form and thus minimize the amplitude of deformations.

An alternative method of resisting lateral loads, used in some offshore terminals, is by the use of batter (raker) piles, sufficiently inclined to develop a substantial horizontal component of their axial capacities. Batter piles must have a reaction in order to be effective; this is usually provided by a mating pile battered in the opposite direction, although the deadweight of the platform may also be mobilized as a reaction force. See Figure 8.1.3.

Special methods and equipment have had to be developed to install the large piles required for offshore structures. Driving with very large hammers

170 Installation of Offshore Piles

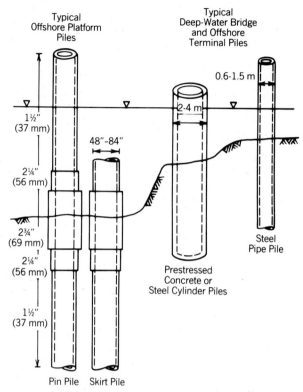

Figure 8.1.1 Typical piles for offshore structures.

is still the preferred method for most cases because it is fast. However, where soil conditions do not permit driven installations and in other special cases, drilling may be employed, with the pile being grouted into the drilled hole. Special foundations such as belled footings have been utilized in the North Sea and Arabian Gulf. Jetting, airlift removal of plugs, and even dredging removal have been used on large-diameter piling for offshore terminals, bridge piers, and a few offshore platforms.

The effect of all such installation operations on the supporting soil must be considered. In some cases, it may be beneficial; in many cases, the results may degrade the performance unless special precautions are taken.

API Standard RP2A warns that piles drilled and grouted may have resisting values significantly different from those of driven piles.

For piles driven in undersized drilled or jetted holes in clays, the skin friction will depend on the amount of soil disturbance, including the relief of stress, which is occasioned by the installation. The strength of dry, compacted shale or serpentine may be greatly reduced when exposed to water from jetting or drilling. The sidewall of the hole may de-

velop a layer of slaked mud or clay which will never regain the initial strength of the parent rock.

In overconsolidated clays, drilled and grouted piles may develop increased skin friction f. If excess drilling mud is present in clays or in soft rock, the value of f may be significantly reduced. In calcareous sands and some silts, driven piles may have very low values of f as compared to those attainable by drilling and grouting.

API Standard RP2A further warns that the lateral resistance of the soil near the surface is significant to pile design (see Fig. 8.1.2), and consideration must be given to the effects of soil disturbance during installation as well as scour erosion in service.

Great strides have been made in recent years in developing hammers and drills for offshore work. As a result, the constructor now has a wide range of effective tools from which to choose in order to meet his particular needs.

8.2 Pile Fabrication

The piles are usually made up of "cans," rolled plate with a longitudinal seam. Individual segments

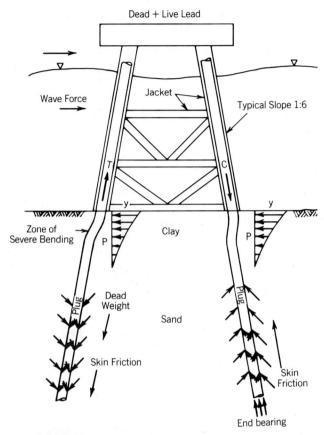

Figure 8.1.2 Typical pile–soil interaction phenomena. Note that bending deformations are purposely exaggerated.

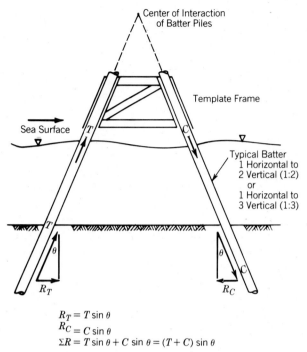

$R_T = T \sin \theta$
$R_C = C \sin \theta$
$\Sigma R = T \sin \theta + C \sin \theta = (T + C) \sin \theta$

Figure 8.1.3 Typical batter (raker) pile installation for offshore dolphin.

(cans) should be 1.5 m (5 ft) or longer in length. The longitudinal seams of two adjacent segments should be rotated at least 90° apart.

A taut wire located at three 120° azimuths should be used to verify straightness of the made-up pile. API RP2A specifies a maximum allowable deviation from straightness in any 3-m (10-ft) length of 3 mm (1/8 in.). For lengths over 12 m (40 ft), the maximum deviation should be 13 mm (1/2 in.).

API RP2A, sec. 4.1.5i, and API RP2B, sects. 4.2 and 4.3, give allowable tolerances for outside circumference and out-of-roundness as well as for beveled ends.

Pile wall thicknesses should vary between adjoining segments by not more than 3 mm (1/8 in.). Where a greater difference is employed, the thicker section should be beveled on a 4:1 vertical-to-horizontal level. See Figure 4.1.2.1.

To ensure proper fitup of splices and to minimize weld time, thick-walled pile segments are best premated or prechecked. Sections can frequently be best matched by rotating one relative to the other.

The surfaces of piles which are to be connected by grout bond to steel tubulars or to the soil should be free of mill glaze and varnish.

Steel piles are usually protected below water by sacrificial anodes. Impressed-current cathodic protection is also often employed but many present problems of reliability, not only due to technical adjustment requirements and possible adverse effects on reinforced concrete members in the vicinity but also to the human factor, namely, that the impressed current interferes with television reception and hence is often turned off.

Above water, steel piles are normally given some additional corrosion allowance—for example, 1/8 in. (3 mm)—and protected by epoxy, polyurethane, or rubberized coatings or by cladding with monel or copper–nickel sheets.

8.3 Transportation of Piling

Large-diameter tubular pile sections of steel or prestressed concrete are usually lifted (or rolled) onto a barge which is then towed to the site. The pile sections must be well chocked and chained so as to prevent any tendency to shift and roll in a seaway. Usually the piles have thick enough walls so that the stack will not locally deform the piles underneath; however, this should be checked and suitable blocking or supports provided as necessary to prevent damage.

Sometimes it becomes practicable and efficient to transport the piles in a self-floating mode, either singly or in a well-secured (chained) raft. This becomes especially attractive when the piles can be subsequently lifted and placed in long sections, for example, skirt piles, which can be entered well below the surface of the sea. The ends of the piles can be closed with steel closure plates or rubber diaphragms. These need to be strong enough to take wave slap during tow to the site. Upon arrival, one end is usually lifted clear of the water by a derrick line to permit cutting out, and then that end is allowed to rotate down to the vertical. In some cases, the trapped air in the other end (sometimes augmented by compressed air) is used to limit the draft when the pile reaches vertical. In such a case that closure must be adequate to resist the internal air pressure. In any event, closures should be provided with a valve so that air or water can be vented and/or allowed to flood in, as required.

In shallow waters, such as those at an offshore terminal, piles have been temporarily stored on the seafloor with recovery pendants and buoys attached.

Removal of a closure plate below water can be a dangerous operation. In one case off Australia, the diver who was cutting out the closure plate was sucked into the pile by the rush of water and drowned. Provision of a valve and prior equalization of pressure on both sides of the closure plate would have prevented this accident.

When a long pile is upended in the water, very large bending moments may be introduced at certain stages of rotation. While these are usually less than those which occur when upending in the air, they are not negligible and must be checked.

For many recently constructed offshore platforms, some piles have been transported with the jacket, either set in the main legs or in the skirt pile sleeves and guides, where they provide additional buoyancy (if closed) as well as additional weight.

The purpose of the pre-installation is to enable several piles to be driven down immediately after seating of the jacket on location so that the jacket will be stable against the action of waves and current.

Typically, as soon as the jacket is seated and leveled (insofar as practicable on the seafloor while resting on the mud mats) the piles are cut loose so that they can penetrate the soils under their own weight. They are then extended as necessary, and all four (or more) are driven down a short distance where they may be temporarily welded to the jacket top. Final leveling of the jacket may then take place. One by one these initial piles may be raised as necessary so as to eliminate any bending stresses, relowered, and driven down to required penetration.

As noted above, the piles when transported in the jacket must be adequately secured against the forces of launching and upending. During the upending of the Magnus platform jacket in the North Sea, a number of piles ran down, hit the seafloor, and were bent. For a while it was feared that the entire project would be set back one year. As it was, mammoth effort plus favorable weather permitted replacement of the piles and installation late that same season.

8.4 Installing Piles in Offshore Jackets

The piles for the typical offshore jacket are delivered on barges, with the first section of each pile being as long as can be handled and placed by the derrick barge.

Pin piles are centered inside the jacket legs and typically extend the full height of the jacket. Skirt piles are encased in sleeves bracketed out from the lower end of the jacket. Many jackets incorporate both pin piles and skirt piles.

Skirt piles must be driven driven either with a follower or an underwater hammer. The piles are typiclly clustered around the corner legs of the jacket and are aligned parallel to them, so that the piles must be driven on a batter of from 1:12 to 1:6. Guides are attached at intervals along the jacket legs to aid in setting the piles through the sleeves.

Some recent jackets have been constructed with vertical sleeves, thus eliminating the guides and enabling a very long initial pile segment to be stabbed into the underwater sleeve. Guidance of the pile may be by means of a tensioned line or, in deep water, by the use of short range sonar and video.

Recently, for very large jackets, some of the piles have been preinstalled in the jacket and secured for transport, launch, and upending. This enables the jacket to be expeditiously pinned to the seafloor soon after the jacket is placed.

"Add-ons" (additional lengths of pile) are "stabbed" onto the top of the previous pile section as it is driven down near to the top of the jacket.

American Petroleum Institute Standard API-RP2A, sec. 2.6.9d, suggests that reasonable assurance against failure of the pile will be provided if static stresses are calculated for each stage as follows:

1. The projecting section of the pile is considered as a freestanding column with a minimum effective length factor K of 2.
2. Bending moments and axial loads are calculated on the basis of the full weight of the hammer, cap, and leads, acting through the center of gravity of their combined masses, plus the weight of the pile add-on section, all with due consideration of the eccentricities due to pile batter. The bending moment so determined should not be less than that due to a load equal to 10 percent of the combined weight of hammer, cap, and leads applied at the pile head perpendicular to the pile centerline.
3. The pile resistance is to be based on normal (elastic) stresses, with no increase for the temporary nature of the load.

One means of reducing the bending in the pile during this stage is to suspend the hammer and leads in a bridle at the proper batter. See Figure 8.4.1. This is especially important in offshore terminal construction where relatively flat batters are often employed (e.g., 1 horizontal-to-2 vertical). This is also very important when driving piles having low bending strength—for example, prestressed concrete piles—on a batter.

Offshore piling are typically large-diameter,

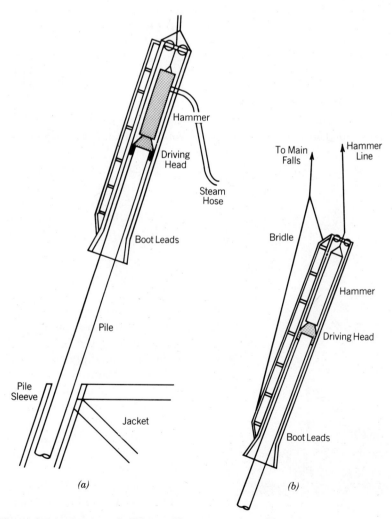

Figure 8.4.1 Driving of offshore piling; suspension of boot leads and hammer.

thick-walled tubulars (pipe) ranging from 1 to 2 m in diameter. They are driven with high-energy impact hammers, either steam, hydraulic, or diesel. As a general rule, the hammer with attached driving head rides the pile rather than being supported by leads (see Fig. 8.4.2). This means that the driving head (helmet) must be secured to the hammer by wire rope slings and that the driving head in turn must seat well on the pile and have a guiding bracket or ring attached in order to keep the hammer aligned with the pile. These latter are called "boot leads." During driving, the hammer line from the crane boom is slackened so as to prevent transmitting impact and vibration into the boom.

For steel piles, there is usually no cushion block used between the helmet and the pile, although an internal cushion is used in certain makes of pile hammer. Because of the tremendous energy required to raise the ram, steam is usually used rather than compressed air. Offshore hammers are generally single-acting with rates of up to 40 blows per minute. Hammer energies of current equipment range from 60,000 to 1,250,000 ft-lb per blow. The larger hammers represent major lifting tasks in themselves, weighing up to 400 tons.

Large diesel hammers are much used on offshore terminals. These have nominal energies in the range 130,000–160,000 ft-lb per blow, but in most practicable driving conditions they can be equated in effectiveness with a steam hammer of 60 percent of that rated energy. The diesel hammer is much lighter to handle and much more economical in fuel consumption, but its effective energy is limited.

Recently new hydraulic hammers have been developed for underwater driving. These are radically new versions of the underwater hammers formerly used in bridge pier construction. The new hammers not only develop several hundred thousand foot-pounds of energy per blow, but are virtually unaffected by depth. They are thus useful for driving skirt piles whose heads may finally be located 100 or 200 m below sea level.

Large vibratory hammers have been used on offshore terminal piling: for example, a quad unit of four large vibrators was used to install the heavy steel piles on the Yanbu, Saudi Arabia, pipe off-loading pier. These actually activated the piles to a resonant amplitude of about 10 mm, and the pile "drove itself" through dense sands and limestone layers.

Experimental driving with vibratory hammers has been carried out on piling for offshore platforms, including the use of hammers of sonic frequency, but so far they have not proven satisfactory for the extreme weights and lengths typical of such piles.

Impact hammers impart an intense compressive wave to the head of the pile, which travels down the pile at the speed of sound in the pile material. This compressive wave is a dynamic stress wave which eventually reaches the tip and extends it into the soil.

The newest large steam and hydraulic offshore hammers are instrumented so that the velocity of the ram can be measured just before it strikes the anvil. A representative pile in a jacket can also be instrumented to measure both strains and acceleration during the hammer blow.

The dynamic portion of the stresses induced in the pile during driving can best be computed by the *wave equation*, which is a one-dimensional elastic stress wave transmission analysis using selected parameters for the response of the hammer, cap block, cushions, pile, and soil strata. (See E. L. Smith, "Pile Driving Analysis by the Wave Equation," *Transactions ASCE*, Vol. 128, 1962, Part I, Paper No. 3306, pp. 1145–93.) Such an analysis is useful in determining the maximum stresses at such critical points as head, tip, and changes in section.

Figure 8.4.2 Driving piling in offshore platform in Gulf of Mexico.

Because of the transient nature of the blow and its very short duration, buckling as a column has been shown usually not to be a problem and can generally be neglected.

It is not always practicable to determine all the parameters needed with accuracy, especially soil parameters, nor is it always practicable to carry out a wave equation analysis, although the widespread availability of computer programs makes this latter a fast and economical practice that is increasingly employed even on moderate-sized projects.

In the absence of reliable calculations of these dynamic stresses induced during driving, an empirical rule is to limit the static portion of the stresses to one-half the yield strength of the pile.

A machined driving head is fitted between the hammer and the pile head. This ensures uniform transfer of the impact blow to the pile and prevents local distortion of the pile head.

The D/t ratios for piles must be limited so as to preclude local buckling at stresses up to the yield point of the pile steel.

Where only moderate driving resistances are anticipated or where the pile will be drilled and grouted (not driven), the pile may be designed as a steel cylindrical member and checked for local buckling due to combined axial compression and bending. This latter is non-critical when D/t is less than or equal to 60. When D/t is greater than 60, then a more in-depth analysis such as that given in API RP2A, secs. 2.5.2a(2) and (3), should be followed.

For piles which will be subjected to sustained hard driving in excess of 20 blows/in. or 250 blows/ft (8 blows/cm or 820 blows/m) the minimum wall thickness of the pile should normally be not less than

$$t \text{ (in.)} = 0.25 + \frac{D \text{ (in.)}}{100} \quad \text{(English system)}$$

or

$$(t) \text{ (mm)} = 6.25 + \frac{D \text{ (mm)}}{100} \quad \text{(metric system)}$$

Pile wall thicknesses are normally varied throughout their length in order to adjust to the in-service axial plus bending requirements. The minimum pile wall thickness should be selected as indicated above to preclude local buckling and also to maximize penetration under the hammer blows.

Maximum bending in-service normally occurs at and just below the mudline. Since designed pile penetrations often have to be modified to some degree in the field based on actual driving resistance, the length of pile with maximum wall thickness should be increased over the theoretically required length so as to permit some tolerance in pile penetration and hence in location of the thickened section.

In selecting pile section (add-on) lengths, the following factors should be considered:

1. The lifting and stabbing of the pile add-on segment. What is the maximum capacity of the crane and boom length to handle the segment? Check bending moment in pile during upending.
2. The capacity of the crane and geometry of the boom when seating the hammer and leads over the top of the newly added-on segment (and often over the intervening corner of the jacket).
3. The possibility that the initial pile section will "run" when it penetrates the jacket leg closure. If allowed to run free, it may drop below the level at which the next add-on may be welded. One solution, of course, is to provide a restraint, such as a preventor sling or cushioned bracket.
4. Stresses in the pile segment when lifting and when the hammer is placed (as noted earlier).
5. Wall thickness at field welds, with consideration of material properties and welding procedures required.
6. Possible interference with adjoining pile segments or structures. This is often critical in offshore terminal construction where there are batter piles which radiate in several directions and whose axes may intersect near deck level.
7. Soil characteristics. It is desirable to plan the segment lengths so that the temporary location of the pile tip is in relatively soft soils, enabling driving penetrations to resume easily when the splice has been made. Conversely, if the pile tip is allowed to sit during splicing in a zone of material with high set-up properties, then excessive resistance may develop when driving resumes.

The head of pile sections on which driving is carried out may be deformed during driving and hence require reheading in order to weld on the next section. Hence API RP2A recommends an allowance of 0.5–2 m for reheading. Modern well-fitting driving heads and some hammers (e.g., Hydroblock) minimize the head damage; so that with heavy-walled piles, reheading may be unnecessary.

When pile add-ons are placed, they are equipped with stabbing guides to facilitate entry and proper alignment. The stabbing guides should have a tight fit in order to provide a proper fitup for the weld. The guides may be designed to support the full weight of the pile during welding so that the crane may be freed from other tasks. Further, support from a floating derrick boom during welding is often unsatisfactory due to the movement of the boom tip and transmitted vibration. However, it is usual practice for the crane to continue to hold on with a slack pile line as a safety precaution until at least one full weld pass has been made.

Pile sections may also be held by temporary supports on guides which extend up from the jacket and provide a support 10–20 m above the deck.

They may also be held by a hydraulically operated clamping and alignment device which is clamped onto the previously set pile section or supported on a temporary work deck on the jacket. This latter allows quick-stabbing guidance and then final accurate alignment of the new section. See Figure 8.4.3.

In addition to connecting pile sections by full-penetration welds, breechblock connector sections have been developed which enable the splice to be effected rapidly by applying torque. Accurate alignment is essential, and hence a hydraulic clamping–aligning device is appropriate. These mechanical connectors have been used for pile followers and have shown generally adequate performance during driving; however, for permanent piles, full-penetration welds are still the standard practice.

For piles to be joined by welding, the add-on section is pre-beveled, ready for a full-penetration weld. After stabbing, the bevel is inspected and, if necessary, ground or gouged to open it up so as to assure a full-penetration weld. Weld procedures and materials should be carefully selected with regard to the pile steel qualities and the temperature at which driving will be carried out, since these welds will certainly be under high impact. This will be especially critical for pile driving in Arctic and sub-Arctic regions which may be carried out at low temperatures. In any event, low-hydrogen electrodes should be used.

Back-up plates are usually built into the stabbing guides. However, where drilling is to be carried out,

Figure 8.4.3 Setting piling in thistle platform using hydraulically operated clamping and alignment device.

internal backup plates cannot be employed, and the stabbing guides must be external to the pile. API RP2A notes that special skills are required for single-side welding or complete-penetration welds where back-up plates are not used.

For long piles with numerous add-ons, great care should be taken to ensure accurate axial alignment of each section, so that the resultant full pile will be as straight as possible. This is especially important for clustered skirt piles, in order to keep their tips properly spaced.

The times required for welding of the large-diameter, thick-walled piles typical of most offshore platforms is significant. On the 54 in. piles of the Hondo platform, the average times were, for 1-in. walls, 3¼ hours; for 1¾-in. walls, 7¼ hours; for 2½-in. walls, 10½ hours.

Welding machines should be properly grounded to prevent underwater corrosion damage due to stray current discharge.

When pile sections are lifted, they are usually provided with lifting eyes. The lifting eyes and their weld details are designed for the stresses developed both at initial pickup and as the pile is rotated to alignment with its final axis. Thus both angle of lead of the sling and the load acting on the padeye will change. An allowance must be made for impact during lifting—normally 100 percent for the lifting eye and 35 percent for the crane.

When lifting eyes or weld-on lugs are used to support the pile sections from the top of the jacket, each eye or lug should be designed to support the entire hanging weight.

Lifting eyes for tubular pile sections must be welded transverse to the initial tensile force; there is no suitable alternative. Hence, it is especially important that a full-penetration weld (not fillet welds), with proper procedures, be used. Similarly, the load is inherently across the thicknesses of the pile plate, requiring adequate through-thickness toughness.

Final removal of padeyes or lugs should be by flame-cutting ¼ in. (6 mm) from the pile surface, followed by grinding smooth.

Holes are often used in lieu of lifting eyes, especially for smaller, less critical piles such as used in offshore terminals. These holes should be burnt undersized and then reamed. They should be located near the top of the section so that they are part of the length cutoff when stabbing a new add-on.

Use of burnt holes alone is dangerous, due to stress concentrations during lifting and adverse effects during driving.

With large and heavy offshore piles, internal cones or running tools are often used for lifting so that the pile section will hang vertically.

The piles for offshore structures are typically heavily loaded in both compression and tension. There are relatively few piles employed; hence each becomes a major structural component. The integrity of the structure therefore depends on each pile being driven to approximate design tip elevation as determined by prior geotechnical investigation and analysis. See Figure 8.4.4.

The pile is also designed to resist lateral forces which are transmitted from the pile to the soil in the upper layers of the soil. The interaction is complex and depends on the stiffness and diameter of the pile and the nonlinear behavior of the soil. Thus the pile wall is usually thickened in the region near the design mudline, and hence it is important that

Figure 8.4.4 Driving piling in jackets of offshore terminal, Kharg Island, Iran. (Courtesy of Santa Fe International.)

the pile be driven in accordance with the initial design so that the thickened wall section ends up at the proper place.

Table 8.4.1 lists some of the more commonly used pile hammers for driving offshore piles, together with their principal characteristics. The "slender" designation indicates that the hammer has a sufficiently small cross-section that it can follow the pile down through the guides of the jacket, thus eliminating the need for followers.

The driving of each pile should be carried out as nearly continuously as practicable in order to minimize the increased resistance which may occur due to "setup." When the tip of any pile has entered a zone of stiff plastic clay, for example, every effort should be made to eliminate or minimize interruptions in driving.

A backup hammer with leads should be available whenever setup is anticipated, as otherwise the breakdown of the hammer may permanently prevent the subsequent driving of a pile to prescribed tip elevation. See Figure 8.4.5.

Note in Figure 8.4.5 the cross-hatched areas. Not only were there additional blows required to break the pile loose, but additional blows were required for all subsequent penetration. In the example shown, some 6000 additional blows were required. For a hammer striking 40 blows/min, this is 150 minutes extra or 2½ hours per pile at a rate which

TABLE 8.4.1 Large Offshore Pile-Driving Hammers Driving on a Heavy Pile

Hammer	Type	Blows per Minute	Weight, Including Offshore Cage, If Any (metric tons)	Rated Striking Energy (ft-lb × 1000)	Rated Striking Energy KNm	Expected Net Energy (ft-lb × 1000) On Anvil	Expected Net Energy (ft-lb × 1000) On Pile
Vulcan 3250	Single-acting steam	60	300	750	1040	673	600
HBM 3000	Hydraulic underwater	50–60	175	1034	1430	542	542
HBM 3000 A	Hydraulic underwater	40–70	190	1100	1520	796	796
HBM 3000 P	Slender hydraulic underwater	40–70	170	1120	1550	800	800
Menck MHU 900	Slender hydraulic underwater	48–65	135	—	—	651	618
Menck MRBS 8000	Single-acting steam	38	280	868	1200	715	629
Vulcan 4250	Single-acting steam	53	337	1000	1380	901	800
HBM 4000	Hydraulic underwater	40–70	222	1700	2350	1157	1157
Vulcan 6300	Single-acting steam	37	380	1800	2490	1697	1440
Menck MRBS 12500	Single-acting steam	38	385	1582	2190	1384	1147
Menck MHU 1700	Slender hydraulic underwater	32–65	235	—	—	1230	1169
IHC S-300	Slender hydraulic underwater	40	30	220	300	—	—
IHC S-800	Slender hydraulic underwater	40	80	580	800	—	—
IHC S-1600	Slender hydraulic underwater	30	160	1160	1600	—	—

Source: Adapted from Heerema, E. P. "An Evaluation of Hydraulic vs. Steam Pile-Driving Hammers," OTC 3829, Offshore Technology Conference Preprints, Dallas, Tx, 1980, with additions by author.

may be $5000 per hour, thus amounting to $12,500 extra per pile. Note also that this could in some cases prevent the pile from reaching its final tip elevation, regardless of the number of blows expended.

The factors affecting pile resistance during driving are many and complex. Thus the mere attainment of a high blow count or "practical refusal" does not necessarily indicate that an adequate capacity has been achieved in either compression or tension. Thus it may be necessary to continue driving either at high blow counts or to use the additional methods, such as cleaning out the pile plug, jetting, or drilling, to be described in a later section of this chapter.

API Standard RP2A, sec. 5.5.9, states, "The definition of pile refusal is primarily in order to define the point where driving with a particular hammer should be stopped and other methods instituted such as using a longer hammer, drilling or jetting, etc." Continued driving at "practicable refusal" may be ineffective, may damage the hammer or the pile, is costly, and may extend the time so much as to endanger completion within the weather window.

The standard further states: "The definition of refusal should also be adapted to the individual soil

8.4 Installing Piles in Offshore Jackets 179

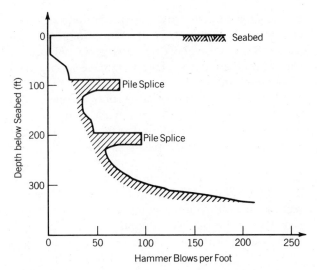

Figure 8.4.5 Typical effects of suspension of driving during splicing of pile in cohesive soils.

the problem of preventing excessive local damage in the pile due to dynamic stresses induced by the hammer and have been determined largely from industry experience in driving offshore piling in the medium to large sizes with moderate-sized hammers.

The empirical rule of thumb used is that the hammer energy per blow, rated in foot-kips, should be equal to twice the cross-sectional wall area of the pile in square inches. This criterion cannot always be met in practice.

When steel piles are driven onto rock, the rebound compressive stress may be almost twice that imparted directly to the pile by the hammer, thus often being above yield. This condition is most frequently encountered in constructing offshore terminals. The pile tip may deform, tear or "accordion." Tip reinforcement is definitely beneficial. See Figure 8.4.6. The wave equation is a useful tool

characteristics anticipated for the specific location." For example, when driving piles in the Arabian Gulf, piles may reach virtual refusal on limestone, coraline, or cap rock layers only to break through into less dense material below after a few hundred blows.

Standard API RP2A suggests a typical definition as follows:

Pile driving refusal is defined as the point where pile driving resistance exceeds either 300 blows per foot (985 blows per meter) for 5 consecutive feet (1.5 m) or 800 blows for one foot (0.3 m) of penetration. (This definition applies when the weight of the pile does not exceed four times the weight of the hammer ram. If the pile weight exceeds this, the above blow counts are increased proportionally, but in no case should they exceed 800 blows for six inches [150 mm] of penetration.)

If the pile driving has been stopped for one hour or more in order to splice a pile or because of equipment malfunction, weather delays, or the like, then the pile should be driven at least 1 ft (0.3 m) before the above criteria are reinstated.

Driving at resistances greater than 800 blows for 6 in. (150 mm) should not be attempted. The pile will be deforming (yielding), and no appreciable penetration will be attained.

Approximate guidelines for selecting hammer size in relation to pile diameter and wall thickness are given in API RP2A, sec. 5.5.10. These do not include other important parameters such as pile length and soil characteristics, but they do address

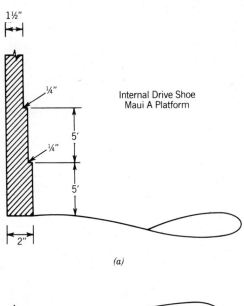

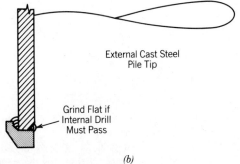

Figure 8.4.6 Tip reinforcement for piles expected to encounter rock or hard layers.

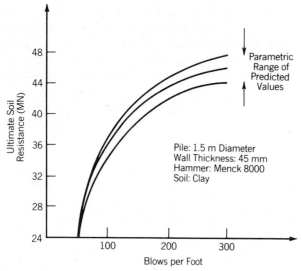

Figure 8.4.7 Typical wave equation analysis prediction of drivability.

for prediction of pile tip stresses in such cases. See Figure 8.4.7.

The trend today is toward the use of thicker pile walls in order to increase the effectiveness of the hammer in obtaining penetration. Heavier hammers are used in order to achieve more effective penetration and driving rates. A somewhat more conservative approach is therefore suggested. In Table 8.4.2 typical values are given which, of course, can be modified based on more specific data or detailed analyses.

When entering an initial section of pile into a sleeve, a longer length than normal may be used, limited only by the moment due to pile deadweight alone and the lifting capacity of the crane. Once entered and run down, an axial force may be applied to run the pile further into the soil. This is done by rigging a line around the head of the pile section, down through a snatch block affixed to the jacket near the pile sleeve, and thence to a winch on the jacket or derrick barge. Exertion of a tension on the line pulls that pile down through the soil to a temporary top elevation safe for mounting of the hammer.

Similarly, on offshore terminal construction, with prestressed concrete piles, by simultaneously jetting at the pile tip (through an internal jet) and exerting a downward tension axially aligned with the pile, the pile may be caused to penetrate through soft clays and sands to a more workable elevation.

With large-diameter cylinder piles, the pile will have to be slung from padeyes on both sides of the

TABLE 8.4.2 Typical Values of Pile Sizes, Wall Thicknesses, and Hammer Energies For Steam Hammers

Pile Outer Diameter		Wall Thickness		Hammer Energy	
(in.)	(mm)	(in.)	(mm)	(ft-lb)	(kN-m)
24	600	⅝–⅞	15–21	50,000–120,000	70–168
30	750	¾	19	50,000–120,000	70–168
36	900	⅞–1	21–25	50,000–180,000	70–252
42	1,050	1–1¼	25–32	60,000–300,000	84–420
48	1,200	1⅛–1¾	28–44	90,000–500,000	126–700
60	1,500	1⅛–1¾	28–44	90,000–500,000	126–700
72	1,800	1¼–2	32–50	120,000–700,000	168–980
84	2,100	1¼–2	32–50	180,000–1,000,000	252–1,400
96	2,400	1¼–2	32–50	180,000–1,000,000	252–1,400
108	2,700	1½–2½	37–62	300,000–1,000,000	420–1,400
120	3,000	1½–2½	37–62	300,000–1,000,000	420–1,400

Note 1: With the heavier hammers in the range given, the wall thicknesses must be near the upper range of those listed in order to prevent overstress (yielding) in the pile under hard driving.

Note 2: With diesel hammers, the effective hammer energy is from one-half to two-thirds the values generally listed by the manufacturers and the above table must be adjusted accordingly. Diesel hammers would normally be used only on 36-in. or less diameter piles.

Note 3: Hydraulic hammers have a more sustained blow, and hence the above table can be modified to fit the stress wave pattern.

pile (opposite ends of a diameter) so as to hang vertically. See Figure 8.4.8.

Upending large cylinder piles presents a problem since it may be practicable to lift only from the top, yet there maybe insufficient bending strength to permit such support. One solution has been to cap the lower end of the pile and to turn it in the water so that buoyancy provides support along the lower half of the pile. This maneuver requires consideration of water depth, pile length, net weight of the buoyant portion of the pile, and pile bending strength. Once turned to vertical, the bottom cap must be removed. The pile should first be filled with water to equalize the head on each side of the cap so that the closure plate may be safely removed.

Other methods of removing caps have been used, in a variety of circumstances. On the Drift River Terminal in Cook Inlet, Alaska, lucite caps were used which shattered when the pile was driven down.

On the Hondo platform off Santa Barbara, California, a chain was welded in spiral fashion to a steel plate; the idea was that when the chain was pulled, it would rip open the plate like the cover of a sardine can. It worked in tests on land; it didn't fully work in the actual installation, with the result that some plates had to be drilled out—a costly and time-consuming operation.

Figure 8.4.8 Upending 4-m-diameter cylinder pile for Collier Chemical Co. offshore terminal, Cook Inlet, Alaska.

A similar failure occurred on an offshore terminal at Inchon, Korea. The steel plate closures were seal-welded and then a chain hinge provided at one side. The concept was that when the pile was driven down, the seal welds would break, the plate break free, and hinge around the chain link so as to move clear of the pile. It didn't work that way. The steel pile broke the plate loose, but it then rode down as a plate cover to the pile, making it a closed-end pile and preventing full penetration until drilled out.

Similar closures are used to keep the jacket legs buoyant during launching, floating, and upending. They are designed to be ruptured by the impact of the pile. Nylon-corded and reinforced neoprene closures are now used almost universally. They are domed so as to resist hydrostatic pressure efficiently, yet they are designed to be easily cut out like cookies by the pile as it is driven through them. The closures are designed to resist the maximum hydrostatic head to which they will be exposed. On a few rare occasions these closures have been made so strong that they could not be cut through by driving. They then had to be drilled out. See Figure 8.4.9.

Rock drills are not efficient in drilling rubber; as one can imagine, the teeth become fouled and the water jets cannot clear the rubber. It requires many trips of the drill stem in and out of the pile in order to finally clear the seal. One solution, then, has been to cut the end of the pile on a scallop, like teeth, so that it slices through the seal progressively.

Reinforced rubber seals remain the state of the art despite the potential problems they can pose—for example, when drilling must be carried out through the pile and cut slabs of reinforced rubber are encountered, causing the drill to be clogged.

Diaphragm-type rubber leg closures are available in sizes from 18 to 144-in. O.D. For deep-water structures and very-large-diameter legs of jackets, sleeves, or cylinder piles, mechanically locked rubber diaphragm elements are available for pressures up to 700 psi in 90-in. O.D. closures and 300 psi in 149-in. O.D. sections.

Unreinforced concrete plugs, both of lightweight and normal-weight concrete, have been employed on the concrete offshore platforms in the North Sea to seal the conductor sleeves. The plug is usually 1.5–2 m thick, with welded shear lugs on the sleeve to transfer shear. These plugs are then drilled out prior to driving the conductors.

Increasingly the piles for deep-water offshore platforms are arranged in clusters around the cor-

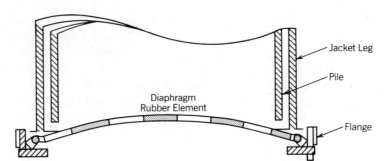

Figure 8.4.9 Standard diaphragm type leg closure. (Courtesy of Oil States Rubber Co.)

ner jacket legs and their loading transferred to the jacket by means of sleeves bracketed out from the sides. The final top of these piles will then be underwater a distance equal to the water depth less the sleeve length. This latter is usually 20–30 m. The pile connection is effected by grout.

To drive the pile so far below water requires the use of either an underwater hammer or a follower. Several types of underwater hammers are now made, two of which can fit inside the pile guides which are bracketed out from the jacket at higher levels. Most common, however, is the use of a follower. This is a moderately thick-walled pile section with a machined driving head on its tip which fits snugly over the head of the pile, transmitting axial compression, while preventing local buckling. For long followers, it may be necessary to join segments with mechanical joints, such as the breechblock connectors previously described.

Occasionally, due to misalignment or minor variances in the pile head, the pile becomes jammed into the driving head and the follower cannot be removed. Then the pile must be cut off, either by divers or else by a drill rig using expander casing–cutter tools. To prevent excessive delays under such a circumstance, the corrective tools should be on board.

Experience shows that with a properly fitting driving head, a square cut on the pile, and a pile wall thickness that is not too small, that is, not less than 1 in. (25 mm), then there is very little loss in efficiency by use of the follower.

Where excessively hard driving is expected—as, for example, when driving though limestone or cap rock strata—a driving shoe should be provided. API RP2A, sec. 2.6.9h, suggests that this be at least one diameter in length and have a wall thickness 1.5 times the minimum thickness of pile section in that pile. Experience in driving through weak limestone containing embedded basalt cobbles has indicated that such a shoe should be two diameters in length to prevent buckling like an accordion. Steel quality should be as high-yield as can be properly welded: Since the weld is made in the shop, it can be properly preheated and postheated as required.

More recently, cast steel shoes have become available for the small- to moderate-diameter piles. These are more easily affixed and welded. See Figure 8.4.6b.

The shoe should normally have the same internal diameter as that of the pile in case it becomes necessary to drill through the pile as a casing; otherwise, the drill may catch. The slightly larger diameter tip will relieve some of the skin frictional resistance on the main pile body.

When grouting piles to the jacket sleeves or when grouting between an insert pile and a primary pile, it is essential that the spaces be completely filled. Experience has shown that grout can trap water and bypass it unless great care is taken.

As noted earlier, the steel surfaces should be free of mill scale or varnish. Mud must be excluded from the annulus; this may require the use of wipers when working in very soft muds. The steel surfaces must also be clean of marine growth (which may form in relatively short periods of submergence) and free from oil or other contamination. Drilling mud should be flushed out with water where this can be done without endangering a drilled hole. If it cannot be safely flushed out, a polymer-based mud or thin bentonite mud should be used, preferably one that has been converted to a calcium base so that it will not coagulate upon contact by cement.

Both neat cements and expanding cements are used; the latter can give much improved bond and shear transfer. API RP2A requires that an expansive, nonshrinking grout be used. Admixtures are employed to promote flow, reduce tendency to segregate or wash out, to provide controlled expansion during the curing period, and reduce bleed. Shrinkage under water is actually not of much magnitude, but bleed is an undesirable property which should be minimized.

The bond is also affected by movements and vibrations of the structure during the setting period. Speed of set and strength gain are controlled by the type of cement and its fineness of grind as well as by the water temperature and by the use of admixtures. The grout should in any event develop at least 1500 psi (10 N/mm^2) within 24 hours.

Special grouts are available which give improved bond strength, decreased bleed, and decreased shrinkage.

Mechanical devices such as shear lugs, strips, and even weld beads can be installed on the inside of the sleeve and the outside of the pile to improve bond. Lugs must be designed to permit proper flow of grout and beveled so as not to trap bleed water under them.

If grout is placed so as to fill not only the annulus but also the body of the pile, consideration should be given to the heat of hydration to ensure that excessive temperatures will not be developed which may destroy the concrete's tensile strength through internal microcracking. For this same reason, the width of the annulus should be limited to about 4 in. maximum, or else a low heat mix should be employed. See section 6.6.8.

Centralizers (spacers) should be used to maintain a uniform annulus between the pile and the sleeve. A minimum width of 1½ in. (38 mm) is required by API RP2A but, 2–4 in. appears optimum. Beyond 4 in., potential shear in the grout may reduce the transfer of loads.

Packers are used to confine the grout and prevent its escape around the tip of the pile. See Figure 8.4.10. The packers must be so installed at the bottom of the sleeve as to protect them during pile entry and driving. Experience shows that packers are often damaged; hence a double set may be a prudent precaution.

Some designs of packers are passive, that is, just flexible rubber. Others are expanded by water pressure or by the grout itself. In the latter case, the grout first fills the packer; then as the back pressure rises, the grout opens a flap valve into the annulus (See Fig. 8.4.11).

When a packer is damaged and the grout is escaping, then all that can be done is to allow the grout to set and try again. Unfortunately, it will

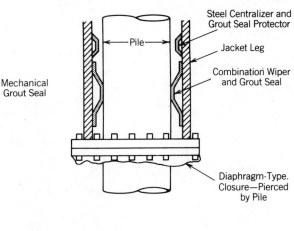

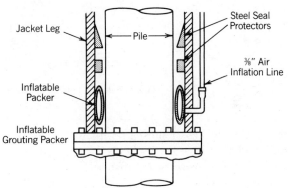

Figure 8.4.10 Wipers seals and packers for pile–jacket leg closures. (Courtesy of Oil States Rubber Co.)

tend to set in the grout pipe also. Flushing out slowly with water at minimum pressure may keep the grout pipe open for the second injection.

For this reason, two grout pipes, with entry ports spaced 2–4 m apart vertically, are often installed. If grout escapes around the pile tip, the first grout pipe can be abandoned. However, water should then be circulated slowly through the second pipe to prevent any possibility of its plugging.

The grout equipment should maintain continuous flow until the annulus is completely filled. If the configuration and relative elevations do not permit grout to be returned to the surface so as to verify complete filling, then suitable means should be employed, such as electric resistivity gages, radioactive tracers, well logging devices, or overflow pipes which can be verified by divers.

Recently, a new method of locking piles to sleeves has been developed in which the pile is "forged" into recesses in the sleeve by means of intense hydraulic pressure. This method, known as "Hydra-Lok," has been successfully employed to fix pin-

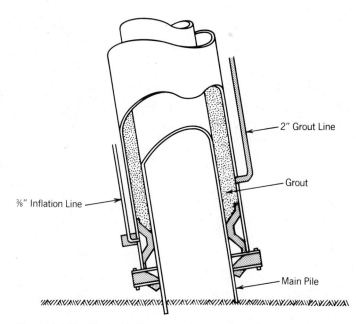

Figure 8.4.11 Typical inflatable packer and grouting arrangement.

piles to the sleeves of subsea templates of the Balmoral and Southeast Forties fields in the North Sea.

Pile installation records must be kept to record the following data:

1. Pile identification.
2. Penetration of pile under its own weight, after penetrating the pile closure.
3. Penetration of pile under weight of hammer.
4. Blow counts throughout driving.
5. Unusual behavior of hammer or pile during driving. For example, a sudden decrease in resistance which is not explicable by a review of the soil profile may indicate a ruptured weld.
6. Interruptions during driving. A record of "setup" time and a record of the blows subsequently required to break the pile loose.
7. Elapsed time of driving each pile section.
8. Elevations of soil plug and internal water surface after driving.
9. Actual length of pile section and length of pile cutoffs as each section is added.
10. Grout mix.
11. Equipment and actual procedure employed.
12. Volume of grout placed.
13. Quality of grout at intervals (tests).
14. Record of delays or interruptions during grouting.

Five examples of recent offshore pile driving experience follow:

8.4.1 Hondo platform, Santa Barbara Channel, California

The Hondo platform was built in a water depth of 264 m. The soils are primarily finely grained, normally consolidated cohesive silts. There are 8 main piles, up to 382 m in length; 12 skirt piles were driven. The main piles are 48 in. O.D., the skirt piles 54 in. O.D. Breechblock connectors were used for the followers for the skirt piles. The main piles required 13 add-on sections due to their long length.

Hammer size was limited by the then-current availability to the Vulcan 3100 hammer, with a Menck 4600 hammer as a backup, both developing about 300,000 ft-lb of energy per blow. Derrick boiler capacity was also a constraint.

Extensive investigations and analyses were made to predict pile-driving performance. Based on prior experience, hammer efficiencies were assumed at between 55 and 80 percent, with 80 percent as the most probable. Pile-driving logs, hammer blow counts versus driving resistance, and pile head forces were all generated for each assumed efficiency rating. Special attention was directed to the last three add-ons.

8.4 Thistle Platform, North Sea

The Thistle platform in the United Kingdom sector of the North Sea, was built in 161 m of water depth. Piles were designed for ultimate axial loads of 35–52 MN (3500-5200 metric tons). These required penetrations up to 140 m below the seafloor into hard clay with multiple sand lenses. Wave equation analyses showed that only about 30-m penetration could be achieved by the available hammers.

Therefore, a two-stage pile solution was selected. In the first stage, 1.37-m piles were driven to 30-m penetration using a Vulcan 560 hammer (300,000 ft-lb energy per blow). The piles typically formed a plug at about 25-m penetration, when the driving resistance would rise from 150–250 to 600 blows/min. Some of the plugs could not be broken free by driving and had to be drilled out to enable further penetration.

A 1.5-m-long driving shoe of steel pipe 12 mm thicker than the normal wall was attached to the tip; this was beveled so as to force the soil out from the pile tip rather than inside.

Piles were driven with a follower, assembled by breechblock connectors. A hydraulic clamping–aligning frame was used to hold the pile add-ons.

From 30-m to 140-m penetration, holes 1.21 m in diameter were drilled. Salt water was used as the drilling fluid (with return to deck level at +30 m) until an 85-m penetration was achieved. Then because of interbedded sand lenses, drilling mud was employed. To prevent hydraulic fracturing in the sand lenses, the mud specific gravity had to be very carefully monitored and controlled.

Holes were drilled with a flat bit, using airlift reverse circulation in a dual-walled drill pipe.

When the holes had been drilled, they were gauged and found to vary from 1.22 m in diameter in clay to 1.52 m in sand lenses.

The insert pile, 42-in. (1.06-m) pipe pile, was sealed at the bottom by a plug, weighted by filling with heavy drilling mud (1.8 sp. gr.) so as to overcome buoyancy and seated in the hole.

Grout of 1.68 sp. gr. was injected through a valve in the insert pile plug so as to completely fill the annulus. The grout had a radioactive isotope admixture added; when it ran out of the grout overflow pipe at the top of the first stage pile, it was detected by a Geiger counter.

The grout was allowed to set 24 hours, and then the weighting mud was pumped out.

8.4.3 Heather Platform, North Sea

For the Heather Platform, also in the United Kingdom sector of the North Sea, large piles were driven to extremely high capacities in hard, sandy, silty clay. Pile design loads were 29.5 MN (6630 K), requiring an ultimate capacity of 44.3 MN (9950 K). Pile penetrations of 43 m were required. Total pile length was 96 m.

Piles were of 1.52-m (60-in.) diameter and used 2½-in. (64-mm) walls throughout so as to enhance driveability.

Connection to the sleeves was by grout. To improve bond transfer, rings of weld beads were run on the piles. The first pile section was 64 m long, the second 32 m.

To expedite add-ons, a hydraulic clamping and aligning device was stabbed onto each first section.

An internal driving shoe, 3½ in. thick and 0.5 m in length, was attached, having the same outer diameter as the pile.

Pile-driving performance was predicted by means of geotechnical investigation and use of the wave equation. Special attention was paid to the mechanism of plug formation. It was predicted that plugs, once formed, would give only partial end bearing under the hammer blows because of the different transmission velocities of the compressive wave. This was partially confirmed by the actual performance, in which blow counts built up to 150–200 blows/ft and then stayed about constant for further penetration. See Fig. 8.4.12.

A pile follower was used, made up of two sections. Gravity connectors were used between the follower sections and between the follower and pile.

To achieve the very high capacities, Menck 8000 and Menck 12500 hammers were employed, the latter developing 1 million ft-lb of energy per blow.

Initial driving, however, was started with the Vulcan 560 and Menck 4600 (300,000 ft-lb/blow) so as to reduce pile-bending stresses.

Pile performance was carefully monitored by means of strain transducers and accelerometers in the pile head. These showed that the losses in gravity connectors were only 2 percent, except in one

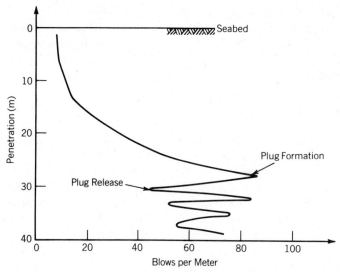

Figure 8.4.12 Typical pile penetration vs. blow count in hard clay.

case of misalignment. Hammer efficiencies ranged from 40 to 62 percent.

8.4.4. Maui A platform, New Zealand

At the site of the Maui A platform, clayey soils were interbedded with a dense layer of volcanic ash. Some piles were able to penetrate this layer; others hung up at refusal. The problem was aggravated by hammer breakdowns and weather delays, leading to high setup. Eventually jetting was employed to break up and remove the plug of ash which had transformed the piles into end-bearing piles.

A similar problem with volcanic ash occurred on a land piling project, where long steel pipe piles were being driven. Here also it became necessary to remove the ash plug, in this case by drilling out the pile plug.

8.4.5 Offshore Terminals, Cook Inlet, Alaska

Like many sub-Arctic and Arctic areas, the recent sediments of Cook Inlet are underlain by glacial till and overconsolidated glacially derived silts. The glacial till is similar to the boulder clay of the North Sea in being very stiff and containing rock fragments of various sizes. The overconsolidated silt is extremely difficult to penetrate under hammer impact alone; the material is so dense and its structure so strong that it cannot displace.

Large cylinder piles from 2–4 m in diameter have been installed in the several Cook Inlet terminals by taking advantage of the rapid breakdown of structure which occurs when water is introduced by jetting. A ring of jets has been built into the pile tip, arranged so that the jets can be continuously operated while the hammer works. This jetting then breaks up the overconsolidated silt into a colloidal suspension. See Figure 8.4.13.

Similar benefit to pile installation has been reported from the East Dock at Prudhoe bay, where jetting enabled ready penetration of 25-m-long steel sheet piles through overconsolidated silt and partially ice-bonded sands.

8.5 Methods of Increasing Penetration

Methods of increasing the penetration include the following:

1. Use a heavier walled pile section; that is, increase the minimum wall thickness. The wave equation indicates a substantial improvement in penetrating ability when a thicker minimum wall is employed.
2. Use a larger hammer, especially one with a heavier ram.
3. Jetting internally, in order to break up the pile plug. In many soils, a plug will form by compaction in the pile tip and transmit bearing

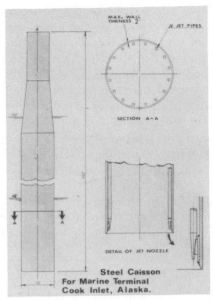

Figure 8.4.13 Details of jet arrangement for 4-m-diameter cylinder piles for Collier Chemical Co. Terminal, Cook Inlet, Alaska.

through skin friction on the inside of the pipe pile. This plug may be broken up by jetting and thus eliminate the end-bearing resistance temporarily. However, the plug may reform in the next 5–10 m or so of penetration, requiring repeated operations. The removal of the hammer and insertion of a jet, sometimes 100 m or more in length, is a time-consuming rigging operation. Quick connectors similar to drill casing can be used to connect the jet pipes so as to minimize the time required.

In some offshore terminal construction, jets have been built into the pile, attached to the inside wall, and connected to a jet water supply by sleeves through the pile wall, below the head. These enable the continuous breaking up of the plug without interruption of driving. See Figure 8.4.13.

Such jets are best permanently attached to the pile and abandoned in place. Attempts to remove them are very time-consuming, usually costing more than the value of the jets. More serious is the fact that loose-fitting sleeves, designed to facilitate later removal, may not adequately support the jets during the combined driving and jetting operation.

Multiple jets are usually installed, spaced at about 1 m or less around the inside periphery. The nozzles should terminate 150 mm or so above the tip.

8.5 Methods of Increasing Penetration

If so installed, the jets must be run continuously during driving in order to prevent plugging.

Typical jet characteristics are:

Diameter	40 mm
Pressure	2 MPa (280 psi) (at pump)
Volume	700 liters/min (175gpm) per jet pipe

When jets prove to be inadequate, then the plug may require removal by drilling. Such a procedure is typically required when driving into hard clay, sandy clay, or weathered rock or through interbedded limestone or sandstone.

In inshore and harbor projects, jets are often run ahead of the pile, as free jets, or alongside the pile, in order to reduce skin friction. Such practice is generally not followed in offshore work because of the tendency to destroy the bearing and lateral capacity of the soil around the pile. Further a free jet is extremely difficult to control as to direction and position when working at depth. However, in cohesionless material such as sands, a free jet may be used ahead of the pile, since the subsequent driving will reconsolidate the sands. Piling should be driven a specified number of blows (say 100–200) after all jetting has ceased in order to insure reconsolidation. Such a free jet "probe" is also useful in determining the character of the soils ahead of the pile tip and in identifying hard layers, for example, coral and limestone in the Mideast and tropical regions.

Standard API RP2A describes the removal of the soil plug inside the pile by jetting and air lifting. The use of a drilling sub or swivel is an even more efficient way of removing compacted material, such as decomposed granite, from the plug. In the rare cases, sometimes encountered in offshore terminal construction, that the pile is plugged with gravel or cobbles, then an airlift-plus-jet combination may be used to clean out the pile.

When the soil plug in the pile is removed to facilitate continued driving and deeper penetration, then it must usually be replaced by a grout or concrete plug, which is placed underwater by tremie means or pumping. The concrete or grout plug must have sufficient length to develop its full capacity by bonding to the inside of the pile.

Excessive heat may be developed in large-diameter (1-m or more) piles with straight cement grouting; this may then result in cracking as the grout cools. Thus the use of concrete, with its lower cement content and hence lower heat as well as higher tensile strength may be indicated. Inclusion of polypropelene fibers in a sand–cement mortar mix will restrain crack development and inhibit crack propagation. Use of a blast furnace slag–cement will reduce the heat of hydration.

When piles encounter hard layers or boulders, it may be necessary to drill them out ahead of the tip. Usually the internal plug is removed by jetting and airlifted down to a meter or so above the tip. Then the pile is reseated with the hammer to prevent a run-in of sand under the tip. The water in the pile should be balanced with the water outside; flow in either direction is undesirable. Then the drill drills out the boulder.

A churn drill will "wander" slightly and hence may drill a hole somewhat larger than the drill diameter, enabling the pile to be driven through. A rotary drill will drill a more regular hole. In very hard rock, it may be necessary to use underreamers in order for the pile to be driven. In the more normal cases, especially where the hard layer is of limited thickness, it may be possible to progressively chip through the sides of the drilled hole by driving after drilling. If such layers are anticipated, the pile tip should be reinforced. Normally, where drilling through the pile is anticipated, the inside should be flush all the way (same inner diameter), which means that a heavier-walled tip or cast steel tip protector will increase the outer diameter. As noted earlier, this may be beneficial in making it easier for the rest of the pile to drive through.

If drilling ahead is required but the drill will not bite on the gravel, cobbles, or sloping rock, one solution is to inject grout below the pile tip so as to form a plug in the base, which after hardening serves as a starter for the drill.

With large-diameter piles of relatively short length such as are often used in offshore terminals, the hard layers may be broken up by chiseling or by drilling several small holes. A chisel can be made from a piece of shafting or from heavy-walled pipe filled with concrete. It may be most effective if driven with the hammer. Care should be taken not to drive too far ahead or the chisel may become stuck.

Drilling and blasting ahead of the pile tip so as to break up coral and limestone layers have been employed. The author's experience and observations indicate that this often damages the pile tip (splitting or curling it), and thus the practice is not recommended.

In hard clays, boulder clays, and glacial till, it may be necessary to drill ahead in order to obtain the required penetration. Generally speaking, the diameter of the drilled hole should be somewhat less than the pile diameter in order to preserve the consolidated soil pressure against the side of the piles to as great an extent as possible. Drilling ahead may be done with salt water or with drilling mud to keep the hole open. Saltwater is easier and cheaper and does not coat the walls of the hole. In some cases it may be possible to keep the water level higher inside than out and thus help to hold the sides of the drilled hole open.

Usually, the length of the hole which may be drilled ahead is limited in order to prevent caving. This means that drilling and driving must be carried out in an iterative manner, repeating one after the other.

API standard RP2A notes that when a hole is formed by drilling an undersized hole ahead of the tip, the effect on pile capacity is unpredictable unless there has been previous experience in the same soils under similar conditions. Jet drilling ahead of the pile tip, while often effective in aiding penetration, is especially unpredictable as to its effects on eventual pile capacity in cohesive soils.

8.6 Insert Piles

When it proves impossible to drive the primary pile to the required tip elevation, another solution is to drive an insert pile. The soil plug is first removed from the primary pile, and then the insert pile is placed and driven ahead. This pile, of smaller diameter, will be free from skin friction over the length of the primary pile and hence can usually be driven to a substantial additional penetration.

When insert piles are preplanned, they usually give good performance, although representing a substantial increase in material cost. When used as an emergency measure, several problems arise. One is due to the fact that the thicker wall section of the pile will be up in the jacket instead of at the mudlines; hence the pile's moment resisting capacity may be less than designed. Grouting in of the insert pile through this zone helps to restore the lost moment capacity.

Another disadvantage is the decreased friction area and end-bearing area per meter of pile penetration. Hence the insert pile may have to penetrate more deeply.

A third problem is the carrying out of the grouting operation to fill the annulus between the insert pile and the primary pile. Since this is usually not a planned operation, it may be necessary to insert a small-diameter (20- to 40-mm) pipe between the two walls in order to grout from the bottom up. Alternatively, a grout pipe may be installed inside the insert pile, with its exit nipple just above the estimated location where the insert pile projects from the tip of the primary pile.

8.7 Anchoring into Rock or Hardpan

With offshore terminals, piles frequently end up on hardpan or rock without having developed sufficient uplift capacity. Several solutions are practicable. One is to drill ahead into the rock a long socket of slightly smaller diameter than the inside diameter of the pile. An insert pile, either a pipe pile or H-pile or cross (+), made up of plates, is inserted in the socket hole with grout pipes attached. It is then grouted up, bonding the insert pile to the walls of the socketed hole and to the primary pile above.

A second method has been to fill the primary pile with concrete, having previously set in a small-diameter tubing (say 150-mm diameter). A hole is drilled through the preformed hole beyond the pile tip, deep into the rock, and a posttensioning tendon is inserted with grout pipe attached. The tendon is grouted to the rock. Then the bar is stressed and anchored. Finally the tendon is grouted to provide corrosion protection and permanent anchorage. While drilling costs are reduced, this method does require care in controlling the levels between first- and second-stage grouting; this point should preferably be below the pile tip. The tendon is in effect a ground anchor used to tie down the pile.

Another solution is to drill the socket, set in a reinforcing cage and then grout or concrete it to the socket and primary pile. The bars of the cage should be well spaced to give ample room for the concrete or grout to flow through them. Vertical bars may be bundled in groups of two, three, or four. Spiral reinforcement should be spaced out; if necessary, it can be made of heavier-gauge steel or bundled also. Spacings and clearances should be at least four times the size of the coarse aggregate.

A modification of the above is to set in a precast plug which contains all the reinforcement and a grout pipe. The walls of the precast plug should be roughened to ensure good bond. Then the annular space between plug and socket and plug and pile is grouted, all in one operation. This method reduces the quantity of grout required and is a very practicable solution.

In a number of cases where sockets are to be used, it may be most practicable first to fill the socket with a fluid grout or concrete mix and then to set in the insert pile, driving it if necessary. In this case, one can be sure that the grout completely fills the annular space.

Attempts to grout between the walls of a pre-driven insert pile and a rock socket through drilled or shot holes in the pile have generally not been too successful, due to inconsistencies in filling. The grout tends to channel, trapping water on one or more sides. Use of greater pressure may fracture the formation.

8.8 Prestressed Concrete Piles

There is an increasing trend toward using large-diameter prestressed concrete piling for offshore terminals and platforms. While these have had a generally excellent record in such places as Lake Maracaibo and the shallow waters of the Gulf of Mexico and the Arabian Gulf for offshore structures, their application has been limited by self-weight and bending moment capacity. However, concrete strengths continue to improve year by year, and recent studies and tests show that the ultimate curvature of prestressed piles, and hence the overload capacity, can be greatly enhanced by confinement, such as by closely spaced spiral.

Driving stresses with large hammers are extremely high. With prestressed concrete piles, the limiting factor may be the rebound tensile stress in soft driving. The probable stresses can be predetermined by use of the wave equation, which will often point up weak points and indicate the need for extra cushion blocks for prestressed piles.

Because of the rather high incidence of rebound tensile cracking of prestressed piles, the basic driving rules are restated here:

1. Use a softwood cushion block between the driving head and the head of the pile. Preferably use green, rough, laminated lumber, of 1- to 2-in. thickness per lamination, total thickness 8–14 in. Use a new block each pile even if there is no apparent damage or crushing of

a previous block. Do not use plywood or hardwood.
2. Use a large ram with small stroke so as to reduce the velocity of impact.
3. Avoid prejetting a void below the tip of the pile. If prejetting is used, minimize the time the jet is held at any one elevation.
4. If a firm surface layer overlies a very soft lower stratum, predrill through the firm surface layer.

Of these precautions, item 1 will solve the large majority of problems.

8.9 Handling and Positioning of Piles for Offshore Terminals

The handling and positioning of offshore piles involves special problems because of the length, the lack of fixed reference points, and the continuous movement of the sea. Therefore, some form of template becomes necessary.

The typical offshore terminal employs steel tubular piles, 0.6–1.5 m in diameter, with 1.0 m diameter most common. Wall thicknesses are 20 to 40 mm. Lengths are of the order of 40–60 m.

Use of a permanent template, frame, or jacket gives the opportunity to make all intersecting welds under optimum conditions. Offshore structures are subject to cyclic dynamic loading and hence, cumulative fatigue. Joints which are field-welded are seldom successful, due to the vibration during welding, the difficulty in keeping the metal dry, and the sudden cooling of the weld with seawater splash. Such joints are also subjected to corrosion-accelerated fatigue.

A template can be set on the seafloor, similar to a jacket, as described in Chapter 9 on the installation of steel platforms, or it can be suspended from an offshore barge or jack-up, or it can be cantilevered out from the previously constructed portion of the structure. In a few cases, self-floating templates have been used. Once one or more vertical piles have been driven, the support of the template may be transferred to these piles.

When piles are installed on a batter through templates near the sea surface, they are of course cantilevered through the water column until they are finally lowered far enough to engage the seafloor and obtain support at their tip. This situation frequently occurs in the construction of offshore terminals where inclined batter piles are used to provide the lateral resistance for mooring and breasting dolphins.

The deflections may be significant and result in significant residual stresses in the piles. Various means of minimizing the deflections have been employed, including temporary evacuation of the water from the pile so as to provide near-neutral buoyancy. Neoprene or frangible caps such as lucite can be installed on the pile tip, or the top of the pile may be capped and compressed air used to exclude water from entry at the tip. Removable floats can be attached to the tip, but these are awkward to install and are likely to break free.

The above systems were all used on the VLCC terminal at Ise Bay, near Nagoya, Japan, because of the deep water (25–30 m).

Another way in which undesirable residual stresses are built into offshore piles is by re-leveling a jacket after some pin piles have been allowed to run in. The bending stresses so imposed may be significant. As noted earlier, good practice calls for initial setting of the jacket as level as possible, driving a few (three or four) pin piles just far enough to permit releveling and temporary fixing. Then after other piles are driven, the first piles are lifted back up into the jacket and then once again lowered and driven. Of course, if the initial setting of the jacket was level and needed no correction later, then the initial piles can be directly driven on down.

Large-diameter cylinder piles are often used for the mooring and breasting dolphins of offshore terminals. They are often delivered by self-flotation, with diaphragm closures. Ballasting may be used to help upend them. See Figure 8.9.1.

Because of their large size, the effect of waves and currents on them may be significant. In Cook Inlet with high-tidal-current velocities, vortex shedding caused oscillating transverse forces to act on the pile. This was in addition to the large direct force of the current. Guiding frames were required, cantilevered out from the derrick barge, with hydraulic positioning devices, in order to enable the piles to be set vertically in their correct position.

Installation methods have included driving accompanied by jetting and drilling as described in the preceding sections as well as special installation methods discussed in Section 8.12.

Figure 8.9.1 Setting a 4-m-diameter steel cylinder pile prior to driving. Collier Chemical Co. Terminal, Cook Inlet, Alaska.

8.10 Drilled and Grouted Piles

Installation of drilled and grouted piles may be carried out in either one or two stages. The piles are to be placed within drilled holes, which are held open temporarily either by seawater or drilling mud. A casing must first be placed through the water and seated into the soils so as to prevent piping due to the imbalance in fluid heads during drilling. Usually this is accomplished by just driving the casing into the overlying soils to a moderate penetration. Then a hole is drilled ahead using either direct or reverse circulation. See Figure 8.10.1.

Since the piles and hence the casings for offshore piles are usually of relatively large diameter (e.g., 1–2 m), normal direct circulation will not develop sufficient return velocity of flow to transport the cuttings to the surface. The annulus between casing and drill stem is just too great in area. By building up the drill stem by means of pipe sections, the annulus can be reduced, thus increasing the return flow velocity to satisfactory velocities.

The head inside may be built up above the outside sea level; if carefully done to a precalculated differential head, this may help to keep the hole open. One means of controlling this is to cut windows in the casing at the prescribed elevation above sea level.

The relatively rapid flow of the fluids through the drill bits help to clean the teeth. Conversely, they may erode the walls of the hole below the tip of the casing. Hence reverse circulation is most often employed. In this operation, seawater or drilling mud is pumped into the casing to keep its level at the desired head. The fluid then moves slowly down the casing and through the teeth and then accelerates to a high velocity up through the drill stem. This high velocity can remove cuttings of high density, such as pyrites. The low velocity along the walls of the hole prevents erosion and minimizes caving.

When holes are drilled ahead of the casing, they are in effect cantilevered beyond the tip. Hence they must have as much guidance as possible within the casing. Centralizers or stabilizers can be used on the drill string to keep it centered. These can also be used below the tip of the casing in hard and firm soils and rock. When holes are drilled on a batter, there is a natural tendency for the hole to droop. This droop can be countered to some extent by the stabilizers and by reinforcement of the drill string to increase its bending stiffness.

Drilling muds can be used to keep the holes from caving and are especially useful in sandy soils. The drilling mud has a higher specific gravity and hence overbalances any inward flow of water. It also penetrates the sands and forms a cohesive layer. While this layer helps to keep the hole open, it also reduces skin friction with grout placed later. This may then require flushing of the hole prior to grouting. A thin drilling mud or seawater may be used for flushing. Alternatively, a polymer mud may be used.

It is essential to monitor carefully the fluid weight and head of drilling mud, since a few pounds per cubic foot can amount to a significant head differential at a depth of 1000 ft.

192 *Installation of Offshore Piles*

Figure 8.10.1 Drilling of offshore piling. (Courtesy of Volker–Stevin Offshore.)

Therefore, care must be taken in all drilling operations not to "fracture the formation" by excessive head of drilling fluid. In this case, piping ensues and the drilling fluid is lost to the sea. One method to prevent this is to "spot mud" only, that is, drill with salt water at a slightly higher hydrostatic head than ambient and then "spot" drilling mud in the socket only (not the full casing) to keep it open long enough to place the pile. If excess head is developed, resulting in formation fracture, it may become necessary to grout the hole with a weak grout, such as a sodium silicate foamed grout. After the grout has set, redrilling in the same location can be quite difficult due to the tendency for the bit to wander off line as a result of differential cementation.

After the hole has been drilled, the pile is placed and grouted. It should be centered in the hole by spacers so as to preserve an annulus for grouting.

As an alternative to inserting a pile, the drill string may be used as the pile by fitting expendable cutting tools to the tip so as to avoid the time required for removing the drill and insertion of the pile. This is especially effective for tension piles, because the pile drill stem can be kept to a workable diameter.

In the case of a two-stage drilled and grouted pile, the primary pile is driven to a predetermined tip elevation. This elevation is selected as one at which there is confidence that the pile can be driven and below which there is confidence that the hole can be kept open. This outer pile then becomes the casing for the drilling of the second-stage pile.

When piles are placed in drilled holes (beyond the tip of the casing), the diameter of the drilled hole should be at least 6 in. (150 mm) larger than the outside pile diameter.

As described earlier, drilled and grouted piles using two stages were used very successfully on the Thistle platform in the North Sea. See Figure 8.10.2. The primary piles were driven through the overlying sands and clays having interbedded sand lenses to seal in a dense clay stratum. This work could be carried out expeditiously during the short summer weather window. Then drill rigs were set on top to work during the winter, drilling ahead, placing secondary piles, and grouting the whole so as to act integrally with the jacket.

8.10 Drilled and Grouted Piles

Figure 8.10.2 Installing drilled and grouted piles for Thistle platform, North Sea. (Courtesy of Varco and Santa Fe International.)

Drilled and grouted piles are especially effective in some calcareous sands. The grout penetrates the crushed shells and develops interlock with the unbroken shells behind. This appears to be the only positive method of achieving good skin friction transfer in calcareous sands. However, there are some calcareous sands, notably those on the Northwest Shelf of Australia, which are essentially impermeable and for which grouting just crushes the sand, increasing the effective pile diameter slightly but not increasing the effective friction significantly.

A grouting shoe may be installed near the tip of the pile in order to permit grouting of the annulus without filling the interior of the pile. A packer, of course, serves a similar function. With closed-end piles of large diameter, it is necessary to check that the pile will not be raised (floated) by the pressure of the grout as the fluid grout is placed.

In soils which may soften or slack upon exposure to water, the pile should be placed and the grout injected as soon as practicable after drilling.

The quality of the grout should be verified at frequent intervals during injection.

Holes for adjacent piles, for example, adjacent skirt piles under one leg, should not be open at the same time unless the soil properties are sufficiently strong and consistent to ensure that grout will not migrate during placement into the adjoining hole through fractures or seams and that the soils will not suffer relaxation.

In some cases, it becomes practicable to drill ahead of the primary pile and then lower it to the full penetration and grout it to the soil. This saves the cost of the overlap of secondary pile into primary pile. It allows the primary pile to be used as the casing, as before. In this case, an underreamer is used, with which the bit expands after it emerges below the tip of the pile and so cuts an enlarged hole.

This system was used on the underwater storage tanks at Khazzan Dubai, in the Arabian Gulf, in which the drilled portion of the holes was in a limestone stratum. It has also been used for a number of major bridges with large-diameter piles of steel and prestressed concrete—for example, the Saudi Arabia–Bahrain bridges and the Ohnaruto Straits and Yokohama harbor bridges.

When large-diameter holes are to be drilled ahead, use of a two-stage bit, such as a pilot bit of $18\frac{1}{2}$-in. diameter followed by a hole opener of 26-in. diameter located 2 or 3 m above the first, may be most efficient.

There are several types of drills available for use, depending on the size and depth of hole and the available supports for the drill. For emergency or limited use, such as arises when a single pile encounters a boulder, a churn drill may be used, suspended from the crane boom if necessary. Another type of drill, a rotary drill, is supported on the pile or casing itself, and gets its torque reaction from being clamped to the pile. Drilling subs and drilling swivels are extremely versatile and easily handled drilling machines. They are suspended from the derrick boom. They get their torque reaction from chains attached to an arm. This arm and the chains are subjected to very severe shock loads and hence must be properly designed and secured for impact. This type of drill is very flexible in use, since from one location of the derrick, several piles can be drilled.

For major pile installations the use of a drilling rig with inclining mast appears to be most applicable, since it can attain faster drilling rates and can make and unmake long drill strings more rapidly. Such a drill rig is usually skid-mounted and is set on a temporary work deck mounted on the jacket. See Figure 8.10.3. These drills use conventional rotary drill bits. They may be operated either in direct circulation or reverse circulation.

Figure 8.10.3 Drilling in of piling for offshore terminal, Ridley Island Terminal, Prince Rupert, British Columbia. (Courtesy of Reidel International.)

An excellent description of drilling and grouting of piles for offshore structures is given in the chapter "Grouted Piles" by Paul Richardson, in the book "Planning and Design of Fixed Offshore Platforms", edited by McClelland, B. and Reifel, M.D., (Van Nostrand-Reinhold, N.Y. 1986).

8.11 Belled Footings

One important development in pin pile installation of offshore platforms has been the bell footing, first used on the Ekofisk Field platforms and since extended to other structures in the Arabian Gulf and elsewhere. In this case, the primary pile serves as a casing through which a drill rig drills a moderate-length hole ahead. Then it employs a belling tool to enlarge the socket into a bell of 4- to 5-m diameter. Reverse circulation is employed, usually with a bentonite slurry (drilling mud) as the drilling fluid. Then a heavy reinforcing cage or steel insert pile is set. The bell and socket and a portion of the casing are filled with underwater concrete, using "fine concrete" aggregates, for example, maximum size 9 mm. As with straight drilled shafts (sockets), salt water may be used as the drilling fluid and the bell "spot-mudded" with polymer mud to hold it open until concreting. See Figure 8.11.1.

API Standard RP2A describes belled piles as they are used to give increased bearing and uplift capacity through direct bearing on the soil. A pilot hole is first drilled below the base of the driven pile to the elevation of the bell base and slightly below so as to act as a sump for unrecoverable cuttings. Then the bell is drilled, using an expander tool. Reverse circulation must be employed in order to gain enough discharge velocity to remove the cuttings.

Then an insert "pile" is run down; this may be a tubular or structural member or reinforcing steel bars assembled in a cage. The bell and pile are then filled with concrete to a height sufficient to accomplish load transfer between the bell and the pile.

Lugs on the insert pile or the deformations of the reinforcement are used to gain high shear transfer in the relatively short height of the bell.

Prestressing tendons may be used for connecting bells to piles which are designed to act only in tension.

Reinforcing steel bars are usually bundled so as to permit the concrete to flow between them and out into the bell. The reinforcing bars are enclosed

8.11 Belled Footings

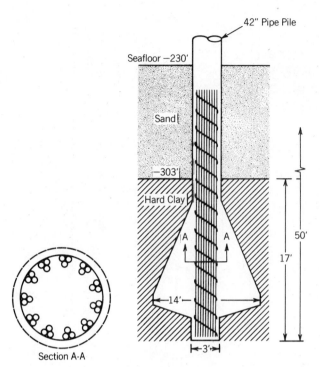

Figure 8.11.1 Underreamed bell footing for Ekofisk platforms.

in spiral; the spirals may also be bundled to facilitate flow.

In the typical installation of offshore belled footings, the primary piles are driven down to seal in the bearing stratum. The holes are then drilled down and belled, reinforcing cages are inserted, and the whole is filled with concrete.

It is not possible to confine the bell concrete with hoop reinforcing. There is no way the hoop steel can be placed. If the bell has been drilled into rock, the rock may confine it; however, bells at Ekofisk were used in soils of relatively low stiffness compared to the concrete. Hence there will be flexural and shear stresses in the concrete under service and extreme loading conditions. This means that the shear and tensile strength of the concrete must be utilized.

Such a confined mass of concrete will get very hot during the period immediately following placement, due to heat of hydration. Subsequent cooling, starting at the outside edge, may produce severe cracking. Cracking may also be caused during the expansion stage. Hence the cement selected should be a low-heat-type cement, such as ASTM Type IV or Type II with pozzolanic replacement of up to 50 percent of the cement. Alternatively, coarse ground slag–portland cement in a 70:30 mix may be used; it has very low heat generation.

The mix should be as cool as practicable at time of placement; aggregates, for example, can be sprayed with water to cool them or liquid nitrogen injected into the mix.

Drilling contractors prefer to use a cement slurry, since they are familiar with it, it is relative simple to handle, and it can be directly pumped. Cement contents will normally be high; hence heat of hydration is potentially even more of a problem. Use of blast furnace slag–cement (93:7) is indicated. However, the tensile strength and shear strengths of neat cement slurry (grout) are relatively low.

Steel fibers have been proposed as one means of enhancing the tensile strength. However, they tend to segregate and bunch and may plug the placement pipe. Even with steel fibers, cement grout tends to have a brittle mode of fracture, with rapid propagation of a few large cracks.

The use of polypropelene fibers gives more satisfactory behavior in all the above regards.

The author prefers a mix which incorporates sand—that is, a sand–cement mortar—or, even better, one incorporating small aggregate (8–10 mm) (3/8 in.) such as pea gravel. This latter has relatively low heat properties and good tensile strength, yet is still able to flow readily.

Plasticizing admixtures should be used. High-range water-reducing admixtures ("superplasticizers") should not be used because of their tendency to rapid and erratic slump loss.

Placement of concrete at the typical depths of 100–200 m below sea level or even deeper requires careful balancing of fluid head due to seawater, drilling mud, and fresh concrete. Friction losses are normally very small.

Placement through a tremie, under gravity head only, is the most reliable means, based on experience at comparable depths with cutoff walls below dams. In this case the concrete can be pumped from the mixer to a hopper at the top of the placement tremie pipe. The tremie pipe will preferably be drill casing of diameter about 10 times the diameter of the maximum size of coarse aggregate used. The concrete surface will normally be about halfway down the pipe, due to the lack of friction and the balance of heads referred to earlier. This level is self-regulating, except in the case of a blockage; thus it needs to be sounded only occasionally just to ensure that there is no blockage.

The mix, having been designed to give maximum cohesiveness, will only partially segregate during its free fall and will remix upon impact.

This, however, is the reason for using a large enough diameter; it is to permit remixing to take place.

Use of valves or restrictions at the tip of the tremie, while capable of functioning in a laboratory, has not proven reliable in service; their use often leads to plugging.

To start the placement, the tip of the casing is closed with a plate wired on with rubber gaskets. The casing is then assembled and seated at the tip of the bell. Then a charge of the fine concrete is added. The casing is now raised a few inches off the bottom, allowing the plate seal to break free and the concrete to flow out as slowly as it can be controlled, until it reaches a balance of heads. Then concrete is continuously fed in at the top.

At greater depths, a polyurethane pig may be employed, being pushed down the casing by the initial flow of concrete. See section 6.6.

The annulus between insert pile (or reinforcing cage) and primary pile must be 6–10 times the maximum size of coarse aggregate so as to permit flow up around the insert pile.

Electrical resistivity or well-logging devices can be used to determine the level of the top surface as it rises up in the primary pile. Obviously, the placement should extend several meters above the theoretical level so as to account for any mixing of the initially placed grout with the drilling mud.

The drilling mud should preferably be converted to a calcium base so as to avoid coagulation upon contact by cement.

Concrete has also been placed satisfactorily by pumping down into the bells, provided that fine, rounded aggregate is used with an adequate sand content (at least 50 percent). The flow rate can be controlled by the friction in the pipe. For example, a pipe 3 in. (75 mm) in diameter may restrict the flow rate to 3–4 m/s for concrete containing $\frac{3}{8}$-in. (10-mm) maximum-size coarse aggregate (pea gravel). Tests should be run with the actual mix to determine the friction developed and the pipe sized accordingly.

To prevent a vacuum forming in the down-leading pipe, a relief valve should be installed at the top.

8.12 Other Installation Methods and Practices

Very-large-diameter piles, 3 to 4 m in diameter, have been sunk by a combination of weighting, vi-

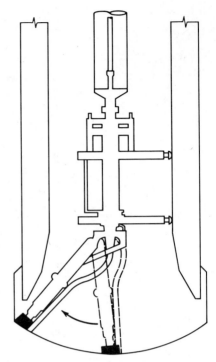

Figure 8.12.1 Mitsui "Aqua-Header" special drilling system for large-diameter cylinder piles.

bration, hammering, and internal excavation. Thus they resemble an open caisson of one cell.

The piles for the Oosterschelde Bridge, 4 m in diameter and 60 m in length, were installed in this manner. In addition to the self-weight of the pile (about 600 tons), another 600 tons was applied by attachment to and hoisting up of the derrick barge through a very ingenious multipart tackle. Excavation was by means of an articulated cutter suction dredge arm suspended vertically from the crane barge boom. A similar means for installation was used to construct the foundations for the new Yokohama Harbor bridge. See Figure 8.12.1.

Weighting is, of course, often used in seating the pile legs of jack-up barges by applying a major portion of the barge weight in succession to the individual piles.

8.13 Improving the Capacity of Offshore Piles

Once piles are installed, it is necessary to evaluate their load-bearing capacity to ensure that the required capacity has been attained. Sometimes the skin friction may be deficient; this is frequently encountered in calcareous soils. Another situation that may arise is where an existing platform is to be upgraded to withstand greater environmental or operating loads.

One method is to clean out the internal soils to a safe distance above the tip (usually several meters) and then to construct a concrete plug inside the pile tip, of adequate length to develop bond transfer to the pile. This converts the pile to a partially end-bearing pile. This method was successfully employed for the previously driven piles in the Kingfish platforms A and B in Bass Straits, Australia, which were driven into calcareous sands.

When inadequate capacity is encountered in the initial driving and the pile fails to develop adequate resistance at the design penetration, then appropriate steps taken promptly may enable the construction of an adequate foundation with minimal additional cost. For example, on one offshore terminal in the Mediterranean, the 2-m-diameter steel piles failed to develop the required resistance with 60-m penetration. On the first pile it was found that the penetration would have to be increased by more than double, at a prohibitive cost. By welding on a tip closure plate which closed 80 percent of the tip, leaving a small central hole for water escape, the piles developed adequate static capacity in the calcereous sands and safely sustained a test load of 4000 tons.

When increased uplift is required, two methods are possible. One is to drill in an insert pile. In one case the drill string itself was used as the insert pile and grouted into place.

Another solution is to weight the pile, similar to placing weights on the legs of a table. The pile is cleaned out, a concrete plug is placed, and iron sand or barites are placed. These should be carefully selected for in-place density, durability, and freedom from corroding effects on the steel. In-place densities of 3.5–4.0 have been achieved with iron sands.

A third solution, applicable in stratified soils where both compression and tension capacities must be increased, is to construct a belled footing, as described above.

Concepts have also been studied for perforating the pile wall with explosive charges and then injecting grout into the surrounding soil. However, they have not been adopted because of concern as to the ability to determine where the grout would actually travel, especially if hydraulic fracturing occurred.

To increase the capacity of the piles of the North Rankin A Platform, it is planned to construct belled footings below the tips of the existing piles.

Insert piles may be driven through existing piling, being made up in short segments, welded as they are installed. They are driven with a follower using mechanical threaded connections for the follower segments. After driving, the insert piles are connected to the primary piles by grouting of the annulus. Alternatively, the insert piles may be installed by drilling and grouting.

Lateral resistance of existing piles may be increased by installation of an insert pile, grouted or concreted in, so as to stiffen the pile in the vicinity of the seafloor. The seafloor sediments themselves may be in some cases strengthened by vibratory densification or pressure grouting, or they may be surcharged with gravel, either normal-density or high-density, so as to consolidate the existing soil, replace any settlements, and fill any gaps that occur under cyclic wave action.

Freezing of the soil around the pile and under the pile tip is another method which has been proposed as a means of increasing the capacity of a previously driven pile, especially in the Arctic. It has not actually been applied, to the author's knowledge, for offshore piles. The concept is that, after driving, the steel pile would be cleaned out by jetting, airlifting, or drilling and then used as the freeze pipe casing. Among the matters which must be considered here are the following:

1. The behavior of saline soils when frozen, the formation of brine lenses and pockets. While most soils show dramatic increase in strength, this is apparently not true of all carbonate soils. Some others may develop weak planes.
2. Required temperature to achieve solid freezing. Due to salinity, this may be $-7°C$ to $-10°C$.
3. Adfreeze from frozen soil to pile.
4. Load transfer from stiff pile to elastoplastic frozen soil.
5. Frost heave.
6. Creep of frozen soil under sustained load.
7. Load transfer properties from strongly-frozen to weakly-frozen to unfrozen soil.
8. Sensitivity of steel pile and especially the welded joints to low temperature. Resistance to brittle failure under cyclic tensile loads, rapidly applied.
9. Rate of warmup (thaw) in case of system failure, especially since the pile acts as a heat conductor.

Lateral resistance of existing piles can be increased by placement of an insert pile and grouting the annulus so as to ensure combined behavior. This was successfully carried out for a platform off Bombay High, India, where the pile was showing excessive deflection under cyclic loads. The soils were calcareous sands.

Alternatively, the seafloor soils themselves may be strengthened by drainage, vibratory densification, or pressure grouting. They may be surcharged by small gravel, either normal-density or high-density such as iron ore, so as to consolidate the existing soil, replace any settlements, and fill any gaps which develop under cyclic loading.

> *And now the storm-blast came, and he*
> *Was tyrannous and strong.*
> *He struck with his o'ertaking wings*
> *And chased us south along.*
>
> SAMUEL TAYLOR COLERIDGE, "THE RIME OF THE ANCIENT MARINER"

9

Offshore Platforms: Steel Jackets and Piles

9.1 General

This chapter addresses the typical offshore platform, originated in the Gulf of Mexico and now spread world-wide. Its range extends from water depths of 40 feet (12 m) to over 1000 feet (300 m) and from relatively benign climates in Southeast Asia to the North Sea. Approximately 3000 such platforms have been constructed. Jackets, the main component of the system, range in weight from a few hundred tons to 35,000 tons and more.

Construction is now underway for steel jacket-pile structures in depths up to 500 meters, in the Green Canyon area of the Gulf of Mexico.

The principal structural components are the jacket, the piles, and the deck. See Figure 9.1.1 and 9.1.2. The concept is very simple: the jacket is prefabricated on shore as a space frame, then it is transported to the site and seated on the sea floor. The piles are then driven through sleeves in the jacket, and connected to the sleeves. The deck is now set.

Jackets are also employed for offshore terminal construction, especially for the loading platform and breasting dolphins.

The typical offshore drilling and production platform does not exist for its own sake but rather is thought of as a necessary but expensive support for the primary functions which are the reason for the project. These functions are to drill wells, produce oil and gas, process it as necessary, and discharge it to pipelines to shore or a loading terminal.

From the platform, conductors are installed, held by conductor guides bracketed out from the jacket. On the deck, derrick and drilling modules are installed, so that the wells can be drilled. Processing modules are installed on the deck, and all the necessary support modules for accommodations, power and water generation, sewage disposal, communication, and heliport. Cranes are installed to handle drill collars and casing, and all consumables from barges or supply boats to the deck. On the deck are stored drilling mud, cement, fresh water, and diesel oil.

Other functions, such as reinjection of water or gas, may also be performed from the platform.

An emergency flare stack is provided in order to flare excess gas.

While diesel oil is used initially to fuel operations, produced gas may be used after production and processing are established.

The construction phases of a jacket for an offshore platform include fabrication (described in Chapter 4), loadout, transport, launching, upending and seating, piling (described in Chapter 8), and deck installation, and module erection.

9.2 Loadout and Tie-Down

The jacket, having been fabricated on shore, must be transported to the site. Typically it is skidded onto a launch barge. See Figures 9.2.1 and 9.2.2. The launch barge is usually grounded at the dock, on a prepared and screeded sand pad. Water ballast is placed, sufficient to hold the empty barge on the bottom even at high tide.

For very heavy jackets and where the water is too deep to ground the barge or the tidal change too great, the barge must remain afloat. In this case the barge must be continuously ballasted so as to remain level and at the proper elevation relative to the skidways on shore, while the weight of the jacket is progressively transferred onto the barge.

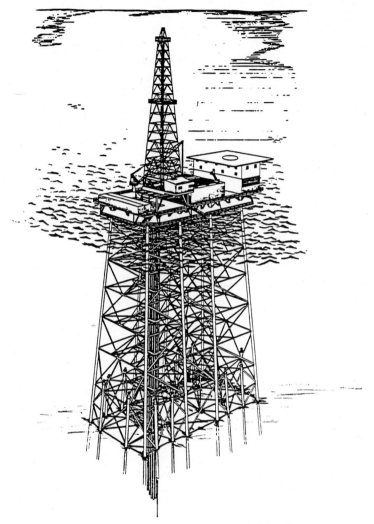

Figure 9.1.1 Steel jacket with pin and skirt piles.

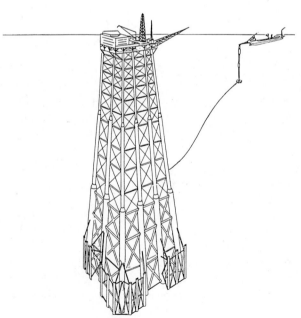

Figure 9.1.2 Steel jacket and clustered skirt pile system.

Figure 9.2.1 Heather jacket, ready for loadout onto grounded barge at Ardesier, Scotland.

9.2 Loadout and Tie-Down

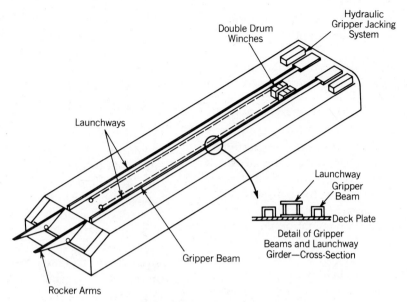

Figure 9.2.2 Launch barge.

API RP2A Sec. 2.4.3b requires that "structures moved horizontally onto the transportation barge by means of ways or wheeled dollies on track supported by cribbing should be checked for the effects of localized loading resulting from the change in slope of the ways or tracks and the change in draft of the transportation barge as the weight of the structure moves onto it. Since movement is normally slow, impact need not be considered."

With the barge positioned end-on to the dock, the jacket is pulled onto the barge, usually by winches on the outboard end of the barge itself, since this automatically holds the barge tight against the dock. See Figure 9.2.3.

Alternatively, the barge may be moored taut against the dock, with lines strong enough to resist the friction force from the jacket. Jacks or screw-rods or on-shore winches are used to pull the jacket out onto the barge.

If spacer struts are used between the on-shore fabrication ways and the launch beams on the barge, these will be placed in heavy compression. They must be heavy enough to take the forces without buckling and supported against "kicking out" sideways.

The barge must be secure against transverse movement induced by current, wind, or the wakes of passing boats. The alignment of the jacket must be accurately maintained while the jacket is pulled out onto the barge. For this reason, it is often pref-

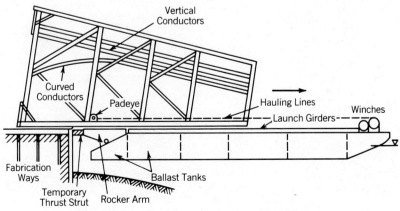

Figure 9.2.3 Load out of jacket onto launch barge.

erable to ground the launch barge before loadout, whenever such factors as water depths make this practicable.

During loadout, the jacket is supported on the fabrication ways, usually on two inner legs of the jacket. These are strengthened by plates so as to act like girders, able to support the jacket weight with some free span between points of contact. These girders are also converted into the bottom chord of a large truss, by using the basic platform bracing, often supplemented by additional diagonals, for example, to enable it to span between points of support, especially when part of the jacket is on the barge and part still on the fabrication ways. See Figures 9.2.4 and 9.2.5.

Initial friction of the jacket on the ways may be as high as 10 percent, especially if the jacket has been erected with its weight bearing on the ways continuously. In many cases, the initial fabrication is carried out slightly elevated above the ways by means of sand jacks. Alternatively, hydraulic jacks are used to permit removal of a filler piece. At time of launching, the jacket is lowered onto the skidways.

To reduce the sliding friction, grease on hardwood, or heavy lubricating oil on steel, or even fiber-filled Teflon-faced pads, are used: values of friction as low as 1 percent are usually attained.

A check list of the operations relating to loadout of jackets follows:

1. Is jacket complete? Has the structure been analyzed for loadout stresses on the basis of the actual structure as fabricated at the time of launch?

2. Are the conductors, both straight and curved, in the same configuration and support condition as has been assumed in the analysis? Conductors, especially curved conductors, are often installed during the onshore fabrication and fixed to the jacket frame, as opposed to the vertical conductors which are often installed offshore, through conductor guides. Since decisions on the number, direction, and time of installation of conductors are often changed during the fabrication process, their support and tributary loads may differ from those used in earlier analyses.

3. Is the launch barge securely moored to the loadout dock, so that it won't move out during the loading? Is the barge properly moored against sideways movement?

4. If compression struts are used between the barge ways and fabrication ways, are they accurately aligned and supported so they won't kick out during launch?

5. Have the pull lines, shackles, and padeyes been inspected to ensure they are properly installed and can't foul during loadout? See Figure 9.2.6.

Figure 9.2.4 Platform Eureka during loadout. (Courtesy of Shell Oil.)

Figure 9.2.5 Load out of platform Eureka. Note temporary buoyancy tanks. (Courtesy of Shell Oil.)

Figure 9.2.6 Multiple-part lines used to pull platform Eureka onto launch barge during loadout operation. (Courtesy of Shell Oil.)

6. Is the barge properly ballasted? If the tide will vary during loadout, are ballasting arrangements made? Will ballast be adjusted as the weight of the jacket goes onto the barge? Are there proper controls?

7. If the ballast correction is to be made iteratively, step-by-step as the jacket is launched, are there clear paint marks so that each stop will be clearly identified?

8. If the loadout is taking place in an active or potentially active waterway, has the Coast Guard been asked to issue a Notice to Mariners to stop all traffic? Has a boat been stationed so as to stop the private power cruiser or tug which may not have received the Notice to Mariners?

9. Are the tugs on station? Are standby tug or tugs available in case of tug breakdown?

10. Has the weather forecast been checked? Squalls are especially dangerous due to their sudden occurrence.

11. Have clear lines of supervision and control been established? Are the voice radio channels checked?

12. Have the marine surveyors been notified so they can be present? Owner's representatives? Verification agent? Have their approvals been received?

Once the jacket is on the barge, the barge must be ballasted for sea. During loadout, many tanks will be partially full only, in order to control deck elevation and trim. Now with the jacket fully supported on the barge, these considerations are no longer active, and the tanks can be ballasted to suit the demands of the sea voyage.

Tanks should normally be either "pressed-up" full or else completely empty, so as to eliminate free surface and sloshing effects.

The draft and freeboard will have been carefully selected so as to maximize stability and especially so as to prevent the out-rigger-like legs of the jacket from dipping into the sea during roll of the barge.

Trim will be adjusted to optimize tow speed and to give directional stability during tow: usually the barge will be trimmed down by the stern.

The above remarks apply when the barge has no restrictions from the loadout dock to the open sea. Many interesting variations arise in inland channels and bays, which have to be dealt with as a special site-specific and jacket-specific operation. Examples follow:

1. Shallow water or a bar may limit draft and necessitate even trim.

2. A narrow channel may require that the overhanging legs of the jacket be high enough to clear dolphins, boat slips, even docks.

3. Fixed bridges may limit the height and necessitate ballasting down to deeper draft and occasionally to severe trim down by the stern, even to the degree of submerging the stern of the barge. Since this reduces transverse stability (the water plane is reduced), this condition has to be checked with extreme care. This procedure was brilliantly executed in connection with the transport of platform Eureka under the Richmond-San Rafael Bridge in San Francisco Bay, with only 1 m clearance below the bridge deck girder.

The tides can be selected so as to give the greatest benefit at these critical stages. Tidal currents must also be taken into account, and adequate reserve tug capacity must be available to abort the tow and pull back if proper conditions are not maintained.

Wind from the side can cause the barge to heel; if the spread of the jacket legs at the bottom is 50 m, then even a 2° wind heel can cause a 1 m increase in height of the jacket leg, or perhaps a half-meter increase in draft. Once past these constraints, then the barge can be ballasted for sea.

The barge, while very stiff, is nevertheless a flexible member. The jacket is typically even stiffer than the barge. Therefore, adjusting of the ballast of the barge should preferably be done prior to tie-down for sea. If one scheme of ballasting was used for the inner channel tow and another will be used for sea, the tie-downs should be freed during the change in ballast so as to prevent imposing bending deformations on the jacket legs.

Tie-downs are therefore installed after loadout and prior to entering the open sea. See Figure 9.2.7. They are major structural systems, subjected to both static and cyclic dynamic loads. Therefore the gravity and inertial forces involved must be calculated for all anticipated barge accelerations and angles of roll and pitch during the design storm adopted for the tow, usually the 10-year return storm for that season of the year and location. Since the loads are dynamic, impact must be minimized

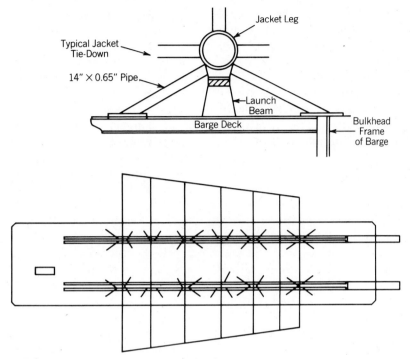

Figure 9.2.7 Tie-down of jacket on barge.

and fatigue in a corrosive environment must be considered. The tie-downs will see approximately 14,000 cycles of fully reversing load for each day at sea.

Inertial forces are due to acceleration in heave, roll, and pitch, and are therefore dependent on the period of response of the barge with the jacket loaded on board. Gravity loads depend on the maximum angle of pitch or roll. Wind loads must also be considered, although they will normally be a much smaller component of the total load.

For typical short-term tows in temperate seas, the following criteria have been used for design of tie-downs:

Single amplitude roll, 20°

Single amplitude pitch, 10°

Roll or pitch period, 10 seconds, double amplitude

Heave force, 0.2 g

The design load will then be the sum of heave plus pitch or heave plus roll.

For long tows, or tows at seasons of the year when storms are likely, special studies should be made as to weather conditions, barge responses, and force combinations imposed. Model tests in both regular and irregular seas are used to measure motions and forces, for various headings of the barge. Typical model scales are 1:50 or larger.

Computer programs have been developed which have been correlated with model tests and which are quite reliable when applied to relatively standard jackets on conventional barges. These give significant values of the responses of the barge for various headings and sea conditions, from which the extreme values can be computed.

If elastic analyses are used, then it is normal practice to use the AISC allowable stresses for the significant value of the responses, without any increase for the short-term duration of the load.

If load factor analysis is employed, then the load factors should be chosen appropriate to the significant or extreme value adopted.

If cold weather will be encountered during the tow, then the seafastenings must be constructed of steels having suitable impact values for the temperatures involved.

Tie-downs are structural members, connecting the jacket to the barge. Therefore the point at which they are connected to the jacket must also be able to resist these forces. Similarly, the structure of the barge to which the tie-downs connect must have proper strength. Usually this means penetrating the deck and making a shear connection to an internal bulkhead or even to the side of the barge. The penetration itself must be sealed so as to prevent water entry. See Figure 9.2.7.

Wire rope is normally not usable for tie-downs; under cyclic loading it stretches and starts to work.

Wedges, even though driven tight, must be welded in place; otherwise they may work free under repeated loads. Therefore, chain is employed where bending flexibility is required. However, in the great majority of cases, fixed structural members such as heavy walled pipe are used.

Punching loads on the barge deck can be spread out by suitable doubler plates, stiffeners, or bearing beams.

Shallow water jackets, such as those employed for offshore terminals, are short and squat. They often may be loaded out and transported vertically rather than on their side. In this case, they may be skidded onto a barge, supported on temporary steel girders under or alongside the jacket legs. Since the weight of such jackets is usually less than 1000 tons, the loadout forces are not excessive. However, because the launch beams on the jacket are temporary, they must be checked for possible eccentricity and also for web buckling; adequate lateral support must be provided.

Once on board, the effects of barge response must be fully checked as to the loadings on the temporary girders, since the loadings will now have lateral components, and the webs of girders will no longer always be vertical.

The third method of loadout is that of the self-floater. See Figure 9.2.8. In this case, the jacket is fabricated in a drydock or shallow basin. The legs on one side are typically made much larger in diameter so as to provide flotation for the entire jacket. Alternatively, extra legs or buoyancy tanks may be provided. Thus for example, if in-service leg diameters of 2 m are required for structural purposes, the legs on one side may be 8 or even 10 m in diameter to provide the necessary buoyancy so the jacket can be self-floating.

This system has been successfully employed on several platforms off the California coast, on the Brent A, Thistle, Ninian South, and Magnus platforms in the North Sea, on the Maui A platform in New Zealand, and also on the Drift River Offshore Terminal in Alaska.

The British Standards Institute Code of Practice, BS 6235, emphasizes it is important to ensure that when self-floating jackets are built in a basin or dry dock, then as the basin is flooded and open to the sea, the structure will not "bottom out" on a subsequent low tide. To assure this, the structure will usually be ballasted to negative buoyancy before the basin or dock has been fully flooded, thus remaining on its supports until the day of float-out. Then it is deballasted on a rising tide and floated out at or near high tide.

Alternatively, temporary buoyancy in the form of a raft of tubular members temporarily affixed to the jacket structure may be provided and later removed after installation of the jacket. This system was successfully employed on the platforms for the BP-Forties Field in the North Sea and the North Rankin A platform in Australia. It has now become almost standard practice to include temporary buoyancy tanks as part of the jacket, for example,

Figure 9.2.8 Self-floating jacket for platform Thistle leaving the Moray Firth in Scotland, en route to installation site. (Courtesy of Santa Fe International.)

as in "Heather" in the North Sea and "Eureka" on the California coast.

The obvious disadvantages of increasing the diameter of the permanent jacket legs are the increased wave and earthquake forces generated in service and in the large number of high stress cycles incurred during tow. The obvious advantages are in eliminating the forces imposed during loadout and launching, as well as the costs of these operations. The requirements for a launch barge and for tie-downs (seafastenings) are eliminated.

Use of temporary buoyancy tanks or rafts eliminates many of the disadvantages of the concept but imposes the new problems of making temporary connections which will not fail due to the dynamic impact forces of transport and installation, nor suffer fatigue in the corrosive salt-water environment, yet still be readily disconnected after the jacket is installed and pinned to the seafloor.

Properly designed, the self-floater tanks can act as the hull of a barge, stiffened by the total jacket framing. Roll response is minimized by the typically wide spread of the legs. The Maui A platform, enroute from Japan to New Zealand, successfully rode out a typhoon with only minor damage.

Self-floaters have also been built on ship-launching ways and launched down into the water, just as is done with a ship. There are of course severe bending stresses induced during launching, since the jacket is partially supported by buoyancy and partly by the ways. Even more serious may be the concentrated bearing forces exerted on the upper end of the jacket and on the ways as the jacket rotates upward at the outer end. Side launching, of course, produces less severe loadings during launch.

A self-floater must be analyzed for the hog-sag and quartering responses to the sea, the first set producing bending and the second, torsion. The large legs on which flotation will take place must be checked for watertightness and internal subdivision provided, just as with a barge.

9.3 Transport

The tow of a launch barge with jacket tied-down or of a self-floating jacket must be planned with great care because of the sensitivity of these awkward structures to sea and wind conditions. Tows have been successfully carried out from Japan to California, Alaska, New Zealand, and Australia, over distances of many thousands of miles. Since towing speeds are inherently slow, a delivery voyage may extend up to 30 days or even more. This makes it probable that at least one major summer storm will be encountered enroute. See Figure 9.3.1.

Tow routes are selected to minimize adverse weather and storms, to avoid complex channels if feasible to do so, and to take advantage of favorable currents. Often two tugs will be employed, as insurance against breakdown and to provide better control of the tow in close quarters. In any event, having one additional boat in attendance while at sea appears wise. The tow route is also planned so as to have good sea room to lee in event a major storm is encountered.

The advice of a competent weather forecasting–ocean routing service should normally be obtained so as to be able to select the most favorable routing for the tow, first from the point of view of safety, i.e., storm avoidance, and second from the point of view of minimum time of transit. Such a service will usually provide daily advice to the towmaster both as to weather to be expected over the subsequent 72 hours and recommended changes, if any, in routing. In turn the tow master advises the ocean routing service daily as to his position, speed, course, weather, and sea state.

Tow boat size is determined partly by its horsepower and concomitant bollard pull and partly by the hull form of the boat itself. For long tows in the open sea, the boat should have substantial length and draft, whereas for working through crowded channels or in close proximity to other structures, a short boat, with minimum draft, often equipped with a bow thruster, will be found best.

For the long open sea tow, the boats should be able to continue along the tow route at about 2–3

Figure 9.3.1 Platform Eureka under tow to Southern California. (Courtesy of Shell Oil Co.)

knots, even when heading into a 15-foot sea with a 40-knot wind and 1-knot current.

Under severe storms, the boat(s) should be able to keep the heading even though the forward speed may be zero or negative. Under some circumstances, the tow boat may cast the tow free, allowing it to drift to leeward while the boat rides out the storm. When the storm abates, the tug picks up the tow again. This is one reason for having a trailing nylon or similar rope which can be picked up by a tug in moderate seas and used to haul the spare bridle and pennant off to make up with the tow line.

In narrow channels, where tidal currents may be adverse and tricky, or in the vicinity of islands or land from which gusty winds may strike, larger boats and sometimes more boats are required.

When the BP-Forties platforms were towed out of the Moray Firth of northeast Scotland, two pulling tugs were supplemented by two stern tugs, being pulled astern through the water, arranged so they could act to slow or turn the self-floating jackets. Such tugs must be selected, or modified, to have high enough stern sheets that they will not take in water over the sterns. The engine room hatch must be battened down securely and provision made to drain the stern well in the event water is shipped.

Stability against capsizing is a major design consideration for a jacket on a barge, primarily because of the high center of gravity of the jacket, but also because of the possible sudden wave slam impact if the overhanging leg of a jacket is engulfed by the crest of a wave. See Figure 9.3.2.

Metacentric height is a valid measure of stability only for small angles of list; it can be considered a first approximation for purposes of determining accelerations and stability. For survival in a storm, for assurance of stability against capsizing, righting moment versus angle of heel charts must be prepared. From computations based on the center of gravity of barge, ballast, and jacket, corrected for the free surface effect of any partially filled compartments, a righting moment curve can be drawn, representing the resistance to overturning.

The angle of downflooding is that at which water will enter the vents in the ballast tanks. To provide extra safety against such an event, if approved by the Marine Surveyor, the vents should be capped after the ballasting has been completed for the sea voyage. A word of warning: the vents must be uncapped during ballasting or else overpressure and structural damage may occur. They are then capped until the barge returns to port, then uncapped. Caps should be painted red.

The wind heeling moment is generally based on the maximum velocity to be expected during the tow: steady state plus gusting. Alternatively 100 knots (50 m/sec) is arbitrarily assumed for a summer voyage.

The above calculations are based on calculated values of the centers of gravity and buoyancy, etc.

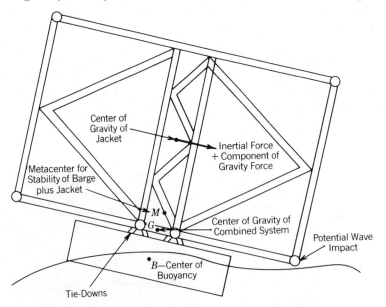

Figure 9.3.2 Launch barge with jacket under heavy roll during storm.

The vertical center of gravity is very sensitive to the actual weight of the jacket plus seafastenings.

With large and important jackets, strict weight control is exercised, both for purposes of transport and especially for launch. These include measurements of actual O.D. of all tubular members and wall thicknesses; the latter will usually be found to be near the upper limit of tolerances. Padeyes, seafastenings, conductor guides, instrumentation panels, mud mats, and so on must all be accounted for. Weld material presents a special problem: calculations should be made on the basis of sampled weld profiles, which again are usually greater than the minimum shown on the drawings.

The reason for such detail is of course the magnitude of the effect that a small additional weight has when it occurs on the upper side of the jacket, perhaps with a $\overline{KG}$ for that element of 50 m.

Before leaving harbor, an inclining experiment will usually be required by the marine surveyor. A known weight is placed on deck and moved a known distance off the centerline. The angle of heel is then used to determine the metacentric height $\overline{GM}$, from which the position of the vertical center of gravity $\overline{KG}$ can be determined. $\overline{KB}$ and the other factors in the equation, I and V, are geometric properties which can be determined by carefully measuring the draft at all four corners of the barge, after assurance that there are no internal compartments with free water surface that have not been fully accounted for.

Increasingly, marine surveyors are concerned about stability under damaged conditions, when one external compartment of the barge may be flooded. This could occur, for example, by collision from a tug boat, or from hitting floating ice or debris, or from rupture of a pipe in a ballast compartment of the barge.

Evaluation of stability and heel under this condition is usually made while neglecting wind, or assuming only a moderate wind, and may use as a criterion that the edge of the deck on the low side should not go below water, i.e., that the peak of the righting moment curve under this condition should not have been exceeded.

Computerized programs have been developed to predict the loads on a jacket-barge combination during oceanic tow and the resulting stresses in these two bodies. One of these, by Noble-Denton and Associates, Inc., known as OTTO, predicts the maximum stresses during a design storm as well as the fatigue due to cyclic loading during the entire tow. It takes into account gravitational and buoyancy forces, wind and current forces, wave-induced inertia forces, and wave-induced hydrodynamic pressure forces.

Fatigue during tow is becoming of increasing concern as tows become longer, through more severe seas, and with larger jackets having greater inertia. During the tow of a jacket from Japan to California, for example, many nodes may use up a significant portion of their total fatigue endurance. This is especially critical for self-floating jackets.

When towing a jacket in semi-restricted waters such as the North Sea, contingency planning should be carried out, including designation of storm lay-by areas, with a plan to proceed from area A to area B and so on only on the basis of favorable weather reports.

Jackets have been lost during tow, a recent case being one under tow from Scotland to Brazil, which sank in the North Sea shortly after the start of the journey. The impact of such a loss far exceeds the value of the jacket; it means the loss of at least a year in placing the field into production, with serious cash flow and interest cost implications.

Most losses of jackets can be traced to one of two causes; either the seafastenings are ruptured, leading to increased inertial forces and shifting of the jacket, or the watertightness of the deck is violated, usually at a seafastening but sometimes at a manhole or vent, resulting in shipping of water, free surface in the hold, and consequent loss of stability.

9.4 Removal of Jacket from Transport Barge; Lifting; Launching

Smaller jackets, designed for shallow water, are often lifted directly from the barge by one or two crane barges and set on the seafloor. The slings are attached and then the tie-downs and connections to the temporary skid beams are cut loose. See Figure 9.4.1.

In most cases, this operation is related to offshore terminal construction and can take place in protected waters. However, in some cases, this operation is in an exposed open coastal location, where long period swells are being amplified and shortened by the shallow water, hence causing significant differential movement between crane barges and transport barge.

Appropriate slack must be left in the lines during

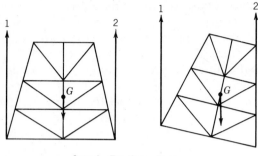

Case A—Two Crane Barges

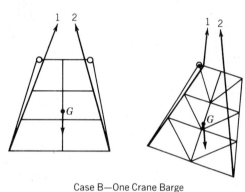

Case B—One Crane Barge

Figure 9.4.1 Lifting off jacket with slings attached below center of gravity.

the period of cutting loose. The cutting loose operation must be carefully preplanned in order to prevent endangering the men, since most of the cuts must be made by hand.

Short vertical guide posts may have been pre-installed at the loadout site to prevent lateral shifting of the jacket once it is cut loose. These braced vertical posts can form part of the tie-down frame; they must be adequately braced for impact.

The jacket may also have chain stoppers acting as supplemental tie-downs. When the primary tie-downs are cut, the chain stoppers still hold the jacket laterally. These chains can then be severed remotely by power-actuated (explosive) cutters.

Slings for the jacket will preferably have been attached above the center of gravity of the jacket, so that the jacket will hang more or less vertically as it is lifted. In this case, it is only necessary to try to catch a group of lower swells or waves, and then hoist as rapidly as possible as the barge starts to rise on a crest.

The dangerous time is the first wave crest after lifting off, when the jacket may once again be contacted by the barge or by the guide posts. These posts therefore should be only the minimum height necessary to prevent lateral displacement during cutting loose and have an inclined protector plate welded across their top end so as to minimize cutting or punching if the jacket leg should contact them on the second rise.

Occasionally the height of the jacket as compared to the length of the boom will prevent direct lift from points above the center of gravity. The slings have to be attached below the center of gravity in order to get a reasonable angle of spread. This is a dynamically unstable lifting mode, since if the load rotates, the righting moment decreases. Recognizing this deficiency, the system can still be used safely, especially if there is a wide spread between the points of attachment of the slings. It must be recognized that a high proportion of the entire load of the jacket may then occur on one sling and one point of attachment, and this in turn severely stresses the jacket frame. Such critical lifts have been made, for example, by two crane barges lifting from opposite sides of the jacket.

The slings or lifting lines can also be run through the jacket legs in such a way that any tilting of the load results in a reactive force from the line itself; this line of course must be guided in such a way that it cannot be frayed or cut on a sharp edge. See Figure 9.4.1.

In one case, for a shallow water jacket in the Gulf of Mexico, the jacket was skidded on its side onto the launch barge. At the site, it was lifted and set on the seafloor. Pre-attached slings then permitted rerigging of the hook so that the jacket could be relifted from its end, causing it to rotate to the vertical for placement.

Note again that certain slings must take the entire load of the jacket during rotation, which in turn reflects on the padeye design and the stresses in the jacket frame.

For any offshore jacket lifts, it will usually be found expedient to pre-attach the slings at the loadout site. Then when the barge arrives at the site, the eyes of the slings may be quickly raised by the whip line and placed over the horns of the hook, ready for the lift to take place.

Tag lines may be used during the initial phase of lifting clear of the barge, to keep the jacket pulled slightly inward toward the lifting barge and thus preventing it from swinging.

Because of the heavy weight of a jacket and complications of such a lift, the crane barge(s) is usually pre-positioned and moored on site. The transport barge is brought in across the stern of the crane barge and secured athwartships. Then the jacket is lifted free. The transport barge is cut loose and pulled clear. Now the load is lowered to the seafloor.

This minimizes the need for swinging either the barge or the boom, and keeps the crane barge picking over the stern where it has highest capacity and minimum roll response.

Most jackets are launched from the transport (launch) barge. Jackets have been launched which weigh over 30,000 tons and which are over 250 meters in length. See Figure 9.4.2. Even larger and longer platforms are now under construction for water depths of 400 m. This is one of the most dramatic operations in offshore construction, yet has been successfully performed many hundreds of times, just as ship launching has also proven successful.

Jackets have also been damaged or even lost during launching, emphasizing the critical, dynamic nature of this operation.

The procedure itself is relatively straightforward. With relatively calm seas, the barge is headed into the sea. It is ballasted down by the stern so that it has an angle of 3° or more. The seafastenings are cut loose. The jacket is then pulled off the stern by lines from the winches rigged around blocks at the stern and back to the bow of the barge. With larger jackets and dedicated launch barges, the jacket may be pushed off by hydraulically operated gripper jacks.

As the jacket moves end-O, off the stern of the barge, it finally reaches a point at which its center of load is beyond the pin of the rocker arms. These rotate to their limit (usually about 30°). The jacket now slides off the rocker arms into the sea.

There is a strong horizontal reaction imparted by the jacket to the barge, causing it to surge forward, at the same time as the stern kicks ups due to the release of the jacket load.

If this is a manned operation, the men, stationed near the bow, should have a safety line so as to avoid being thrown off the barge by this rather violent reaction of the barge. In most modern cases, this operation is carried out by remote control, unmanned, so as to avoid the danger. An umbilical cord from the tug or radio may be used to actuate the launching system.

The jacket, leaving the barge, has combined downward and rotational momentum. It will therefore usually plunge, with some jackets plunging even deeper than nominal diagonal length, before slowly returning back to sea level in a horizontal attitude. See Figure 9.4.3.

Most jackets are designed to ride, self-floating,

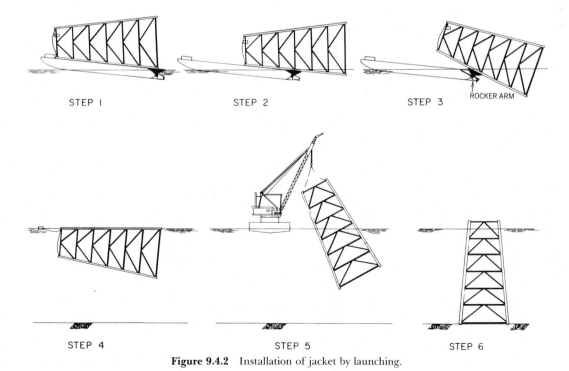

Figure 9.4.2 Installation of jacket by launching.

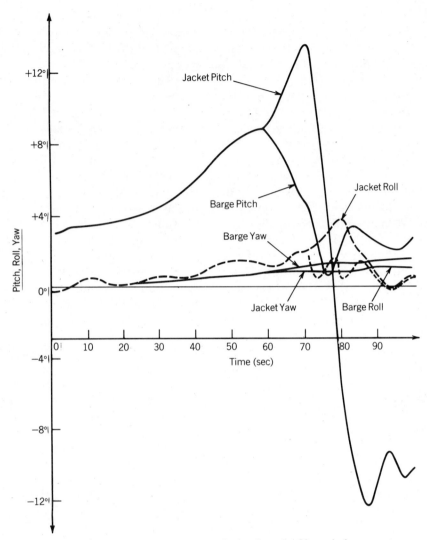

Figure 9.4.3 Motion responses during launch of large jacket.

on the upper side legs, with these about half immersed. This means a freeboard of only half the diameter of a jacket leg.

To return to the launching operation, starting friction may be relatively high. This will require the use of high jacking or pulling forces, opposite in direction to those applied in loading the jacket.

As the jacket moves down the launching ways, its weight is imposed progressively on a smaller and smaller length of the two central jacket legs, until finally all the load is that at the rocker arm. The jacket now rotates partially into the water so that it is supported over two zones: the water and the rocker arm. The jacket continues to slide, the two legs still carrying a high portion of the total load until the jacket finally slides free.

Note that while the vertical load at the time the rocker arm rotates will normally be the maximum, there is in addition a friction force acting parallel to the jacket leg.

The worst thing that can happen during a launch is for the jacket to skew sidewise and thus not only tilt the barge so as to cause the jacket to roll but also cause loads on the jacket frame at points and in amounts for which it is not designed. The proclivity to roll is in part due to the raise of the bow of the barge out of the water as the center of gravity of the load, i.e., the jacket, moves aft, thus reducing the waterplane moment of inertia.

During initial movement of the jacket on the barge, moving astern until its center of gravity reaches the rocker arm pin, the jacket can be kept properly aligned by controlling the jacking/pulling system and by steel plate side guides on the two center legs of the jacket, which hold them on the launching ways. Feedback instrumentation must be installed to verify that the jacket is correctly aligned as it reaches the rocker arms.

As the jacket rotates into the water, there are impact forces (similar to wave slam) on the legs and cross-members and any temporary buoyancy elements, tending to tear them lose.

There are inertial forces acting on any piles which were preinstalled in the jacket legs or sleeves, tending to cause them to plunge downward.

Jackets have been loaded and launched with the lower end (base) launched first, and also the reverse, where the top is launched first. Present practice appears to favor launching with the top of the jacket first.

As the jacket enters the water, many of the tubular members will have been made watertight and empty, so as to provide the needed buoyancy to cause the structure to float properly. These are subjected to hydrostatic forces, principally hoop stresses but also complicated by axial compression due to hydrostatic force on the end. See Figure 9.4.4.

API RP2A suggests a formula 2.4.6b for computing the design hydrostatic head; this is not intended to be used as the head to use during launching as it is unnecessarily refined and may be inadequate to take care of the short-term loading.

The tubular members and temporary buoyancy tanks may also experience ovalling due to drag as the water rushes past the bracing and legs during launching.

The design against buckling must also consider initial out-of-roundness of the tubulars.

It is obvious that control of the jacket weight and its distribution are very critical for the launching process. This is why detailed accounting must have been carried out as to variations in wall thickness and diameters, weld material, temporary attachments, mud mats, conductors, piles, and so on. On the buoyancy side, outside diameters must be thoroughly checked; usually circumferences are more readily measured and can form the proper basis for buoyancy calculations.

For the larger jackets involved in many current projects, temporary buoyancy tanks are necessary, to aid in controlling flotation depth and attitude. See Figures 9.4.5 and 9.4.12.

In a few cases in the past, improper and inadequate calculations have led to jackets ending up floating upside down, then imploding due to excess hydrostatic pressure acting on members which were never intended to be deeply submerged. Figures 9.4.6 through 9.4.11 show the sequence of such an unfortunate accident.

The jacket must be launched in sufficient depth of water so that there is no danger of it hitting the bottom, again taking into account the momentum of launch and the diagonal length across the jacket. See Figure 9.4.13.

On several occasions, jacket legs have been damaged by hitting the seafloor.

Computer programs have been developed that portray the entire launching process graphically. They also give the stresses on the jacket members and launch barge rocker arms during launch and enable the entire dynamic process to be examined in detail. Note, however, that such programs are only as valid as the input data. The actual behavior

Figure 9.4.4 Launching of large jacket. Note temporary buoyancy tanks attached to guides for skirt piles.

Figure 9.4.5 North Rankin A jacket, destined for Northwest Shelf, Australia, with temporary buoyancy tank attached.

Figure 9.4.6 Sequential phases of an unsuccessful launching of a jacket in the Gulf of Mexico in early 1960s. Phase A: Commencing the launch (Courtesy of Bruce Collipp. Shell Oil Co.)

Figure 9.4.7 Phase B: Jacket rotates into water. (Courtesy of Bruce Collipp. Shell Oil Co.)

Figure 9.4.8 Phase C: Jacket continues rotation. (Courtesy of Bruce Collipp. Shell Oil Co.)

Figure 9.4.9 Phase D: Jacket turns over, upside down. (Courtesy of Bruce Collipp. Shell Oil Co.)

Figure 9.4.10 Phase E: Base of jacket awash. (Courtesy of Bruce Collipp. Shell Oil Co.)

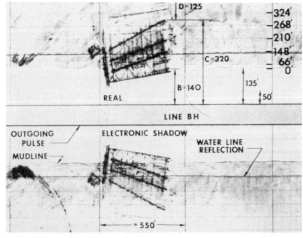

Figure 9.4.11 Phase F: Jacket sinks to seafloor. Position ascertained by side-scan sonar. *Note:* jacket was subsequently successfully recovered and installed. (Courtesy of Bruce Collipp and Shell Oil Co.)

Figure 9.4.12 Launching of North Rankin A jacket, Northwest Shelf, Australia.

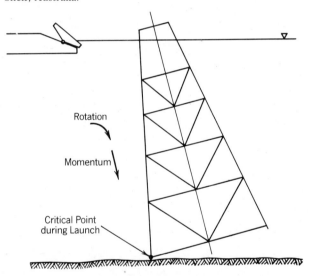

Figure 9.4.13 Potential impact of jacket on seafloor.

Figure 9.4.14 Controlled launching of platform Eureka off coast of Southern California. (Courtesy of Shell Oil Co.)

is very sensitive to relatively slight variation in the amount and distribution of weights and buoyancies. Such programs permit controlled launching, which is becoming increasingly important for deep water platforms. See Figure 9.14.14.

The tubular members designed for buoyancy must be positively sealed. Cross-bracing will normally be welded closed. Filling holes should be left and vents provided for those members which are to be free-flooded: plugs can be installed for those which are to be temporarily buoyant and flooded after installation.

Attention is called to the fact that the location and details of the holes must be determined and/or approved by the jacket designer; otherwise, these may become stress raisers or initiate cracks under cyclic loading in service.

The ends of sleeves or legs through which piles are later to be driven are usually sealed with reinforced neoprene jacket leg closures, so that they can be penetrated by dropping of the first pile section.

The above descriptions have been addressed to launching off the end of the barge, "end-O launching," which is the common method employed because of the typical jacket configuration.

However, just as side launching of ships imposes less severe structural demands on the hull, so sideways launching of jackets, when applicable, imposes less severe forces. For guyed towers, with their rectangular profile, giving a uniform cross-section throughout, or for the loading and breasting platform jackets of offshore terminals, which often are rectangular in profile, side launching is very practicable. Recent Japanese analyses indicate that side launching may also be practicable for tapered jackets, provided guidance is provided while still on the barge.

Several short athwartships launching girders can be used, thus reducing the load acting on any one point of the jacket frame. The barge can be heeled 5 to 7° by differential ballasting.

Relatively small rocker arms can be employed, pinned at the side of the barge. The jacket is pushed or pulled to the down side of the barge; then all restraints except one on each end are released. These two must then be cut simultaneously. The jacket slides and rolls off, in turn kicking the barge sideways. This concept of sideways launching is especially attractive for very long jackets, such as the 500-m jackets proposed for use in the guyed tower system.

Brown and Root has developed a system for launching a jacket from two barges, in which the barge supporting the lower end of the jacket is towed end-O, until the rocker arms release the support of the lower end. This now drops into the water, rotating and therefore allowing the upper end of the jacket to slide off the second barge.

This system places a heavily concentrated load at the lower end of the jacket, and hence both this location on the jacket and the rocker arms on the first barge must be heavily reinforced. See Figure 9.4.15.

Recent schemes have been proposed in which the heavy end of the jacket is to be supported on a launch barge and the upper end is to be made self-floating by means of temporary buoyancy tanks. Since in this case, as with the two-barge concept, the jacket does not rotate as it leaves the first barge, the lower end supports of the jacket and the rocker arms must carry very heavily concentrated loads. The rocker arms must be arranged so they can rotate through 90°. Sliding shoes are built into the ends of the jacket support. The temporary buoyancy tanks at the upper end are easily removed after upending.

A significant variation of the above is to provide self-flotation at the lower end of the jacket and support the upper end on a conventional launch barge. Now launching can be combined with up-ending, and the upper end of the jacket will rotate in conventional fashion. Since the enlarged legs or tanks are now deep below the surface, they attract minimal wave forces and may be left permanently in place, filled with water. See Figure 9.4.16.

McDermott has proposed joining a second barge astern of a conventional launch barge, using an articulated connection. This second barge would then rotate downward as the jacket moves aft, providing support as the jacket tilts into the water.

Since proposals are being made to install conventionally framed single-piece jackets in depths of 500 m or more, and guyed towers up to perhaps

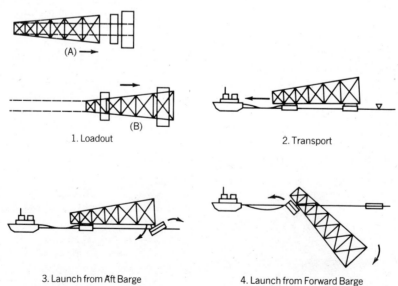

Figure 9.4.15 Twin-barge launching of deepwater jacket. (Adapted from "Installation of Deep Water Platforms" Blight, G. J. and Tuturea, D. P. *Petroleum Engineer International*, Nov. 1978.)

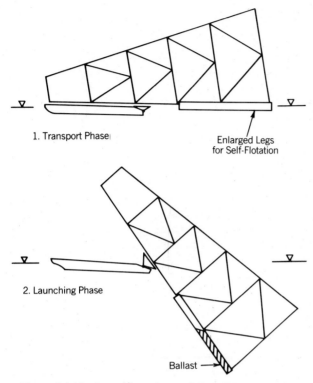

Figure 9.4.16 Launching of a partially self-floating jacket.

800 m, it appears that innovative concepts of launching such as those outlined above must be developed.

9.5 Upending of Jacket

The upending of smaller jackets has been often accomplished by a combination of differential ballasting augmented by the lift from the crane boom of an offshore derrick barge. While this provides excellent control, it involves several potentially dangerous dynamic aspects.

First, the jacket, having a large actual mass plus an added mass (hydrodynamic mass) of almost equal magnitude, cannot respond to the accelerations induced in the boom tip by the heave and pitch of the barge. These latter have a typical double amplitude period of 6 seconds, which means that it is the boom and derrick barge that are pulled down when the wave crest passes, rather than the jacket being pulled up. There is, of course, elastic stretch in the wire rope falls, hence use of as many parts as practicable is desirable. There is also the flexibility in the boom and the stretch in the boom topping lift lines. Nevertheless, this procedure is safe only in a very calm sea.

The slings for this upending should have been preattached, so as to be readily accessible, above water, for hooking on.

This crane boom can provide control of the jacket's attitude; but the primary upending moment must come from differential ballasting in which water is flooded into the lower portions of the jacket legs on the high side. As the jacket rotates, water may be drained out of upper bracing.

API RP2A provides in part: "Generally, the upending process is accomplished by a combination of a derrick barge and controlled or selective flooding system. This up-ending phase requires advance planning to pre-determine the simultaneous lifting and controlled flooding steps necessary to set the structure on site. Closure devices, lifting connections, etc. should be provided where necessary. . The flooding system should be designed to withstand the water pressures which will be encountered during the lifting process."

Large jackets have extensive ballasting and control systems installed, to permit flooding and venting, as well as hydraulic lines with which to operate valves.

The bending moments and forces induced in the jacket during upending must be determined, in order to prevent overstress in the jacket frame. Any tubular members which are empty or partially empty during the upending process must be able to withstand the combined hoop forces and axial forces induced by the water pressures at the depths involved; these conditions and forces may not necessarily be the same as those during launching or in service. Failure to recognize the effect of combined stresses is believed to have been partially responsible for the collapse of the temporary buoyancy tubes on the Frigg DPI Platform, which resulted in loss of the jacket.

Note especially that self-floating jackets will first experience significant hydrostatic pressures during upending. See Figure 9.5.1.

One means of countering high hydrostatic pressures is through internal pressurization with compressed air. On the BP Forties platforms, nitrogen gas, released from liquid nitrogen, was used to internally pressurize the temporary buoyancy tanks.

The British Standards Institute Code of Practice for Fixed Offshore Structures, BS 6235, states, "Whenever possible, the use of internal balancing pressurization should be avoided due to the con-

218 *Offshore Platforms: Steel Jackets and Piles*

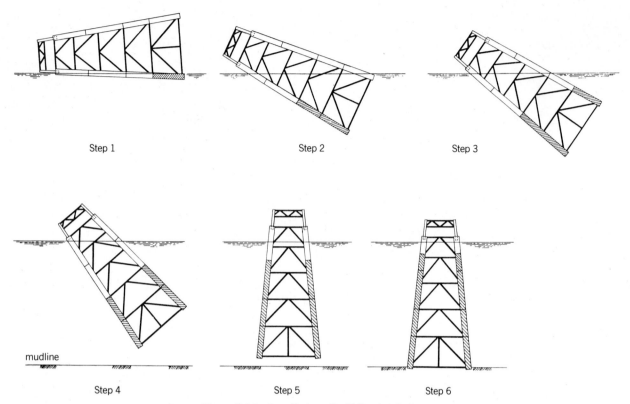

Figure 9.5.1 Installation of self-floating jacket.

straints upon design and handling that it produces. If it is used, the following should be noted.

a) The rate of pressurization should not exceed the structure's ability to withstand stresses induced by the increased temperature due to compression of internal air.
b) The process has to be capable of being arrested at any stage without the need for power."

Control of relatively small jackets has usually been by umbilical (electric-hydraulic) from the derrick barge, actuating the opening and closure of valves, and feeding back information on progress of flooding. Usually the valves are equipped with spring closures, so as to automatically close in event of power or hydraulic failure. Screens are provided over intakes to prevent entry of debris which might prevent closure of valves.

Pressure sensors, sensing the rise in pressure of the air compressed at the top of a member as it is flooded, or of the water at the bottom, provide necessary information. As more sophisticated and larger jackets are installed, valve-position indicators may also send the signals back to the control station on the barge.

As jackets have become larger, the upending process has usually been carried out remotely, without involvement of the derrick barge for lifting control. Three-legged jackets usually roll during upending, making it unsafe to have a line from the boom, while deep water jackets usually traverse too great an arc for the boom to follow. See Figure 9.5.2.

Remote control has been exercised as before, through an umbilical. However, umbilicals have been broken by the extended sweep of the upper end of the jacket as it rotates. Radio control has thus been found more reliable and is now the state-of-art for major jackets.

As a back-up, there is usually a station on the upper end of the jacket, where manned controls can be activated in an emergency. The men are usually not on board the jacket during the initial part of the upending but may be transferred by helicopter or boat. For this latter purpose, a rope ladder is arranged to hang down from the control station.

Large jackets, designed for deep water, obviously

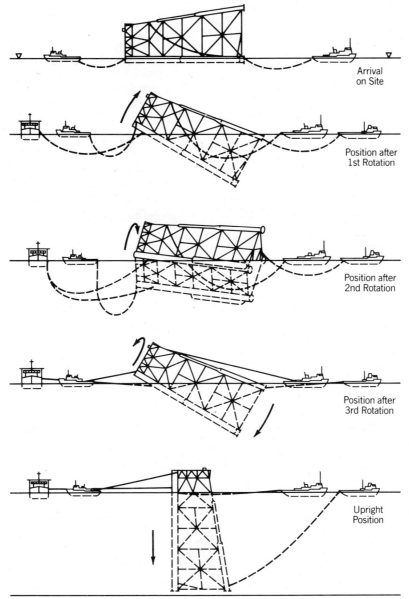

Figure 9.5.2 Upending of the Thistle platform jacket. Ballasting sequence not shown.

require a more sophisticated plan for upending in order to avoid overstressing of the jacket frame. The large legs of self-floaters can be subdivided both in plan and length.

Similar subdivision can be carried out for those jackets in which only skirt piles are employed, thus permitting the legs to be divided by transverse closures. Large legs and temporary buoyancy tanks may be pressurized internally to resist hydrostatic pressure.

On the BP-Forties platforms, liquid nitrogen was carried on board; it was released as gas into the temporary buoyancy raft and the legs as the jacket was upended.

Note that expanded gas from liquid is extremely cold and may freeze valves, preventing their operation. Compressed air, conversely, can get very hot and interfere with controls or minicomputers.

Upending is usually planned by means of a computer program which takes into account the constantly changing configuration of submergence and the changes imposed by ballasting. Once a suitable plan has been developed, physical model tests are almost always run. These serve two purposes: one, to verify the behavior of the jacket during upending, and two, to acquaint and train the key people—barge superintendent and offshore engineer—in this complex dynamic operation. See Figure 9.5.3.

Shell's "Eider" platform in the North Sea, 180 m in length, is currently being designed to be self upending after launching. This is a revolutionary new concept which may significantly reduce costs and improve reliability. It is scheduled for installation in 1988.

The jacket, now in vertical attitude, and with a draft only 3 to 5 m less than that at the installation location, is towed slowly to its final site location. Wherever feasible, the upending is of course carried out at or in the immediate proximity of the final site. However, the jacket's diagonal depth may exceed the final draft. Where seafloors are very uniform in depth over a large extent, as in parts of the North Sea, this may necessitate upending some distance from the site and then towing it to final location.

Such final tows will have a bridle preattached near the center of rotation of the jacket, so that the jacket will remain vertical in the final tow. Towing force and speed is purposely reduced to a minimum.

To eliminate or reduce this extra step (final positioning tow), the present trend is to use temporary buoyancy tanks which will enable uprighting in the depth of water that exists at the final site.

Currents during up-ending may displace or tilt the jacket relative to the derrick barge.

9.6 Installation on the Seafloor

To ensure that the jacket will be installed in its proper location, an offshore derrick barge is nor-

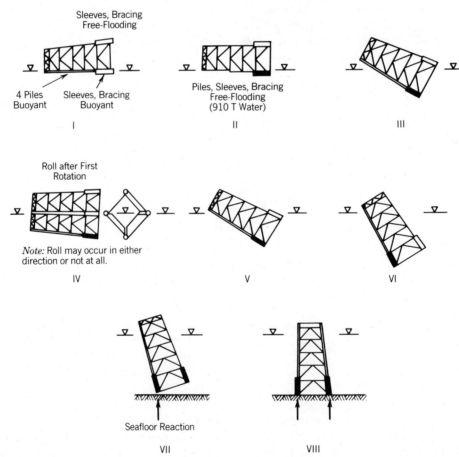

Figure 9.5.3 Steps in the upending of self-floating platform (Ninian southern). (Adapted from M. Metcalf et al., "On the Installation of a Self-Floating Offshore Platform," *Marine Technology*, July, 1978.)

mally moored on location. In shallow water, this mooring is accomplished by the derrick's own anchoring system, with the anchors being carried out by anchor handling boats. Once set, a pull is taken successively on each anchor line to ensure the anchor is properly seated. The final location and orientation of the derrick barge is then established by means of survey, principally electronic survey, but often keyed in to any preset acoustic transponders on the seafloor.

API RP2A Sec. 5.4.2 requires that the anchor lines be of sufficient length for the water depth at the site and that the anchors and lines be of the proper size (weight) and shape to hold against the maximum combination of wind, current, and waves.

In deep water, the derrick barge's own anchoring system may have inadequate length of lines and hence mooring buoys are preset to which the derrick barge's lines can then be run.

API RP2A also contains a rather curious section 5.4.2c suggesting that, where holding ground is poor or the mooring system cannot be made fully adequate, the derrick barge should be located so that if the anchors do slip, the barge will move away from the platform. This provision may be appropriate for small platforms being constructed with marginal equipment, or may have been intended primarily for application at later stages of construction when the jacket is firmly seated in place.

However, for installation of major jackets, it would seem more appropriate to orient the barge so that it would have the minimum boom tip motion. Further, as piles are driven and later, as deck sections are erected, the barge has to locate itself within the limiting radii and sectors.

Fortunately the second set of criteria will sometimes match the first, in that the derrick barge will have its stern to the platform. The jacket's location will then be guided by lines from the derrick barge, controlling not only its location but also its orientation.

To install a platform over an existing subsea well template, great precision and care are required in order to prevent damage to wells. The template will normally be held in place by piles although a gravity base could conceivably be employed. Two of the piles (or spuds from a gravity base) are fitted as guide posts, with tapered tops to engage cone-shaped funnels from the jacket. These guide posts will usually be decoupled from the template at this stage. Independent "bumper" piles may be used to protect the template during final positioning.

The jacket is brought into proper position, floating with several meters of clearance between the bottom of the jacket and the top of the guide posts. To provide full control of the jacket's position during this operation, a second derrick barge will normally be moored on the far side of the jacket.

In early development of this technique, guide lines from the jacket top were attached and tensioned so as to give a visual indication of location and verticality. Today, with sophisticated sonar (acoustic) locators and transponders, plus video cameras which can be mounted on the legs, it is feasible to dispense with the guide lines and place full reliance on the instrumentation. Redundancy in instruments must of course be provided in the event of malfunction of any instrument. The jacket is now slowly ballasted down to engage the guide posts, and then on down to seafloor contact with the mud mats.

Whether set independently on the seafloor, or over a well template, the jacket now must derive temporary support from the mud mats bearing on the soils at or just below the surface. The jacket must be self-supporting until pin piles can be driven. It is important that the jacket be level and remain so within a small tolerance until the piles are installed. Levelling of the jacket after pile installation produces generally unacceptable bending stresses in the piles.

A careful evaluation must be made of the soil loadings during this phase. The jacket will be bearing at this stage either on the bottom bracing or on mud mats or a combination of both. The weight of the jacket must include any piles or conductors which are being supported by the jacket during the installation.

The bearing pressure on the soil must be within allowable limits under the combination of direct load and that due to waves and current during the piling phase. API RP2A Sec. 5.5.4 allows a one-third increase in allowable soil bearing values during this phase if wave action is considered. This may be roughly acceptable in smaller installations. A much more thorough analysis is required for major structures, taking into account short-term consolidation, settlements, the effect of cyclic lateral and vertical strains, and so on. Scour around and under the mudmats must be prevented.

All structural elements bearing on the soil or supporting the mud mats must be adequate for the maximum bearing loads anticipated, including those due to storm. While API RP2A allows a one-third increase in allowable stresses for this condition if wave action has been considered, however, in the event of a sudden storm, a local failure of the soil which is self-limiting in extent, for example, tilting of the jacket, is obviously preferable to structural failure. The design of the mud mats should also address the failure mode, to be sure that structural failure will take place in the mat proper, rather than by damage to the permanent jacket legs or braces.

Mud mats were originally timber planks affixed to the bottom bracing so as to increase the bearing area. With major jackets, these mud mats are now structural steel, heavily reinforced flat plates, carefully designed to provide proper bearing. They are frequently tailored to fit the bottom contours; in the case of the Hondo platform off the Southern California coast, there was 20 m difference in elevation from the deepest to the shortest leg. This means that the jacket must be accurately oriented as well as positioned.

The effective weight of the jacket on the bottom may be controlled by ballasting. This permits moderate adjustment of level of the jacket, which may be supplemented by moment induced by lines from the controlling derrick barges.

The jacket must also have resistance to lateral displacement. In competent soils, this may be increased by added ballast. Especially if a storm comes up or the derrick barge has to suspend operations before the jacket is adequately secured by piles, then the addition of ballast may be indicated.

Another means of increasing lateral resistance is by having the pile sleeves or jacket legs extend below the mud mats, to act as spuds.

In mud slide areas, in areas of sand waves, and in very weak soils, jackets are being designed to penetrate well below the seafloor, so as to provide frame action at depths up to as much as 15 m. This may be required in the future in unconsolidated sands which may be subject to liquefaction in an earthquake.

To enable jacket legs and pile sleeves to penetrate into soft soils, the addition of water ballast is normally sufficient.

Bracing, however, is more difficult to penetrate because of its large area. Jets can be preinstalled, with nozzles acting along the underside of horizontal bracing so as to wash out material from under the bracing and lubricate the sides.

For self-floating jackets, which typically have two enlarged legs, or where pile sleeves are of large diameter, supplemental means may also be necessary to cause them to penetrate. Jets can be arranged inside to break up the plug and an airlift or eductor system employed to remove the material. These can be designed to operate below the pile closure.

For the Maui A platform off New Zealand, jet and airlift systems were built into the two enlarged legs, so as to enable the material within the large diameter legs to be progressively removed.

In poor soils, one other way around the dilemma of providing vertical and lateral support to the jacket during this early phase is by driving four temporary piles to a short penetration only. The jacket can then be levelled by jacking, lifting, or ballasting, and temporarily welded off to these four piles.

The "temporary piles" may be four of the permanent piles, driven initially only to a small penetration. They would typically have been transported with the jacket, so as to expedite their release and installation. Sometimes these short piles are made permanent and used only as spuds for (P/y) support. In most cases, after the remaining permanent piles have been driven, these "temporary piles" are cut loose, and raised as necessary to release any bending stresses. Add-ons are welded on and the lengthened piles are now driven to final penetration.

9.7 Pile and Conductor Installation

The jacket, now temporarily supported on the seafloor, is ready for pile installation. Some pile sections may have been transported with the jacket. The initial add-ons are welded on and the piles driven, as described in detail in Chapter 8.

In some cases, only a few piles are driven from a floating derrick. A work deck may have been preinstalled on the jacket or may be now placed. On this deck, cranes may be set, so that all further operations may be carried out from the platform itself. See Figure 9.7.1.

Fully self-installing platforms have been designed; these have a stiff leg derrick pre-attached to a work deck, the whole built into the jacket, so that upon upending, the stiff leg may erect itself, then pick and drive piles. Whether such a solution is an acceptable one in any specific situation will depend on the remoteness of the location, availability of offshore derrick barges, sea and weather states during the installation period, and ability of the soil to support the temporary loads of the jacket, with work deck, stiff-leg derrick, and live loads.

Figure 9.7.1 Jacket for Thistle platform incorporated a "work deck" on which to set drilling rigs and crawler cranes for use in pile installation.

The piles will penetrate the jacket closures as they are initially dropped. The grout seals at the base of the sleeves will keep mud out of the jacket leg as the piles are driven to final penetration. The driving of the piles and the subsequent grouting are described in detail in Chapter 8. Grouting is currently the accepted means for transfer of load between skirt piles and sleeves and/or jacket legs.

Grouting of piles in jacket legs is also an effective way of stiffening the gross section and preventing local buckling, as, for example, at nodes where the bracing members intersect the legs.

Piles can also be secured to the jacket by welding; this system has been much used in the past where piles extended up above water, inside the jacket legs. It is also currently employed in offshore terminal construction where jackets and pin piles are employed.

The transfer of high cyclic axial loads from the top of the pile into the jacket leg requires careful consideration as to weld details, since the welding will have to be carried out under adverse conditions of wetness (spray), perhaps low temperature, and while the jacket may be vibrating under wave action.

Steel shims are used to center the pile in the jacket leg; these are usually one-quarter or one-third segments of steel pipe of the proper radius. The welds are best designed as shear welds, from the pile to the shims, then from the shims to the jacket leg. A developed section of this detail is shown in Figure 9.7.2.

Currently under development are hydraulically forged connections, in which the pile is expanded against the sleeve to provide full shear transfer by friction.

The lateral resistance of the installed platform is developed by the P/y (lateral load-deflection) of

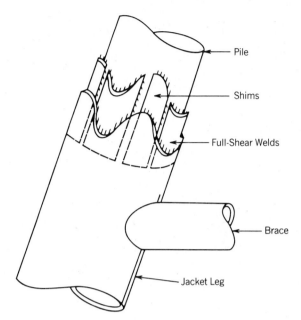

Figure 9.7.2 Scalloped welded connection for transfer of load between pile and jacket.

the pile-soil system, which normally takes place over the top 5 to 8 m of the soil. Since this is normally the zone of weakest soils, lateral resistance may be critical. Several schemes to enhance the properties of the soil have been proposed, including built-in drainage systems at the tip of the legs (or sleeves). To prevent annular gaps from forming around the pile under cyclic wave loads, pea gravel has been dumped on the seafloor around the pile: as a gap opens on one side, the pea gravel works down and wedges the pile, preventing progressively increasing displacements.

For the same reasons, it is important to prevent scour around the piles and bottom of the jacket, both when the jacket is temporarily supported by the mud mats and in service. Some prospective areas for platforms, for example, Sable Island off Nova Scotia, have sandy seafloors and high bottom currents, and have shown rapid scour behavior around the legs of jack-up drilling platforms. Shallow water areas, with sandy bottoms, where wave action may be severe, are especially suspect, since scour due to eddy action may be augmented by the pumping action of the jacket vibrating and rocking under the waves.

Scour protection around jacket legs can probably be most expeditiously and practically accomplished by the placement of graded rock through a long tremie pipe. Obviously the depth of practicability is limited, but fortunately so is the depth at which scour action usually occurs. Alternatively, controlled dumping from the surface has been utilized with generally satisfactory results.

Conductors are now installed, in much the same manner as piles. See Figure 9.7.3. The lower section of some of the conductors may have been carried out with the jacket, but for the most part, they are transported by barge, threaded in through the conductor guides, extended by add-ons, and driven to the required penetration. Since they are usually of smaller diameter than piles, that is, about 30 in. diameter × 1 in. walls, and usually penetrate to less depth than the piles, they are easier to drive, and smaller hammers can be used. Their penetration requirement is determined primarily by the ability to seal off flow during drilling, so that drilling mud will not escape to the sea. They must also be driven to a sufficient depth to prevent escape of shallow gas, which could form a flow path for future release. Conductors also must support the wells.

Alternatively, they may be installed by the drilling rig, which may use either a pile hammer or drilling-jetting techniques to install the conductors.

In mud slide areas, the conductors may be enclosed within a larger diameter tubular, for example, 5 m̊ diameter, which provides the strength and stiffness to resist the lateral forces from the moving mass of mud.

In other areas, e.g. Cook Inlet, Alaska, the conductors were drilled through the supporting piles, using the latter as protection against ice.

9.8. Deck Installation

Standard API RP2A, sec. 5.6.1, requires that the deck elevation be within plus or minus 3 in. (76 mm) from the design elevation and shall be level. The degree of level is usually limited to about 12 in. (300 mm) differential height across the longest dimension of the platform, but in any event should ensure proper drainage and proper operation of processing equipment.

Deck sections are now to be lifted on. With smaller platforms, the "pancake" concept was often adopted, in which some of the permanent equipment was preattached to decks, with each deck of the platform being lifted on in succession. After each deck was erected, remaining equipment for that deck was set.

With larger platforms, the "deck" now consists basically of module support structural frames, consisting of girders and trusses onto which large modules of assembled and integrated equipment are set. The initial sections have legs extending below them, with stabbing guides to fit into the piles or jacket legs. The stabbing guides are so configured that they also act as back-up plates. Since the mating leg is the same diameter and wall thickness as the extension to which it is to be joined, a full penetration girth weld is made, similar to the splice in a pile. See Figures 9.8.1 and 9.8.2.

To aid in stabbing four legs into four sleeves, the stabbing guides may be made slightly different in length, so that one can be entered first, then the module rotated so the second can be entered and then the whole lowered to fit the remaining joints.

Transport of a large deck module will usually

Figure 9.7.3 Installation of conductors on platform Eureka. (Courtesy of Shell Oil Co.)

Figure 9.8.1 Setting deck units for platform Eureka offshore Southern California. (Courtesy of Shell Oil Co.)

be by barge, although smaller modules and equipment may be transported by supply boat or on the derrick barge itself.

The weight of modules has grown in recent years, to 500 tons, then 1000, and most recently, 5400 tons. At least one derrick barge is now under construction that will lift 10,000 ton modules. The purpose is to enable more complete assembly on shore and reduce the time and cost of offshore hookup.

These monstrous lifts require calm seas and a derrick barge with minimum response to the seas.

Figure 9.8.2 Setting large module on deck of jacket.

The latest generation of heavy lift derrick barges is of the semisubmersible type. This of course is a trade-off between the reduced motion response of the barge and the concomitant reduction in stability, especially in roll, as the load is lifted.

For this reason, heavy lifts are generally made over the stern, with the swing being used only for minor adjustments in position to engage the stabbing guide. In fact, for heavy lifts such as these, a shear legs crane barge would be equally suitable, with the minor positioning adjustments being made by the deck engines. The largest offshore derrick barges in the North Sea now are fitted with two huge cranes, one on each stern quarter, so that their combined capacity may be used.

For heavy lifts, boom tip response is very critical. On-board minicomputer programs have been developed to optimize the heading and boom angle.

The derrick barge is pulled back from the platform, and the cargo barge, with the large deck unit on deck, is pulled in across the stern. The lift is made and the cargo barge pulled clear. The derrick barge now pulls astern, to the platform, where it sets the deck unit.

Because of the numerous parts of line needed for such heavy lifts, getting rid of the load after having landed it on deck is often a problem, even with free overhaul release. The stabbing guides must have a length equal to the heave at that point, otherwise on the next heave cycle after entry they may be disengaged again.

Load out and transport of these large deck units and modules on a cargo barge requires procedures and seafastenings similar to those of the jacket, with added complications. The unit, with its four or more legs extending downward, is difficult to support. Large reinforced plate-bracket assemblies are needed to distribute the load (static plus dynamic) over the deck so the leg will not punch through, and can be supported both on the skidway and on the barge. The center of gravity of the unit is high above the deck; the tie-downs must provide adequate lateral support to resist the lateral forces due to roll angle and accelerations during transport.

A typical deck section tie-down is shown in Figure 9.8.3. This shows the support of a leg. Supports may also be built-up so as to provide direct support to the module frame itself. Further description of module erection and hook-up is given in Chapter 14.

A major development along different lines has been the Hi-Deck concept, in which a fully integrated deck, complete with all or most of its equipment modules installed, is floated in between the legs of a previously installed jacket and lowered down to mating. For details, see Chapter 14.

Decks have thus undergone a series of major evolutionary developments, spurred by the recognition that the greatest portion of the total costs of an offshore platform is generally in the processing and support equipment and the greatest labor demand is in hook-up and testing. Operating with high volumes (50,000 to 300,000 barrels of oil per day), high pressures, high volumes of gas, and so on requires precision assembly and thorough testing. Hook-up is a major demand on skilled manpower and on a large platform can require 2 to 2.5 million man-hours. Significant savings in costs and time can be achieved by carrying this work out under favorable conditions at inshore shipyards.

At the same time, the total deck "payload" has grown from 7000 tons to over 40,000 tons, partly because of increased requirements such as gas reinjection, water flooding, and so on, partly because of more remote and demanding environments requiring weather protection which in turn requires

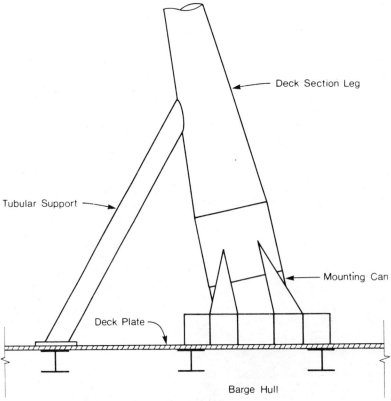

Figure 9.8.3 Sea-fastening for leg of deck section.

more ventilation, large helicopter services, and greatly enlarged quarters and support. The typical jacket-pile structure is very sensitive to total deck load. One way to reduce the total is by integrating the deck or at least the modules so as to make more efficient use of the deck structure.

The evolution then has been from individual deck sections and individual pieces of equipment, to module support frames and large integrated modules, to completely integrated decks.

9.9 Examples

In the following pages, a description is given of the installation procedures employed for three landmark offshore platforms: Hondo, Cognac, and Cerveza. Each of them employed important new techniques which will have an influence on future platforms, especially those in deeper water.

Example 1—Hondo

The Hondo platform installation off the coast of Southern California in 270 m of water was carried out in a unique fashion, by launching the jacket in two halves, then mating them afloat, and upending as a single jacket. See Fig. 9.9.1 through 9.9.5. The two halves of the jacket were constructed on one long fabrication ways at Oakland, California, to ensure exact match of the huge space frame. At the juncture between the two sections, mating cones were provided, each with a series of hydraulic ram connectors.

Each half was separately loaded onto a large launch barge, towed to a semiprotected site east of Santa Rosa Island, and launched in conventional fashion. The two halves now floated on their upper legs, with a freeboard of only about 1 m.

The two halves were aligned and pulled together, the mating cones engaged, and the hydraulic ram wedges activated. See Figure 9.9.6.

Full-penetration welds were run on the inside of the corner legs of the platform. Access to the two legs 50 m below water was through special tubes, which also provided ventilation, power, and light for the welder.

A similar mating procedure has been planned for the guyed tower concept when applied to water depths beyond those practicable for transporting and launching a single-piece jacket.

For the Hondo platform, pile sleeves were sealed with a double set of heavily reinforced neoprene pile sleeve closures designed for the hydrostatic head at 270-m depth. When the piles were later released, they did not penetrate under their own weight and had to be driven through the closures. Subsequent piles were cut with a serrated bottom edge which facilitated penetration of the closure.

Figure 9.9.1 Fabrication and erection of platform Hondo. Note connectors at mid-length for future mating of the two halves.

228 Offshore Platforms: Steel Jackets and Piles

Figure 9.9.2 Lower half of platform Hondo en route from San Francisco Bay. (Photos courtesy of J. Ray McDermott and Exxon.)

Example 2—Cognac

This platform was the first truly deep-water platform, breaking the 300-m depth limit. The jacket was constructed in three sections: base, middle, and top. See Figure 9.9.7. The installation of this platform also served as a proving ground for many advanced deep-water techniques: acoustic and video positioning devices, deep diving, underwater hydraulic hammers, and so forth. The sequence of construction is shown in Figure 9.9.8.

At the site, 12 large mooring buoys were set, each with three anchors. Two offshore derrick barges were positioned, one on each side of where the platform was to be built. See Figure 9.9.9.

Figure 9.9.3 Launching lower half of Hondo.

Figure 9.9.4 Joining the two halves of Hondo to form a single jacket for 270-ms water depth.

Figure 9.9.5 Setting deck section for platform Hondo. Note stabbing guides.

The lower section was built vertically (in the same orientation as its final position), transported, and launched. It was then positioned between the derrick barges, given slight negative buoyancy, and lowered to the seafloor by means of four 3½-in. and four 3-in. wire lines. Control of the structure was by selective ballasting, by means of the electric–hydraulic riser. See Figure 9.9.10.

Intensive studies were made of the dynamic responses during this lowering, and strict limitations imposed on the sea conditions which would affect the derrick barges during this critical operation. All lowering lines were equipped with motion compensators.

The piles were now transported in full length, self-floating. Each was upended by the derrick, then inserted in the sleeve almost 200 m below water, guided by sonic and video devices and entered into the funnel. The hydraulic hammer, an HBM Hydroblok hammer rated at 800,000 ft lb per blow, was then lowered onto the pile and guided by ten-

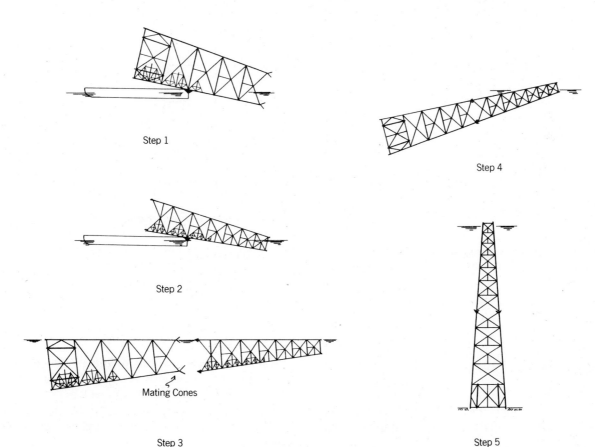

Figure 9.9.6 Assembly of jacket for Hondo platform.

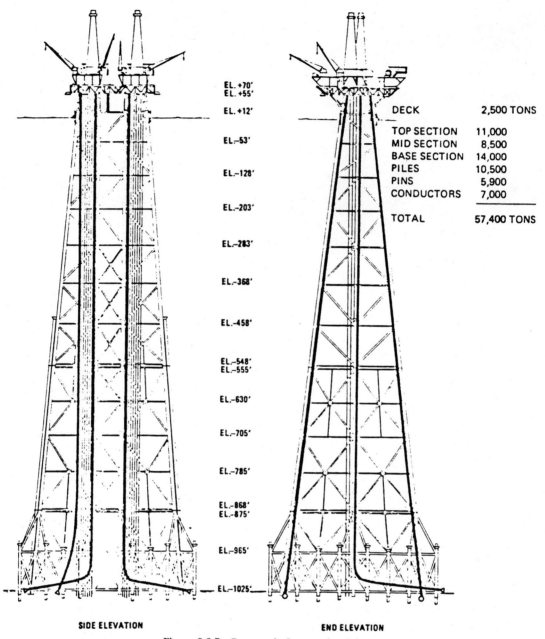

Figure 9.9.7 Cognac platform and weights.

sioned guide lines and acoustic transponders, and the pile was driven to full penetration. There were 24 piles, each 84 in. in diameter and 625 ft long, weighing 465 tons. See Figure 9.9.11.

Piles were then grouted to the sleeves of the lower section using a drill pipe. This completed one season's work.

The next season, the middle and top jacket sections were transported, launched, and upended.

Each was then lowered to mate, with mating cones, into the previously set jacket section. Guidance was by acoustic transducers, video, diver visual reports, and a mating funnel. Hydraulic clamps were activated to temporarily fix each section to the other.

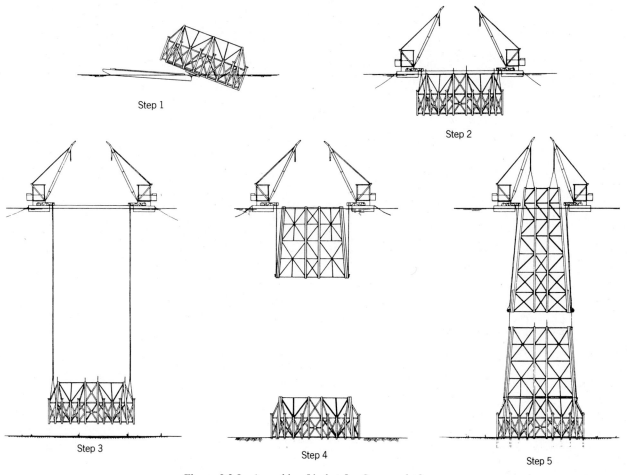

Figure 9.9.8 Assembly of jacket for Cognac platform.

To join these sections together, 10 large pile-like tubular dowels, each 72 in. diameter × 1025 ft long, were inserted through the jacket legs and grouted to all three sections.

Observation and monitoring of all underwater operations was carried out by a remote-controlled vehicle (ROV) and by divers, using TV as well as direct visual observation.

Example 3—Cerveza

This jacket-pile type platform was located in water almost as deep as Cognac, yet was fabricated and installed as a single piece. The jacket was 954 ft high and weighed 24,000 tons.

For loadout, the launch barge was ballasted down to the draft of 31 ft so that the top of its launchways was 9 in. below the top of the fabrication skidways. Then as the jacket was pulled onto the launch barge, the barge was progressively deballasted so as to pick up the load of the jacket. See Figure 9.9.12.

Once the jacket was fully on the barge but still overhanging the fabrication skidways, the barge was further deballasted to lift the jacket clear.

During loadout, closely spaced paint marks on the jacket served to guide the ballast controller. The rocker arms of the launch barge were locked during this operation.

During the loadout, transport, and installation, the following were preinstalled and temporarily fixed in the jacket:

16 curved conductors
4 skirt piles, each 560 ft long

232 *Offshore Platforms: Steel Jackets and Piles*

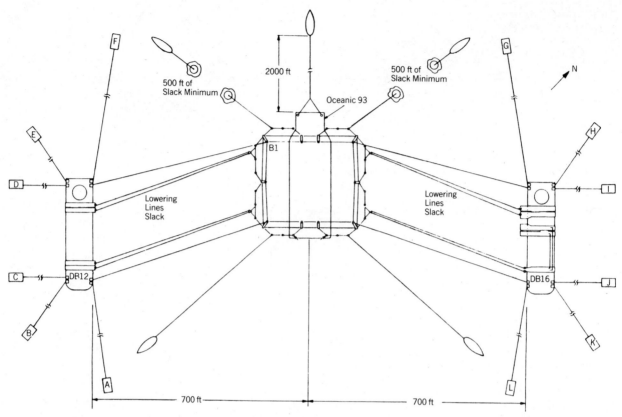

Figure 9.9.9 Launch configuration of mooring spread.

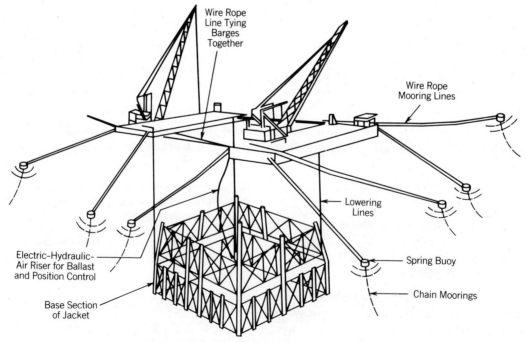

Figure 9.9.10 Installation of Cognac base section.

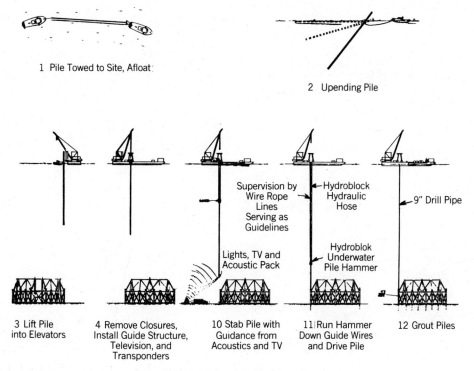

Figure 9.9.11 Piling installation at Cognac.

6 pull tubes (for future risers)
4-275 ft long skirt piles, used as temporary buoyancy tanks

The jacket was positioned on the barge for minimum draft and level trim during tow, with a 280-ft overhang on one end. The launch barge was 650 ft long × 170 ft beam × 40 ft depth.

In preparation for launch, the barge was ballasted with 29,000 tons of water which trimmed it down by the stern 3°. The static coefficient of friction of the jacket on the launchways was 0.11 and

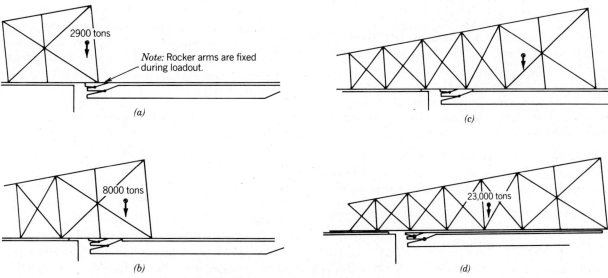

Figure 9.9.12 Load out of platform Cerveza.

234 *Offshore Platforms: Steel Jackets and Piles*

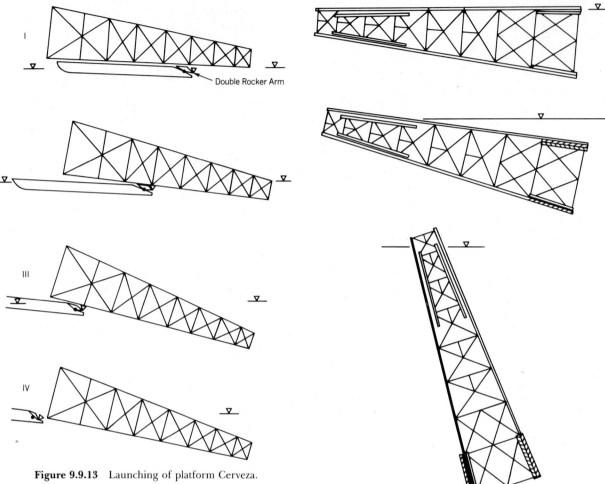

Figure 9.9.13 Launching of platform Cerveza.

Figure 9.9.14 Upending of platform Cerveza.

required a jacking force of 1400 tons to initiate launch. The dynamic coefficient of friction was 0.05. See Figure 9.9.13.

The jacket attained a launch velocity of 10 ft/s, a maximum dive angle of 13°, and a maximum dive depth of 265 ft (80 m). During this launch, the maximum submergence of the barge keel was 77 ft (24 m), an indication of the need to ensure that launch and similar barges can withstand excess hydrostatic heads without imploding.

Personnel on the launch barge were kept clear of the outboard side of the launchways; in between they were safe as the jacket moved over their heads.

The key feature of the launch was the use of the double rocker arms, which reduced the forces on the jacket legs and on the barge.

The jacket when launched had a reserve buoyancy of 10 percent, that is, 2400 tons.

Upending was carried out by ballasting two pile sleeves with tops closed so as to trap air and then two more which were open to the atmosphere. The jacket then upended to a stable vertical attitude. See Figure 9.9.14.

The controls for upending were on board the jacket, on a temporary control skid. The jacket was upended against the current, at a site 1½ miles away, in 400 m of water, giving the tugs time to correctly orient the jacket as it drifted back to location.

Control of final lowering and touchdown was by the use of sonic depth sounders located on each corner leg. All skirt sleeves had one flooding valve and one venting line. The grout lines were available to inject compressed air to deballast if necessary.

Valves were hydraulically operated, with a spring return. Indicator tubing from each compartment led to the control board, enabling the state of flooding to be monitored.

A contingency plan was prepared so as to be able to deal with all postulated emergencies during up-ending. In addition to the air pressure indicator tubing, inclinometers gave continuous readouts on attitude. It was recognized that variations in the current with depth might affect verticality.

Any damage to pile sleeves or in-leakage was able to be offset by ballasting, valve closure, or air pressure. In case of other damage, for example, to bracing, the jacket would be brought back to horizontal for repair.

*Thus this mysterious divine Pacific
zones the world's whole bulk about;
Makes all coasts one bay to it; seems
the tide-beating heart of earth.
It rolls the mid-most waters of the
world, the Indian Ocean and
the Atlantic being but its arms.*

HERMAN MELVILLE,
Moby Dick

10

Concrete Offshore Platforms (Gravity-Base Structures)

10.1 General

Offshore platforms of the gravity-base category are designed to be founded at or just below the seafloor, transferring their loads to the soil by means of shallow footings. Such gravity-base platforms have usually been constructed of reinforced and prestressed concrete but may be built of steel or of a hybrid design of concrete and steel. See Figures 10.1.1, 10.1.2, and 10.1.3. These gravity-base structures are often extremely large in both dimensions and mass: the Gulfaks C platform, designed for a water depth of 217 m, will require 240,000 m^3 of reinforced and prestressed concrete.

These platforms are almost always constructed in their vertical (final) attitude, enabling much or all of the deck girders and equipment to be installed at an inshore site and transported with the substructure to the installation site. These structures are usually self-floating, although when necessary additional lift forces may be developed by temporary buoyancy tanks or special lifting vessels.

To minimize soil bearing loads, these structures usually have a large base "footprint." To provide buoyancy, they usually have large enclosed volumes. They thus generate much greater inertial forces under waves and earthquake, 50,000–100,000 tons of lateral force being typical, with special structures developing twice these amounts. Thus sliding tends to become the dominant mode of failure, at least for water depths up to 150–200 m.

To transfer this lateral load into the soil and thus prevent sliding, steel skirts and steel dowels are employed, designed to penetrate and thus force the failure surface further below the seafloor. Such skirts also provide protection against scour and piping. While the skirts are typically fixed to the base of the platform during fabrication, in special cases where shallow water limits draft, skirts or spuds may be installed through sleeves after the structure has been seated on the seafloor.

Figure 10.1.1 Raising the slipforms on the shafts of the Statfjord B platform.

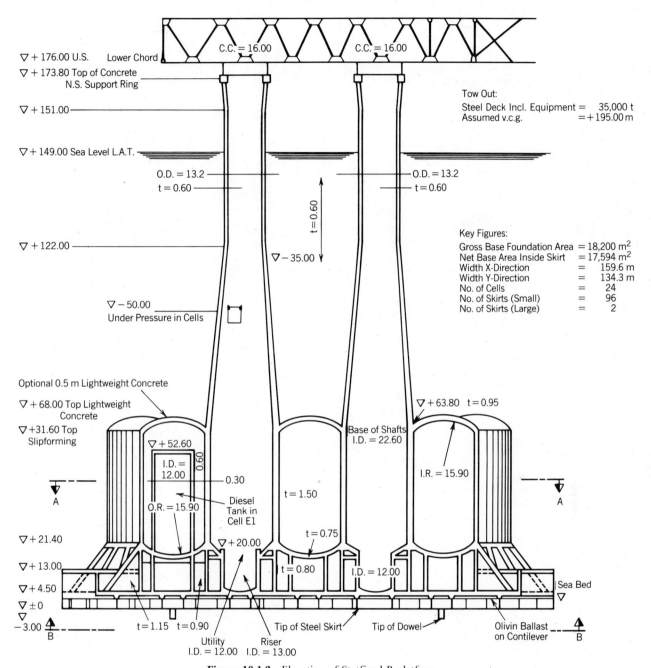

Figure 10.1.2 Elevation of Statfjord B platform.

10.2 Construction Stages

The construction of a typical concrete gravity-base platform takes place in a well-defined sequence of stages. For each stage, there are several important criteria which must be met.

1. The structure must be watertight and have stability and freeboard at all stages of construction.

2. The loading conditions and combinations acting on the structure are significantly different from one stage to the next. Structural integrity must be assured at each stage.

3. Ballasting and compressed-air systems (if these latter are employed) must be carefully and positively controlled at all stages.

To meet the above criteria, it becomes necessary to control weights and dimensions with great care.

238 *Concrete Offshore Platforms (Gravity-Base Structures)*

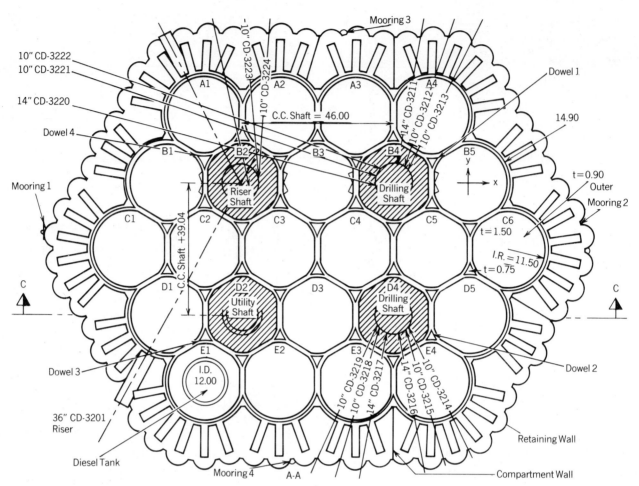

Figure 10.1.3 Plan view of Statfjord B platform, Norway, showing struts which support cantilevered base slab.

These structures are very large and massive, extending up to almost 200 m on each of the three axes.

The sequence shown in Figures 10.2.1 through 10.2.6 inclusive shows 15 stages of construction. The number of stages has been purposely abbreviated in order to give the overall pattern of construction. There are numerous sub-stages in each main stage. Each such stage must be carefully analyzed to be sure all criteria are met from the beginning to the end of that stage.

Most errors to date have been due either to overlooking an intermediate stage or to combining two or more stages to save computational effort. Detailed sketches of each sub-stage, along with evaluation of the pertinent hydrostatic, hydrodynamic, and structural loadings, must be prepared so as to enable visualization by both design and construction engineers.

While these structures are extremely massive, they are flexible structures supported on an elastic foundation and hence are subjected to deformations. These deformations become built into the structure by the subsequent construction. Residual stresses therefore must be considered, with appropriate allowance for their relief by creep.

The internal subdivisions of the structure are subjected to differential pressures, primarily due to the different ballast water heads acting on each side. Compressed air may be used on occasion to pressurize a compartment; the pressures occasioned thereby must be considered.

Accidental conditions must also be considered: the loss of compressed air from under the base skirts on one side, rupture and flooding of one compartment due to collision from a boat, a broken ballast pipe, or a failed penetration. Under these accidents, the structure may be permitted to suffer

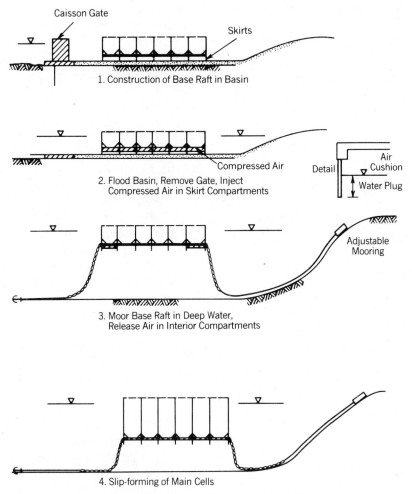

Figure 10.2.1 Stages of construction: 1–4.

minor local distress so long as its integrity, stability, and buoyancy are maintained.

Progressive collapse—for example, where one compartment floods, overloading the adjoining bulkhead which in turn fails, and so on—cannot be permitted.

A more detailed description of the special requirements and considerations at each step and stage will now be presented.

10.2.1 Stage 1—Construction in Basin

In stage 1 the basin must be capable of being safely dewatered against the maximum high tide plus storm surge and under the maximum rise in water table and rain runoff.

In a basin constructed in Queensland, Australia, the initial gate was made of steel sheet piles, supported by a sand berm. Some of the sheet piles had been driven out of interlock. Piping through these gaps eroded the sand berm and led to failure of the gate wall under extreme high tide, which in this case was 7 m above MLLW.

Similar dikes of sand have been protected in the Netherlands by well points and rock slope protection in addition to steel sheet piles. Dikes used in the basin for the construction of the Condeeps in Stavanger, Norway, have been of steel sheet piles backed by a substantial berm of rock.

The base course under the raft must be free-draining; perforated pipe or vitrified clay pipe drains are usually installed under a crushed rock base. The surface itself may be a concrete slab, provided with adequate bleed holes in case of a rise in the water table. This same provision of free water movement will later be important when the basin is flooded, so as to assure equal water pressure act-

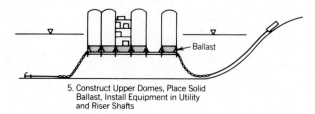

5. Construct Upper Domes, Place Solid Ballast, Install Equipment in Utility and Riser Shafts

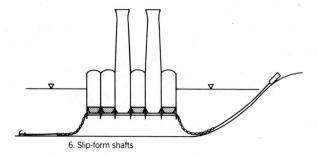

6. Slip-form shafts

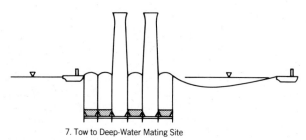

7. Tow to Deep-Water Mating Site

Figure 10.2.2 Stages of construction: 5–7.

ing upward on the base raft of the gravity-base structure (GBS).

Where the basin has been constructed in clay soils, as in the Concrete Technology Corporation basin in Tacoma, Washington, a pile-supported slab, with a full underdrainage system, was constructed.

Side slopes must be protected against excessive erosion under heavy rain and, more importantly, against slope failure. Horizontal drains, rock berms at the toe, or well points can be used to prevent slides; gutters at the top and plastic bitumastic or shotcrete coverings applied to the slope will serve to prevent erosion.

Access is frequently underestimated. Two well-surfaced roads should lead into the basin or trestle access provided at the top.

The roads around the base raft, on the bottom of the basin, should be well drained; a basin tends to collect large quantities of water and mud.

Caisson gates are used where several uses of the basin are planned; they can be removed in one day and replaced in one day. Such gates were installed after the piping incident in Queensland, and they are installed at the Highland Fabricators basin in Nigg Bay and Howard–Doris's basin at Loch Kishorn, both in Scotland. A new basin under construction by Mitsui Engineering and Shipbuilding at Oita in Kyushu, Japan, will similarly use caisson gates. Two details need to be verified: first, the provision of drainage from under the gate caisson so that no uplift can occur; second, the connections at each side, where a sheet pile wall is usually extended well back into the bank.

The base raft of the GBS caisson is constructed. The first items installed are the skirts, which are constructed of either steel or concrete depending on the foundation soils at the installation site. See Figure 10.2.1.1. In some cases, such as at McAlpine's yard at Ardyne Point, Scotland, the precast concrete skirts are set down into slots. Then the base slab can be directly supported on the basin's foundation slab.

Conversely, in Norwegian Contractor's basin at Stavanger, where longer steel skirts are employed, the skirts are supported on the basin's foundation slab and the structure's base slab is constructed on falsework scaffolding.

In the latter case, where anodes for cathodic protection, filtered drains to relieve the pore pressures in the surficial sands, and grouting nozzles all have to be installed under the slab, access by personnel must be considered. This is usually provided by temporary "doors" through the skirts. Later, the "door plates" can be replaced and welded.

Some of the larger GBS base rafts have had almost 100 skirt compartments, all more or less identical, covering a total of perhaps 20,000 m^2 of area and 500 m of perimeter. In order to ensure an orderly flow of materials and efficient access of personnel and equipment, these skirt compartments must be clearly marked; otherwise half of any worker's time may be spent hunting for the right location.

Concrete base slabs are usually 1–2 m thick, they typically generate high heat of hydration, leading to thermal expansion. If the subsequent cooling is restrained, as by the skirts, scaffolding, or adjacent slab pours, substantial cracking may occur. To minimize this problem, the sequence of pouring

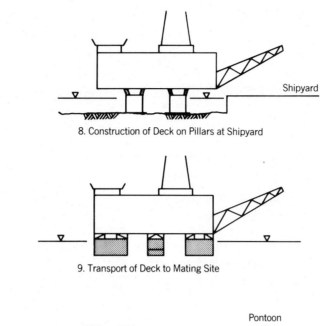

8. Construction of Deck on Pillars at Shipyard

9. Transport of Deck to Mating Site

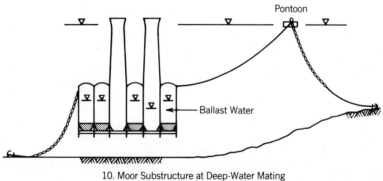

10. Moor Substructure at Deep-Water Mating Site, Ballast Down to Minimum Freeboard

Figure 10.2.3 Stages of construction: 8–10.

and the time between pours needs to be carefully determined, based on thermal analyses. Thermal probes may be used to monitor the actual temperature in the interior of a slab.

Chapter 4 discusses many of the practical problems of congested reinforcement, prestressing ducts, and concreting.

Base slabs are usually posttensioned with long tendons; once again the concrete slab must be free to shorten as it is not prestressed until the shortening actually takes place. If restrained by friction, this will not occur until the base raft is floated. For this reason, when there are no long skirts to deflect, special means are provided to reduce friction—for example, a sand layer between polyethylene sheets, on which a plywood soffit is laid. This was the method used during the construction of the Super CIDS concrete structure at NKK's shipyard dock in Tsu, Japan.

Where there are skirts, these may be supported on thick neoprene pads so as to permit outward and inward shear deformation. Long steel skirts may have adequate flexibility in themselves to accommodate the expansion and contraction.

The lower raft walls are now constructed, forming cellular partitions, capable of developing the necessary shear when the base raft is floated.

If there is an upper slab on the base raft as an integral part of the design configuration, it will provide the deck of a bargelike structure. Then any hog–sag moments during floating will be resisted by the top and base slab.

Unfortunately, having such a top slab on the base raft is the exception rather than the rule. Usually

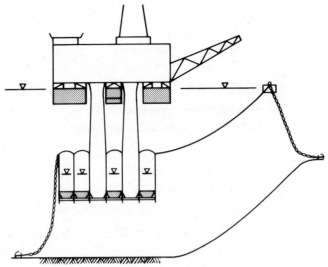

11. Maneuver Deck over Substructure and Transfer Deck to Substructure

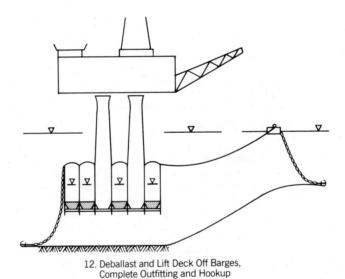

12. Deballast and Lift Deck Off Barges, Complete Outfitting and Hookup

Figure 10.2.4 Stages of construction: 11, 12.

there are only the multitude of intersecting cell walls. There is no upper flange, no "deck," for moment resistance during floatout.

Unequal moments arise, for example, because of cantilevered extensions to the base slab, which will be submerged during floatout and/or mooring. Because the center of gravity of a semicircular annulus lies outside that of a complete semicircle, a significant hogging moment may be induced. This may be aggravated by the weight of the skirts, since these are usually concentrated around the outside, and by cranes, mooring chains, and so on.

Reduction of the hogging moment may be attained by increasing the air cushion under the outer skirts and by adding ballast to the center cells. However, this latter usually conflicts with the requirements to take all possible steps to reduce draft during floatout.

Another solution, often adopted, is to prestress the top of the cell walls and/or add reinforcing steel so as to offset or resist the hogging moment and reduce residual stresses.

Note that since many of the larger GBS base rafts are not symmetrical, it is usually necessary to check

10.2 Construction Stages   243

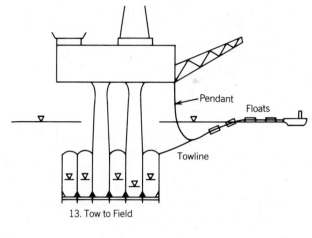

13. Tow to Field

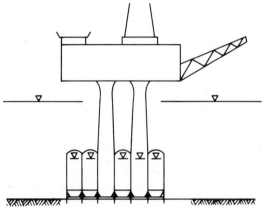

14. Ballast Down to Seafloor, Add Ballast so as to Penetrate Skirts, Grout under Base Within Skirt Compartments

Figure 10.2.5 Stages of construction: 13, 14.

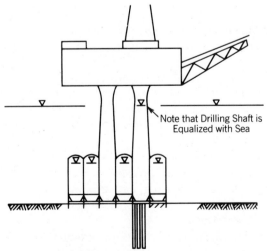

15. Drill out Conductor Plugs, Drive and Drill Conductors

Figure 10.2.6 Stages of construction: 15.

Figure 10.2.1.1 Steel skirts under base of Statfjord B Condeep will later be forced into seafloor at site. Note sacrificial anodes for cathodic protection.

bending around a number of possible axes, both orthogonal and inclined.

With the base raft structure nearing completion, the mechanical systems are installed, principally consisting of the salt-water ballast piping, underbase grouting system, skirt drainage and skirt venting, instrumentation such as strain gauges on skirts, bottom clearance acoustic sensors, and base mat pressure cells.

A careful inspection is now made of all skirt compartments to remove the accumulated debris, forms, scaffolding, and so forth that inevitably are left. The doors in the skirts are now welded closed. Ballast water is pumped into the cells to hold the raft on its foundation slab.

10.2.2 Stage 2–Floatout

With the structure ready for floatout, final checks are made of weight and displacements to ensure that flotation can be accomplished within the design tolerances of draft and heel. Note that for the typical 120-m-diameter raft, it takes only 1° heel to increase draft by 1 m. A minimum underkeel clearance of 0.5 m is usually required, from 3 hours before high tide to 3 hours after.

To reduce draft, air cushions are usually pressurized under the skirts. Compressed air may be introduced through the underbase grouting system, for example.

To prevent the air from escaping under the tip of the skirts, a "water plug" about 0.75 m high is usually left. If the skirts, for example, are 4 m long, then the use of the air cushion can reduce the draft by 3 m or so, depending on the area covered by skirts which are capable of holding an air cushion.

In the case of the Andoc Dunlin base raft, draft limitations were so critical that it was desirable to eliminate the water plug so as to gain another 0.75 m of bottom clearance. Therefore, large rubber inflatable "rafts" were placed in the skirt compartments, allowing complete filling by compressed air and preventing any unwanted escape of the air caused by the drag effects from the water due to currents, waves, and the towing speed in shallow water.

The basin is now flooded through sluice gates. See Figure 10.2.2.1. When the flooding is complete, the air cushions are tested for each set of skirt cells to be sure there are no leaks. The ballast water is removed, and the base raft floats up to its floatout draft. All systems are checked.

Meanwhile the exit channel has been re-sounded to ensure adequate draft, and dragged or profiled (with an electronic–acoustic profiler) to ensure against any rocks, debris, piles, or other obstructions above grade.

If all is OK, the gates are now removed. If the "gates" consist of sheet piles and a rock dike, this area must also be carefully swept to ensure against obstructions.

Navigational aids will have been established to guide the exit and to mark the route to the deep-water construction site. Laser or light ranges are useful in monitoring the effect of side currents. Electronic distance-measuring and position-plotting systems are usually installed on a temporary control deck on the base raft, supplemented by a sextant or theodolite.

Moorings will have been set at the deep-water site, secured to mooring buoys.

All boat traffic in the area will have been stopped. Weather reports will have been double-checked, especially with regard to wind. Tidal predictions will have been verified for the specific exit site.

Floatout from the basin has to be very carefully controlled because the typical raft is very unstable and weak in bending and hence sensitive to accidental loads and events. Winches mounted on the walls or sides of the basin control lateral movement, while highly maneuverable tugs pull the base raft out into the channel. The intact freeboard will be a minimum of 1 m above the wave crest height with allowance for runup.

Because mooring lines can jam in a sheave or otherwise be fouled, contingent means are available to cut the lines if necessary.

The base raft is now towed to the deep-water site. A harbor crane barge will aid in securing the mooring chains from the mooring buoys to the base raft.

The air cushion will still be contained within the skirts. The release of this air must follow a carefully calculated procedure to ensure against overstressing the base raft in bending. In general, the air cushion under the central portion can be released, while that under the periphery cells is kept.

Control of the air pressure under skirt cells is very tricky. If pressure gauges are used in different segments of the base, the readings may lead inherently to the wrong action.

When the Statfjord A base was floated out, pressure gauges were the only instrumentation used; see Figure 10.2.2.24. As the base raft tilted slightly,

Figure 10.2.2.1 Base raft of Ekofisk caisson afloat in construction basin. (Courtesy of C. G. Doris.)

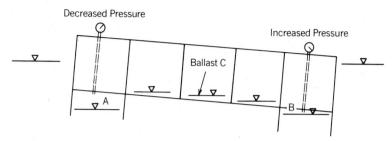

Figure 10.2.2.2 Effect of platform heel on air pressure in skirt compartments. Note how the intuitive response of reducing pressure in B will further increase the platform's list.

the pressure on the downside (B) rose, and that on the upside (A) fell. So the natural reaction was to bleed air from the downside (B) and add air to the upside (A). This caused increased heel, continuing to show higher pressures on the downside and lower pressures on the upside. So in effect, the "corrective" action was slowly jacking the platform to ever greater heel.

Earlier reference was made to the low stability of the base raft. This is largely due to the air cushion during floatout which creates a free surface in each skirt compartment, reducing the gross waterplane moment of inertia by the sum of the individual waterplanes.

Thus in the case shown in Figure 10.2.2.2, we have a relatively low stability, actions being taken (injection and later release of air) which is having the opposite effect to that intended, and a minimal freeboard.

Any ballast water in an internal compartment such as C will also shift to the downside, adding to the overturning moment.

The proper means of control, therefore, is not pressure but volume. Water level indicator gauges—for example, based on electrical resistivity—or hinged floats should be installed in the skirt compartments so as to give a measurement of the height of the water plug relative to the base slab and hence a measure of the air volume.

Base raft level indicators of suitable sensitivity should also be installed at the control station.

During floatout, mooring, and initial construction operations at the deep-water site, there is always the possibility for loss of air pressure under one set of skirt compartments. The subdivisions must therefore be carefully selected so as to minimize the adverse effects of loss of air pressure or of local flooding.

10.2.3 Stage 3—Mooring at Deep-Water Site

Either three or four mooring legs will have been pre-installed, terminating in mooring buoys arranged around the GBS deep-water construction site. These will typically consist of a heavy clump anchor on the seafloor or drilled-in anchor on the beach, several shots of chain, and then a wire rope mooring line or chain to the buoy. The buoy is held in approximate location by a separate small clump anchor and pendant.

From the buoy, chain is run to the base raft and connected by shackles. Now the chain forms a continuous mooring from base raft to anchor, with a float supporting its seaward end.

At least one line will have provision for progressive takeup as the construction proceeds and the mooring sinks lower in the water. See Figure 10.2.3.1.

In addition to moorings, a water line and a power line are usually run out from shore. By mating these lines to a wire rope, adequate strength can be provided.

10.2.4 Stage 4—Construction at Deep-Water Site

Shortly after the base raft of the Ninian central platform was moored, a storm arose. Continuous wave action beating against temporary bulkhead closures near the waterline caused working and then fatigue of the bolts holding the closures on. The external compartments on that side flooded. The water pressure inside that compartment ruptured the temporary internal closures to two adjoining compartments. Fortunately, all other closures held and the raft did not heel further.

However, the concrete batching and mixing barge had been moored tightly to the far side. Its

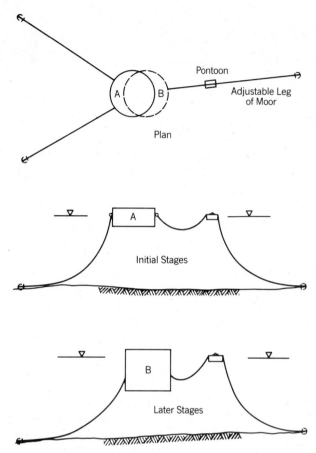

Figure 10.2.3.1 Moor of substructure at deep-water construction site.

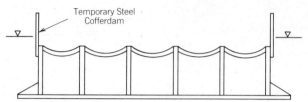

Figure 10.2.4.1 Extending peripheral wall so as to increase freeboard during float out and early stages of construction afloat.

manholes on deck were not secured. When the base raft listed into the waves, the far side of the base raft rose, lifting the adjacent side of the concrete barge, consequently submerging its deck on its far side. The water poured into the manholes, and the concrete barge broke loose, turned over, and sank.

On all temporary closures, therefore, the bolts should be designed to resist fatigue. Double nuts with lock (spring) washers can be used. Large bearing plates can be installed to prevent local crushing of timber struts.

A procedure which can be usefully employed to increase the freeboard during the critical stage of floatout and initial caisson construction is to extend the outside walls higher than those on the inside. On the Ekofisk caisson, for example, the outside walls were kept built up above the inner concrete.

This can also be accomplished by extending a steel cofferdam up from the base raft, cantilevered from the base raft so as to give adequate freeboard.

In turn, this may allow floatout with a less deep base raft and hence less draft. See Figure 10.2.4.1.

During construction afloat, the platform is required to maintain its 1-m freeboard under the 100-year storm wave plus runup, for that location and season of the year, as well as one-compartment stability, that is, to be safe with any one compartment flooded. If this cannot be maintained, then the intact freeboard must be increased to about 6 m. The possibility of internal flooding due to ruptured valves must always be considered.

When the base structure is too large for construction as a single structure (in plan), then it may be constructed in two or more smaller sections; each is floated out of the basin or building dock and then joined afloat. Such joining afloat has been carried out for both steel and concrete structures, notably a steel VLCC in Rotterdam, the Hood Canal Replacement Bridge in Washington, the Valdez, Alaska, floating container terminal, and several U.S. Navy floating seaplane docks, as well as the steel jacket structure for platform Hondo.

The structures must be accurately aligned by use of mating cones. Their contact areas should be cushioned by a crushable material such as wood, fiber, or neoprene. They are then rigidly connected by external steel beams or by temporary posttensioning so as to prevent joint rotation in the vertical and horizontal planes.

The edges of the contact areas are sealed, usually by the compression of the seating and cushioning seals. The space can now be dewatered by drainage into one of the sections. Posttensioning, bolting, and/or welding may now be carried out in the dry, along with epoxy injection and/or cement grout fill.

Details are of course all-important, including the details at the end of the joint section reinforcement, where stiffnesses and moment and shear capacities are reduced from their normal values.

Construction can now take place at the deep-water site. See Figure 10.2.4.2 If the GBS is being

Figure 10.2.4.2 Construction afloat at deep-water site in Loch Kishorn, Scotland. Ninian central platform. (Courtesy of C. G. Doris and Chevron U.K.)

constructed of reinforced concrete, as has been most common to date, then transport and installation of reinforcing steel, prestressing ducts, and concrete must take place in substantial volume and on a more or less continuous basis.

Typically slip-forming is used to raise the walls, using retarding admixtures in the mix because the large volumes involved necessitate a very slow rise, perhaps only 1 m per day. See Figures 10.2.4.3, and 10.2.4.4.

A typical base raft will have 1200 linear meters of 80-cm-thick walls, so that a continuous production of perhaps 1000 m^3/day is required to raise them at the rate of 1 m/day. Meanwhile, perhaps 300,000 kg (300 tons) of reinforcing steel must be placed, along with prestressing ducts and embedded plates. As many as 600 workers per shift are required, around the clock, for up to 60 days of continuous operation.

There are many variants in the construction scheme, all of which have applicability under special circumstances.

Where labor supply is limited, where there is an unusually large concentration of reinforcement, or where embedments must be set with great accuracy, then slip forms may not be applicable. Panel forms, cantilevered up, or flying forms are then employed, with each pour being defined as to extent and duration. This often necessitates the use of window boxes and tremies ("elephant trunks") to place the concrete. It also means that there will be a large number of horizontal construction joints. Both these requirements can and have been successfully met, not only in base raft construction for offshore platforms, but also in the rather similar walls of nuclear reactor containment structures.

Slip forms have also been employed in such cases, subdividing the structure into segments, with vertical construction joints. The Ninian central platform, with its seven concentric walls, was so subdivided.

Precast concrete elements have been extensively incorporated into base raft construction, the Ninian central platform and Global Marine's Super CIDS being prime examples. See Figure 10.2.4.5.

These should be cast in as large segments as practicable; 200-ton segments were used on the Ninian central platform. They can be incorporated into the remainder of the structure so as to act in

Figure 10.2.4.3 Construction of Ninian central platform afloat in Loch Kishorn, Scotland. Note combined use of slip forms and precast concrete shell elements.

248 *Concrete Offshore Platforms (Gravity-Base Structures)*

Figure 10.2.4.4 Slip-forming the main cells of the Statfjord A platform, with the structure moored at the deep-water construction site.

monolithic fashion by means of dowel extensions and cast-in-place concrete joints or by posttensioning.

Match-cast jointing techniques appear applicable in the case of many precast assemblies; these can follow the technology now perfected for long-span bridges, where units up to 20 m wide have been successfully joined by this method.

Match-casting was employed with great success for the breakwater wall of the Ninian central platform, where horizontal joints were employed. Match-casting not only led to rapid construction but also assured perfect alignment of the complex posttensioning ducts.

Other construction expedients include the use of precast concrete soffit forms to support hori-

Figure 10.2.4.5 Erecting precast concrete shells on the Ninian central platform.

zontal slabs and domes during construction of the base caisson. The Seatank platforms used precast concrete pyramidal forms which were domed shells constructed of shotcrete.

Steel soffits with internal support beams built in can be similarly used, the whole designed for composite action with the cast-in-place concrete.

When extremely thick walls are required, as for some designs for ice-resistant structures, precast concrete hollow box units may be employed, to be filled later with cast-in-place concrete.

Examples of the effective use of combined precast concrete and cast-in-place concrete are the outer walls of the Ekofisk caisson and the Ninian central platform.

With steel GBS structures, large prefabricated segments are being lifted, guided into place, aligned, and welded. They must be given temporary support against wind. Welding locations must be protected from rain and spray. Columns and shafts must be adequately stiffened to prevent out-of-roundness distortions.

For all structures being constructed afloat, access must be provided for the large number of personnel. This matter has often been underestimated in the past. Men must move from shore to a barge or floating dock alongside, up and over the outside walls, and across open cells and the internal walls to their site of work.

These walls are constantly being extended up, with closely spaced reinforcing bars, ducts, and climbing rods (for the slip-form yokes) all projecting above the walls. The freeboard of the structure is also changing. To add to the difficulties, distances between cells and walls are large, extending the reach of cranes to their limits and sometimes necessitating rehandling of reinforcing steel and other materials.

Clear identification of cells and walls is essential so that there will be no confusion of newly hired workers as to where they are to go. Access routes must be planned and safe walkways provided. Reinforcing steel bars may purposely be spread at certain specific locations; this should be incorporated in the working drawings and checked by the designer rather than being improvised in the field.

Freeboard can be kept constant by selective ballasting of the cells. However, these operations must be carefully planned and checked at every stage and substage. On the Beryl A platform, main cells were ballasted with water up to 40 m deep. No ballast was placed in the small star cell interstices between the primary cells. No notice was taken of this because these interstices were covered over by temporary timber work platforms that rose with the slip forms. Small in plan as they were, they had never been designed nor checked for a hydrostatic head inside the main cells, placing tension and shear on the connecting walls. Only after conclusion of concreting of the base caisson to a height of about 52 m, when the work platforms were removed, was the large and extensive cracking in the walls made apparent. The structure's integrity was finally restored, but only through a mammoth effort of concrete repair.

This illustrates the need in planning to portray each of the many incremental steps of construction, stage by stage, so as to reflect the step-by-step addition of concrete, which changes weight, trim, and draft, the progressive changes in ballast, change in stability (usually not critical), the changes in hydrostatic loading, and the structural response to the different load combinations.

In particular, the matter of differential hydrostatic head on the internal and external walls must be addressed. In the usual case, the net load is the difference between two large numbers, each the square of the water head. Mere application of a load factor to the differential may be totally inadequate to cover tolerances in these heads. Conversely, application of a load factor to the larger of the two heads of, say, 1.3 and 0.9 to the lesser head, can result in impracticably, almost ridiculously, great design loads, especially as the total head (depth) increases. Therefore, it is necessary to establish realistic tolerance ranges for the water ballast in the various compartments.

During construction afloat, the base structure's draft gradually increases. This changes the scope of the mooring lines. Usually, the necessary tension can be maintained by taking in on one, or at the most two, of the mooring lines. This is facilitated if one of the lines can be run to shore, assuming it is nearby. If not, adjustment means can be provided by a large mooring pontoon, moored over the bight of the line, taking up vertically to shorten the effective scope. Such a system is essentially a large spring buoy and may be excessively flexible; this is why a large pontoon is desirable. See Figure 10.2.3.1.

Taking up tension in one leg of the moor by hauling in at the base structure presents practicable difficulties because of the underwater connection and the large forces involved. However, this has been done on bridge pier caissons, which present a similar problem, by using a cable- or chain-grip hydraulic tensioner, working through a watertight sleeve built into the structure. From time to time, the mooring line has to be stopped off, the sleeve extended upward, and the tensioner relocated.

The moorings must be designed to hold the substructure during any storm wind plus waves and currents having a significant probability of occurrence during the construction period. Usually this is selected as the 10-year return storm, but considering the seriousness of the consequence of a rupture of the moorings and the delay and loss to the total project, a longer-return-period storm should perhaps be adopted.

The substructure almost always has several large barges moored to it, and it may be impracticable to move these away prior to a storm that arises suddenly, as was the case at Ninian. Therefore, these should be included in the calculations.

Crew changes are another practical problem due to the large number of workers per shift and the need to carry on work continuously. During rough weather and high winds, there arise the problems of getting men on and off the boats, seasickness, location and course in fog, rain, or snowstorm, and so on. These must all be addressed in the planning stage.

Provision must also be made for emergencies. What is done if a man is injured? Are there pallets or basket stretchers available which can be handled by a tower crane to hoist the injured man to a boat?

What about man overboard? A continuous safety boat has been provided 24 hours a day around some of the larger platforms. Life preservers, trailing floating lines, and the like should be provided.

Life jackets should be worn by all men working over the side and during transport. It is unnecessary and perhaps counterproductive to try to have men wear them while working in the cells; they may catch on the reinforcing steel. However, if there is ballast water in the cells, life jackets will be appropriate.

Cells eventually are 60 m or so deep. Safety nets may be needed where men are working across their top and to catch falling debris when men are working underneath. Usually these nets are provided at least for the principal shafts.

Projecting reinforcing bars are a puncture hazard and an eye hazard. They should all be fitted with a red plastic cap.

Ducts become convenient receptacles for tools, bottles, aggregate, and the remains of lunches. They should all be covered with red plastic caps, which will also serve to keep out the rain. On one project, a heavy rain was followed by freezing weather; the water in the ducts froze and split the concrete walls in multiple fractures.

Fresh water and electric power must be supplied to the structure during construction afloat. If the deep-water site is relatively near to shore, a hose and power cable may be run, attached to a wire line. Otherwise a water barge and generator barge must be moored alongside.

During the final installation process, many stages later than those being presently discussed, the platform will be ballasted down onto the seafloor. As it nears touchdown, the water trapped under the caisson tends to cause the platform to skid laterally, more or less uncontrollably. While venting of the skirt compartments and reducing the rate of descent will minimize this problem, another method that has been developed to prevent dislocation uses three or four dowels which engage the seafloor while there is still 2–3 m of bottom clearance.

These dowels could not be installed in the basin because of draft limitations, so they are lowered and fixed in position at the deep-water construction site. A typical dowel is a steel tubular 78 in. (2 m) in diameter, with 3- to 4-in. (75- to 100-mm) walls, extending 4 m below the skirts. In any specific case it is designed to provide adequate lateral resistance to hydrodynamic skidding by developing passive resistance in the soils, which of course produces shear and moment in the dowels. The dowel is designed so that failure, if it occurs, will first be by plowing through the soil and then by bending/buckling of the dowel so that the structure itself will remain undamaged.

As soon as the structure is moored at the deep-water site, each dowel is lowered and secured in place by grouting within a sleeve. Also during this phase, major mechanical system installations take place, primarily the crude oil piping and oil level indicators. Several complex pumping and ventilating modules may be placed in the utility shaft. In addition, risers for flow lines from satellite wells and for crude oil transfer lines are installed in or on the periphery of the base caisson. The risers may include J-tubes and, in some cases, curved conductors. In other cases, entry tunnels may be constructed, through which the pipelines will eventually be pulled.

Experience has shown that the mechanical installations can occupy a disproportionately long period in the overall construction schedule. To minimize this time requirement, as well as to permit more efficient assembly, modularization should be performed to as high degree as practicable, with large modules being set into the utility and riser shafts. Similarly, conductor guides and support frames should be preassembled for rapid installation by a large crane barge.

Modularization and prefabrication permits coatings to be applied under shop conditions. At the site, conditions of moisture and temperature are usually adverse, and hence coatings at the site should be limited to touchups and welded splices.

10.2.5 Stage 5—Completion of Base Caisson

The next substage at the deep-water site is the construction of the domed roof of the base caisson. The construction of this domed roof, covering 10,000–20,000 m² of area, covering over the 15–30 cells, often requires more time on the construction schedule than the construction of the walls of that base caisson, even though the ratio of concrete may be as low as 1:3. See Figure 10.2.5.1. The delay, of course, is due to support of the forms. Prefabricated dome soffits of concrete have been extensively employed as well as the more conventional trusses and plywood. A steel soffit, containing all the necessary reinforcing steel and having multiple studs for composite action, could conceivably be placed as a single lift, permitting rapid concreting thereafter.

Figure 10.2.5.1 Forming, reinforcing, and concreting of domes on the base caisson of the Statfjord B platform.

Once concreted, the domes are often covered with a meter or so thickness of lightweight concrete to absorb the impact of objects dropped from the deck during drilling and production operations.

Over-dome pipe supports are also provided for any crude oil transfer piping which may have been installed on the exterior of the base caisson and which now must be led horizontally into the riser shaft.

Since the substructure may later be towed to another site for deck mating, towing attachments and notches are fitted around the periphery of the walls where they intersect the domes.

Circumferential prestressing is also usually required at this intersection between domes and exterior walls. Proper staging should be applied to facilitate this operation.

Most offshore platforms require solid ballast in order to lower the center of gravity as much as possible. Solid ballast may consist of concrete placed above the base slab, within the cells. Mass concrete typically generates high heat of hydration and hence leads to thermal expansion which may adversely load the adjoining walls and crack the slab. Therefore, the mix should be selected for low heat and low modulus of elasticity. High density is desirable; this may be enhanced by selection of aggregate of high specific gravity. Blast furnace slag–cement, low-heat cement, or a portland cement–pozzolan mix should be employed. Strength is normally not a criterion; hence total cement content can be kept relatively low. To prevent undesirable structural interaction between the ballast concrete and the walls, sheets of crushable material (polyurethane foam) are often placed on the boundary walls.

Iron ore has also been used effectively as solid ballast, placed by pumping in as a slurry. Provision must be made to decant the slurry by drainage.

Finally, where solid ballast is to be placed on exposed slabs, such as cantilevered base slabs, high-density rock can be placed.

There are several methods available for placement of rock on exposed underwater slabs: placement through a large-diameter (e.g., 1-m diameter) tremie tube, placement in a skip or bucket that is lowered to the slab before discharge, and discharge from the surface in a mass, as from the pocket of a small bottom dump or side dump barge. In the latter case, impact force has to be considered.

Tests both on offshore platforms and on similar operations have shown that discharge in a small but

Figure 10.2.6.1 Using slip forms to construct the tapered shafts of the Beryl A platform. (Courtesy of Norwegian Contractors.)

coherent mass is usually quite satisfactory, without excessive segregation, especially if the mass is flooded before discharge so that it doesn't contain large quantities of entrapped air. The alternative, applicable in deeper water, is placement through a tube.

10.2.6 Stage 6—Shaft Construction

The next stage is typically the construction of the shafts, which typically may be from one to four in number. These are usually tapered, with varying wall thickness, necessitating the use of sophisticated adjustable slip forms. See Figure 10.2.6.1.

Shaft slip forms, like chimney (stack) slip forms, tend to rotate. To correct this after it has started is difficult. Chain jacks may be used to react against the climbing rods, or a more rigid steel beam stub may be embedded in the wall and used as a reaction point.

The large single shaft of the Ninian central platform was constructed using match-cast precast concrete segments later post-tensioned to act monolithically. See Figure 10.2.6.2.

Control of verticality is by lasers, set accurately on the base slab. Unlike chimneys on land, the vertical axis of a floating structure changes with list and trim of the structure, so all points must be rel-

Figure 10.2.6.2 Outer shaft of Ninian central platform is completed with match-cast precast concrete segments posttensioned together.

ative to the base slab. Control of tolerances is discussed in Section 18.9.

On some gravity-base platforms (e.g., the Andoc Dunlin platform) the upper portion of the shafts was made of structural steel in order to reduce weight and overturning moment. The large "cans" of the shaft were set by use of a large floating derrick after the base structure had been ballasted down to reduce the height above water.

Gravity platforms have been constructed of steel: examples are the four small Loango platforms off the mouth of the Congo River and the large Maureen platform in the North Sea. Their construction generally follows that of the concrete platforms except that, being light, most of the structure can be constructed in the construction basin. See Figures 10.2.6.3 and 10.2.6.4.

Large "bottles," cylindrical tanks attached to the edge of the base, provide both buoyancy and stability during tow and installation.

An especially daring hybrid concept has been proposed by Saga Petroleum of Norway. The mating sequence proposed involves joining the top section of steel tubular framing to the lower section of concrete cellular structure while both are inclined at 30° to the horizontal. Obviously, control of attitude, orientation, and rotation are all required, yet this same operation has already been successfully carried out for an articulated loading column of somewhat similar concept. See Figure 10.2.6.5.

Mechanical outfitting must continue within the shafts. Temporary doors may be left to facilitate personnel entry and provision of services. The "doors" subsequently have be to concreted closed and must then be able to carry the stresses imposed by hydrostatic and gravity loads and by the global bending of the shaft under wave action. They must of course be watertight. Detailing of the closures needs to be carefully worked out by the constructor's engineers in coordination with the design engineer. Particular attention has to be devoted to the upper horizontal joint where bleed water and settlement may allow water to permeate under high hydrostatic head. A secondary injection of grout or epoxy may be utilized. Consideration must be given to residual stresses and differential prestress.

There is typically a large, heavy ring beam or girder to be constructed on top of the shafts. This may be of reinforced and prestressed concrete, as with the Condeep platforms, or of fabricated structural steel, as used at the Ninian central platform. In either case, tolerances; level, distances, and so on must be controlled to within a very few millimeters. Since at the time of their installation the structure will probably have ballast distribution different from that at the time of deck mating, correction factors must be calculated and applied to the different inclinations and hence distance apart of the shafts.

10.2.7 Stage 7—Towing to Deep-Water Mating Site

At this stage, the structure may be towed to a different site for deck mating, since this requires even deeper water than at the final installation site. The structure will still be floating with waterline near the top of the large base caisson. Since the tow will be partially or wholly within restricted waters, the boats will tow with short scopes of their towlines. Unfortunately, the thrust of the propellers will react against the base structure and the net forward speed will be significantly reduced or even prevented. In the case of the Statfjord B platform, the

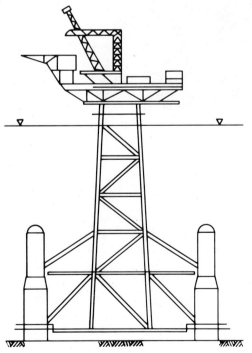

Figure 10.2.6.3 Loango steel gravity-based platform. (Courtesy of Tecnomare.)

Figure 10.2.6.4 Steel gravity-based platform under fabrication in Scotland, for Maureen Field in North Sea. (Courtesy of Tecnomare.)

tow had to be aborted and rearranged. Use of two pusher tugs then proved far more effective.

Control of the structure within narrow channels has been effectively carried out by having one or more lead boats, auxiliary boats on each side, and either pusher tugs or stern boats.

For such a tow, a temporary control platform is mounted. A generator must be provided for power and lights, with fuel for the trip, fresh water supplies, radios, and navigation gear. A life raft is provided, plus fire extinguishers.

The Statfjord platforms were towed about 100 km to their deep-water mating site; the Ninian central platform was towed only 30 km, but the Andoc substructure was towed from Rotterdam to Bergen, Norway.

The deck mating site should be selected so as to provide adequate depth to enable the substructure to be almost fully submerged by ballasting yet provide good protection from winds and good mooring positions. An ideal location would be at the head of a relatively narrow and deep inlet so as to be protected from high winds.

Moorings are installed both in the seabed as necessary and onshore, the latter facilitating the needed adjustments in lengths of line as the structure is ballasted down and deballasted up.

The substructure is moored upon arrival.

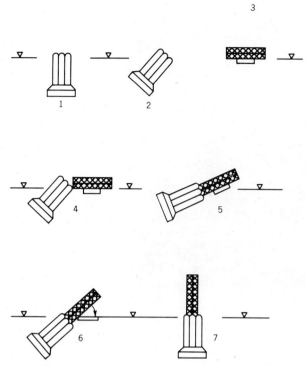

Figure 10.2.6.5 Mating sequence proposed for Saga's steel–concrete gravity-based structure.

A second set of moorings is established in order to permit mooring of the deck–barge complex upon its arrival.

10.2.8 Stage 8—Construction of Deck Structure

While the substructure is being thus constructed, the integrated deck is erected. This may be done in a shipyard, on girders or trusses spanning a graving dock or overhanging a trestle (jetty). These support conditions may be different from those that will be acting when the deck is finally mated on the substructure, and hence deflections and twist deformations must be calculated and accounted for. These can amount to as much as 10–15 cm in extreme cases and affect the equipment and piping as well as the girders themselves. Most recently, therefore, pillars have been constructed in shallow water near the shipyard. The pillars are short segments identical to the top of the shafts of the substructure, enabling erection of the deck under exactly the same support conditions as it will have when mounted on the substructure. It is therefore essential that these pillars be correctly located, with a tolerance in plan of only 2 or 3 cm and in level of perhaps only 1 cm.

The deck structure will typically start off with large, heavy structural steel ring girders, perhaps designed to be filled with concrete later, after the deck has been finally mounted on the substructure. Then very heavy steel trussing or plate girders are built up with prefabricated sections. Because of the spans and loads involved, plate thicknesses often run up to 100 mm. Welding procedures will probably require both preheat and postweld heat treatment. Since these joints are subjected to cyclic loading, fatigue considerations will probably require grinding of the weld profiles and checking not only of the welds by NDT but also of the heat-affected zones adjacent for hardness.

As the support structure is built up, various modules of equipment are installed. Scheduling is obviously a very complex matter, since equipment modules take time to assemble, yet the support structure must also be completed. Careful planning will enable equipment modules to be lifted over the support structures, lowered, and in some cases skidded to final location. See Figure 10.2.8.1.

The deck, now supported on several pillars or other temporary supports, is a very large structure which will undergo expansion and contraction with the temperature and will experience rotations as weights are installed. Therefore, it is usually supported on heavy laminated neoprene and steel pads, allowing shear deformation and rotation.

At the same time, the deck must be adequately secured against a windstorm. Stops must be provided so that the structure can in no way slide off. Wind forces can reach 500–1000 tons or more under a severe storm. Provision must also be made for jacking if the combination of temperature cycles and wind cause the structure to "crawl" sideways. Bearing pads should therefore be regularly inspected visually.

During this period, a great many men will be working in confined spaces and with possibly conflicting demands for scaffolding, cranage, ventilation, lighting, and access. Since the deck is almost always on the critical path and the labor costs are high, careful planning and scheduling can produce significant savings.

Appropriate provision must be made for fire protection. Also, since the work is over water, either floats or nets should be installed to protect a man who falls.

Proper provision must be made for grounding

Figure 10.2.8.1 Steel deck for Statfjord C platform is constructed on four pillars located adjacent to shipyard quay.

of electric welding equipment. Welding cable insulation should be maintained in good condition and repaired or replaced when damaged by abrasion. See API RP2A, sec. 5.7.

Housekeeping and cleanliness, along with proper lighting, will have positive benefits on safety and efficiency. Cleaning of insulation, sandblast sand, and debris is especially difficult but of course is no different in nature than with large ship construction. Some operations will require protection from rain and wind; fireproof canvas, plastic, or similar shelters may be required.

Detailed weight control of the superstructure must be maintained, as it must for the substructure, with the added impetus that weights on the deck are high above the center of buoyancy and hence will have a disproportionate effect on the stability. For example, to counter an extra 100 tons on deck may require adding 1000 tons of solid ballast in the hull.

When the deck is ready for transfer, it is essential that the weight and its distribution be accurately known, since the deck structure will be transferred by floating equipment. This structure may differ from the final configuration, since one or more modules may have been delayed. Temporary materials and equipment may be on board and must be accounted for.

10.2.9 Stage 9—Deck Transport

The deck is now to be lifted off the pillar supports by means of large barges. These "barges" may be conventional offshore barges, halves of surplus tanker hulls, or special pontoon combinations designed to float in under the deck, between the pillars. They are then deballasted so that seafastening supports may be affixed between the barge and the deck girders at appropriate lift points.

A careful analysis and evaluation has to be made as to where to use fixed connections, where to use pinned connections, where to allow flexibility, and where to use ties or struts. See Figures 10.2.9.1, 10.2.9.2, and 10.2.9.3.

The two or three barges, together with the deck, will form a complete structural system and must be analyzed as such, under both static conditions and the dynamic conditions (waves) expected during this particular movement.

When all barges have been properly ballasted and the seafastenings connected, and the weather forecast is favorable, they are progressively deballasted until just before liftoff. Load cells may be used to verify forces in key members. Then the three barges are further deballasted, raising the deck clear of the pillars.

During this operation, the barges are moored to

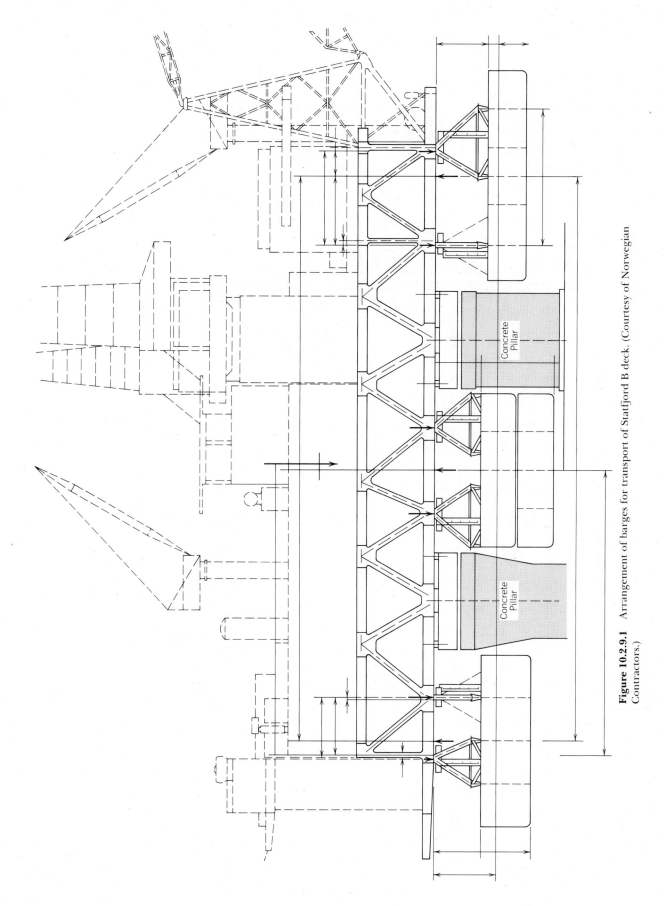

Figure 10.2.9.1 Arrangement of barges for transport of Statfjord B deck. (Courtesy of Norwegian Contractors.)

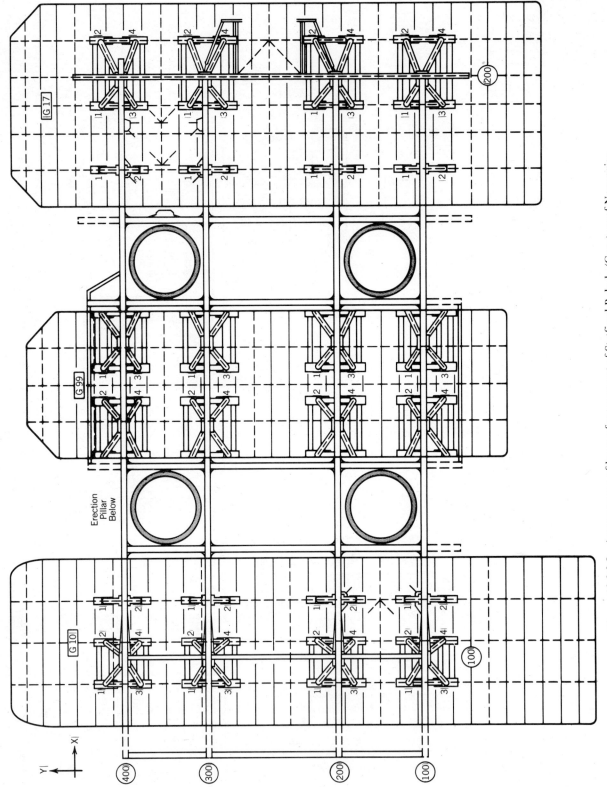

Figure 10.2.9.2 Arrangement of barges for transport of Statfjord B deck. (Courtesy of Norwegian Contractors.)

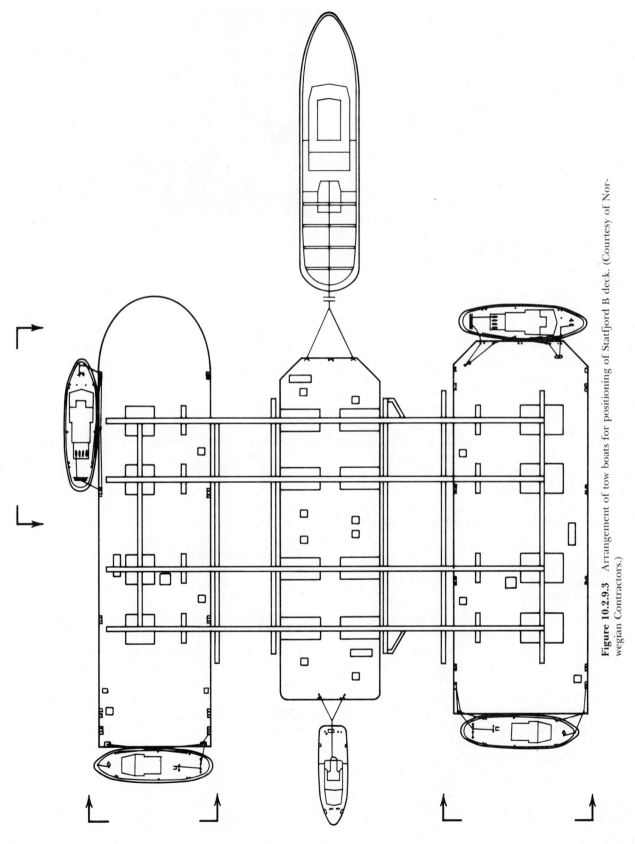

Figure 10.2.9.3 Arrangement of tow boats for positioning of Statfjord B deck. (Courtesy of Norwegian Contractors.)

the pillars by taut lines of sufficient length to accommodate the vertical movement. Nylon lines may be used to absorb shock. If blocking is used between barges and pillars, it should have Teflon coating or similar provisions so as to enable sliding to take place.

The boats then take the structure in tow, providing both pull and lateral guidance. See Figure 10.2.9.4.

The major concern during such a catamaran or trimaran tow is wind across the channel. Boats must have adequate power. Bow thrusters are desirable if the tow must be made in narrow channels; this will enable the boats to operate on very short scope if necessary.

Contingent plans should be made in case a wind storm is suddenly forecast; lay-by areas should be selected.

Temporary facilities will have been mounted for the tow similar to those provided for the substructure in order to provide for control, communication, lighting, and emergencies.

Upon arrival at the site, a decision must be made, depending on weather forecasts, as to whether to moor the deck–barge complex or to proceed directly to the mating.

When the deck–barge complex is finally committed to the mating, many of the seafastenings can be loosened so that while support is still provided, there are as few tension connections between barges and deck as practicable.

10.2.10 Stage 10—Submergence of Substructure for Deck Mating

Before arrival of the deck–barge complex, the substructure will have been subjected to two tests. One is a standard inclination experiment, in which a known weight is moved a known distance. The resultant angle of heel is measured. From this, the true position of the metacentric height can be determined and hence the location of the center of gravity.

The second is to test the ballasting and deballasting systems and control systems and to verify watertight integrity of the substructure. It also is an excellent opportunity to train the crew in this all-important operation. In some cases, watertightness depends on temporary closures; in the case of Ninian, several hundred plugs were used to close the holes in the perforated breakwater. These must all be double-checked to ensure they are secured

Figure 10.2.9.4 Trimaran tow of Statfjord B deck passing Stavanger, Norway. (Courtesy of Norwegian Contractors.)

tightly; the failure of only one could cause the loss of the entire platform.

The substructure is submerged in a series of steps, checking weight of ballast water added versus resultant draft.

In a number of cases, compressed air has been introduced into the main cells during deep submergence in order to provide additional safety against implosion due to the high hydrostatic forces. It is recommended by both the Federation Internatiónale de la Precontrainte (FIP) *Recommendations for Concrete Sea Structures* and the American Concrete Institute ACI 357 *State of Art Report on Concrete Sea Structures* that the structure have a safety factor of at least 1.05 in the event of loss of air; thus the use of compressed air is limited to that needed to further raise the safety factor.

Care must be taken to ensure that when compressed air is introduced into a cell, it is done in a step-by-step pattern matching the external hydrostatic pressure during both deep submergence and subsequent deballasting. The air pressure is constant over its full height, whereas the hydrostatic pressure not only decreases due to deballasting but also varies in a triangular diagram along the height of the cell. Uncontrolled air pressurization could result in an outward explosive force on a structure designed only for external load.

Compression of air raises its temperature which will later cool back to that of the external water, in turn reducing the pressure.

The typical structure is very large. It is not unusual for the compressed air volume to add several hundred tons to the total weight of the structure during the filling process.

Figure 10.2.11.1 Statfjord C deck being maneuvered over deeply submerged concrete substructure. The top of the shafts can be seen in the lower right, with the base now 165 m below water.

10.2.11 Stage 11—Deck Mating

The scene is now set for the deck transfer operation. The weather forecast must of course be favorable, with minimum wind. All boat traffic in the vicinity is stopped. Water density measurements at depth are made to be sure that calculations are based on actual densities.

The substructure is ballasted down so that only 3–5 m of shaft extend above the waterline. The deck–barge complex is slowly winched in around the shafts. Clearances are typically only 30–60 cm (12–24 in.). See Figure 10.2.11.1.

Hydraulic winches can control the structure within a few centimeters, but are subject to even minor surges. Blocking is therefore sometimes employed, arranged with hydraulic jacks, to effect the final horizontal control. The blocks are fitted with Teflon so they can slide vertically. See Figure 10.2.11.2.

When horizontal positioning has been verified, it is time to deballast the substructure. There will

Figure 10.2.11.2 Adjusting final position.

Figure 10.2.11.3 The mating takes place.

usually be 1–2 m of clearance. The substructure is brought up just short of contact; all points are checked. Then deballasting continues. See Figure 10.2.11.3. When 10 percent of the load has been transferred, a final position check is made. All remaining tension seafastenings are now disconnected so that they provide vertical support only.

During this stage, a very complex interaction takes place. As the substructure starts to raise the deck, the barges also rise due to relief of weight. They tend to follow the deck up as it is picked up by the substructure.

Since the substructure will now have the weight of the deck on it, the height of ballast water in the cells will have to be further lowered, which means that the net external hydrostatic heads are now the maximum which the structure will ever see. This condition typically is the most critical in the structure's life, as far as implosion is concerned.

Finally the substructure is picked clear of the barges and on up to a safe height. See Figure 10.2.11.4. This "safe height" is one selected to balance safety against implosion (with all compressed air removed), stability, and access for further work. The barges can now be pulled clear.

Since the conditions of support of the deck during transport must be different than those on the pillars or on the substructure, there will be differential vertical and horizontal deflections and twist. These must be accommodated either in the flexibility of the deck and shafts or else by the use of bearing devices which will permit lateral movement on the bearings. Sand jacks, neoprene pads, and sliding plates are some of the systems used. Whatever is used, positive stops should be provided.

Vertical tolerance adjustment between the deck and the ring beams on top of the shafts has been provided by a number of means. C. G. Doris has employed flat jacks, which permit measurement and equalization of load. After such adjustment, the water in the jacks is replaced by cement grout. Norwegian Contractors have employed soft iron pipe sections, thick-walled, which deform under load concentrations, equalizing the load onto adjoining pipe sections.

The deck now rests on the shafts by gravity alone. A check must be made to ensure safety in the event of accidental flooding, which would produce a heel. A typical requirement is that the deck not slide off even if a significant heel (e.g., 7½° or even 15°) is experienced due to accident.

Such an accidental heel did occur at a later stage with the Statfjord A platform. During testing of the ballast control system, valves were left open, causing

Figure 10.2.11.4 The substructure is deballasted, raising the deck.

an unplanned shift of water to one side, resulting in an increasingly severe list. Alarms sounded and the workmen started to abandon the structure. Fortunately, the supervisor on the platform had the wisdom and courage to descend deep into the utility shaft, correct the valve errors, and ballast the structure back to vertical.

The deck is now usually additionally secured to the shafts by prestressing. Prestressing is especially effective in eliminating cyclic fatigue and in providing safety against uplift under accidental conditions. These prestressing tendons are short, and hence a system must be used that permits adjustment for initial seating losses, since in a 4-m-long tendon, a seating loss of 6 mm amounts to a loss of prestress of 350 N/mm^2, almost half that input by the jacks!

These prestressing tendons must also be protected against corrosion, since this is one of the most vulnerable zones. In particular, sealing and drainage must be assured to prevent salt-water spray sitting in anchorage pockets.

During the entire deck-mating operation, careful checks are made of the ballast water in all the cells to detect any unexplained variances in water level or quantity pumped. Even relatively small discrepancies should be checked by visual examination within the cells.

A relatively small in-leakage in one of the cells of the Statfjord A platform, showing only about 300 liters/min inflow, turned out on more detailed investigation to be due to a large laminar crack in a cell wall, caused by secondary bending of the large caisson as it was subjected to the high hydrostatic forces of the deep submergence. Extensive repairs were necessary, including epoxy injection.

The Ninian deck transfer was carried out in a different but very ingenious manner. The deck structure, weighing 7000 tons, was constructed on land, at Inverness, Scotland, then skidded onto a single large barge for tow around the north end of Scotland to the mating site near the Isle of Skye. In a sheltered inlet, the deck was transferred to two barges which supported it only along its edges. See Figure 10.2.11.5.

Meanwhile the concrete substructure was ballasted down. In suitable weather, the twin barge assembly was brought to the substructure. By means of lifting rods supported from towers on the centerlines of each barge, the deck was raised so that it would clear the top of the single central shaft. The twin barges now were guided past the shaft, straddling it. The lifting rods now lowered the deck onto the shaft.

Temporary supports were flat jacks on neoprene pads. With the change in support conditions, the deflections of the deck trusses also changed: the lateral movements were accommodated by the pads. The flat jacks were then used to equalize the load at each of eight support points. Then they were pumped full of grout.

10.2.12 Stage 12—Hook-up

The next operations are those of hook-up of all equipment and its testing. These can require many hundreds of thousands of man-hours. While the site of work is protected, nevertheless all access must be by water, and the deck is now 30–50 m or so above the water. Hundreds of men must move on and off each shift. See Figure 10.2.12.1.

It is necessary to provide boat landings, adequate crew boats, elevators and stairways, lighting, and water and power supply. As soon as possible, the permanent cranes on deck are made operable so they can handle supplies and fittings on and off as required.

264 Concrete Offshore Platforms (Gravity-Base Structures)

Figure 10.2.11.5 Deck erection on the Ninian central platform takes place behind the protection of the Isle of Skye in northwestern Scotland. (Courtesy of Tokola Offshore and Chevron U.K.)

Figure 10.2.12.1 Inshore hook-up of mechanical and process systems on Statfjord B Condeep. (Courtesy of Norwegian Contractors.)

Safety of the workers must be assured; this requires life rafts, a safety boat, and suitable temporary railings. Fire protection must be provided.

10.2.13 Stage 13—Towing to Installation Site

The structure is now placed under tow to the site. Its displacement may be several hundred thousand tons; Statfjord B displaced almost 700,000 tons. Four to six of the largest tugs in the world are typically employed, each being rated near 20,000 IHP and having a bollard pull in excess of 150 tons. These are arranged with towing pendants and retrieving lines as described in Section 6.1. A more in-depth discussion of towing is given in that section.

Within restricted waters, two stern tugs are also employed, to prevent yaw and to keep the GBS platform fairing true behind the boats. See Figure 10.2.13.1.

When the tow emerges into the open ocean, the scope of towlines is lengthened and the stern tugs are cast off. See Figure 10.2.13.2.

A pilot boat may lead the way during exit from the channel, available to warn other shipping and in some cases to verify clearance across relatively shallow waters.

A safety boat may run alongside, for use by supervisory personnel, interboat transfers if needed in an emergency, and to pick up a man overboard.

The GBS platform is outfitted for the tow with a navigational bridge, radios for communications

Figure 10.2.13.1 Tow of Statfjord A platform from Boknafjord into the North Sea. (Courtesy Norwegian contractors.)

to boats and to the shore base, electronic position-finding equipment, safety and firefighting equipment, quarters for the riding crew, diesel and water supply, and generators for power and lighting. The heliport on the deck may be activated for emergency use. Duplicate radar, gyros, fathometers, and forward-searching sonars are installed, as well as emergency generators.

The route will have been carefully surveyed using profiler equipment so as to identify any pinnacles, wrecks, ridges, or shallows which may have been missed by conventional fathometers.

At least one of the boats may have been equipped with satellite position-finding equipment.

Lay-by areas are provided along a long tow route, where the structure may safely ride out a severe summer storm.

In the event of a storm, the structure may be ballasted down more deeply so as to increase stability.

Data concerning tows of gravity-based offshore platforms are given in Section 6.1 and in Tables 10.2.13.1 and 10.2.13.2. Fig. 10.2.13.3 shows graphically the total tug horsepower in relation to displacement for large gravity-based structures in the North Sea.

During the tow, stability will probably be the controlling parameter. The draft, ballast, and pay-

Figure 10.2.13.2 Ekofisk oil storage caisson under tow in open sea.

TABLE 10.2.13.1 Towage and Emplacement of Concrete North Sea Structures

Unit	Date of Tow and Departure Point	Tow Displac. (tonnes)	Draft with/ Without Dowels (meters)	Distance (nautical miles) Channels or Fjords	Distance (nautical miles) Open Sea	Total Tow Time (days)	Mean Open Sea Speed (knots)	Total Tug I.H.P.
Brent A (Condeep)	July 75 Stavanger	338,000	82	44	118.4	6	2.36	68,000
Brent B (Condeep)	August 75 Stavanger	384,500	76	44	184	7	2.20	68,000
Frigg CDP.1 (C.G. Doris)	August 75 Andalsnes	209,000	67	44	268	8	1.89	44,500
Frigg MCP01 (Doris)	June 76 Kalvik	206,000	66.5	25	350	11.0	1.6	43,000
Brent D (Condeep)	July 76 Stavanger	382,000	116/113	48	172	8	2.30	72,000
Frigg TP.1 (Seatank)	May–June 76 Ardyne	166,000/ 209,000	35 inc. to 64	147	620	11.5	2.78	54,000
Statfjord A (Condeep) (Base Tow)	August 76 Stavanger	370,000	65.2/60.5	75	40	3	2.70	78,000
Statfjord A (Condeep)	May 77 Stord	457,000	119/114.3	24	165	5	2.06	68,000
Dunlin A (Andoc) (Base Tow)	July 76 Rotterdam	232,000	25.2	65	508.5	7.10	2.98	78,000
Dunlin A (Andoc)	June 77 Stord	419,000	131.2/129 (locally 151.2/149)	32	136	5	2.3	68,000
Frigg TCP 2 (Condeep)	June 77 Andalsnes	292,760	91.54/ 86.04	44	307	14	2.15	68,000
Brent C (Seatank) (N. Channel)	July 77 Ardyne	298,000	38.4/38.4	199	738	13.12	3.42	84,000
Cormorant A (Seatank) (N. Channel)	June 77 Ardyne	346,500	37.5/37.5	173	687	8.58	3.34	95,000
Ninian Central (Howard Doris)	May 78 Raasay	601,220	84.2	—	499	12	1.8	76,500
Statfjord B (Condeep)	August 81 Vatsfjord	825,000	130/127	70	164	5.5	1.7	86,000
Statfjord C (Condeep)								

load of equipment on deck will all be selected to give 1–2 m of positive metacentric height. While this may seem small, it must be remembered that the initial righting moment is the product of the GM and the displacement, and the displacement in this case is very large.

Damage control provisions may also govern design. Usually the structure will be required to be able to suffer flooding of any one exterior compartment. Internal subdivision of a shaft and fendering of a shaft are both rather impracticable in most cases; hence these requirements are usually

TABLE 10.2.13.2 Towage and Emplacement of Concrete North Sea Structures

Unit	Total Static B.P. (tonnes)	No. of Towing Tugs	Speed of Immersion (m/hr)		Time to Emplace (hr)	Conditions at Touchdown: Wind Force/ Wave Height, Period	Water Ballast Added to Emplace (tonnes)[b]	Distance Off Target (meters)
			Max.	At Seabed				
Beryl A (Condeep)	584	5	8	8	45	SE'ly 2 0.5 m 5 sec	90,000	32
Brent B (Condeep)	584	5	12	6	39	S'ly ⅔ 0.6 m 6 sec	124,000	25
Frigg CDP.1 (C.G. Doris)	372	4	5.6	2	39	WSW ¾ 1.0 m	6,000	14
Frigg MCP01 (Doris)	335	4	6	1.6	13	NW ¾ 2.0 m 6 sec	70,000	8
Brent D (Condeep)	545	5	6	6	11	WSW 3 1.2 m 5 sec	12,500	8
Frigg TP.1 (Seatank)	440	4	10	Minimal[a]	11.7	SE 4 1.5 m	4,500	1
Statfjord A (Condeep) (Base Tow)	605	6						
Statfjord A (Condeep)	565	5	8	8	7	SW 1–2 1.3 m 8 sec	13,000	10
Dunlin A (Andoc) (Base Tow)	560	6						
Dunlin A (Andoc)	632	6	10.5	5	4	NE 3 2 m 8 sec	2,500	12
Frigg TCP 2 (Condeep)	565	5	6.2	6.2	6	NNW 3 1.5 m ⅘ sec	5,000	2
Brent C (Seatank) (N. Channel)	645	5						
Cormorant A (Seatank) (N. Channel)	730	6						
Ninian Central (Howard Doris)	585	5	—	—	19			10
Statfjord B (Condeep)	715	5	7	1	6	W ½ 1.5 m 7 sec	11,000	15

[a] Speed of immersion of TP.1 at contact with seabed was related to moving at slow speed onto a sloping seabed.
[b] Calculated from commencement of tug rearrangement to point where unit no longer considered in transit condition.

waived if the shaft is specially thickened and reinforced to withstand the impact of a boat near the waterline at the towing draft. Note that the towing draft is normally less than the installation draft.

Dynamic response of the platform's motions must be considered under the action of the design storm. This is usually taken as the 10- or 25-year return storm for the period of year involved. With its relatively low metacentric height, the structure will typically roll and pitch in a relatively long natural period, perhaps 60 seconds, developing maximal accelerations of about 0.20 to 0.25 g. The accelerations of course affect the deck equipment and require that all attachments be capable of resisting the lateral forces which develop.

It is important to calculate righting moments for

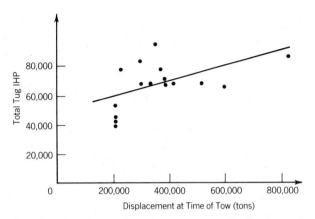

Figure 10.2.13.3 Total tug horsepower vs. displacement of North Sea gravity structures as ballasted to towing draft.

the various angles of heel and the heeling forces due to wind and waves, since a typical platform does not have the response of a conventional ship hull and metacentric height is a measure of stability only at very small angles of heel. Usually, a limit of 5° will be placed on the maximum heel during the 10-year storm, assuming the towline is slack or has been dropped.

10.2.14 Stage 14—Installation at Site

Upon arrival at the site, the structure is to be ballasted down onto the seafloor. The several boats fan out in starlike pattern so as to hold the GBS platform in location. Here is where bow thrusters are very helpful, enabling the boat to select its heading without necessarily exerting excessive pull.

Witt and Meurs in their paper "The Positioning of Offshore Constructions. Research and Training by Simulation" read at the European Offshore Petroleum Conference in Stavanger, Norway, 1978 suggest the following rules to govern tug manipulation and positioning:

1. Wait to give a new order until the former order has had noticeable effect.
2. The structure moves with great inertia and little drag, so all braking must be done by tugs.
3. Use as few tugs as possible.
4. Redirect tugs back to their original position after each action.
5. Use little steps in changing power and course as opposed to radical changes.

It would appear that rule 1 will need to be interpreted in context, since braking action may sometimes have to be initiated soon after a thrusting force is applied.

An alternative method to the use of the boats for positioning is that of mooring lines, generally using the inverted catenary mooring system, with wire lines or chains leading from the anchor to a spring buoy. From winches mounted on the deck of the platform, wire lines are run down to submerged fairleads, then out to the spring buoy.

This system of moorings is likely to be increasingly employed as offshore structures are installed over subsea wellhead templates with predrilled wells.

One or more large derrick barges may take the place of several spring buoys, serving as control vessels during the lowering, as was done in the case of the Cognac steel jacket platform.

Control during installation is carried out from a control room within the structure. The instrumentation will generally include the following:

Echo sounders, to give bottom clearance at four or six corners

Pressure transducers to read draft at the same locations

Pressure transducers to read internal ballast water levels in each cell

Strain gauges to read axial forces and moments in dowels and in selected skirts

Differential pressure transducers to give water pressure in each skirt compartment

Biaxial inclinometer for tilt

Earth pressure transducers to give contact pressures on base slab

Strain gauges to give strain of reinforcing steel in base slab or domes

Pressure transducers in a closed hydraulic system to give skirt penetration

The installation manual will contain a description of all the systems involved, background data, and drawings. It will give a detailed set of guidelines for installation, including:

Positioning and orientation
Touchdown
Dowel penetration
Ballasting and drainage (venting)
Concrete skirt penetration
Base slab (or dome) seating
Underbase grouting

Ballasting down is carried out by pumping water in or by controlled free-flood. Stability generally becomes greater during submergence, at least with structures having generally vertical walls. However, conical structures such as those proposed for the Arctic, lose waterplane rapidly as they are submerged more deeply and may become unstable. For these, temporary buoyancy tanks may be employed.

Another system, extrapolated up from the smaller barge-mounted facilities in the Gulf of Mexico, is "tipping down." One end is purposely tilted down to engage the seafloor; then the entire facility is rotated down. In such a case, stability will initially depend on the inclined waterplane and, later, on the support from the bottom as it engages the lower end and prevents sideways rolling.

While such maneuvering has been successfully carried out for smaller structures, its use on large structures should probably be limited to relatively minor angles of tilt and be well verified by model basin tests before field implementation. Consideration must be given to the loading on the structure's edge during this landing and the disturbance done to the seafloor. Skirts may suffer overstress in bending.

For large structures and major offshore platforms, the use of temporary buoyancy would appear to be more conservative and appropriate.

Temporary buoyancy tanks will have to be relatively large in diameter (e.g., 10 m) and high (e.g., 40 m) so as to raise the center of buoyancy without significantly affecting the center of gravity. Therefore, they will usually be made of steel, with heavy ribs, although one study has shown that a hybrid steel–lightweight concrete shell was even more efficient. Temporary buoyancy tanks also give a minor improvement in the moment of inertia of the waterplane but unfortunately also increase the displaced volume, so that the net effect on BM is not usually significant.

Temporary buoyancy tanks give their major contribution by raising the center of buoyancy. Therefore, they do not have to be symmetrically placed around the structure. For example, one such tank can be installed on one side only, if that is desired, and verticality maintained by selective ballasting.

These tanks must be attached to the structure at some point on the enlarged base caisson. These attachments are subject to the dynamic cyclic forces of the waves during the tow; these can lead to fatigue, especially in the corrosive environment.

Prestressing of the tanks to the base appears one good means of attachment, using posttensioned unbonded tendons prestressed from the top of the tank at a stage when that top is above water. These can then be released after installation, again working from the top of the tank. Alternatively, they can be cut by explosives, hydraulic shears, or underwater burning.

To release such tanks, they should be ballasted to a slight negative buoyancy and then released. A towline to an attachment at the center of rotation of the tank will prevent the tank from rotating up, with possible damage to the shafts, deck, or tank.

Temporary buoyancy tanks were employed on 10 caissons of the offshore terminal at Hay Point, Queensland, Australia. During the installation of the last caisson, one connection failed due to corrosion-accelerated fatigue; fortunately, this happened when the structure had almost touched down, so there were no serious consequences.

Returning to the structure at the site, it is being held in position by the tugs (or mooring lines) and being ballasted down by pumping or flooding in of seawater. All events and conditions are being monitored in the control room. As it nears the seafloor, the rate of descent is slowed so as to allow the water to escape.

Attention is directed to the five last columns of Table 10.2.13.1 which pertain to installation data. It will be noted that the gravity-base structures in the North Sea have typically been installed within an accuracy of 10–15 m by use of tugs in a "star formation."

Where greater accuracy is required, Noble-Denton Associates have suggested the following alternative methods.

1. Tugs in star formation using pin pile system on structure: ±5 m
2. Tugs in star formation using pin pile system on structure and anchor winches on tug: ±4 m
3. Anchor winches on board structure leading to preinstalled anchors on seafloor: ±2 m
4. Anchor winches on board structure leading to seafloor anchors, with pin pile system: ±1 m
5. Anchor winches on board structure leading to anchors, using pin pile system plus control lines to moored derrick barge: ±0.5 m
6. Anchor winches on board structure, leading to anchors, plus pin pile system and vertical taut line to template on seabed: ±0.5 m

Ten concrete caissons for the offshore terminal at

Hay Point, Queensland, Australia, were all installed within 0.5 m by the use of anchor winches on board, with lines leading either to preplaced seafloor anchors or to previously placed caissons. Similar procedures were followed with the four concrete caissons for the Tarsiut Caisson Retained Island in the Canadian Beaufort Sea.

At touchdown, the dowels plow into the soil. Bending stresses, converted from strain gauges on the dowels, are read out in the control room. The short-range, high-frequency echo sounders on each corner give the distance between base slab and seafloor.

The skirts now engage. These have been designed not to buckle even if they hit a subsurface boulder; they will displace it laterally through the soil. Ballast is continued to be added until the desired penetration is achieved. The initial penetration rate is kept slow, 10–15 cm/hr, so as to avoid overpressure in the skirt compartments and consequent piping. Once skirts are well embedded, the rate of penetration may be increased to about 1 m/hr. Meanwhile, water is being vented out from the skirt compartments.

Eventually, penetration is achieved and the structure comes to rest on the base slab or on concrete skirts or sills purposely designed to halt further penetration and thus prevent excess local pressure on the slab.

Without such arresting, the structure will continue to penetrate until it bears directly on the slab. High spots of clay will be displaced and squeezed out. Even a surface boulder will be forced down into the soil. A high spot of sand, however, may give very high local bearing resistance, thus overloading the slab.

Pressure cells on the base slab and strain gauges may be used to monitor slab–foundation interaction.

If the skirts do not penetrate the required distance, even when the structure is fully ballasted, then the water trapped inside the skirt compartments may be bled off into the utility shaft at an underpressure, thus increasing the effective force causing penetration. Such selective ballasting and underpressuring allows very accurate control of level. The degree of underpressure must not be so great as to initiate piping under the skirts.

For the Gullfaks C platform, 22 m long concrete skirts will penetrate the soft clays with interbedded lenses of sand. By reducing the pressure inside the cells, the necessary driving force will be developed.

A large scale field test demonstrated the validity of this concept and the enhanced effectiveness of cycling the internal pressure.

For other platforms in similar soils, where deeply-penetrating skirts are required, it has been proposed to progessively remove the soils from within the skirt compartments by air lifts, eductors, or pumps.

Once the structure has been penetrated to its designed embedment, the spaces remaining beneath the base slab are often filled in order to ensure equal bearing on the soil. Grout fill has been used in many cases.

Thixotropic admixtures can be added to the grout to reduce segregation and prevent excessive intrusion of the grout into interstices of rock and grout escape through minor openings into the sea. For example, for offshore caissons in Australia which were seated on a prepared base of crushed rock, tests showed that normal grouts were much too fluid and penetrated too far into the rock, whereas a thixotropic grout gelled as soon as the resistance to flow increased.

Grouts for underbase fill must always be placed under a very low head to avoid development of piping under the skirts and to avoid lifting the caisson. Use of a gravity feed through an open hopper at the correct elevation needed to just overcome the hydrostatic head by about 15 psi (0.1 MPa) is the best procedure. Provision must be made for venting of the displaced water.

When the void under the base is thick, for example, 0.5–2 m, as it is in many North Sea installations, then means must be taken to prevent excessive heat of hydration from developing in the underbase grout. Some full-thickness tests showed that the temperature with normal cement mixes could rise to 100°C and more, creating undesirable strains in the base slab.

In addition to selecting a grout mix for low heat, consideration must be given to the logistics involved: How will the materials be delivered, batched, and mixed out at the site, taking into account remoteness, sea conditions, and so on? While some operations such as mixing can perhaps be performed on the platform, other solutions involve prebatching with dried materials and transport in water-resistant containers of steel or plastic.

The installation of 14,000 m^3 of underbase grout on the ELF-TCP2 platform required nine days.

Finally, each installation needs to be evaluated as to what is desired. In many cases a low modulus of elasticity is desirable. Strengths need be little better than that of the foundation soils: 1–2 MPa may be a reasonable value in many cases. Bleed should be minimal.

In the North Sea, underbase grout mixes have been developed which are a slurry of cement and seawater, with a foaming admixture such as sodium silicate, from 4 to 10 percent by weight of seawater, along with retardation and stabilizing admixtures. This presents minimum logistics problems and develops a low-strength, low-modulus mix that flows well and tends to fill up underbase cavities.

Cement-based grout fills have been flowed into the underbase gap for most of the North Sea concrete gravity-based platforms as well as under the 10 offshore terminal caissons in Queensland, Australia, and are being intruded under the 66 gate-support caissons of the Oosterschelde Storm Surge Barrier.

In cases where the platform must be relocated, for example, for an exploratory drilling structure for the Arctic, or where a permeable underbase fill is desired, properly graded sand may be slurried into the underbase spaces. The sand slurry is flowed in through piping, either internal or external, having minimum bends; those bends that are necessary should be of large radius. The sand tends to drop out of suspension and build up a little dam; this causes a slight rise in pressure, and a rivulet of sand slurry breaks through to create new fill and successively new dams. Spacing of discharge points is usually 3–5 m on center, assuming a 30-cm depth underbase void is to be filled. See Figure 10.2.14.1.

Sand flow methods have been thoroughly developed by both Danish and Dutch engineers for use in their submerged tube and harbor construction. Offshore, they have been successfully applied under Arctic exploratory drilling caissons.

In all cases of underbase filling, whether by grout or sand, the feed should be at low head and well controlled so as to prevent raising the structure by excess pressure.

The next item of work is that of scour protection, if required. Skirts which penetrate are one means of scour protection; however, these are not always practicable due to soil conditions or draft constraints.

Rock may be dumped around the periphery of a caisson, using a long discharge tremie from a rock hopper barge. This system was developed by a Dutch dredging contractor for covering the Ekofisk–Emden gas pipeline through Danish waters.

In sandy soils a filter fabric of some sort is desirable. Fabric mats can be preattached to the edge of the base slab and then rolled up along the sides of the lower walls. After landing, they may be cut loose and spread out radially from the base. To hold their outer edges, divers may place sandbags or drive a headed steel pin through them to anchor them in the sand. This procedure was employed on the Ekofisk caisson. The filter fabric was then covered by dumped rock. Shortly after completion of this work, the structure encountered a major storm. See Figure 10.2.14.2. The scour protection proved to be fully adequate.

Filter fabric for this application should be heavily reinforced with nylon or even stainless steel wire. A heavily reinforced fabric, sewn together with a finer mesh geotextile, has been used extensively in coastal applications in the Netherlands.

However, dumped rock may be impracticable where there are flow lines and other mechanical

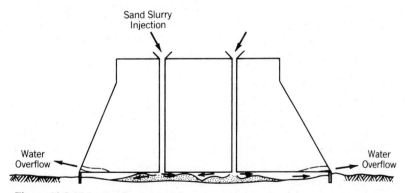

Figure 10.2.15.1 Statfjord A platform in service during North Sea winter storm.

Figure 10.2.14.2 Ekofisk oil storage caisson is subjected to severe North Sea storm shortly after installation. (Courtesy of Phillips Petroleum.)

fittings near the base or where it is impracticable to locate such a vessel in the vicinity of the platform, keeping in mind the many other activities which must be carried out during these first few critical months. The primary effort must of course be the spudding in and drilling of the wells.

Hence one solution is to place prefabricated mats made up of heavily reinforced filter fabric to which concrete blocks have been preattached. Such mats can be placed by derrick barge, using spreader beams; large mats would have been laid out and stacked on a barge for easy handling. These were used to provide scour protection for the caissons of an offshore terminal in Queensland, Australia and successfully performed through a major cyclone shortly after installation.

Such mats can also be affixed to the periphery of the GBS base and stopped off up along the sides. After seating of the GBS, they can be cut loose and laid out on the seafloor with the aid of cranes or work boats.

Such mats could probably not stand the action of the sea on a long open sea voyage and hence would not be attached until at or near the final installation site. They do appear well suited to use in the Arctic, where they might be attached during the wait for ice to clear at a specific installation site.

Articulated concrete block mats were planned for use on a large concrete caisson structure for the Alaskan Beaufort Sea. They were to be transported with the structure. On arrival, they would be attached to the base and laid out over the peripheral soils.

In the construction of the Oosterschelde Storm Surge Barrier in the Netherlands, articulated concrete block mats, incorporating reinforced filter fabric backing, were made up on huge rolls and then rolled out onto the seafloor to prevent scour.

10.2.15 Stage 15—Installation of Conductors

The final construction operation at the site is to install the conductors. These will usually have been transported with the GBS structure and already set in the conductor guides. In the typical platform base, conductor sleeves will have been concreted in the base slab and then sealed against water entry during transportation.

While a double set of reinforced neoprene pile closure sleeves might be utilized to seal these conductors, the closures to date have been a 1.5- to 2-m plug of unreinforced concrete, which is later drilled out by the drill rig.

The conductor sleeves, penetrating the bottom slab, have both internal and external mechanical shear strips or keys to prevent any possibility of failure, as this could cause the loss of the platform during tow.

When the sleeves are later drilled out, the drilling water level inside the drill shaft should first be equalized with the external sea level. Once the drill penetrates, any significant head differential will either cause the foundation soils to flow in or piping to take place to the outside.

In the case of Beryl A, when the head inside had been pumped down about 15 m to permit installation of fittings in adjoining cells, as soon as the drill penetrated, several hundred cubic meters of sand flowed into the shaft. Piping in from under the outer cells occurred. Satisfactory support and filling was restored by extensive underbase grouting and external filter plus rock protection.

For subsequent structures, skirts have been provided around the drilling shafts and water heads have been kept balanced during this operation. Skirts around the drilling shafts will also limit the flow of drilling mud which may spill over from the wells into any adjoining underbase voids.

The conductors are now driven and drilled down to the prescribed penetration, following which the surface pipe is run and the wells drilled. The annuli between well casing and surface pipe and often that between surface pipe and conductor are cemented.

The wells remain free to move vertically, independently of the structure, as the platform settles over time. Figure 10.2.15.1 shows a typical offshore gravity platform in service.

10.3 Enhancing Caisson–Foundation Interaction

In some installations at sites with very weak soils, it may be necessary to augment shear transfer from the base to the foundation in order to prevent sliding. This need can arise with platforms in the Arctic where very high lateral ice forces must be resisted yet skirts are not practicable due to draft restrictions for transport, or where there is inadequate weight of the caisson structure, even when fully ballasted, to drive the skirts into the soil.

Spud piles have been proposed for the as-yet unbuilt Sohio Arctic Mobile System (SAMS). These would be driven after founding of the platform. They would be jacked or driven through steel sleeves so as to engage firmer soils such as partially ice-bonded dense sands at some depth below the seafloor. See Section 20.11.

A typical spud might be 78–96 in. (2–2.5 m) in diameter, with wall thicknesses up to 4 in. (100 mm), designed to penetrate 40 ft (12 m). It will probably extend a further 70 ft (20 m) up into the spud well. Ten to 40 such spuds might be used on a platform.

The spuds could of course be jacked in, in the same manner as the legs of a jack-up rig. The internal plug can be broken up by jets, which are permanently attached to the sides of the spud. The material can then be removed by airlift.

Alternatively, the spuds can be driven in with a

Figure 10.2.15.1 Statfjord A platform in service during North Sea winter storm.

large vibratory hammer or even a diesel or steam hammer. Penetrations are very short as compared to piles, and hence installation time should be minimal.

Several of these spuds can be lowered a meter or two below the base as it nears the installation site, with the spuds being held by jacks or slips, so that they can act as dowels to hold position as the structure is founded. Then, immediately on founding, the remaining spuds can be released and successively driven and jetted to the required penetration.

If subsequent removal is required due to the desire to relocate the platform, the spuds can be pulled up using the following procedures:

1. If there is concern that the spuds have been frozen in by permafrost, jet, and airlift out any remaining plug inside, then use steam to heat the water.
2. Remove top spud supports.
3. Raise the water level inside the spud as high as possible, well above sea level, so as to raise the pore pressure in the adjoining soil.
4. Activate jacks to remove spuds, or use a vibratory hammer to break them loose for lifting.
5. An alternative method of removal, used on similar-sized but longer piles or spuds, is to cap the piles and then apply water pressure to jack the spud out. Since platform relocation is a planned event, the caps can be welded on long ahead of time. Alternatively, attachment of the caps can be by high-strength bolts. See Figure 10.3.1.
6. If a spud has become buckled or distorted so badly by overload that it cannot be pulled, then a diver can always go down in the heated water inside the spud and cut the pile off. Such major distortion will presumably have been known well ahead of the move, giving adequate time for cutting.

Similar-sized damaged piles have been readily cut off in this fashion at an offshore terminal in Cook Inlet, Alaska.

Spuds can also be used effectively in seismic areas where shear transfer to the foundation soils is required yet the mass of the structure is limited and hence there may be insufficient friction between the base and soil. At the same time, the low mass

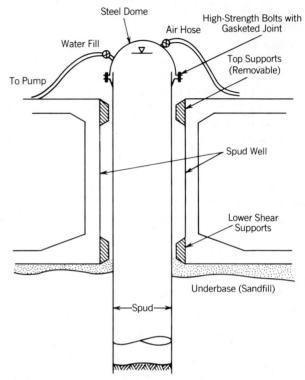

Figure 10.3.1 Spud removal.

reduces the seismic forces. Spuds can then be used to supply the needed shear resistance.

Spuds do not transfer vertical load. Hence the caisson is free to settle. The weight of the caisson on the soil increases the passive resistance of the soil acting against the spud and hence the efficiency of the spuds.

As opposed to spuds, which are useful only in transmitting shear forces to the soil, pilings have also been proposed for some cases, in order to provide increased resistance to bearing and uplift and to prevent excessive settlement in very weak soils. Such a proposal has been developed for the Condeep Tripod 300 platform in the Troll Field of the Norwegian North Sea; as to whether piles will finally be found necessary or not remains undecided at the time of this writing.

Piles have also been considered in planning for the support of caisson-type structures in the Navarin Basin, with its deep water and soft soils.

Uplift-resistant piles were drilled and grouted into the foundation soils in order to hold down the three large underwater oil storage tanks off Dubai, in the Arabian Gulf.

When piles are used, provision must be made to

transfer the pile loads into the structure, which will usually be done by grouting within sleeves. This typically requires a relatively long sleeve length, say 20–30 m, although use of mechanical shear transfer lugs or ribs will reduce the necessary length.

Skirts may also be used to develop supporting capacity for the structure. For example, it has been reported that Norwegian Contractors proposes to use 35-m-long skirts on their newest concept for the Troll platform, which is located in over 300 m of water depth, with extremely soft surficial soils. The skirts will be designed to develop both skin friction (cohesion) and end bearing in the soils, thus minimizing long-term settlements due to compressibility. Such a solution could also be applicable to the similar conditions in the Navarin Basin of the Bering Sea.

The penetration may be able to be attained by ballast weight alone, provided provision is made to vent the trapped water and semifluid mud. Alternatively, techniques previously used on the caissons for deep bridge piers may be adopted, namely jetting and removal of the soil from under the base while increasing the driving weight by controlled ballasting.

To remove the material from the skirt compartments a combination jet–airlift or jet–eductor system may be built into the structure. The peripheral jets slurry the material which is then pumped up and out over the base for disposal. Vents must be provided to the skirt compartment to prevent piping in the case severe under- or overpressure develops as a result of imbalance in the flows.

In weak soils such as clays, the increase in weight must be carefully monitored to prevent sudden shear failures of the soils.

Removal of material from underneath is best accomplished within skirted compartments or open-bottom cells extending below the base slab. Multiple or rotating jets are used to break up the soil.

As an alternative to jets and eductors, an open sleeve may be run to the top of the base structure through the domed roof and a portable dredge pump lowered down—for example, a Pneuma Pump, a Marconaflo System or a hydraulic dredge pump—to remove material successively from each of the many skirted compartments underlying the base.

External jets around the periphery may be installed so as to permit lubrication with water or bentonite slurry (drilling mud) if dense sands must be penetrated.

A number of other approaches to the matter of structure–foundation interaction have been proposed, especially for Arctic and sub-Arctic regions where lateral forces and, in deeper water, overturning moments, are very high. See Figure 10.3.2. Many of these have been described in Chapter 7.

1. In relatively shallow water, with only a moderate depth of weak soils, these may be removed by suction dredge prior to arrival of the structure. It should be noted that this increases the depth of the structure and hence increases its cost, weight, and draft.

2. Dredging as in item 1, above, may be followed by filling the excavation with sand or gravel. Such refilled soils must have adequate consolidation to minimize settlements and to give adequate strength in bearing and shear. If necessary they may be densified by dynamic compaction (repeated dropping of a very heavy weight) or by vibratory consolidation, such as the Tomen, Vibroflotation, or Dutch vibratory compactors.

3. Wick drains may be installed under the base area, to enable weak, claylike soils to drain as the structure weight is applied on the surface. A blanket of free-draining material should first be placed so as to provide lateral escape for the water pushed out of the soil.

    Installation of wick drains may be carried out by barge-mounted equipment ahead of the arrival of the caisson or by working through the caisson itself immediately after founding. This latter has been proposed by Brian Watt Associates for their BWACS Arctic Caisson Drilling Structure.

    Drainage as a means of consolidation is time-dependent; several months are usually required for full strengthening to resist design loads. This may be acceptable in some regions of the Arctic and sub-Arctic where the design ice load is not expected until 3–5 months after installation.

4. Embankments may be placed at the site in the seasons preceding installation, while the GBS is being built. Especially if these are extended over an area larger than the structure's footprint, then they will help to consolidate the clays

276 *Concrete Offshore Platforms (Gravity-Base Structures)*

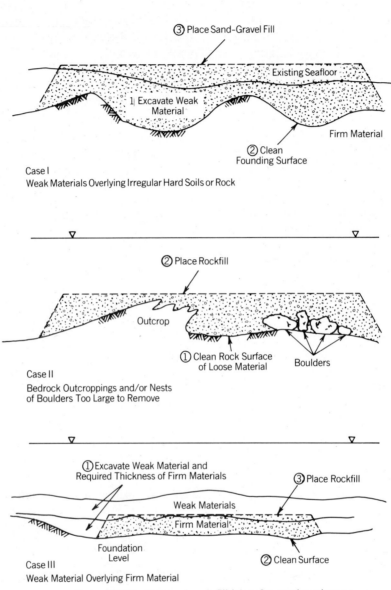

Figure 10.3.2 Prepared rock fill bases for gravity caissons.

and increase their shear resistance, as well as provide external berm resistance to shear failure in the soil.

Of course, external berms can also be installed after the caisson has been founded in order to increase the effective resistance of clays to bearing failure and sliding.

5. Multiple sand piles or stone columns may be driven in a closely spaced pattern under the area on which the structure is to sit. Such sand piles, if closely spaced, are effective in providing direct bearing and shear resistance. If a blanket of permeable material has been placed over the seafloor, then these sand piles will also act as drains when the structure's load is imposed.

6. Loose sands can be predensified by vibratory compaction. A major operation of this type was carried out to strengthen the foundation for the 66 piers of the Oosterschelde Storm Surge Barrier in the Netherlands. Multiple spud piles 1 m in diameter were jetted and vibrated to 30 m depth in the sands. Then intense horizontal

and vertical vibrations were applied as the spuds were withdrawn.

7. Chemical injection of granular materials has been proposed, to be injected after the structure has been installed. The soils must be sufficiently permeable to be penetrated effectively.

   "Jet grouting" has been proposed for relatively impermeable soils. A high-pressure injection of grout fractures the soil formation, allowing lenses of grout to penetrate.

8. Artificial freezing of the soil has been proposed; it has appeared attractive in the Arctic.

   Natural freezeback stabilization of embankments has been used for dikes and levees on land and in shallow water in the Arctic, with varying success. Further discussion of the technique is described in Section 20.7.

9. To cover rock outcrops or irregular rock seafloor areas, stone fill may be placed as a blanket, compacted, and roughly leveled so that the gravity-based structure may have a uniform bearing. This procedure can also be applied to an irregular rock surface which has been purposely cleaned off by removal of soft and compressible material.

   Consolidation of such underwater stone fills can be achieved by dynamic compaction, explosive compaction, vibratory compaction (for fine material), or simply the successive dumping of masses of broken rock.

   In other cases, the rock surface will be pre-leveled, as by the use of shaped charges followed by grinding. To provide a tight seal under the edge of the structure as well as prevent "hard spot" bearing, crushable material may be preinstalled under the peripheral edge. On the piers of the Honshu–Shikoku bridges, collapsible neoprene cushions were employed. Another method of sealing is to drape canvas curtains from the base, these to be weighted at the bottom with chain so as to conform to the actual rock surface. Then underbase concrete or grout can be flowed under the base.

10.4 Sub-base Construction

In some installations planned for gravity-base platforms (often called "offshore caissons"), it becomes desirable to place the structure in two halves, a so-called sub-base being placed first, followed by placement of the platform proper on top of the sub-base. The sub-base will thus be large in plan, but must be submerged well below the sea level. Thus considerations of stability during installation arise. It might be tipped down onto the seafloor as previously described; however, this develops severe stresses in the structure and may deform the seafloor.

Therefore, temporary buoyancy "shafts" extending above the sea level are preferred. They permit visual control of position and orientation and give stability and draft control during installation.

Once the sub-base is founded on the seafloor, the platform proper can be ballasted down onto the sub-base.

The temporary buoyancy shafts, if properly located, can serve as guides for the positioning of the platform. The platform can therefore be accurately positioned as it is ballasted down onto the sub-base.

To provide cushioning and equalize bearing between the sub-base and structure, a number of concepts can be employed. If this is a permanent installation, then polyurethane pads can be located at three or four spots around the periphery and underbase grout injected for permanent bearing.

Alternatively, the subbase may be ballasted down but kept still floating by means of the temporary tanks. The platform caisson is then positioned over the sub-base, guided by lines from the tank. The sub-base is deballasted, rising up to effect the juncture.

Another material that can be used to fill the horizontal joints between units is rubber asphalt; this was used on the Super CIDS platforms between the three segments of the structure.

Sand or sand and gravel have been proposed as a founding material on top of the subbase; in this case it should be carefully screeded and compacted before submersion or else screeded after the sub-base is in place. If screeded before submersion, means must be found to ensure that the sand or gravel fill is not displaced by wave wash during submergence; for example, use of an asphaltic binder.

Once the subbase is guided down by the shafts, vertical mating cones can be engaged so as to result in exact mating.

A structural lip can be provided around the periphery of the platform extending up from the sub-base so as to provide positive resistance against displacement.

Tubular steel dowels or spuds can be entered through sleeves in both segments and fixed in place by compacted sand or grout, depending on whether the installation is temporary or permanent.

10.5 Platform Removal

Some gravity-base platforms are designed for use as exploratory drilling structures, taking advantage of the relative ease with which bottom-founded structures may be broken loose from the soil, de-ballasted to towing draft, and then reinstalled at a new location. This is especially attractive for exploratory structures in the shallow to moderate-depth waters of the Arctic.

"Permanent" offshore platforms are required to be removed at the end of their useful life.

A detailed description of removal of gravity-base platforms is given in Chapter 17.

"Woulds't thou," so the helmsman answered,
"Learn the secret of the sea?
Only those who brave its dangers
Comprehend its mystery."

Till my soul is full of longing
For the secret of the sea,
And the heart of the great ocean
Sends a thrilling pulse through me.

HENRY WADSWORTH LONGFELLOW, "THE SECRET OF THE SEA"

11

Other Applications of Offshore Structures

In Chapters 9 and 10, typical offshore platforms employed in drilling and production of offshore oil and gas were used as a basis for describing the construction procedures required. In this chapter, a number of other applications and other types of offshore structures will be evaluated as to their special construction requirements. No attempt, however, will be made to repeat in detail the "standard" construction procedures previously described.

The several other applications to be addressed are the following

1. Offshore terminals
2. Single-point moorings
3. Articulated columns and loading platforms
4. Guyed towers
5. Tension leg platforms
6. Seafloor well templates
7. Underwater oil storage vessels
8. Offshore bridge piers
9. Cable arrays and moored buoys
10. Subaqueous tunnels (tubes)
11. Moored floating energy plants, e.g., OTEC
12. Storm surge barriers

Most but not all of the above are employed in the development of offshore petroleum resources. However, most of these structures are also used by other industries and for other functions. Offshore terminals are employed for all types of bulk commodity transfer, including especially coal and iron ore. Single-point moorings have been employed for slurry transfer of iron ore and coal, and tension leg platforms for a variety of military installations.

11.1 Offshore Terminals

Offshore terminals are typically built in water depths exceeding 20 m in order to accommodate very large crude carriers (VLCCs) and deep draft ore carriers. Wherever feasible, of course, these have been located in protected or semiprotected waters, but on many continental margins adequate water depth is found only offshore, in a partially or fully exposed location. Therefore, the construction operations must be carried out in the ocean environment, subject to the normal waves, wind, and current for that season, and with suitable precautions for possible storms.

The typical offshore terminal consists of a loading platform, two (or four) large breasting dolphins, and four mooring dolphins. See Figure 11.1.1.

Catwalks usually join all these structures and may require intermediate supports. A trestle may connect the loading platform to shore, or submarine pipelines may be used instead.

The initial structural concept employed was an extension of the harbor-type structure of independent piles supporting a deck and fender system, adapting dimensions to the more severe design conditions of the open sea. Such structures have been extensively used in Japan, in the Arabian Gulf, and along the coasts of Brazil and Australia, for example.

Pilings are typically large-diameter steel pipe piles, perhaps 1 meter in diameter, 40–60 m in length, having wall thicknesses of ¾–1½ in. (18–38 mm). High-yield steel [f_y = 50,000 psi (350 MPa)] is normally employed. Both vertical piles and batter piles are employed, intersecting at the deck level so as to react against each other under lateral loads. Batter piles are also called raker piles.

280 *Other Applications of Offshore Structures*

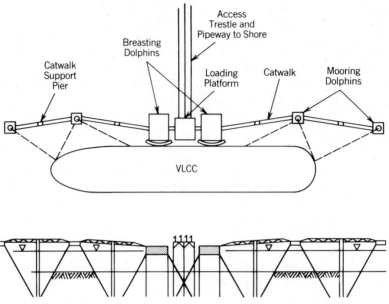

Figure 11.1.1 Offshore loading terminal.

As the size of ships and the exposure conditions become more severe, the proportion of batter piles has increased so that typically they dominate the construction. Relatively high axial loads are used as the basis for design, 400–600 tons in compression and 50–100 percent of that in tension. Therefore, it is important that the piles be driven to their design penetration and that they be accurately located so that the axes of a batter–vertical pile group intersect at a single point, this latter so as to minimize bending. The connection at the intersection must be adequate to develop the flow of forces.

The construction problems for offshore terminals arise principally from the following:

1. The offshore site is usually moderately remote, involving logistical, personnel transfer, and survey problems.
2. The piles are large by harbor standards (although not if compared to those of deep-ocean production platforms).
3. Very close tolerances are required for positioning of the piles at their heads.
4. The relatively shallow water (20–30 m) causes refraction of waves and changes the wave and swell characteristics, making operations difficult even in moderate sea conditions.
5. Last but not least, the economic constraints and the relatively short distance offshore tempt the contractor to utilize harbor-type equipment and methods, which often prove too small and limited for the exposure conditions and work involved.

Figure 11.1.2 shows a typical construction scene at an offshore terminal. The contractor needs to study carefully the bathymetry, which may be changing rapidly on a steep coast; the geotechnical information, which may indicate the presence of hard or firm strata that may be difficult to penetrate; the wave refraction and breaking patterns, which may influence the work at the site; and the currents, which may affect the route of supply services.

Among the many problems which have been encountered are:

1. Sand waves, resulting in slowly changing bottom bathymetry
2. Refraction of waves causing intersections and concentrations of wave energy at certain zones, with consequent pyramidal waves and confused seas
3. Effect of currents on waves, giving excessively steep waves during every ebb (or flow) tide
4. Cap rock on or near the surface that is difficult to break up and to penetrate yet is underlain by loose, almost liquid, sands and silts

Figure 11.1.2 Offshore terminal under construction. (Courtesy of Volker–Stevin Offshore.)

5. Weathered rock, which gives widely varying resistance to piles, even to adjacent piles
6. Boulders on and under the surface
7. Sloping hard surfaces, on or below the seafloor, causing the pile tips to tend to run downhill during driving
8. Calcareous sands, requiring special methods in order to develop uplift and bearing capacity
9. Overconsolidated silts and clays, which are extremely difficult to penetrate

An initial construction requirement is to set up a shore base for support of the construction. Ideally, there will be an adjacent harbor with dock and craneage facilities. Unfortunately, this is usually not the case; therefore, such a support base must be established.

In the case of most offshore terminals, there are a very large number of structural and mechanical elements to be transported out to the construction site. An analysis of the number of barge loads and lifts will usually indicate that efficient transfer at the shore base is essential. Some protection must be provided against the breaking waves. Swells not only lead to much larger breaking waves, even in calm winds, but also can lead to severe surges in harbors and channels, where mooring lines may be suddenly snapped.

Adequate draft must be provided not only for the barges but also for the tug and crew boats.

Services (power, water, fuel) must be provided to the shore base, as well as communications, both local and to centers of supply, and especially to weather forecasting services.

Unfortunately, in the past history of offshore terminal construction, the contractor has usually failed to set up an adequate shore base initially and has had to progressively improve it as the job went on, meanwhile suffering from the inadequacy.

The second step is to set up survey control. Horizontal control may consist of ranges, using focused brilliant lights visible for several miles at sea, even in the daytime, or lasers.

Electronic-positioning systems are usually also set up: The shore stations are established on the coast. Tidal gauges are installed, preferably in protected wells. Levels can be run by laser, corrected for curvature of the earth if the structure is distant offshore.

As so often happens with all types of offshore structures, the geotechnical investigation made by the design engineer for his purposes may be inadequate for the constructor's needs. Therefore, additional site information may need to be developed as to the seafloor sediments, boulders, and obstructions. Through use of a "sparker" survey, jet probings, and borings, more information can be obtained on the upper soil strata through which the piles must be driven.

For example, the contractor may want to handle and set a 60-m-long steel pile as a single piece on its designed batter. He needs to have a reliable estimate as to how far the pile will run down under its own weight. Will jetting be required?

At the construction site proper, the contractor will now set up moorings, to facilitate moving his floating vessels, derrick barges, and so forth along and around the terminal. Preset moorings minimize the time of moving and the problem of handling and resetting anchors. In many cases they eliminate the problem of crossed anchor lines when two pieces of floating equipment must work in close proximity.

Another early step may be the construction of an offshore survey tower, which will provide visual reference for close-in surveying. The availability of

282 *Other Applications of Offshore Structures*

electronic-positioning devices of high resolution has minimized the need for such towers in recent years.

The next decision is what to do in case of storm. Presumably the boats will run to safety in a harbor. If reasonable weather forecasting services exist, with proper planning and judgment the contractor may avoid being caught with supply barges at the site.

But what about his major floating equipment, for example, his large derrick barge? This will always be at the site, and hence may be vulnerable to a sudden storm. It will not always be practicable nor safe to try to tow it to a harbor; the harbor entrance may have breaking seas, shoals, or cross currents which make it exceedingly dangerous to enter in a storm. The tugboats may not have enough power and size to handle a large derrick barge under severe sea conditions.

Experience has shown that it is often safer to ride the storm out at sea on a preset storm mooring. Such a mooring will consist typically of a single long wire from the barge to a mooring buoy (spring buoy), which in turn has a line or chain leading to a heavy anchor. The anchor is usually two anchors, piggybacked (one behind the other joined by a half-shot of chain), or one large anchor, with a shot of chain to ensure that the pull is horizontal. See Section 6.2, "Moorings and Anchors."

During any season when a sudden storm is possible, the storm mooring line is kept connected, but usually slack, so as to permit normal maneuvering on the short, taut operating lines.

Now construction proper can begin. The piles are delivered, either on a barge or self-floating, and are picked up for driving. The first pile in each dolphin should preferably be one of the vertical piles in order to give a point of reference and support for subsequent batter piles. See Figure 11.1.3. Yet many mooring dolphins are designed with all batter piles, no vertical piles!

A second problem is that the piles must be driven alongside each other and then cut and pulled to final position. With heavy-walled, large-diameter pipe piles which are driven, for example, through hard seafloor material, they may be quite inflexible and difficult to position. Pulling the heads may impose bending stresses that will be permanently fixed in the piles.

There is another serious problem with the subsequent connecting of the piles. If these tubular members have to be cut, fitted, and welded at their point of intersection, this may prove impracticable to carry out in a satisfactory manner. The piles may be vibrating due to waves and vortex shedding from the current. The joint area may be wet with spray, the steel below optimum welding temperature. Weld positions will be unfavorable. See Figure 11.1.4.

These joints are subject to cyclic loading and dy-

Figure 11.1.3 Driving piling for offshore terminal at Septiba Bay, Brazil. (Courtesy of Bechtel Corp.)

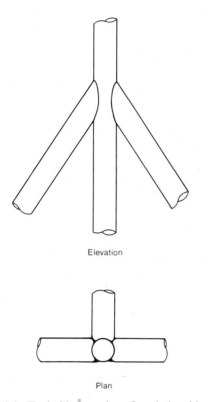

Figure 11.1.4 Typical intersection of vertical and batter piles of mooring dolphins indicating difficulties involved in making proper connections.

namic loads; thus they often fail from fatigue.

As a result of the twin problems of positioning batter piles and connections, the prefabricated template has been evolved. In this case, a template, typically extending from high-tide level up to the top of the dolphin (a height of 5 m or more), is prefabricated of tubular steel members. See Figure 11.1.5. All welding is done in the shop, where conditions are optimum and proper procedures and nondestructive testing can be carried out. The template will have sleeves, at the proper angle, through which the piles may be set and temporarily supported.

The template is then held over the stern of the derrick barge in proper plan location, and one or more vertical piles driven through their sleeves. They are not normally driven to final penetration at this stage but only as needed to provide lateral and vertical support to the template. If the final design does not contain a vertical pile, then the contractor adds one, with its sleeve, as a temporary support pile, usually in the center of the dolphin.

Now the template is raised vertically to proper elevation and temporarily welded off to the vertical piles. Its position and orientation are checked.

Now batter piles are successively set through the sleeves, alternating directions so as to avoid dislocating the template. As they are set, they are driven to grade. Finally, the vertical pile is cut loose, spliced, and driven to its final penetration. If the vertical pile is not in the design, but was only supplied by the contractor for temporary purposes, then it is cut loose and removed.

The connections between piles and sleeves of the template are usually by a combination of shear welds, on a scalloped profile, as previously described (Fig. 9.7.2), plus cement grout injection. This latter is needed in any event to prevent vibration.

Where a jack-up construction barge is used, as has been the case for a number of offshore terminals in Japan and elsewhere, then the jack-up platform can provide the initial support for the template. However, the template should not be rigidly attached to the jack-up platform but be free to slide as dynamic lateral loads are applied by the driving of the batter piles; otherwise the stability of the jack-up rig may be endangered.

Breasting dolphins must resist the heavy loads imposed during docking. Thus they are often full jackets, similar in concept to although smaller than the offshore oil drilling and platform jackets described in Chapter 9. See Figures 11.1.6 and 11.1.7.

Similarly, the loading platform may be a jacket, often with all vertical piles, since none of the loads from the vessel are transmitted to it. See Figure 11.1.8.

During the construction of an offshore terminal, with its many independent structures, the derrick or construction barge will have to move many times in order to be able to handle the batter piles at the many different angles. In turn, it must have its mooring lines out at various angles, which change as the barge is moved. It is, of course, undesirable to have a line run around a dolphin, as it may apply a high lateral force in a direction other than that for which the dolphin was designed, displacing it. More likely, it will break the line just at the time it is most needed. See Figure 11.1.9.

Jack-up construction rigs in clay soils must be careful not to reset their legs too close to a previous hole. On the Ise Bay terminal, near Nagoya, Japan, some 65 positions were required for the jack-up. An extremely careful layout was required to pre-

284 *Other Applications of Offshore Structures*

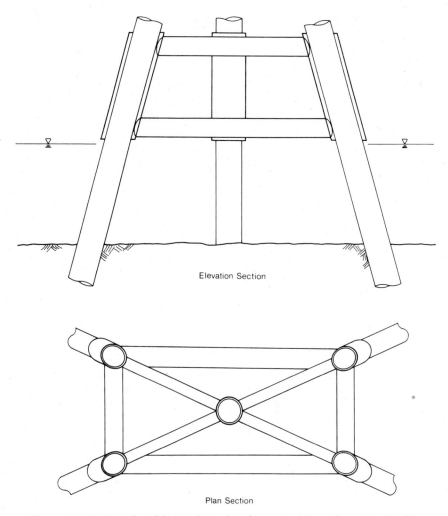

Figure 11.1.5 Use of prefabricated template for construction of mooring dolphins.

vent overlap of leg holes which might cause bending of the legs or loss of support.

These many positions, orientations, crane reaches, and mooring arrangements all need to be laid out in sequence on construction drawings.

The positioning of the mooring dolphins is usually not critical; a meter or two each way is acceptable.

The relative positioning in and out of the breasting dolphins and loading platform is very critical, because when the ship berths and lies against the breasting dolphins, it must not hit the loading platform, even with temporary deflection. At the same time, it must be close enough to allow hose connections to be made or to stay within the shiploader's radius. Hence great care must be taken in establishing the front face for the breasting dolphins and the setback for the loading platform.

Massive fenders are provided on the breasting dolphins to absorb the impact energy during docking. Typically, the 250,000-DWT tanker docks at 15 cm/sec (6 in./sec), and all this energy must be absorbed by the fenders plus the elastic distortion of the dolphins.

These fenders therefore are large, massive, energy-absorbing devices with a predetermined load-deflection response. Many different types have been developed, utilizing deforming rubber fenders, springs, hydraulic rams, the deflection of high-strength steel tubulars in bending or torsion, or the potential energy of gravity weights.

Regardless of these details, the fenders must be

Figure 11.1.6 Setting jacket for Kharg Island offshore terminal No. 5, Iran. (Courtesy of J. H. Pomeroy and Co. Inc.)

Figure 11.1.7 Driving piling through jackets of large offshore terminal, Kharg Island, Iran. (Courtesy of J. H. Pomeroy and Co. Inc.)

properly and accurately set and installed by the contractor, who has frequently underestimated the tolerances and man-hours required.

Fender units are prefabricated in the largest segments which can be conveniently handled. Temporary guides should be installed which will automatically position them in proper position for bolting or welding.

The previous remarks about the difficulty of welding in the splash zone apply here. Because of the impact forces involved, for example, 1,200,000 lb (5.4 MN), very extensive shear welds are required.

High-strength bolts are therefore usually preferable. To aid in fitup, slotted holes should be provided, slotted in one direction on the jacket frame, the other direction on the fender bracket. If proper

Figure 11.1.8 Installation of Keiyo II Seaberth Terminal in Tokyo Bay. Note use of crane barges of 2500- and 3000-ton capacities. (Courtesy of Kajima Corp.)

Figure 11.1.9 Piles for offshore terminal at Valdez, Alaska are installed by combinations of driving and drilling. (Courtesy of Reidel International.)

bearing plates are used under the bolts, a good connection can be rather quickly made. Tolerances are very important, in that the face of the fenders on the two or more breasting dolphins must line up so as to be engaged simultaneously and equally by the ship's hull.

The main emphasis by the constructor, therefore, must be on maximum prefabrication, provision for tolerances, and adoption of all practicable expedients to aid in installation.

The superstructure of the loading platform is now installed. To the greatest extent practicable, it should be prefabricated in large modules. The hydraulic loading arms are especially time-consuming to set, because of the accuracy required and their awkward shape. Prefabrication (preassembly) of these in a modular frame will save many hours of time at the site. See Figure 11.1.10.

Similarly, erection of a ship loader on the loading platform is a major task due to the heights involved and weights that must be lifted. It may be necessary to erect a stiff-leg derrick on the loading platform in order to set the higher segments. Special planning is required for the rigging when installing a telescoping ship loader boom; not only is the boom heavy and awkward and required to be lifted high above the deck and the sea, but it must be traversed into the closely fitting housing. In turn, the lifting slings tend to foul on the housing frame and may require relocation during erection. See Figure 11.1.11. For these reasons, complete prefabrication and assembly of a shiploader in the shipyard or harbor is of course preferable to assembly over water whenever this is practicable.

Caisson-type (GBS) offshore terminal structures have been built in order to permit complete outfitting in a protected harbor, including shiploader, conveyor stacker, fenders, and the like and the transport and setting of this complete terminal as a single unit in manner similar to that described in Chapter 10. For the Hay Point Terminal in Queensland, Australia, the terminal consisted of three berthing caissons, one carrying the shiploader completely erected, another the conveyor stacker. These caissons were each towed out and installed in a single day, the actual setdown taking only a few hours. Mooring lines were run from each caisson to preset mooring buoys. Winches on the caisson pulled the structure to exact location and held it there during setdown. See Figure 11.1.12.

Figure 11.1.10 Erection of access bridge, Port Latta offshore terminal, Tasmania.

One short mooring line was required, leading to the previously set structure. This was affixed to a rubber cushioning device that was designed to absorb shock loading. A nylon line could have been similarly used to accept the dynamic force variations.

The structures were designed to arrive at high slack water, be positioned during the fall in the tide, ballasted down to seat at low slack, then ballasted to stay on their pads during the subsequent high tide. (Tidal range was 5–6 m.) See Figure 11.1.13.

Dolphin caissons had superstructures of tubular structural steel; they required substantial temporary buoyancy tanks to be attached to ensure stability and control during seating. See Figure 11.1.14.

After seating, the spaces under the caissons' bases

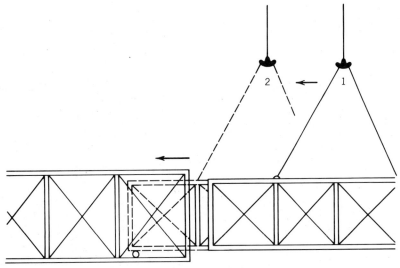

Figure 11.1.11 Entering telescopic boom of ship loader.

288 *Other Applications of Offshore Structures*

Figure 11.1.12 Prestressed concrete caisson is towed to site with shiploader mounted. Hay Point Terminal No. 2, Queensland, Australia. (Courtesy of Rendel and Partners and Utah International.)

were filled with grout containing a thixotropic admixture. Scour protection, in the form of articulated concrete block mats, was placed on the periphery of the caissons.

Another type of design for offshore terminals is that which employs large-diameter cylinder piles 2–4 m in diameter, usually of thick-walled steel. This type has been extensively employed in Cook Inlet, Alaska, where ice loading dominates the design criteria. They have also been used in Iran and in Saudi Arabia.

This type of structure consists essentially of all vertical piles. Each pile represents a major installation process, by jetting and driving, or by drilling and grouting, as described in Chapter 8.

Very large driving heads must be fitted in order

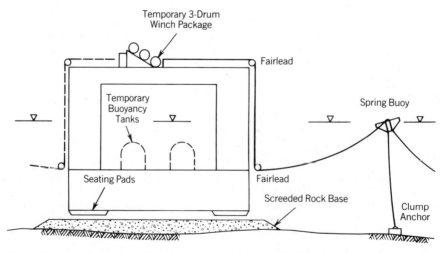

Figure 11.1.13 Positioning of offshore terminal caisson, Hay Point, Queensland, Australia.

11.1 Offshore Terminals

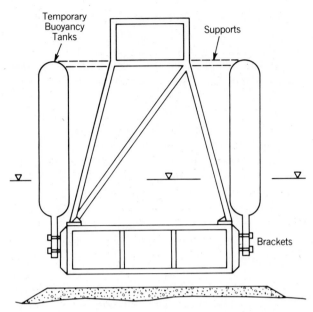

Figure 11.1.14 Use of temporary buoyancy tanks to provide stability and control during installation.

to distribute the hammer blow over such a large-diameter cylinder. Often these are specially fabricated, with stress relieving of welds to prevent cracking under impact. Jets, if required, are usually preinstalled in the piles so as to permit their operation simultaneously with the impact driving.

In many cases, the installation has been carried out using large offshore-type impact hammers. In other cases, multiple vibratory hammers have been employed.

A major construction problem for this type of construction is the handling and positioning of such a large cylinder pile, say 50–60 or more meters in length, 4 m in diameter, weighing perhaps several hundred tons. Tidal currents, such as those in Cook Inlet, which range up to 7 knots (4 m/sec), tend to displace the cylinder pile both in the direction of the current and laterally due to vortex shedding. See Figure 11.1.15.

The pile must have special slings fitted so that it will hang vertically. Once set into the soil, the problems of current are reduced.

The next problem is that of driving such a large pile in the heterogeneous soils which are typically encountered: glacial till and overconsolidated silt in Alaska, cap rock and limestone strata in the Arabian and Persian Gulf areas. The procedures required are described in Chapter 8.

The subsequent superstructure erection is carried out in similar manner to that for the more conventional terminals, except that prefabrication is made easier by the all-vertical pile arrangement. Often it may be practicable to carry out the completion work over the top rather than by floating equipment.

Alternatively, very large prefabricated deck sec-

Figure 11.1.15 Setting 14-foot-diameter (4-m) piles through deck of offshore terminal, Drift River Terminal, Cook Inlet, Alaska. (Courtesy of Pomeroy–Gerwick–Bechtel.)

tions or bridges may be set by a crane barge. For the offshore iron ore terminal at Port Latta, Tasmania, the conveyor bridge trusses were sized to use the maximum capacity of the available crane barge. See Figure 11.1.10.

On some offshore terminals, for example, the LPG terminal at Ju'Aymah and the petrochemical terminal at Jubail, both in Saudi Arabia, large-diameter cylinder piles have been set in predrilled holes and then driven to develop proper bearing and lateral support. These have varied from 1.6 to 4 m in diameter.

Trestle connections from shore are relatively standard in their construction operations. Because they typically cross through the surf zone and over shallow-water areas, part or all of their construction is carried out over the top. By such methods, the lifting, driving, drilling, and framing are essentially independent of the sea state and current. See Figure 11.1.16.

Typically the upper works of a large crawler crane are mounted on long girders, for example, double wide-flange beams designed to span one bay and having extensions that cantilever out to the next bay. See Figures 11.1.17 and 11.1.18.

Spud piles are dropped and the girders are jacked to grade. The crane then sets the piles through the template and drives them. See Figure 11.1.19. The template is now welded to the piles. Longitudinal stay beams are dropped into place and bolted. The rig can now skid forward to the next bay.

Where pile driving is not excessive, this entire cycle can be completed in 4–6 hours by a trained and skilled crew.

Another crane follows up behind, placing prefabricated deck sections, completing all bracing and framing.

The piles in the template need to be fully fixed to prevent excessive vibration and assure proper interaction. Shims are inserted to fix the pile head, and a seal made at the lower end of the sleeve, followed by grout injection and welding at the top. Accelerators may be used in the grout to achieve early strength.

A long trestle will require an anchor bent every 500 m or so; this should consist of a double bent, adequately braced. By shortening the span, a properly prefabricated template may be hung. Because of weight limitations, it may have to be set in two or more segments and bolted together.

Launching forward of the rig is done by pulling against the longitudinal girders. To guard against accidental sideways displacement, lateral stops should be installed on the end of the cap girders of each template. Such a sidewise runoff actually occurred when launching the crane rig forward on the access trestle to the Hay Point terminal, Queensland, Australia; fortunately, the rig did not roll off, and no damage nor injury occurred.

Such over-the-top methods have been used with spans of 20 m: designs have been prepared showing its practicality for spans as great as 40 m or more. In planning for a very long access to a terminal off the Ivory Coast, where heavy swells from the Southern Ocean would make work afloat very difficult, spans of 30–40 m showed significant economies.

The major difficulty with this type of construction is that all materials must be delivered over the trestle, rehandled by the deck construction crane to the stern of the pile-driving rig, then swung around by it to the next bent. Proper packaging and planning of such deliveries, perhaps using a night shift, will simplify this problem.

Even when working on top of a trestle or offshore terminal platform, provision must be made

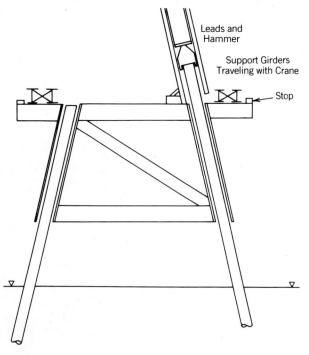

Figure 11.1.16 Template and bent frame for piled access trestle to offshore terminal.

11.1 Offshore Terminals

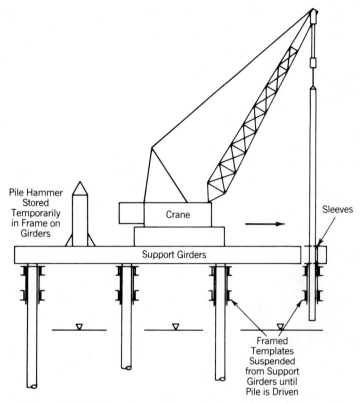

Figure 11.1.17 "Over-the-top" construction of access trestle to offshore platform. Entire crane plus support girders skids forward over each cap as it is completed.

for safe access by personnel. Walkways need to be incorporated into the design, safe stepdowns when the crane swings, adequate lighting, and so forth.

Water safety also needs special consideration. A man may fall overboard. Life jackets should be mandatory. Lifelines of nylon or similar floating material should be strung and a rescue boat on call or patrolling as indicated. From a modern terminal, with deck at +10 or +12 m, and nothing but tubular piles in the water, it can be very difficult to rescue a man if he falls overboard in a choppy sea with a current running.

Figure 11.1.18 Traveling crane on converted Bailey Bridge constructs access trestle to Hay Point Terminal No. 1, Queensland, Australia. (Courtesy of John Holland Constructions Pty.)

292 *Other Applications of Offshore Structures*

Figure 11.1.19 Leading end of access trestle under construction. (Courtesy of John Holland Constructions Pty.)

The discussion of personnel access and transfer in Section 6.4 is especially relevant to offshore terminal construction.

An alternate to trestle construction, usually adopted for offshore terminals that are far offshore or where the waterway cannot be impeded, is the use of submarine pipelines. Their installation is generally described in Chapter 12. However, there are available a number of special options: the pipelines can be pulled from shore, with the pulling winches on the terminal structure itself, or by a pulling barge located seaward of the terminal structure. Small-diameter lines can be stacked and welded vertically on the terminal, fed down through J-tubes, and pulled to land. Where the terminal is far offshore, standard offshore lay barges can be used, laying the end alongside the terminal.

The lines may be laid in a predredged trench: even if it partially silts in, it will be relatively easy to sink it down later with a jet sled operation. Near to the terminal itself, at the very least, the trench should be buried and backfilled to prevent damage to the lines from boat anchors.

11.2 Single-Point Moorings

Single-point moorings have been developed over the last several decades. The initial concepts were designed to enable tankers to moor and offload (or unload) crude oil, and as such they represented an economical solution for transfer of oil wherever sea conditions permitted a reasonable usage, for example, 65 percent availability.

Single-point moorings have subsequently been improved and extended to enable them to handle several different petroleum products and even to load iron ore slurry into ore carriers.

From a construction point of view, they generally consist of a base and anchors, underbuoy risers or hoses, and a floating buoy moored to the base by flexible lines or an articulated strut. The buoy contains swivels which allow a ship moored to it to weathervane with the wind, current, and to some extent the seas. The ship may moor to the floating buoy or be yoked to it by a rigid or articulated yoke. See Figure 11.2.1. Many variants on the basic concept have been developed. See Figure 11.2.2.

The principal construction operations include the installation of the base structure and the anchors. The connection of hoses and so on requires a support rig with hoisting capabilities and divers. The installation of the buoy is mainly a question of positioning, followed by the more tedious job of equalizing the tension in the several anchor legs.

Bases have been usually constructed of steel and are initially buoyant. They are sunk to the seafloor by ballasting down in calm water, with descent controlled by lines from a barge working over sheaves. The elastic stretch in the lines is counted on to absorb the dynamic loads, which are principally due to wave action on the support barge. Ballast water is of course confined within small compartments in the base in order to avoid free-surface effects.

Bases have also been constructed of reinforced concrete and delivered to the site afloat.

Bases are usually relatively small, 10–20 m in diameter, 3–5 m in height. Once on the seafloor, they are usually filled with slurried sand or slurried iron ore or with cement grout pumped through a hose. Grout-intruded aggregate has also been used, even

11.2 Single-Point Moorings

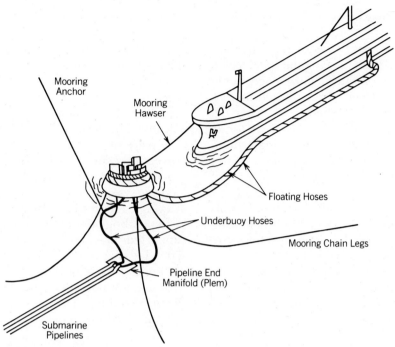

Figure 11.2.1 Typical "Calm" installation for single-point mooring for offshore oil transfer.

though there may be practical difficulties in placing the aggregate and in the multiple grout injection points required. Grout-injected aggregate was successfully used on the Polaris Missile test pads off San Clemente Island, California, which were of similar size and depth and contained even more complex inserts.

Underbase grout may be placed through a hose to fittings on the base structure, in a manner similar to that of gravity-based structures. Since the skirt length on a base structure is necessarily limited, there being very little net weight to cause penetration, the grout should be highly thixotropic so as not to escape out to the open sea. Drop curtains of

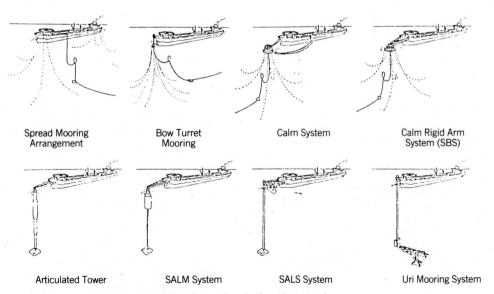

Figure 11.2.2 Alternative single point mooring systems.

294 *Other Applications of Offshore Structures*

canvas or sandbags may be used to seal the bottom edges.

Scour protection may be required. Since the water depth may be less than 30 m, the combination of currents plus wave-induced movements could cause serious erosion. Filter fabrics covered by rock are typical, but other clever schemes have been developed, including "artificial seaweed" (closely spaced nylon ropes) hanging from a ring around the structure so as to slow the water movements. This is intended to cause deposition instead of scour.

The base often is fitted with a manifold on its top, with fixed curved pipes (the "PLEM") leading over the edge of the base, through which hoses are connected to the pipelines from shore.

Bases may also be pile-supported in soft soils. Piles are driven through sleeves in the base and connected by grout. These piles will be relatively short and lightly loaded; their main function is to prevent excessive settlement, tilting, and sliding.

Anchors for the several types of single-point moorings can be large gravity blocks or driven or drilled-in piles. In several cases, in hard seafloors, holes have been drilled and heavy anchor chain run into the holes, after which they are filled with tremie concrete, using very small aggregate, for example, ⅜ in. (8 mm), in order to permit placement through a flexible hose guided by a diver. A steel transition flare piece, like a funnel, is set at the top so as to prevent excessive wear on the chain at that point.

Installation of a typical "CALM" single-point mooring system is shown in the attached sequence of drawings; Figures 11.2.3 through 11.2.8 inclusive.

Depending on soil conditions, the PLEM may be supported on a base structure similar to those described earlier in this section.

Single-point moorings are increasingly being used to moor tankers for storage and especially for floating production systems. In the latter case, they are combined in the field development with subsea templates.

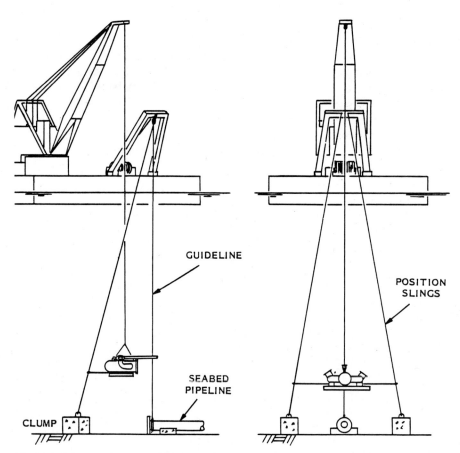

Figure 11.2.3 Installation of Calm single-point mooring for offloading oil: guideline mooring system. (Courtesy of Imodco.)

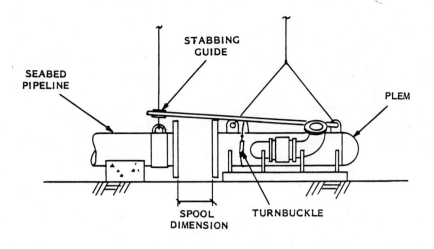
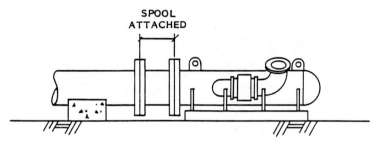

Figure 11.2.4 Plem installation. (Courtesy of Imodco.)

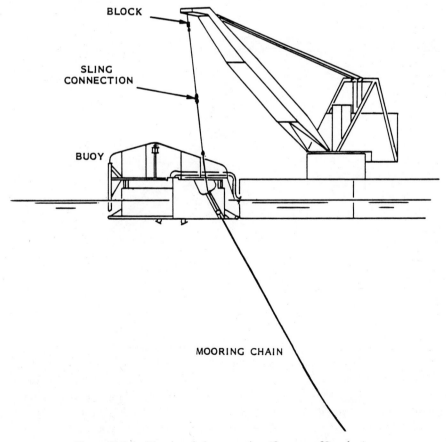

Figure 11.2.5 Mooring chain connection. (Courtesy of Imodco.)

296 *Other Applications of Offshore Structures*

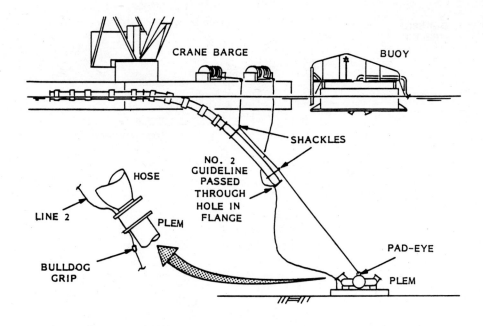

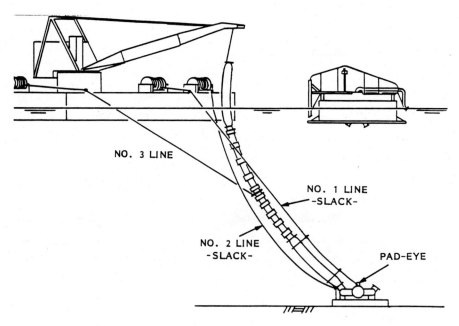

Figure 11.2.6 Underbuoy hose connection to plem. (Courtesy of Imodco.)

When used for semipermanent moorings, the systems must of course be made heavier and more reliable so as to safely hold the vessel, even during storm conditions.

While the great preponderance of current development is directed toward the use of floating production systems for marginal oil fields, the basic concept is also suitable for offshore processing, storage, and transfer vessels for other commodities.

The installation procedure for a "SALM" articulated mooring system, for the semipermanent mooring of a production and storage vessel, is de-

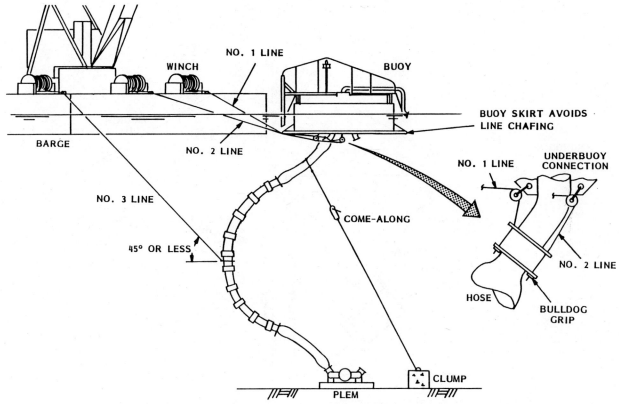

Figure 11.2.7 Underbuoy hose connection to buoy. (Courtesy of Imodco.)

scribed below by way of example. This procedure, and the accompanying schematic drawings, were made available through the courtesy of IMODCO. See Figures 11.2.9 through 11.2.21 inclusive.

Prior to the start of the tow of the SALM to the installation site, an installation barge is moored at the site, using an accurate electronic navigation system such as SYLEDIS. This surface position fix is then transferred to the seabed and an array of transponders is set up so as to permit positioning of the base of the SALM and for determining its azimuth.

A seabed survey by divers or ROV and a side-scan sonar survey should be carried out prior to installation.

11.3 Articulated Columns

A natural extension of the more advanced single-point mooring systems has been the development of the articulated columns or articulated loading platforms, such as those installed in the North Sea for the offloading of crude oil. These differ from the single-point mooring (SPM) systems such as CALM primarily in structural size, offloading rate, and depth of water. See Figure 11.3.1.

These platforms consist of a base, a column, and a deck structure. The base is anchored to the seafloor by gravity weight or piling. The column is buoyant, and its center of buoyancy is well above its center of gravity. The base and the column are joined by an articulated hinge, called a *cardan*, which allows articulation on both axes but is restrained against torsion.

While all but one of these structures built so far have been of steel construction, a number of those planned will be built of both steel and concrete, employing each material in the zone for which it is best suited. For example, the one hybrid structure built to date, that for the Maureen Field in the North Sea, had its central column constructed of prestressed concrete, with upper and lower sections of steel.

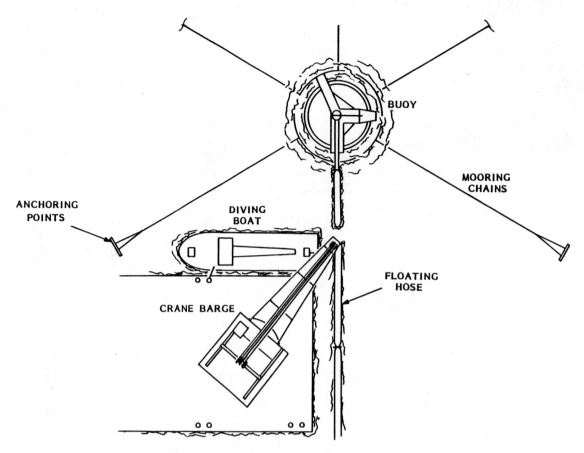

Figure 11.2.8 Plan view of floating hose hook-up for CALM. (Courtesy of Imodco.)

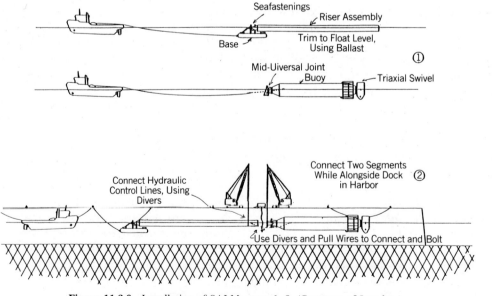

Figure 11.2.9 Installation of SALM: steps 1, 2. (Courtesy of Imodco.)

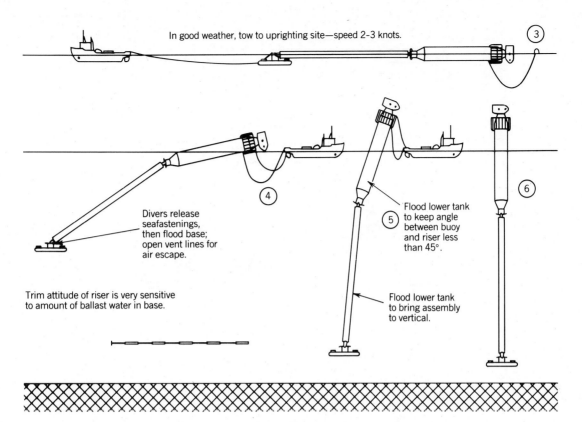

Figure 11.2.10 Installation of SALM: steps 3–6. (Courtesy of Imodco.)

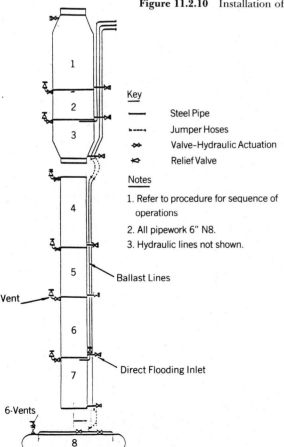

Figure 11.2.11 Installation of SALM: bilge and ballast system. (Courtesy of Imodco.)

The articulation ensures that no moment is transferred to the seafloor. The reduction this makes possible in structural dimensions then reduces the wave force which is attracted and the lateral force tending to produce sliding.

The concept has been applied to flare stacks, offloading terminals, and the deep-water mooring of vessels for floating production systems. An extension of this concept has been proposed by C.G. Doris, for example, for deep-water (500–800 m) drilling and production platforms. The installation of these systems can be quite complicated and requires a very sophisticated application of advanced hydrodynamics and construction engineering. See Figure 11.3.2.

An initial construction concept was to install the base first, ballasting a steel or concrete cellular structure to slight negative buoyancy.

As has been seen with other fully submerged structures, the dynamic response of a moderately large base is very significant, especially when coupled with an offshore derrick barge which is being accelerated in heave, roll, and pitch by the waves.

A large flat cylindrical base exerts the inertia of its own mass plus contained ballast and also a large vertical added mass as it tries to lift the column of water above it. Thus almost all dynamic surge must

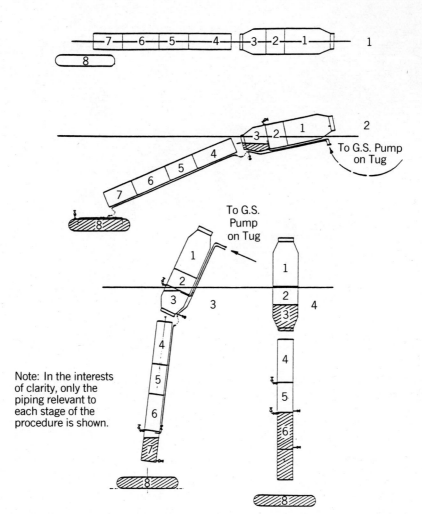

Figure 11.2.12 Installation of SALM: uprighting sequence and flooding diagram. (Courtesy of Imodco.)

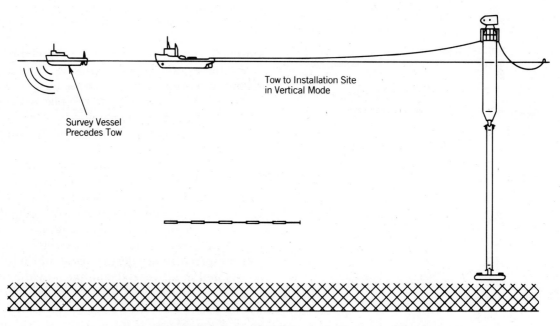

Figure 11.2.13 Installation of SALM: step 7. (Courtesy of Imodco.)

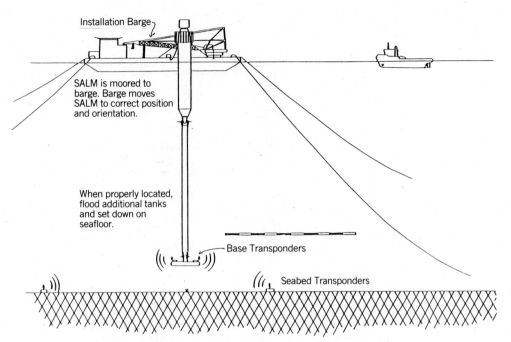

Figure 11.2.14 Installation of SALM: step 8. (Courtesy of Imodco.)

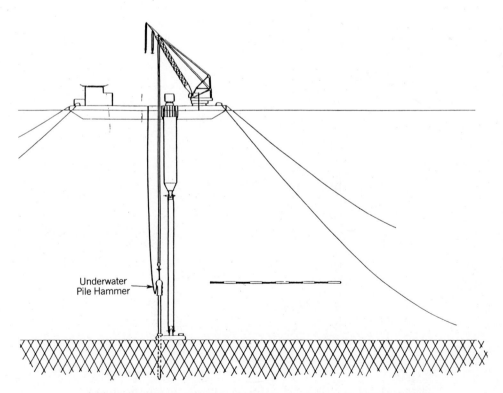

Figure 11.2.15 Installation of SALM: step 9. (Courtesy of Imodco.)

301

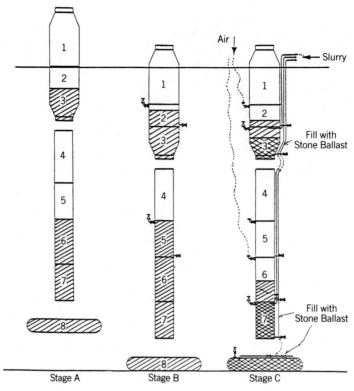

Figure 11.2.16 Installation of SALM: setting sequence. (Courtesy of Imodco.)

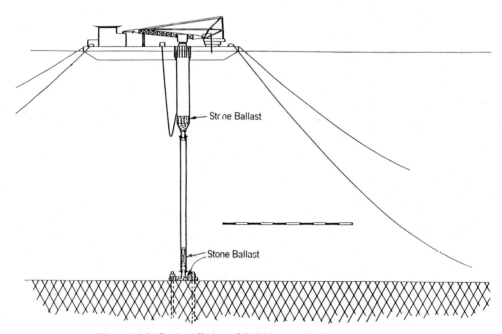

Figure 11.2.17 Installation of SALM: step 10. (Courtesy of Imodco.)

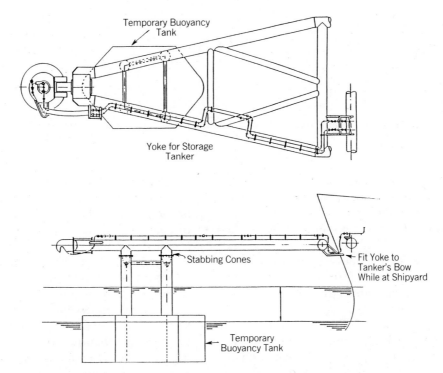

Figure 11.2.18 Installation of SALM: mounting of rigid arm yoke. (Courtesy of Imodco.)

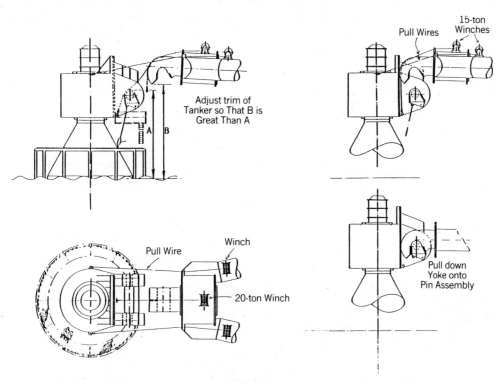

Figure 11.2.19 Installation of SALM: details of yoke connections. (Courtesy of Imodco.)

303

304 *Other Applications of Offshore Structures*

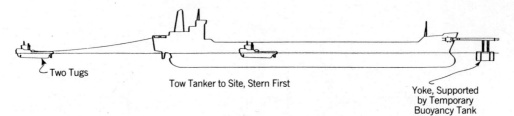

Figure 11.2.20 Installation of SALM for moored floating production–storage vessel. (Courtesy of Imodco.)

be taken out in the elastic stretch of the wire rope and flexibility of the boom. Lifting over the side will perhaps give more cushioning by the vessel, but unfortunately it also produces more roll dynamics.

Among the several systems developed to minimize the dynamic portion of the load has been that of the spar buoy which limits the range of the variable force. See Figure 11.3.3.

Once the base is seated, the column is brought to the site floating horizontally, upended by ballasting to the vertical attitude, and then guided and ballasted down to a mating of the hinge. Diver intervention is required at the mating to drop in the securing pins and activate the hydraulic locks. Tensioned guidelines have been used to control the mating; See Figure 11.3.4. Often the final connection is made by a pulling line which directs and pulls the mating cone into position.

Another method of lowering the base has been to use a "controlled free-fall" system in which buoys are attached at intervals to lines leading to the barge or boats. As the base structure is progressively ballasted, it descends until restrained by the buoyancy of the next set of buoys. Then more ballast is added, either to the base structure or to one of the buoys, so that the base may descend another step. In this way, velocity of descent is kept low, dynamic forces are accommodated, and control is maintained, even to the extent that the process is reversible by ejecting the water from a buoy by compressed air.

As water depths have increased and structures have become larger, with greatly increased functional requirements, other methods have been developed in order to give proper control and reduce the amount and complexity of underwater operations, especially that of mating the critical cardan joint.

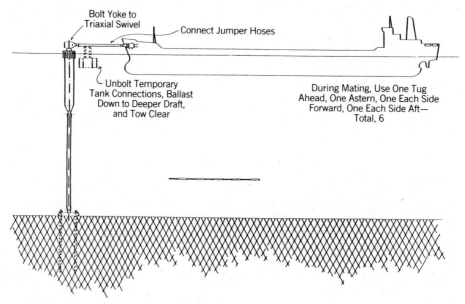

Figure 11.2.21 Installation of SALM: mooring of floating production–storage vessel. (Courtesy of Imodco.)

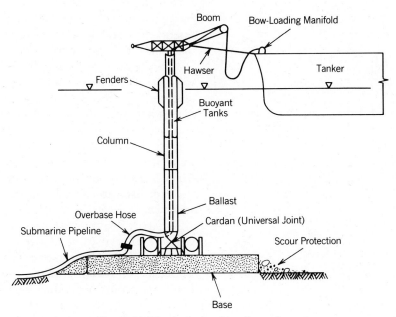

Figure 11.3.1 Articulated loading column.

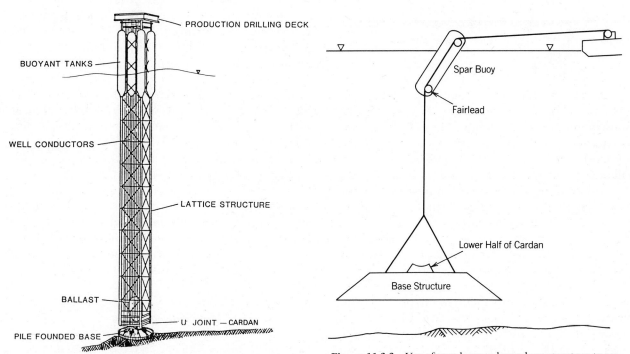

Figure 11.3.2 Buoyant drilling and production tower.

Figure 11.3.3 Use of spar buoy to lower base structure to sea-floor.

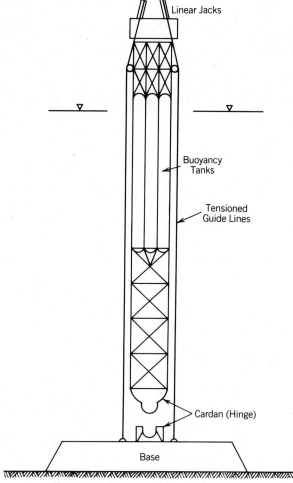

Figure 11.3.4 Mating articulated column with base by use of tensioned guidelines.

One such method is to attach the column while the base is floating at the surface. This enables the subsequent upending and descent to be fully controlled as to stability and depth. It also ensures that the cardan assembly is carried out properly and facilitates inspection. See Figures 11.3.5 and 11.3.6.

Control of orientation is of course one of the major problems in the initial assembly. Since it can be carried out in shallow, protected water, lines to mooring buoys and to anchors can be employed.

Once the articulated column is mated, it can then be towed to the site in either the horizontal or vertical mode. Smaller structures, with minimal facilities, such as mooring structures for floating production systems, are usually towed in the horizontal mode, whereas structures with extensive topside facilities will be upended to the vertical while still in the harbor, the topsides installed and hooked up, and the structure towed to the site in the vertical mode, to be ballasted down onto the seafloor. To upend the combined base–column structure, iron ore ballast may be slurried into the base.

One of the more demanding of such installations was that of the Beryl A flare stack, where the articulated structure had to be not only upended and seated properly, but had to have an exact distance and orientation from the production platform in order to enable the prefabricated flare stack bridge to be set. Lines were therefore run from the platform to the base structure and a boat with bow thruster used to extend them as the base reached setdown.

Where a floating production vessel (e.g., a converted tanker) is to be moored, a rigid but hinged yoke is usually fitted to the bow of the ship in a shipyard. The mating of the yoke and the single-point mooring riser is then carried out at the site, by use of a pull-in wire system, and connection made by a tension-bolted flange. See Figures 11.2.20 and 11.2.21.

Another scheme involves vertical construction of the articulated column. The base is constructed first and moored in a deep-water protected site. The halves of the cardan are preassembled, joined with temporary fixing, and then set on the base. Temporary buoyancy tanks are attached so as to enable the base to be submerged and still remain stable.

Then the column is constructed above the base, allowing the whole slowly to sink as weight is added. Eventually the structure will have been submerged to the point where the stability and buoyancy can be supplied by the column alone. The temporary buoyancy tanks are then removed and the structure completed afloat. The deck can then be mounted, lifting it on if it is a small structure, or transferring by floating in over the column and deballasting if it is a large deck.

The whole can then be towed to the site in the vertical mode. After seating on the seafloor, the cardan is freed, so as to permit flexible articulation. Ballasting is added so as to cause the base to penetrate. In firm soils, underbase grout may be adequate; in other soils, underwater piling will be driven through sleeves in the base and connected by grouting of the annulus.

11.4 Guyed Towers

The guyed tower concept, like the articulated column, transfers no bending moment to the foundation soils. Reactions to the wave loads are taken

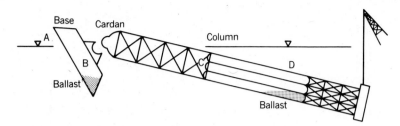

Figure 11.3.5 Assembly of articulated column with base while afloat.

by the guyline legs, usually 16 in number, secured to anchor piles, with a clump anchor in the system. A spud can or pile support at the base of the tower provides shear transfer to the soil and provides the entry point for the conductors. See Figures 11.4.1 and 11.4.2.

The guyed tower is rectangular in profile, typically 40 m or so square. It is designed for deep water, 200–700 m or even deeper.

Fabrication of the square tower is carried out on a ways, as with any jacket. Since this structure is primarily for deep-water use, it will often be fabricated in two halves, just as was done for the Hondo platform (see Section 9.9). Preferably the two halves will be built as one, then later separated, so as to ensure perfect match.

However, if sufficient yard space is not available, a short section at the juncture, incorporating both mating sections, can be constructed first. Then this section is skidded to the far inshore end of the ways, separated, and the lower half fabricated in a normal manner. The mating section of the upper half can be skidded sideways onto a parallel launching ways, then down to the outboard end, and the upper half fabricated concurrently.

There are obviously several variations of the above, depending on yard layout.

Each half is now transported to a protected deep-water site, as was done with Hondo, and then launched. Note that the launching can be either end-O, or sideways, since the cross-section is uniform.

Each section is provided with temporary buoyancy tanks so that it floats horizontally on its upper legs. See Figure 11.4.3.

After mating and the welding of the mated legs so as to form an integral structure, the auxiliary buoyancy towers at the lower end of the guyed tower are ballasted to slightly negative buoyancy. Now the structure, having a slightly inclined attitude, is towed to its installation site.

Ballast added to the spud can section and lower jacket legs causes the structure to upend.

The upper temporary buoyancy tanks are designed so that the structure will float vertically at the site. Further ballasting of the jacket legs and spud can cause the structure to touch bottom and then penetrate the spud can into the soil.

The pile anchors may be drilled in, using a drill ship or semisubmersible, and then cemented. The first segment of the guy line will be attached to each anchor pile as it is set. Then the drill ship will lay out the segment, lower the clump anchor, attach the second segment, and then lay it down. A pen-

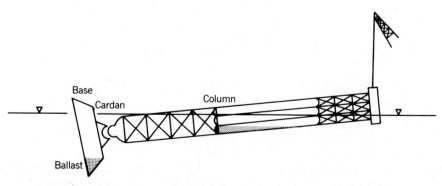

Figure 11.3.6 Alternative method of assembly of articulated column with base.

308 Other Applications of Offshore Structures

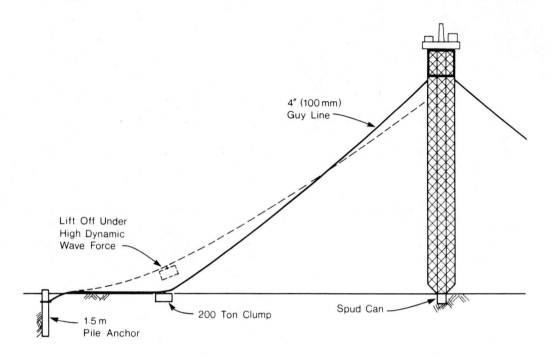

Figure 11.4.1 The Guyed-Tower concept.

nant will be attached, with a marker buoy, so that the guy line can be retrieved when the tower is moved into position.

Once the tower has been set and the spud can has achieved initial penetration, each guy line is fed in through a swiveling fairlead and run up to the deck and stopped off. A cable grip hoist is attached and initial tension taken up gradually around the series of 16 guys.

Then the upper temporary buoyancy tanks are ballasted to a slight negative buoyancy and removed.

To help the penetration, weighted drilling mud, for example, with barites added to bentonite slurry,

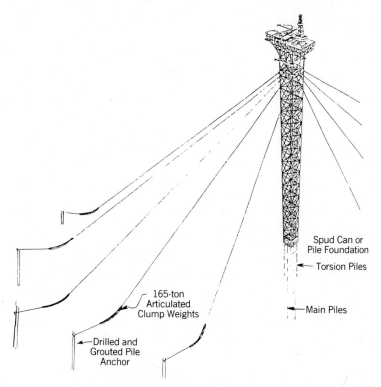

Figure 11.4.2 Guyed-Tower production system.

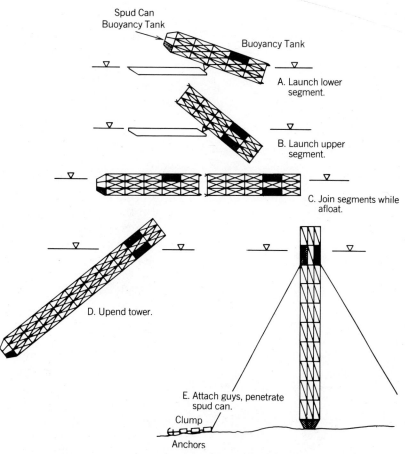

Figure 11.4.3 Proposed procedure for installing Guyed-Tower.

can be used to replace the water ballast in the spud can. After penetration has been achieved, the heavyweight ballast can be replaced by seawater. Spud can penetrations can be 2–15 m, depending on the soil stiffness.

Following final penetration, the tension in each leg is readjusted with the cable grip hoists. The deck structure and modules are then set by derrick barge.

In other installations of guyed towers, a piled base structure will be used, with the piles driven through sleeves in the guyed tower. This was the solution adopted for platform Lena, constructed in 1000 ft (300 m) of water, described in the following section.

Not only was platform Lena a unique structure once in place, but its installation embodied a number of new ideas which proved successful. The jacket was 330 m (1080 ft) long by 36 m (120 ft) square. Launch weight, including main piles and torsion piles, was 27,000 tons. The jacket was loaded out from the fabrication yard in conventional end-O fashion and then lowered onto transverse skids. Launching near the site was sideways, using four launch runners with guides plus rocker arms. Holdbacks were used to restrain the jacket while the barge was ballasted to heel 7° to starboard. Then the 3½-in. frangible nuts of the holdbacks were severed by explosive detonation and hydraulic jacks activated to overcome starting friction.

The jacket launch took only about 10 seconds, one-fourth the time normally required for a stern launch. The jacket had a maximum roll of 53°, the barge a maximum roll of 15°.

Twelve long buoyancy tanks, 20 ft (6 m) in diameter by 120 ft (36 m), were built into the upper portion of the jacket so as to enable upending. High-density iron ore slurry was placed into the base to assist upending.

A derrick barge was used to control the upending, seating, piling, and guyline attachment. This derrick barge maintained position by means of four computer-controlled thrusters.

The lower sections of the main piles of the structure, 54 in. in diameter, were carried out with the jacket. After upending of the jacket and its seating, they were extended and driven to 560 ft (170 m) penetration, making the completed piles almost 500 m in total length.

For the torsion piles, whose upper end terminated in the base, a system of latches and lugs was used to connect pile and hammer together for lowering as a single unit. One 4-in.-diameter multistrand wire rope was used to lower the combined unit, using a 600,000-lb capacity linear winch. Air, electric, and hydraulic lines were lowered with separate constant-tension winches and lines.

Initially hammer efficiency was reduced by the cushioning of compressed air below the hammer ram, but a change in the air exhaust system overcame this problem. After driving to full penetration, the latches and lugs were hydraulically released, enabling the hammer to be retrieved.

Earlier, the 20 guys and their anchors had been installed. Drilled-in pile anchors were placed and grouted, each with the guy line preattached. These guy lines had articulated clump anchors attached. A barge laid out the guy with its clumps and then secured it to a buoy temporarily held in position by a small, taut gravity anchor. Once the tower was installed, connection was made with four lines from the tower, which led out through underwater fairleads. These were then pretensioned with linear jacks. Then the additional 16 guys were completed and tensions equalized. The guy lines were 5⅜-in. diameter wire lines, each 1800 feet in total length and sheathed in polyethylene. These guys had a breaking strength of 1525 tons each and were designed for 500–600 tons maximum load. The clump anchors were each 200 tons, consisting of articulated weights attached to the guy line segments.

11.5 Tension Leg Platforms

The first commercial installation was the Conoco Hutton tension leg platform (TLP), installed in 1984 in the United Kingdom sector of the North Sea. See Figure 11.5.1. The concept appears to be suitable to very deep water and hence is being developed by many petroleum companies and engineering firms.

Typically, the deck is constructed on temporary support pillars, just like the Statfjord B and C decks (see Section 10.2.8). Then it is picked up by one or more large barges, floated in over a partially submerged semisubmersible substructure, and mated.

The tension leg tethers can of course be of any very-high-strength, high-fatigue-life material, such as very large-diameter wire rope. In the case of Hutton, thick-walled steel tubing with connectors similar in principle to drill casing was selected because of its high modulus of elasticity (stiffness) and

Figure 11.5.1 Hutton Tension-Leg platform. (Courtesy of Conoco U.K. Ltd.)

the ability to connect up segments progressively into strings. An articulated joint is fitted at the lower end of each tether. See Figure 11.5.2.

To provide the reaction base on the seafloor, several systems have been developed. In one case, piles are set through the seafloor template which has been accurately positioned by an acoustic positioning system. The piles in turn have been guided into the template by sonar and video. They are driven by underwater hammer and the connection for load transfer made by grouting the annulus between pile and sleeve.

While the TLP is being fabricated, a well template will have been installed on the seafloor and some or all of the wells will have been drilled.

The floating TLP is now towed to the site, moored in position above the seafloor template, and preinstalled bases. One tether from each corner is successively made up and lowered down to the mating joints on the templates. These are run in and locked.

Meanwhile, the TLP has been free to heave. Now the four tethers are connected and the TLP deballasted to tension the tethers. This critical operation must be carried out in a very calm sea and must be completed within a few hours.

Now the other tethers are run down and connection made. Equalization is by hydraulic systems within the TLP corner columns, while primary tensioning is again done by ballasting control.

In the case of a gravity-based anchoring system, the concrete and steel box bases are previously affixed under the hull so that they can be carried to the site by the TLP with initial tether segments connected. The boxes are now given slightly negative ballast and lowered down as the casing strings are made up. Once on the seafloor, the boxes are fully ballasted and the tethers connected and tensioned by deballasting of the TLP.

The dynamic response of the coupled platform–base system will constantly change as the base (or bases) descend. Care must be taken to prevent resonant amplification in heave at critical lowering depths. The most severe dynamic stresses in the tethers are usually encountered during the initial stages of lowering.

During each of these many stages of operations, attention has to be given to stability and draft of the floating semisubmersible platform. Neither the accidental flooding of one compartment nor the decoupling of one tether should lead to the loss of the platform.

The installation of the Hutton TLP was an example of excellent engineering, planning, and execution. While the platform substructure, the tethers, and deck were being fabricated, the well template was placed and the wells were predrilled. Then the foundation (base) templates were installed. To achieve the required tolerance in position of the foundation templates of 250 mm in distance, 2° in orientation and 0.5° in level, a 900-ton steel guiding frame was lowered and positioned on the previously placed well template. The frame was leveled by jacks operating against the frame's mud mats. Foundation templates were supported temporarily by the frame and a single 30-m-long pin pile driven through each to fix its position. Then the frame was removed.

Then through each template, the eight main piles, each 1.8 m in diameter, were driven to 60-m penetration. A Menck MHU 1700 underwater hydraulic hammer was used. Piles were entered into their 7.5-m-long template sleeves and the hammer

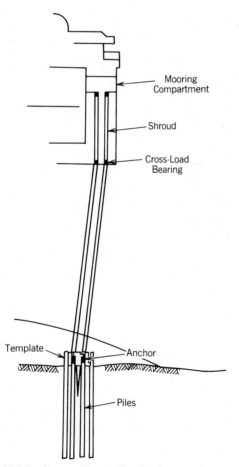

Figure 11.5.2 Conoco Hutton Tension-leg mooring.

was set on them, using an acoustic positioning system and ROV–TV camera. After driving, the piles were grouted in the sleeves.

Installation of the platform itself was carried out in calm seas.

The semisubmersible TLP was held in position by lines from two large semisubmersible derrick barges, which in turn were moored with 12 point anchor systems.

The first leg was run down in each corner and the anchor connector at the lower end latched into the foundation template by hydraulically activated locks. The legs were forged steel hollow tubes, 260 mm O.D., of 92.5-mm wall thickness proof-tested to 115,000 psi. Tapered threaded joints were used to connect the 9.5-m long segments. See Figure 11.5.3.

The installation sequence follows. The acronym *TMCs* stands for the four combination leg tensioner and motion compensators which were operated by pneumatic–hydraulic machines and were located in the mooring chamber of each shaft of the platform's hull.

1. The tension leg platform was moored to the two semisubmersible crane barges and positioned 40 m to one side of final position.
2. One leg was lowered at each corner, using special tapered threaded joints. See Figure 11.5.3.
3. The leg was transferred to the TMC and raised to the top of the stroke.
4. The platform was moved over the templates to final location and the first round of legs was stabbed into the seabed cones. Connectors were latched and 10-ton tensions applied with TMC compensating for motion.
5. TMC valves were closed to suppress heave.
6. The platform was pulled down to the 32-m operating draft, with 500-ton tension in legs.
7. Leg load was transferred from the TMCs to the permanent load block by means of a locking collar.
8. The platform was deballasted to increase tension to 1300 tons in each corner leg.
9. The remaining 12 legs were stabbed and latched.
10. Tension was equalized with jacks in all legs.
11. Tension was adjusted to 815 tons per leg by deballasting.

11.6 Seafloor Well Templates

The use of subsea production templates, working as part of a subsea production system, is growing rapidly. See Figure 11.6.1. Even with fixed structures (e.g., BP's platform Magnus) the use of a template through which to predrill the wells while the platform itself is being fabricated shows significant cash-flow advantages by enabling the platform to be brought into production at an earlier date.

Subsea templates have therefore been developed by a number of firms; they have grown in size and weight to as much as 2000 tons or more. These are typically loaded out on a barge for transport and then launched or lifted off to the self-floating mode. See Figure 11.6.2.

They then must be lowered to the seafloor, with equipment and rigging adequate to sustain the high dynamic forces involved. Stability during submergence is maintained by designing the structure so that the center of gravity is below the center of buoyancy.

In one system, the lowering is done by a drilling vessel, either a drill ship or semisubmersible. Lines

Figure 11.5.3 Tapered threaded joint for Hutton TLP. (Courtesy of Conoco Inc.)

314 *Other Applications of Offshore Structures*

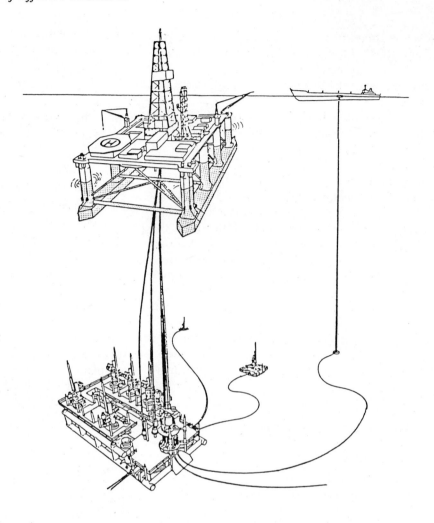

Figure 11.6.1 Multi-well template and semisubmersible field development.

are run from the derrick hoist, down through the moon pool, and then up and attached to the top of the template floating alongside.

A floating derrick barge now hooks onto the template, which is given negative ballast, allowing it to sink below the keel of the drilling vessel. By slacking the floating derrick barge's lines and taking in on those of the drilling rig, the load is transferred to the drilling rig. The barge lines are now disconnected. See Figure 11.6.3.

This transfer can also be done dynamically without the use of the offshore derrick barge, provided the template has inherent stability, that is, center of gravity (CG) below center of buoyancy (CB).

Buoyancy tanks attached to the upper portion of the template can be used to give this stability. For buoyancy tanks which must remain intact during descent into deep water, consideration should be given to pressurization by filling with a light fluid such as gasoline or solvent or by filling with syntactic foam.*

Other means are available for transporting and setting templates, especially the larger ones. A barge may be specially equipped with large cable or casing grip hoists. In an inshore harbor, in relatively shal-

---

*In some engineering articles, this word is spelled *syntectic*. *Syntactic* is believed to be more accurate.

to lower the template is that the heave compensator of the drilling vessel can minimize the dynamic loadings due to heave.

For the larger templates of the future, large spars can be used to provide inherent heave compensation as the template is lowered. In this case, the actual lowering can be carried out by linear winches on the spars, while the spars will be able to minimize differential heave because of their small waterplane area.

Templates can either be gravity-base-supported or pile-supported. In the first case, grouting under the base, between skirts, may be employed to ensure level support. In the second case, piles are set through sleeves, using sonar and video guidance, driven with an underwater hammer, and the connection made by grouting of the annuli.

If a structure is to be later set over the subsea template, bumper piles (actually seafloor "fender piles") may be installed with the subsea template and driven in, as for the support piles. Then they are disconnected from the template so as not to transfer loads into it during the structure's installation. See Figure 11.6.4.

Flow line installation to subsea templates may be carried out by laying the flow line down beside the connection point. The line may then be pulled in using a wire line through a base plate at the connection point. Cameron, Vetco, and Hughes all have developed connection systems suitable for use to 1000-m depth. See Figure 11.6.5.

Alternatively, one end of the flow line can be lowered vertically using a drill rig and tensioned guideline. The flow line is landed on and inserted into the base which is on the seafloor. The connection permits a swivel in a vertical plane, using tools run down from the drill rig. The line at the surface is now transferred to a lay barge, which pulls away, keeping tension on the line.

Connections of flow lines to the subsea template may also be made using remote-controlled manipulators. An overall subsea production system is shown in Figure 11.6.6.

11.7 Underwater Oil Storage Vessels

Underwater oil storage vessels differ from other subsea installations primarily because of their large displaced volumes. In this case, a structure is being installed which is capable of containing 500,000 to over 1,000,000 barrels (80,000–160,000 m^3 of oil.

Figure 11.6.2 Subsea well templates. (Courtesy of Vetco.)

low water, the template is seated on the seafloor. The barge now floats in over the top, attaches the cables or casings, pulls the template up snug under itself, and thus transports it to the site. Once at the site and properly positioned, the template is lowered to the seafloor.

Disconnect is effected by remote-operated devices, for example, acoustically, hydraulically, or explosively activated disconnects.

One advantage to the use of the drilling vessel

316 *Other Applications of Offshore Structures*

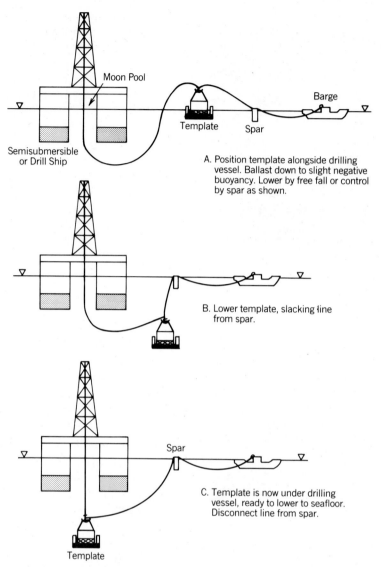

Figure 11.6.3 "Keel-hauling" method of transferring subsea template from afloat on surface to underneath drilling vessel.

Therefore, we are now dealing with very large dimensions in relation to the waves and with very large inertial forces.

Chicago Bridge and Iron successfully installed the Khazzan Dubai offshore oil storage vessels of steel in the Arabian Gulf. Initial construction was in a shallow, dewatered basin. When the tank was sufficiently complete so that it could float as a single unit, using compressed air, the basin was flooded, and the tank, a bottomless hemisphere, was moved laterally into a deeper basin and seated on its floor by release of internal air pressure. The structure was then fully completed. Made to float once again by filling of the tank with compressed air, it was towed to the site and positioned by mooring lines, and the air was gradually released. As the air bubble became progressively smaller within the tank, the tank took a significant list (almost 30°!), until its righting moment equaled the dynamic listing moment. It was allowed to slowly sink further, eventually returning to vertical and seating on the seafloor. This initial list, of course, had been shown in model tests and hence was anticipated.

Through sleeves in the periphery of the tank, piles were set and seated by hammer. Using the piles as casings, holes were drilled and enlarged in

11.7 Underwater Oil Storage Vessels   317

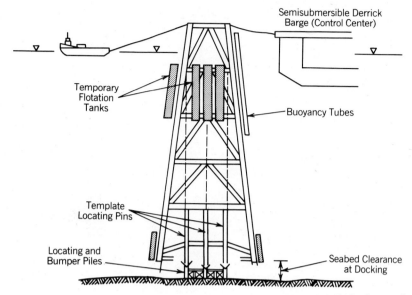

Figure 11.6.4 Docking arrangement for placing jacket over pre-drilled subsea well template. Beryl B platform. (Courtesy of Mobil North Sea Ltd.)

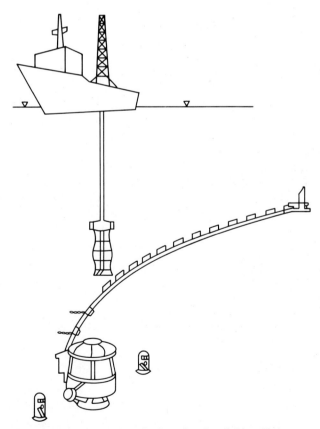

Figure 11.6.5 Installation of subsea flow line linking Christmas Tree to manifold. From B. Renard, M. Lerique, J. Tinchon, "Specific Features of a New Concept for Deepwater Flow-line Laying and Connecting and Its Sea Trial in 250 meters" OTC 4577, Offshore Technology Conference, Dallas, Tex., 1983.)

the limestone strata; the piles were lowered to place and grouted so as to serve both as vertical support and, more importantly, to give resistance to uplift when the tank was filled with oil.

A number of conceptual studies have been carried out for large tanks, both of concrete and steel, for placement on the seafloor in such areas as the Gulf of Alaska, the Navarin Basin, offshore China, the northern North Sea and the deep Arctic.

Conceptually, these tanks would be seated on the seafloor, with sufficient ballasted weight or hold-down piles to prevent uplift when filled with oil. The passage of long-period waves can also give significant uplift forces in the shallower water depths, but these, of course, diminish with depth. Consideration also has to be given to the very long tsunami wave and its potential for uplift.

The tanks are filled with oil, using the natural pressure from the gas. Discharge is accomplished by the differential water versus oil pressures at the depths involved, supplemented as necessary by pumping.

A large tank of this type is initially manufactured, transported, and submerged in a manner similar to that used for large gravity-based structures.

During initial submergence at the site, as the upper surface disappears below water, there is a zone of great dynamic instability. This zone has been frequently noted with semisubmersible drilling vessels.

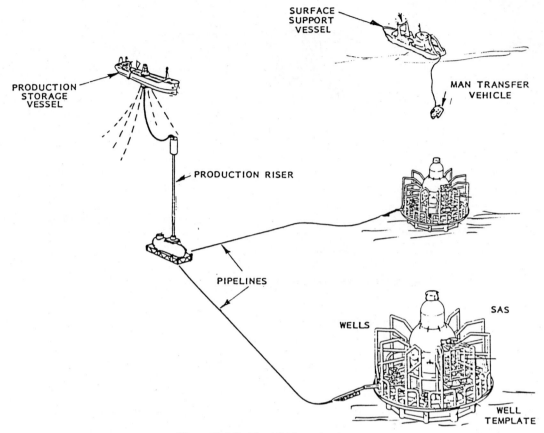

Figure 11.6.6 Total SAS production system.

With the sudden loss of waterplane, the righting moment depends almost wholly on the vertical distance between CB and CG. At the same time, the waves no longer move in fully orbital fashion but break and swirl over the top, creating a Venturi uplift effect, often denoted as the "beach effect."

This problem area is best overcome either by submerging one end first so as to maintain some water plane righting moment or else by the use of columnar buoyancy tanks, temporary or permanent, to give stability until substantial submergence has been achieved. See Figure 11.7.1.

A large underwater storage tank can be pulled down to the seafloor, under positive buoyancy, using tethers or lines secured to a preplaced base (similar to that proposed in the previous section for the TLP) or else lowered down from buoyant vessels on the surface.

Stability and attitude of the unit, once fully submerged, is determined by $\overline{KB}$, $\overline{KG}$, the free surface of ballast water in compartments, and the pull of the lines. A known and definite trim is preferable to an oscillating, indeterminate pitch.

The second problem is that of dynamic response. Large spars, floating above the storage vessel, have the ability to respond to induced heave with a spring response and minimize the dynamic heave due to passing waves.

The spar buoys may either be independent of each other or joined by means of an articulated space frame which maintains the relative positions of the spars in the horizontal plane as well as providing a support for operational control.

In relatively deep water (100–300 m) internal pressurization may be used to prevent implosion of the tanks. Air pressure can be used provided consideration is given to:

1. The weight of air added (often several hundred tons)
2. The temperature rise as air is compressed
3. The reduction in pressure as the air subsequently cools
4. The differential pressure gradient between external hydrostatic head and internal air pressure over a vertical distance

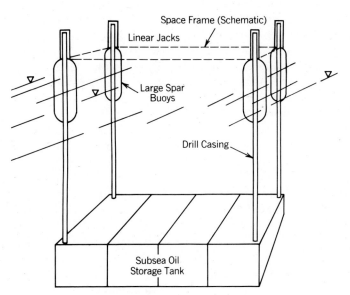

Figure 11.7.1 Using drill casing and large spars to control undersea oil storage tank during descent to seafloor.

The external buoyancy and head are functions of water density, which is not constant but which increases with lower temperatures, greater depths and, in some cases, greater salinity.

The buoyancy may also be affected by the reduction in displacement, (reduction in gross dimensions) due to the increased head.

At greater depths, other methods must be employed. Some potential solutions are described in Chapter 19.

11.8 Bridge Piers in Offshore Areas

Recent advances in the construction of offshore platforms have re-stimulated the imagination of many nations to envision bridges across hitherto unbridgeable waterways. Bridges have significant aesthetic, philosophical, and political appeal as well as strictly economic utility. Thus we see bridges proposed to cross open-sea stretches such as the Great Belt of Denmark, the Skaggerak between Denmark and Sweden, the English Channel between France and England, the Knik Arm of Cook Inlet, Alaska, the Strait of Messina, the Strait of Gibralter, and now, as a reality, the Inland Sea between Honshu and Shikoku.

Bridge piers in deeper waters of the open seas or exposed to severe environments are essentially offshore platforms, exposed to similar extreme environmental forces, with perhaps increased emphasis on ship collision and long-term durability.

The bridges in such deep water are most likely to be suspension bridges of very long span, of the order of 2000–3000 m. Where several suspension bridges are placed in succession, intermittent anchorage piers must be provided, capable of sustaining very high unbalanced pull from one span alone in order to meet the criterion of safety under damaged conditions.

Selection of the founding sites may be more constrained, and indeed at some currently studied sites, exposed bedrock and considerable irregularities are to be expected.

This appears to restrict the potential concepts to a gravity-base structure founded on a preplaced rock fill, perhaps augmented with drilled-in piles or rock anchors and a steel jacket with drilled-in and grouted piles. Belled footings may be used to increase bearing and uplift resistance of the piles. Belled footings were used on the bridge to the industrial terminal at Jubail, Saudi Arabia.

The ship collision and durability–fatigue endurance criteria may require a more massive structure, of either reinforced concrete or hybrid steel–concrete design, floated into place and installed in much the same fashion as the gravity-based structures described in Chapter 10.

First, returning to the foundations, initial data on the site can be given by sparker, echo sounding, and side-scan sonar. More precise information can be obtained by acoustic profiling, by geotechnical borings, with both in-situ and laboratory testing, and by visual–video exploration by submersible. Particular care should be taken to determine the presence of soft, almost fluid, sediments lying in pockets above the more competent materials, since these may not show up in the normal acoustic reflections. High frequency-sonar, water column sampling, and grab sampling may be needed.

The constructor may be called upon to support these investigations and to provide survey control for positioning.

Boulders on the seafloor site should in most cases be dragged clear. With clay seafloor soils, it may only be necessary to remove boulders larger than 0.75 m diameter. This is the practice in the North Sea, where smaller boulders are just forced down into the hard clay.

Soft material may need to be removed, either prior to founding by dredging or else after founding, by jetting and airlift or eductor removal from within skirts. Alternatively the softer material may be displaced by dumped rock.

As noted in Sections 7.7 and 7.8, rock can be placed at depth either by discharging through a large tremie pipe or by direct dumping of a pre-saturated rock mass from a bottom-dump or side-dump barge. Dutch dredging vessels have been fitted for rock placement at depths over 100 m by discharge through a modified dredge ladder from a ship-shaped hull (i.e., a modified trailer suction hopper dredge). At greater depths, a flexible tremic tube is used. A graded rock embankment placed in the above manner will have minimum segregation but will also have an uneven surface with typical variations in height of 1 m or so. Compaction may be achieved by the impact of successive masses of rock, by dynamic compaction by dropped weights and explosives, or by very intense surface vibration.

Concepts for screeding at depths of 100 m and more have been developed, based on use of a vertical dredge ladder hung in the moon pool of a drilling vessel. Alternatively, a screeding frame can be set on the seafloor and operated by remote control. See Section 7.4.

Excessive efforts at screeding to close tolerances do not seem warranted, however. By topping off the compacted rock embankment with smaller-size crushed rock, steel skirts may be employed so as to penetrate and seal off the base area. Bearing pads, three or four in number, can be built into the base. When these land, selective ballasting can be used to level the pier exactly. This is a question of designing the pad area so that it will support the structure with minimum ballasting but will penetrate a short distance into the crushed rock under maximum ballasting. Jets can be built in to facilitate this penetration.

Then the underbase is filled with grout so as to provide permanent support. If desired, the upper portion of the crushed rock embankment may also be grouted.

To meet the damage criterion of unbalanced loads or ship collision, if overturning is critical, tension anchors can be drilled through sleeves in the pier, down into bedrock, where posttensioning tendons (active) or drill casing (passive) can be installed and grouted.

An alternative solution for sites where the overlying material may be removed to bedrock is to drill and shoot the bed rock and then grind it off to a uniform seat for the gravity-based structure. This procedure was carried out for Anchorage Pier 7A, the most critical pier on the Bisan–Seto route of the Honshu–Shikoku bridges.

The final depth was 50 m. The rock was drilled and blasted before removing the overburden. Then the broken material was removed by a 99-m^3 clamshell bucket. A very large jack-up rig was modified so as to provide guides for drilling and grinding just above the seafloor. A rotary drill operating from the jack-up rig ground the peripheral seat areas on which to seat the caisson. A crushable rubber cushion was attached to the caisson's cutting edge so as to seal the periphery during underbase grouting. See Sections 7.4 and 10.3.

The construction sequence for Pier 7A is illustrated in Figures 11.8.1 through 11.8.8 inclusive.

For the Akashi Strait Bridge main pier, one scheme for the large caisson is to support it on a peripheral ring of closely-spaced, large-diameter (about 2m) steel piles drilled and grouted into the underlying rock. Load transfer will be by grout between the sleeves in the caisson and the piles.

For shallow-water bridge piers such as that at the Ohnarutu Straits, which resembles the Straits of Messina with its tidal whirlpools, even larger-diameter (4-m) piles have been drilled, concreted, and grouted. Because these piles furnish the entire compression, tension, and shear support for the pier, intimate contact of the pier as a whole with the irregular rock seafloor is not required.

Many other concepts are technically feasible, especially for the intermediate piers, based on the

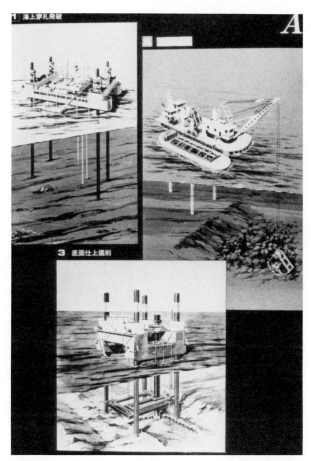

Figure 11.8.1 Construction sequence for Anchorage Pier 7A of Honshu–Shikoku Bridge on Koyama–Sakaide route. (Courtesy of Kajima Corp. and Honshu–Shikoku Bridge Authority.)

Figure 11.8.2 Construction sequence (*cont.*)

Figure 11.8.3 Caisson being towed to site.

Figure 11.8.4 Positioning caisson by means of tugs.

concepts which have been developed for offshore platforms. The principal limitation is usually that of ship collision, which is basically a design rather than a construction consideration.

11.9 Cable Arrays, Moored Buoys, and Seafloor Deployment

It is often necessary to deploy anchored buoys in the deep ocean. These generally consist of a clump anchor, a large buoy, and a connecting line of either steel wire or more usually of fiber such as nylon, polypropylene, or Kevlar. To protect against damage to the line from fishbite, polyurethane or polyethylene coatings may be extruded onto the fiber lines.

To prevent accidental sinking of the buoy from flooding, polyurethane foam or syntactic foam should be used as fill. Buoys have frequently been lost due to rupture of tanks due to boat collision, fatigue of welds, or gunshots from passing fishermen.

Generally the scope of the buoy line is set at 1.2–2.0 times the depth. Two methods of deployment are used: "anchor-first" and "anchor-last."

Figure 11.8.5 Controlling caisson during final lowering to prepared seat.

11.9 Cable Arrays, Moored Buoys, and Seafloor Deployment

Figure 11.8.6 Underwater concreting by the grout-intruded aggregate method is used for all the main piers of the Honshu–Shikoku Bridge. Mortar barge *Seki* is at upper right of photo.

In the "anchor-first" case, the anchor is hoisted over the side and lowered to the seafloor, and then the buoy is deployed. This requires considerable time, during which position must be maintained, as well as requiring a powered drum to control the lowering.

In the "anchor-last" method, the buoy is launched, the boat travels forward a distance equal to the scope, and the anchor is dropped, free-fall.

Both calculations and field measurements indicate that dynamic forces are relatively minimal, that the anchor falls only at 3–5 m/sec due to the drag restraint of the line, and that impact forces of the anchor do not lead to excessive dynamic stresses in the line. In general, the maximum force in the line is equal to the weight of the anchor plus line.

When deploying cable arrays, relative position control is very important. In some cases, use of marker buoys (articulated spar buoys) with short scope may be helpful, although today with acoustic-positioning systems even more exact control can be attained.

In some cases it is necessary to cut a line free—for example, a line used solely for deployment—in which case a hydraulic cutter can be run down the line, either controlled or free-fall, and activated when it reaches the desired depth. Alternatively, an ROV equipped with a cable cutting device may be employed.

Proving of the mooring capacity of a line may be carried out in relatively shallow water by mooring a barge, such as an offshore derrick barge, between the legs of the moor, then reeving wire lines to blocks and thence to the deck engines, so that two legs react against each other. In deeper water,

Figure 11.8.7 Setting structural steel anchorage grill in Pier 7A.

Figure 11.8.8 Completion of concreting of Pier 7A using floating concrete plant.

and with short scopes, the vertical component of force becomes dominant, so that it may be practicable to test the vertical capacity only.

Deployment of objects to the seafloor may be done by lowering down, as described in the section on seafloor templates, with adequate consideration for the dynamic response of the lowering vessel. Alternatively, controlled free-fall deployment may be used, in which buoys are used to reduce the net weight, increase the drag, and provide vertical stability. The buoys will thus also be descending to deep water, so that they must be filled with either light-density fluid or syntactic foam, so as not to suffer significant decrease in volume with depth. Once the object impacts the seafloor, the buoy or sphere is released, either automatically on release of load (pelican-hook arrangement) or by an acoustic-actuated release mechanism.

11.10 Subaqueous Tunnels (Tubes): "Bottom-Founded" and "Submerged–Floating"

Subaqueous tunnel sections have long been used to cross beneath harbors. Some of these have involved individual segments of 35,000 tons displacement and 100 m or more in length. They thus resemble bottom-founded oil storage vessels, except that the demands of vertical and horizontal positioning control of the tubes are much more severe. The sunken tube concept has been proposed in recent times for the crossings of the English Channel, the Skaggerak and the Great Belt.

The difference between harbor crossings and those in the open sea is primarily that of transient dynamic forces during transport, lowering, and positioning. Depths may be significantly greater.

During transport, the flotation will usually be provided by temporary bulkheads recessed within the ends of the segment. These must be designed to resist the maximum hydrostatic head during submergence and in addition must be able to resist wave slam during tow.

Moorings will have been predeployed, enabling the tube section to be moored upon arrival. The tube section may then be lowered to its final position by lines from surface floating structures. See Figure 11.10.1. However, in locations subject to swell, dynamic forces may prove excessive. In such cases, submergence may be carried out by ballasting alone, with stability being maintained by temporary vertical tanks which penetrate the waterplane. These enable the descent to be controlled; they provide a platform for control of the mooring winches, and they give a waterplane moment of inertia which provides stability while minimizing the response to the waves.

The trailing end of the tunnel segment can be guided down by taut wire lines and/or by sonic position indicators so as to engage mating cones and seat on the lower lip of the joint of the preceding section. The elevation of the leading edge can be controlled by jacked spuds, extending from the tube, until such time as the backfill can be placed underneath and sufficiently consolidated to provide support.

The rear temporary tanks are now ballasted to slight negative buoyancy, detached from the tube, and the tanks then deballasted and towed away for reuse. The leading tanks are left in place as a guide

Figure 11.10.1 Artist's rendering of the lowering of a steel and concrete tube segment, San Francisco Bay Area Rapid Transit Project.

for the subsequent mating of the next tunnel segment.

The trench will have been predredged, probably using a trailer suction hopper dredge. Cleanup dredging to remove loosely consolidated infill must be carried out just ahead of the tunnel segment setting.

Backfill can be placed under and around the tube by hydraulic means and consolidated by vibratory probes working alongside the tube. Care must be taken not to raise the segment due to excessive hydraulic pressure from the temporarily fluidized backfill material. The jacked spuds are now retracted hydraulically, transferring the underwater weight of the tube to the fill.

Over the top of the tube, graded rock material may be placed to protect against scour and the whole covered with an articulated concrete block mattress to protect against dropped anchors.

While the above describes the general process, there are a number of alternatives at each stage which may be adopted in specific cases in order to speed installation and reduce costs.

First, after the trench has been dredged, rock fill may be placed and screeded to grade so as to allow the tube to set directly on the rock without backfill beneath. Such screeding requires the use of a jack-up screed barge, which will provide direct elevation control. In harbor projects, such as the San Francisco Bay Rapid Transit tube and the Hong Kong sunken tube, a tension leg semisubmersible barge has been used as a screed barge; such a barge could be used offshore as well, although having more inherent lateral movement (surge and sway) than a jack-up. See Figure 11.10.2.

If the rock material is left slightly below grade, and if skirts are used to penetrate the rock fill, then sand or grout can be injected underneath, working from within the tube itself.

Instead of using jacked spuds at the leading end of the tube, polyurethane pads on concrete slabs may be preplaced so as to crush when full load is applied and thus transfer bearing to the entire base of the tube.

To seal a tube segment into the preceding tube, a horizontal thrust is required. Hydraulically operated winches may provide an initial closure and partial compression of seals. If the space between the rear bulkhead of the tube being set and the leading bulkhead of the tube previously set is now unwatered, then the full hydrostatic head acting on the leading bulkhead is available as thrust to force the sections together, squeezing the rubber joint seal tight.

Then the bulkheads can be cut out and the joint made more positive, if required, by welding and grouting from inside.

As opposed to submerging a tube by ballasting down, it may be pulled down against preset anchor blocks or drilled-in piles. These latter will give position control as well and thus may eliminate the need for surface moorings. Such a pulldown system was employed for the tube across the St. Lawrence River near Montreal, where high river currents prevented lowering directly from floating barges.

A very large construction jack-up barge has also been proposed for setting the tube segments of the English Channel crossing. Such a barge would not only control the tunnel segment during lowering but provide a work platform for underbase filling. The problems with a jack-up are, first, that of bringing the floating tube in between the jack-up's legs without endangering them by impact and, second, the need to relocate the jack-up for each segment, necessitating jacking down to flotation, then

Figure 11.10.2 Tension-leg screed barge enables accurate screeding of gravel bedding for the segment at depth of 50 m. San Francisco Bay Area Rapid Transit Project. (Courtesy of Parsons, Brinckerhoff, Quade, and Douglas, Bechtel Corp; Tudor Engineering.)

back up again, an operation which is very sensitive to and limited by the sea state.

The placement of sunken tube (tunnel) segments is closely related to the installation practices required for gravity-based structures and underwater oil storage vessels, and hence consideration must be given to many of the same aspects, especially:

1. Hog–sag bending under tow
2. Wave slam against bow
3. Free-surface effects during ballasting down
4. The "beach effect"—dynamic instability, as waves first wash over top during submergence
5. Hydrostatic pressures during immersion
6. Stability when fully submerged beneath the waterplane
7. Draft control
8. Erosion of seabed due to escape of water from underbase
9. Displacement during placement of backfill

Several submerged but floating tunnel projects have been proposed recently for the crossing of very deep channels, especially in Greece and Norway. In these cases, the tunnel segments would be submerged only to the depth necessary to permit ship passage overhead and to minimize lateral environmental forces acting on the tube. These will be essentially tension leg platforms, held down by mooring wires to preplaced or predrilled anchors.

Segments can be lowered, controlled, and joined by the same means discussed for bottom-founded tubes with the use of temporary buoyancy tanks. This will minimize bending moment and dynamic forces during submergence. Because their depth of submergence will be limited, temporary columns can be extended up through the waterplane. The pulldown wires or tethers will be tensioned from inside the tube by cable grip linear jacking systems.

Strict weight control will have to be maintained and checks made against actual ballast added versus draft as the installation proceeds. Weight and attitude adjustment of individual segments can be by filling of side compartments with sand, either normal silica sand or high-density iron sand.

Dynamic response studies should be carried out to ensure against resonant action occurring as one segment is brought into contact with a previously tethered segment.

11.11 Ocean Thermal Energy Conversion (OTEC) Systems

Extensive studies and initial test installations for ocean thermal energy conversion (OTEC) have been carried out under contracts with the U.S. Department of Energy as well as by engineering organizations in other countries, notably Japan and the Netherlands. In general, all systems employ a cold water pipe of large diameter, raising cold water from a depth of 1000–2000 m to a surface plant where the warm waters are used to provide the heat source. The surface plant may be a very large floating structure, in which case it is usually planned to be moored in deep water (2000–4000 m). Alternatively the plant may be constructed on the continental shelf in perhaps 200 m water depth, or even in shallow water.

The principle unique construction aspects to be discussed in this section have to do with the deployment of the cold-water pipe. In test installations of relatively small-diameter pipe, several have been lost during deployment, while two have been successfully installed.

For the floating plant system, the pipe is usually conceived as being about 30 m in diameter, 1000 m in length, weighted at the lower end, and suspended from the floating vessel at the upper end. Pipe materials considered have included lightweight concrete, steel, fiberglass, polyethylene and hybrid designs. The principal problems both in service and during construction are dynamic ones, resembling those successfully surmounted in the installations of the Cognac platform and Hutton tension leg platform. Here, however, the sizes and weights are an order of magnitude greater.

The major construction problem is that of transport, upending, and transfer of the cold-water pipe to its location hanging beneath the floating vessel.

The pipe is usually conceived as being articulated in order to reduce its in-service response to internal waves and currents. During installation, these joints may be of use to the constructor or may require temporary locking. Alternatively, polyethylene pipes have been tried, which are able to be buckled during deployment yet subsequently regain their shape.

In one proposed system for deployment, long segments will be assembled in shallow water and towed to the site in the horizontal mode. Upon arrival, they will be upended by selective ballasting. This system suffers from the need to have temporary bulkheads within the pipe in order to enable selective ballasting, since otherwise the joints would develop extreme angular changes exceeding the capacity of the joints.

With a cluster of small-diameter polyethylene pipes, however, the flexibility of the pipes, combined with their inherent low buoyant weight, enables them to buckle temporarily at the overbend and then expand out to their original shape.

Alternatively the joints may be designed with structural stops or restraints of adequate capacity to resist the moment as each section fills with water. This would resemble in principle the Shell articulated pipeline stinger, but on a much larger scale.

Another deployment concept is that of on-site assembly in the vertical mode. In one such system, vertically floating segments perhaps 100–200 m in length would be towed to the site. Lines would be run from under the moon pool of the vessel to the pipe. It, in turn, would be gradually lowered until the lines from the vessel took the load, enabling the pipe to swing in under the vessel. The pipe is then raised up and secured to the vessel, using linear jacks with large heave compensators to offset the differential heave caused by the response of the vessel to the seas. See Figure 11.11.1.

In principle, subsequent segments can also be added below the original section, the only problem being that each such jointing must be carried out at increasing depths.

Mating cones, with guidance by acoustics and TV, are proposed to enable this jointing.

Another approach involves the adding of additional segments progressively at the top, so that the joints are effected at or near the surface. For this, the pipe as a whole is suspended by external lines leading down to an enlarged base section so that new segments can be fed in through the space in the lines.

As with articulated columns and single-point moorings, the trend in development is toward the transport and assembly in the vertical mode to the greatest extent practicable.

Currently, attention is being directed to shelf-mounted plants, located near steeply sloping margins such as those found in Hawaii and the Caribbean. For the "MINI-OTEC" plant on Hawaii, a small polyethylene line was run down the slope to a depth of 600 m. Since it is positively buoyant, it is held down at intervals by tethers to clump weights. The line was assembled in quiet water near shore, towed afloat to the site, and submerged by progressive flooding. As it submerged, it progressively buckled, but resumed its circular shape after the critical bending passed. Reinforcement was provided at the connection of each tether. Chains may be used in lieu of clump anchors so as to drag over the seafloor and adjust to varying profiles.

Deep Oil Technology has proposed a system for pulling a large-diameter line down a slope:

1. Using a drilling vessel, a pile anchor will be drilled and grouted into the rock at the lower end.
2. When the pile is set, it has a terminal gimbal base attached which contains a large sheave. Now, after the pile is in place, and cemented, a line is paid out from the drill ship, around the underwater sheave, and back to a surface boat. This end is towed to shore.
3. A landing base for the cold-water pipe termination is now lowered onto the pile, using tensioned guide lines.
4. When the line reaches shore, it is attached to the prefabricated pipe. The pipeline is now pulled out to the termination pile. A small vessel is moored vertically above the end of the pipeline, moving out with it. The vessel supports a jetting hose and TV camera and provides buoyancy to support a drag scraper or plow.
5. The guide lines pull the end of the pipeline into the termination cone.
6. Finally, a retrievable filter and entry port are lowered down the tensioned guidelines and fitted to the termination structure.

Another proposed approach involves the pre-installation, by pipeline-pulling methods, of a steel pipeline, extending down the slope to the lower termination. This then acts as a guide rail down which the shore-assembled pipe will be run.

In all shelf-mounted installations, adequate holdback capacity must be provided at all stages.

Because the upper portions of the line will be exposed to wave action, the shore crossing, out to a depth of 50–100 m, may be pulled through a preconstructed tunnel. For a similar installation of large-diameter gas lines across the coastal zone in Norway, precast tunnel segments were installed, supported on intermittent piers of underwater concrete. Then the gas lines were pulled through the tunnel.

328 *Other Applications of Offshore Structures*

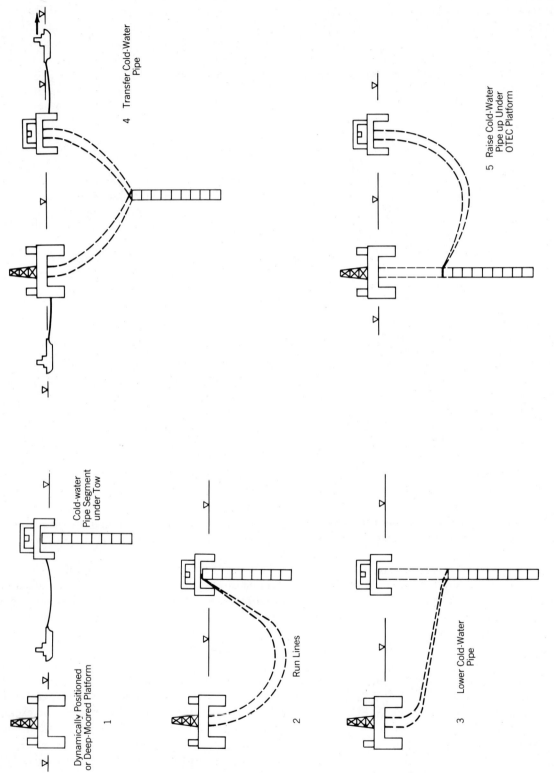

Figure 11.11.1 One concept for deployment of OTEC cold-water pipe.

11.12 Oosterschelde Storm Surge Barrier

The Oosterschelde Storm Surge Barrier, across the mouth of the Eastern Scheldt Estuary in the Netherlands, now nearing completion, must rank as one of the major offshore engineering and construction achievements of the decade. See Figure 11.12.1. Although constructed in water depths only 20–40 m deep, it is exposed to waves and winds from the North Sea and to high tidal currents, so as to require the development of new techniques for offshore construction which will have application on future projects. The project is also notable because of the innovative development of specialized construction equipment. See Figure 11.12.2.

Sixty-six mammoth concrete gate piers are being installed, seated on a prepared foundation on the sands of the river delta. Initial operations offshore commenced with construction of an island near the center of the project from which all construction work was carried out. Three large basins were excavated, diked off, and dewatered to enable fabrication of the concrete piers in the dry.

Simultaneously, extensive slope protection work was carried out on the beaches and dikes adjoining the barrier proper. Aprons of asphalt-filled stone, sand–asphalt, and articulated concrete mats were laid.

Meanwhile, at the site, large-diameter dolphin and anchor piles (steel cylinder piles) were driven to serve as moorings for the extensive floating construction operations to come. Lines from the anchor piles were run up to mooring buoys.

The loose sands in the top 10–20 m of the foundation under the barrier, were then compacted by vibratory penetration. A special floating rig, the

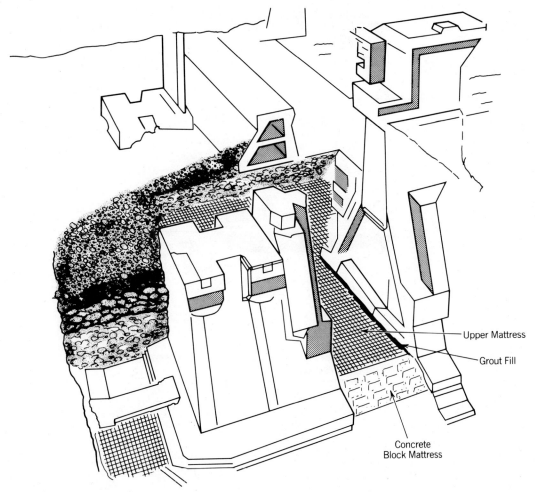

Figure 11.12.1 Isometric view of gate piers and foundation protection for Oosterschelde Storm Surge Barrier, the Netherlands.

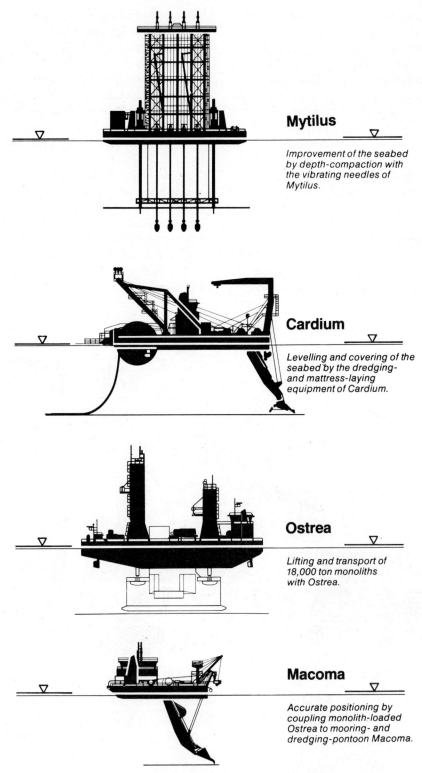

Figure 11.12.2 Specialized offshore construction equipment used to construct the Oosterschelde Storm Surge Barrier in the Netherlands.

Mytilus (see Fig. 11.12.3), jetted and vibrated four large-diameter steel tubes down to a depth up to 50 m below sea level and then actuated heavy internal vibrators as the tubes were withdrawn. Spacing of the vibrated probes was 6 × 6 m each.

The surficial sands of the seabed were then removed by a "dustpan" dredge, which also dragged a screeding compactor and laid out a heavy mattress behind. This mattress, consisting of reinforced geotextile fabrics and graded stone layers, was prefabricated in an onshore plant and reeled up on a huge floating reel to be floated out and hooked up to the dredge. See Figures 11.12.4 through 11.12.7.

An underwater tracked inspection vehicle then crawled over the mattress while being tended by a floating survey vessel above. Sonic and electronic instrumentation plus feeler probes enabled an extremely precise survey of each mattress to be made. See Figure 11.12.8.

From the information so determined, an articulated concrete block mattress was tailored so that when it was laid over the lower mattress, the surface was level within a few centimeters.

Meanwhile, when all concrete piers had been completed in a basin, that basin was flooded (see Fig. 11.12.9), the dikes were removed, and a giant catamaran crane barge, *Ostrea*, moved in over a pier, raised it, and transported it to the site. See Figure 11.12.10. The crane barge was mated to the mattress-laying barge, which was already properly positioned, and the pier lowered into place.

Figure 11.12.3 The *Mytilus* densified and compacted the underlying sands to serve as a foundation for the Storm Surge Barrier.

After seating, grout is pumped in under the base, working from a compartment inside the pier. See Figure 11.12.11. Extensive scour protection is placed outside each pier. Larger stones are run down an inclined ladder so as not to damage the concrete by impact. The pier is filled with sand to provide stability. The nearly completed barrier is shown in Fig 11.12.12.

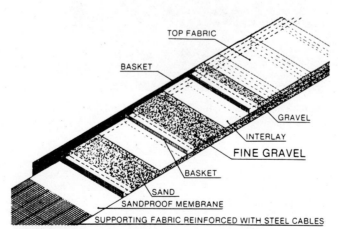

Figure 11.12.4 Foundation mattress for Oosterschelde Storm Surge Barrier.

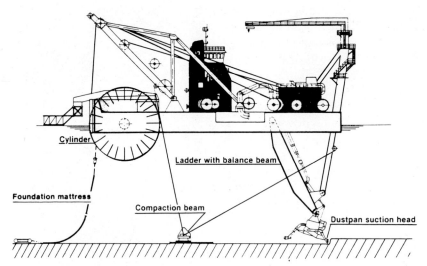

Figure 11.12.5 Cleaning, compacting, and laying of foundation mattress. (From *World Dredging and Marine Construction*, Jan. 1984.)

Figure 11.12.6 The *Cardium* laying the articulated concrete mattresses for the Oosterschelde Storm Surge Barrier.

Figure 11.12.7 Laying of articulated concrete block mattress.

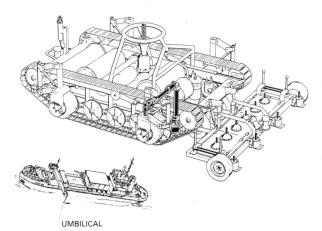

Figure 11.12.8 Underwater inspection vehicle used to ascertain variations in mattress elevations so that filler blocks can be accurately sized and placed. (From *World Dredging and Marine Construction*, Jan. 1984.)

Figure 11.12.9 Piers for Oosterschelde Storm Surge Barrier await their turn for installation.

Figure 11.12.10 12,000-ton capacity lifting barge *Ostrea* setting huge concrete gate piers for Oosterschelde Storm Surge Barrier. (Courtesy of Hendrik Boogaard B.V.)

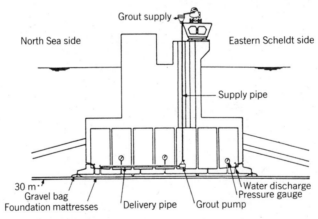

Figure 11.12.11 Undergrouting system.

11.12 Oosterschelde Storm Surge Barrier

Figure 11.12.12 Installing piers for the Oosterschelde Storm Surge Barrier. (Courtesy of Dutch Construction Consortia.)

Whisp'ring wind, soaring bird, gently rolling sea:
Dancing wave, flying fish, beckoning to me.
Shining sail, steady ship, heaven is my chart;
Guiding star, silver moon, call me to depart.
The rolling sea is keeper of my heart.

"KAHUNA KAI" (TRADITIONAL HAWAIIAN SONG)

12

Installation of Steel Submarine Pipelines

12.1 General

This chapter will address the installation of steel submarine pipelines used for the transmission of petroleum products, gas, water, slurries, and effluents.

The diameter of steel submarine pipelines typically runs from 3½ in. up to 54 in., with occasional lines running 72 in. Diameters of pipe worldwide are usually expressed in inches, even though all other units are metric; this is due to their historical tie to the oil industry.

The steel for these lines is usually of relatively high yield strength, 350–500 MPa (50,000–70,000 psi), and is selected for weldability. Wall thickness will normally run from 6 to 25 mm (0.25–1.0 in.), with the upper limit again being constrained by weldability.

Almost all steel pipelines have been joined by full–penetration welds, especially in the petroleum industry, where pressures typically run 1500 psi (10 MPa) and leakage of oil or gas is unacceptable. Consideration is being given, however, to the use of mechanical joints, for example, joints similar to those used with well casing. Developmental work continues on explosively—and hydraulically—expanded connections. In a few cases, flanged connections are used, but these are usually seal-welded.

Since most submarine pipelines are installed empty, they are subjected during installation to high hydrostatic pressure, along with whatever bending may be taking place, and are laid under axial tension; thus buckling under combined loading becomes a principal design consideration. Tolerances are consequently of great importance, out-of-roundness, best-fit circle, and wall thickness being the most critical.

The steel is protected from external corrosion by coatings such as bitumastic or epoxy supplemented by cathodic protection, usually sacrificial anodes. Internally, the line may be uncoated if it is to be in petroleum service, or it may be internally coated with epoxy, polyurethane, or polyethelene or cement-lined when it will carry water, salt, or corrosive substances.

The external coating may be further protected from abrasion by concrete or fiberglass wrapping or the like.

To give stability to the line when in service, especially those lines which must be emptied at some stage of their life or which carry a low density material like gas, the line must have a net negative buoyancy. This is usually supplied by concrete weight coating (which can also serve to protect the anticorrosion coatings) or by increasing the wall thickness of steel. See Figure 12.1.1.

Recently, a number of pipelines in the North Sea have experienced "floating up" off the seafloor due to shedding of their concrete coat. This indicates that the reinforcing mesh may have been underdesigned or that the pipelines may have been subjected to excessive overstress during installation. The latter is within the purview of this book. One can hypothesize that such damage may have occurred during the more severe sea states when the pipelaying barge was subjected to severe dynamic surge. If the coating was then not only cracked but delaminated from the pipe, then transient pore pressures under the storm waves could break the coating off in progressive failure.

This type of failure has previously occurred during pipe-pulling operations. The most obvious solution is to increase the amount of circumferential reinforcing in the coating. Since the coated pipe is

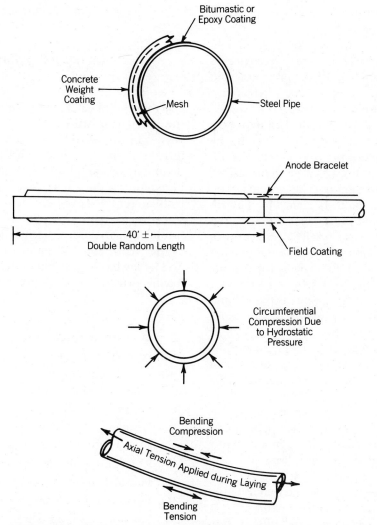

Figure 12.1.1 Typical steel submarine pipeline, showing stresses incurred during installation.

usually furnished by the oil company, this obviously presents a contractual problem to the pipeline installation contractor. Nevertheless, he may often find it in his best interests to verify the amount of circumferential reinforcing and, if necessary, to request (or pay for) its augmentation.

Pipelines are basically designed to lay on the seafloor or in a trench in the seafloor, with more or less continuous support. However, unsupported spans may occur in rough, rocky seafloors or where the sands move under the action of currents and waves. The designer will have set limits on the unsupported span lengths, which the contractor must not exceed; this may require either prior seafloor leveling or post-installation support.

Lines are buried beneath the seafloor in many areas of the world to protect them from fishing trawl boards, from dragging anchors, and from fatigue due to oscillation in a current. Sometimes the trenches are backfilled artificially, but in most cases natural sedimentation is counted on to fill the trench.

As noted earlier, the pipeline usually sees its most severe stresses during installation; thus very close integration is required between the designer and the installation contractor. The designer needs to be aware of and address the needs of the contractor during installation. The contractor conversely must be aware of the limitations and constraints imposed by his installation procedures, taking into account

338 Installation of Steel Submarine Pipelines

the sea state (waves and current), the varying water depths, and the varying seafloor.

In addition, both parties must be cognizant of other pipelines and facilities in the area, recognizing the tolerances in location both of the previously laid lines and facilities, and the tolerances which are inherent in the contractor's procedure.

Submarine pipelines are typically laid in a "corridor" whose centerline and width are given by the client and shown on the approved permit. The installation contractor must have an adequate survey system to enable him to comply. This system is usually an electronic positioning system but may include lasers, ranges, and preset spar buoys.

The installer must verify to the satisfaction of the client and the regulatory body that the line has been satisfactorily installed. Externally, this is done by side-scan sonar and ROVs, using video or acoustic imaging. Internally, the line is pigged and then tested with hydrostatic pressure to a pressure in excess of the design pressure.

A pipeline "pig" is a short cylinder, of slightly less diameter than the pipeline, with several sets of squeegee wipers. When the pig is entered in the pipeline and excess pressure is applied, it travels along the pipeline. The diameter of the pig and its length verify that there is no dent, crimp, or buckle more than the small annular space. The squeegees hold the pressure so that the pig will move.

The pig is usually equipped with an acoustic transponder or radioactive marker so that if it does get stuck, its position can be determined.

Guidance for the design and installation of submarine pipelines is given in the DNV *Rules for Submarine Pipeline Systems*. Excerpts relating to installation are given in the appendices to this book.

Many methods of pipe laying have been employed, selected on the basis of environmental conditions during installation, availability and cost of equipment, length and size of line, constraints of adjacent lines and structures, and so forth. The following are those most commonly employed:

1. Convential lay barge
2. Bottom-pull barge
3. Reel barge
4. Surface float
5. Controlled below-surface float
6. Above-bottom pull
7. J-tube from platform
8. J-tube catenary lay barge
9. S-curve with breather floats

These will be described in the following sections.

12.2 Lay Barge

The offshore lay barge has grown up from the specially modified cargo barge of the 1950s to one of the most sophisticated, efficient, and expensive vessels in the world. Lay barges are often characterized as first-, second-, and third-generation to denote major quantum jumps that have been made in extending the ability to lay lines in deep water, with current achievements being the successful installation in depths over 600 m (2000 ft) across the Strait of Sicily and in such adverse environments as the North Sea. See Figure 12.2.1.

The lay barge is a system that comprises the following principal operations:

1. Seaborne work platform
2. Mooring and positioning of barge
3. Pipe delivery, transfer, and storage
4. Double-ending of pipe, conveying to lineup station, and lineup
5. Welding of joints
6. X-ray
7. Joint coating
8. Tensioning of line during laying
9. Support of line into water (e.g., "stinger")
10. Survey and navigation
11. Anchor-handling boats
12. Communications
13. Personnel transfer—helicopter and crew boat
14. Diver or ROV for underwater inspection
15. Control center
16. Crew housing, feeding, etc.
17. Power generation
18. Repair facilities and shops

A typical second-generation lay barge is shown in Figure 12.2.2. The layout of equipment is shown in figures 12.2.3, 12.2.4 and 12.2.5.

The basic operations of the lay barge can be outlined as follows:

1. The lay barge is positioned on its anchors, 8 to 12 in number, holding it aligned with the pipeline route, with a "crab" or slight orientation angle as needed to accommodate the effects of the current. Its position is deter-

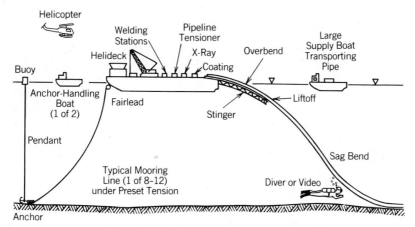

Figure 12.2.1 Typical lay-barge operation.

mined by an electronic positioning system, augmented by laser in some cases. Its orientation is by gyroscope.

2. The anchors will be progressively moved forward as the laying takes place, usually in 500- to 600-m (1600- to 2000-ft) jumps. One anchor-handling boat on the starboard side will move each anchor ahead in succession; another anchor-handling boat will move each of the port anchors ahead in succession. See Figure 12.2.6.

Typically, the anchor-handling boat maneuvers close to the anchor buoy so as to enable the deckhand to hook an eye in the end of the pendant. He attaches a wire line from the deck engine of the tug, which either pulls the buoy aboard or pulls the pendant through the buoy, thus lifting the anchor clear of the bottom 5 m or so. The boat then runs forward, setting the anchor as directed in its new position, releasing the buoy. The boat turns outboard and goes back for the next anchor in the cycle.

The new position of the anchor is given by voice radio command from the control house, which is based on radar, gyro, and the reading on the remote mooring line length counters, reading the line length paid out by the winch.

The proper paying out and taking in of each mooring line on the winch drum is monitored by video in the control house to ensure against crossed lines on the drum or fouling of the line.

3. From a supply boat or barge alongside the

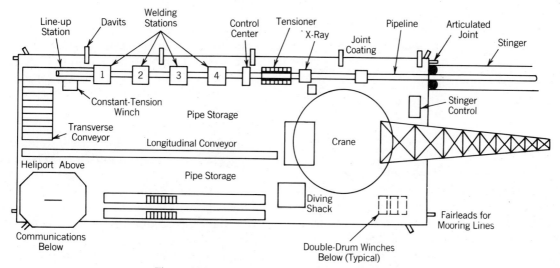

Figure 12.2.2 Second-generation pipe-laying barge.

340 *Installation of Steel Submarine Pipelines*

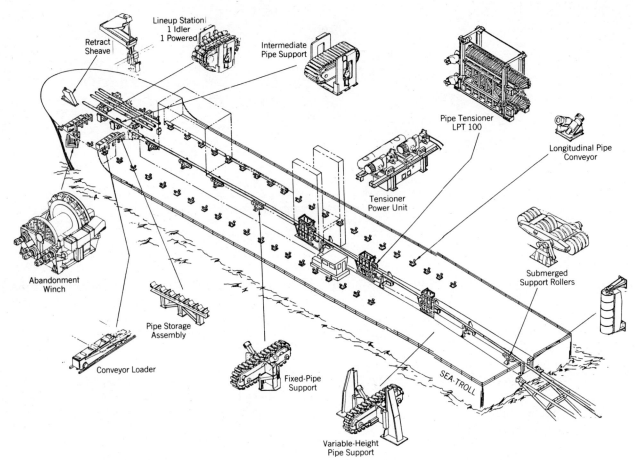

Figure 12.2.3 Layout of equipment on ship-shape pipelaying vessel. (Courtesy of Western Gear Corp.)

port side, the crawler crane on the lay barge snags (picks) one pipe length (±40 ft) at a time, turns, and sets it in storage.

4. From storage, the crane picks a pipe length and sets it on the end-O conveyor which moves it to the transverse conveyor at the bow.

5. This conveyor feeds it onto the lineup station, where it is positioned, usually semi-automatically, in correct alignment and then run forward to the end of the preceding segment.

6. The internal lineup clamp positions it in exact spacing and holds it for the hot-pass weld.

7. The hot-pass weld is made and gouged.

8. The segment moves forward successively to weld stations 2, 3, 4, and so on with one or more passes being applied at each station, and is then chipped or gouged.

9. The fully welded line now passes through the tensioner, where it is gripped by polyurethane cleats on caterpillar-like treads. Hydraulic rams push the pads against the coating, adjusting their pressure so as not to deform the pipe nor crush the coating, while still developing frictional resistance. The tensioners run on torque converters or similar devices so as to pay out under a set tension. This tension will have a rather wide tolerance; for example, it could be 70,000 lb ± 20,000 lb (300 KN ± 100 KN).

10. The joint now goes to the X-ray station where it is X-rayed and the films are developed and checked. If a flaw is found, it must be cut out, rewelded, and re-X-rayed. For a cutout, the barge must be moved astern and the line brought back up on board one or two lengths so that the cutout is forward of the tensioner.

11. The pipe section now moves astern again, where the joint is coated with the specified corrosion-protective coating. A zinc–alumi-

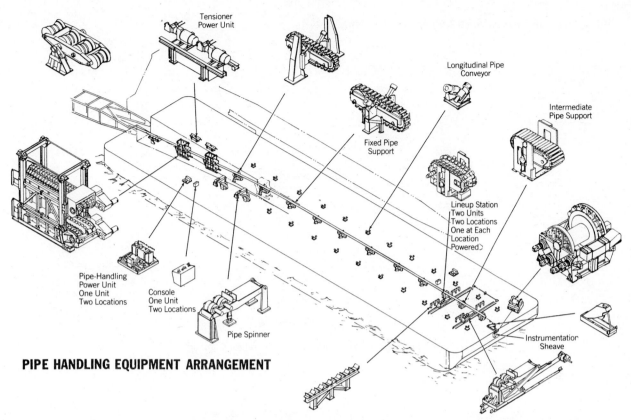

Figure 12.2.4 Equipment on modern pipelaying barge. (Courtesy of Western Gear Corp.)

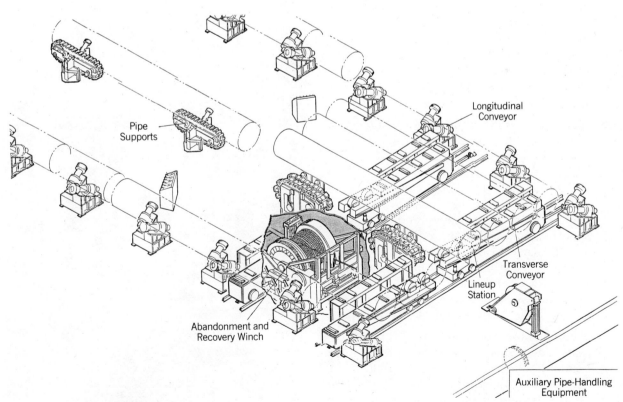

Figure 12.2.5 Arrangement of conveyors and winches in vicinity of line-up station. (Courtesy of Western Gear Corp.)

342 Installation of Steel Submarine Pipelines

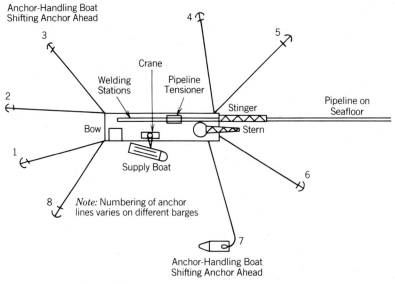

Figure 12.2.6 Typical lay-barge operational spread.

num bracelet or other anode is affixed. Concrete mortar coating is applied so as to protect the corrosion-protective coating. This fresh concrete is protected by a sheet metal wraparound. See Figure 12.2.7.

12. The completed pipeline now passes down the ramp and over the stern of the barge and bends downward. This downward bend is called the "overbend." See Figure 12.2.8.

13. The line rides down the stinger to a point of departure, where it leaves the stinger due to the tension in the line.

    The stinger in this case has a hinged connection to the barge. It has built-in flotation so as to support the pipeline while still allowing a downward inclination and some flexibility to accommodate surge. The stinger may be articulated so as to permit continuous curvature or may have a built-in vertical curve. Load cells on the roller supports, plus depth indicators such as bubble gauges, enable the stinger to be ballasted for optimum support.

14. The line now moves downward through the water and bends back to the horizontal at the seafloor. This bend is called the "sag" bend. At this bend, the pipeline is usually subjected to its maximum stresses due to the combined axial tension, vertical bend, and circumferential hydrostatic pressure.

Figure 12.2.7 Coating joint is last operation before pipe descends down stinger.

Figure 12.2.8 Pipeline descends down stinger, guided and supported by rollers. This is the critical "overbend" zone for the pipe.

15. As the line lays out on the seafloor, its integrity is checked either by divers or video, the latter either riding the pipe or by ROV.

From the above sequence, it can be seen that the typical lay barge system described at the beginning of this section has the following physical components:

Anchor-handling boats (usually two)
Supply boats (usually three) or supply barges (usually two) with tug
Helicopter service
Crew boat
Shore base
Lay barge, equipped with:
Crawler crane
Pipe storage racks
Pipe conveyors
Lineup station
Internal lineup clamp
Welding stations
Tensioner
X-ray equipment
Joint-coating equipment
Constant-tension winch for abandonment and recovery
Stinger and stinger control
Winches with mooring lines
Control room
Radio circuits to shore and boats
Voice and indicator circuits to welding stations, stinger control, X-ray
Gyrocompass
Radar
Electronic-positioning device
Tensioner force readout
Mooring line tension readout
Mooring line length-out readout
Diver shack
Decompression tank
Winch video screens
Heliport
Quarters for crew
Mess hall and kitchen
Office
First-aid and medical facilities
Owner's quarters and office
Repair shop
Power plant
Fuel and water storage
Stores room

The crew required to operate an offshore pipe-laying vessel may be 150 or more men per shift. Normal operations use two 12-hour shifts. A third shift will be off on leave. Work schedule is usually two weeks on, one week off, or even one on, one off.

Tension is maintained in the pipeline from the barge to the seafloor in order to reduce the vertical bending and the tendency to buckle. Values of applied tension range from a low of perhaps 20,000–30,000 lb (100–150 KN) in shallow water and calm seas to 400,000 lb (200 KN) in deep water and rough seas.

The lay barge is subject to dynamic surge motion, depending on the relationship between wave length, barge length, and depth of water. This surge is usually too fast for the tensioner and the welder to follow. Thus the pipe is locked in fixed position in relation to the barge. Therefore, the tension is cyclically reduced and increased about the steady-state force. Typical ranges of tension are of the order of 20,000 lb (100 KN) each way in a moderate sea. Heave and pitch also have some effect on the tension, but generally to a much lesser degree than surge.

This tension must also be introduced and maintained during the startup and laydown of the pipe.

Welders are critical to the operation. They are working on a rolling and heaving barge, yet must produce essentially perfect welds. They must be protected from spray and rain and must have adequate light and ventilation.

If the X-ray discovers a flaw in the weld, the resultant cutout repair stops the entire operation until it is completed.

The actual performance of the welds is also of serious concern to the pipeline installation contractor, due to his responsibility to ensure a sound, leak-free pipe on completion. The combination of axial tension and overbend stresses on the weld are very severe, especially since the latter are dynamic. Not only the toughness of the weld itself is involved, but also that of the heat-affected zone (HAZ), which in turn is influenced by the parent steel quality as well as the welding procedures. The constructor may therefore find it prudent to test the pipe steel and welding procedures under dynamic tension loads prior to finalizing procedures.

In a typical offshore operation, the barge will move one pipe length every fifteen minutes. On the most modern third-generation barges, using advanced welding techniques, rates are even higher. This means that all the work must be completed at each station within that same time frame. This translates to 100 or more 40 foot lengths per 24 hour day. See Figures 12.2.9 and 12.2.10.

These performances have been exceeded by top-notch crews on good days, even with manual welding.

Stresses in the pipe in the laying operation are controlled not only by axial tension but by the net submerged weight of the pipe. This latter is the difference between two large numbers, the one being the air weight of the pipe, the other being the buoyancy due to the displaced volume. The major variable is the thickness of mortar coating, which affects both air weight and displacement, but not equally.

Figure 12.2.9 ETPM's modern pipelaying barge laying 46 in. O.D. pipeline off Northwest Shelf, Australia.

In a typical case, a pipe may have an air weight of 1000 lb/ft (15 KN/m) and a displacement of 950 lb/ft (14.3 KN/m) leaving a net (buoyant) weight of 50 lb/ft (0.7 KN/m). If the coating increases the weight by 5 percent, the displacement may increase by only 2 percent. These numbers may sound small, but they develop an increase in net buoyant weight of 30 lb/ft or a 60 percent increase in the force causing the bending.

Thus while weight control is normally not as critical with lay barge operations as it is with bottom pulls, it nevertheless is of great importance and must be monitored.

The pipe is generally furnished in double-random lengths, which are normally 40 ft. Most sections will run 38–42 ft. However, generalized pipe procurement specifications allow some sections which vary widely from the norm, as short as 16 ft

Figure 12.2.10 Submarine pipeline at Esmond Field moves down stinger into North Sea. (Courtesy of Bechtel Corp.)

and long as 56 ft. While this may be accommodated on land pipelines, it is unworkable at sea. Such sections should be cut or spliced to the normative length of 40 ft at the shore base, or else the procurement order should exclude these variances.

Rates of progress often reach one and even two miles per day. This means that 100 to 200 or more sections must be loaded out each day from the shore base, transported to the site, and then unloaded to the deck of the barge. This last is a critical operation when the seas are running high and may, along with anchor handling, be the controlling operation.

The transfer at sea of the pipe is a typical case of operations involving two vessels alongside each other, of different characteristics, each responding in its own way to the seas, in each of six degrees of freedom.

The relative positions in plan can usually be maintained in a moderate sea state by tying a transport barge alongside the lay barge, with suitable fendering, so that the major individual responses are limited to heave, roll, and pitch. In heavier sea states, barges can no longer be kept alongside, and so supply boats are used. By running a stern line from the boat and keeping power on, a good skipper can hold his boat in reasonably close position, although now he will develop some relative sway, surge, and yaw motions.

The typical laybarge is restrained from lateral motion by the mooring lines; it is also moved periodically one pipe length ahead. These lines, while catenary in scope in deep water, are kept under tension by the winches. The line tensions are measured by tension meters on the wire rope or on the winch drum or both. In the typical second-generation lay barge, the tension may be 80,000 lb (400 KN) with a variance in a moderate sea of ±20,000 lb (±100 KN). This variance is due to the long-period sway plus surge built up by the waves, storing energy in the wire lines as the barge gradually moves to one extreme of its lateral range. The lines on the far side gradually become more taut, so that eventually the barge changes direction and starts its sway excursion to the other end of the range.

The acceleration at the end of its excursion causes a shudder effect in the overall system, which translates into a severe horizontal whip of the stinger and of the pipe. The surge excursions cause cyclic bending in the pipe at the overbend and in high sea states can lead to low-cycle fatigue in the pipe.

The mooring lines must provide the horizontal restraint against wave drift, wind drift, and current drift. They also react against one another and especially must counter the tension on the pipe, which in effect is like a mooring line of relatively equal tension, leading directly astern. Balancing out the tensions in 8–12 mooring lines plus one pipeline is a complex problem, especially when these line forces are not steady but subject to the significant ranges introduced by the long-period excursions.

Typically, the tensions in the mooring lines are set so that under the maximum design surges, the force will not exceed 50–60 percent of the guaranteed minimum breaking strength.

To offset the pipeline tension requires additional mooring line forces in the lines leading forward. The system must be balanced up, which is difficult enough with one positioning of the anchors, but which is rendered more complex due to the constant lifting and relocation of anchors. The system can be satisfactorily resolved by preparing calculations of typical and extreme positions for each permutation; it of course lends itself to the use of an on-board mini- or microcomputer, which can then solve for intermediate situations.

The cost of pipe laying is directly related to the progress, since the cost per day is more or less the same whether any pipe is laid or not. The rate of progress has until recently been controlled by the time required for welding. There is a specific amount of weld metal which must be applied. Only two welders (one each side) can work at any station. Therefore, the rate of progress depends on the number of stations. Typically these are placed one pipe length (40 ft) apart. There is only room enough for a certain number of welding stations on a barge; therefore, the longer the barge, the greater the rate of progress. This explains why prior double-ending of the pipe does not speed the operation.

Another means of accelerating the welding is by the use of microwire welding, but this is usually only acceptable in hot climates because of the dangers of cold lap at lower temperatures.

The biggest jump in pipelaying progress has come with the introduction of automatic welding of one type or another.

Second-generation lay barges are limited by the sea state. When the significant wave height exceeds about 8 ft (2.54 m), operations must shut down. The specific limit of course depends on the relative direction and the period of the waves, as well as the barge length and width. The limiting item is usually control of surge and the interaction between stinger, pipeline, and barge. The working limits can be increased by using a wider and longer barge, by using more powerful tensioners, and by using an articulated or fixed cantilever stinger.

When seas reach 10–12 feet (H_s = 3–3.5 m), then other constraints arise. Anchor-handling boats can no longer pick up the anchor buoy, although this limit has been extended by clever arrangements enabling the boat to run past the buoy and snag it rather than having to back down for the deck hand to make fast to the pendant. Pipe transfer from a barge alongside went out with an H_s of 2–2.5 m, but a supply boat can be used to extend this operation to the 3- to 4-m range.

The barge motions in roll and long-period sway (snapback) become too severe, especially with a beam or quartering sea, and the welders are unable to produce quality welds. The pipe starts to jump out of the stinger, and there is danger of buckling the pipe.

At this stage, a decision must be made as to whether to hold on or to initiate abandonment procedures. The major factor here is the weather prediction. If improvement is forecast within the next few hours, it may be practicable to hang on, maintaining tension. Another factor is whether or not the anchors will hold in the seafloor soils or are likely to drag; a dragging anchor will almost always lead to a buckle.

When abandonment is decided, a bull plug (cap) is welded onto the pipe. A line from the constant-tension winch is attached. A buoy and pendant are also attached to the bull plug. It is a good precaution also to attach an acoustic pinger to the bull plug. The barge then moves ahead, paying out on the line, until the pipe is fully laying on the seafloor. The end of the constant-tension line is buoyed and run off.

The barge can now pick up its anchors to move to a sheltered location or decide to ride the storm out at sea, on its anchors, but turned now so as to head into the sea.

When the storm ends, the barge moves back to location and resets anchors. While one hopes to find the two buoys, it is not unusual for them to have been torn away by the storm. That is when the acoustic pinger helps.

The constant-tension line is now pulled on-board and the tension applied. The barge slowly moves astern, bringing the pipeline back up onto the stinger. A line from the crane may have to be hooked on (by diver) so as to help guide the line back onto the rolls of the stinger without fouling. Now the pipe is pulled on board, through the tensioner, until the bull plug reaches the lineup station; the bull plug is cut off and the pipe end rebeveled, and the laying operation can recommence.

An important point is that abandonment procedures are almost always carried out under extreme conditions, at or above working limits, whereas recovery operations will normally be carried out in good sea conditions.

The start of pipe laying also requires special procedures. Assuming the work will start at the platform, an anchor is set at a point astern of the platform and brought up the stinger onto the barge, where it is welded to a bull plug on the leading end of the pipe. Tension is applied by the tensioner and the pipelaying operation commences, with the lay barge moving progressively ahead and the completed pipe line being pulled off the stinger.

The alignment is set as close as physically and safely possible to the platform so that when the bull plug reaches the seafloor, it will lay alongside the legs of the platform.

Another method is to pull the end of the line off the lay barge. This can be done by running a line, attached to the pipe end, from the barge to a sheave at the preset anchor and back up to a winch on the barge. This enables the end of the line to be pulled to the exact position desired. The potential problems here are fouling of the line at the sheave (e.g., the sheave flips over, jamming the line) or fouling of the line in the jacket bracing.

The jacket may have been pre-fitted with a J-tube and a winch on deck so that it can pull the line off the lay barge. While this system can only be used with small-size lines, up to about 12 in., it is very economical and eliminates the need for a separate riser operation.

If a large boat is available and the seas are calm, it can pull the line off of the lay barge and lay the end down adjacent to the platform. The efficacy of this method depends on the bollard pull being at least equal to the required tension.

When the line terminates at a platform, it is customary to lay past the platform on one side and then follow abandonment procedures, laying the line down on the seafloor.

Earlier it was stated that the most serious problem in pipelaying is a wet buckle. In the case of a dry buckle—that is, where the line does not take on water—the pipe can be just pulled back up on board. In the case of a wet buckle, however, the pipeline has been flooded and cannot be brought back on board without creating continuing buckling. For this reason, at startup at least one pig was placed in a pig chamber at the startup end, along with air fittings. If a wet buckle occurs, compressed-air lines are connected and the pig run along the pipe to the point of buckle. This empties the line so that it can be recovered.

Actually, there is one even more serious case, known as a propagating buckle. This is the case where the ovaling of the pipe at the point of initial buckle reduces the collapse strength below the resistance to the external hydrostatic pressure so that the buckle travels back along the pipe. While this case is usually within the province of the designer, the constructor must make sure that this cannot occur, else he could lose his entire line. Where calculations show this to be possible, buckle arrestors in the form of thicker pipe or reinforced pipe are installed at intervals of 1000 m or so. For example, wrap-around plates may have been pre-installed.

Occasionally a line may be damaged after it has been successfully laid down. Often this is due to an anchor dragging into the pipeline. It may even be an anchor from your own spread, that is, the barge or boats, or it may be from another contractor working on the same platform. The line must be repaired. A similar requirement ensues when two sections of the line must be welded together, a riser attached, or the like.

One method is to use a hyperbaric chamber (a "habitat") lowered down over the line and centered on the junction or repair point. Compressed gas is used to expel the water, and divers descend to make the weld in the gas atmosphere. The selection of the appropriate gas mixture is critical in order to ensure the proper weld quality.

Such a repair in 100 m of water, for example, may require several days. It may be tended by the lay barge, but if the sea state permits, a smaller support vessel may be used.

The repair procedure consists of accurately cutting the lines and beveling their ends. A template is made to ensure an exact fit of a "pup" (a short, specially cut pipe section), which is fabricated on board and lowered down for welding. After the welds are completed, the joint is coated for corrosion protection. X-ray is usually not practicable, and reliance must then be placed on visual inspection and magnetic-particle or other NDT techniques to verify the quality of the weld.

Wet-welding techniques have been under development for many years; the problem of course is ensuring a weld that will be safe under the working pressures, which typically are 1500 psi (10 MPa) or so. At the present time, there is not universal confidence in the quality obtainable by wet-welding techniques, but development continues.

In shallow water, the damaged section can be brought to the surface for a dry weld. The line should be empty, if possible, and a long length brought up so as not to exceed curvature limits in the pipe. For this reason, many lay barges are fitted with davits along one side, enabling lifting from the entire length of the barge. The derrick crane may also pick from the stern, the pipe transfer crane from the bow. Where this curvature is still too great, floats or buoys may be attached to give positive buoyancy along appropriate lengths of pipeline.

As the line is brought up, there are of course length-compatibility problems in all but very shallow water. It is usually necessary to cut the line, thus flooding the pipe. Once brought to deck level, the ends are beveled, a pup fabricated and installed, and the line laid back. Now the new line is longer than required, so it must be laid back so as to lie in a horizontal curve on the seafloor.

For further discussion on repairs, see Section 15.5.

Installation of risers at platforms is another special operation requiring careful preplanning of each stage. There are a number of methods which have been used successfully (see Fig. 12.2.11):

1. The riser is preattached to the side of the platform. The end of the line, prelaid on the seafloor but still empty, is pulled over to that same side of the platform by means of a line so rigged as to maintain axial tension. Then divers make a template of the intervening space and a pup is prefabricated and installed using a hyperbaric chamber lowered over the joint so that welds can be made.

    Alternatively, with large-diameter pipe (e.g., 42 in. or greater), flanged connections may be used. The line is cleared of water, using a pig if necessary, and a welder descends in the riser to weld the joint from the inside.

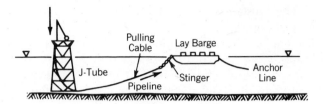

Reverse J-Tube Method

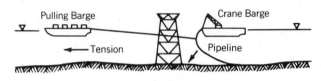

Tension Method

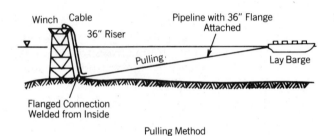

Pulling Method

Figure 12.2.11 Riser installation methods. (Adapted from *Offshore Platforms and Pipelining,* Petroleum Publishing Co., Dallas, Texas, 1976.)

Hydrotech has developed a two-piece diagonally flanged coupling which can be sleeved over the two pipe ends and then rotated so as to accommodate a difference in angle. A three-piece coupling can accommodate up to 15° misalignment. After the joints are bolted up, they are seal-welded, using a dry chamber filled with inert gas.

Vickers has developed an explosive-welding method which is especially suitable for tie-ins between pipelines and risers. The explosion is initiated inside the pipe, forcing it out against the sleeve so as to give a solid intermolecular bond. The reliability of this method has not yet been fully accepted.

2. In shallow water, the line is picked up by the davits along the starboard side. The derrick picks the riser so that it hangs just off vertical, at the proper angle to the pipe. The welded joint is made. The riser and line are then low-

ered back down to the seafloor, the riser coming into position along the jacket to which it is now clamped.

In moderately deep water, it may be necessary to add on to the riser from time to time, the so-called "stovepipe" operation. As the riser and pipeline are being lowered, the riser is stopped off from the platform and a new length of riser added.

3. For smaller lines, such as flow lines, J-tube risers are built into the platform. The laying is started from the platform; the pipe is pulled off the lay barge and up into the riser tube by a line from the pipe and to a winch on the platform deck. The J-tube bends the pipeline in a permanent but controlled deformation.
4. For deeper lines, risers are preinstalled on the platform. Alongside is a riser pull-in tube. In some cases, a line may be led out through the riser. This line is then run to the lay barge. As the laying starts, the pipe end is pulled off the platform and to the mating joint, using a winch on the platform to pull in the line. Initial connection is made by bolted flange, followed by internal or external welding as described earlier.

For lines to be run ashore, there are also several alternatives:

1. A line may be separately pulled out from the shore through the surf zone. The lay barge now moves in just seaward of the end of the pulled line. With a line from the barge exerting axial tension, the shore line is pulled on board into the tensioner and the new pipe sections welded on. Now the standard laying can commence.
2. The lay barge moves in to as shallow water as is safe. A wire line is run ashore to a winch on shore. As the lay barge makes up pipe, the winch on shore pulls the end to the shore. Then the lay barge proceeds with its standard pipe-laying procedure.
3. The lay barge lays from the platform toward the shore. When it reaches shallow water, it lays the end of the pipeline down, then turns itself around and resets anchors. It now pulls a line out from shore. Using the davits, it picks up the end of the previous line, joins the two ends by welding, and relays the line on the seafloor in a horizontal curve to accommodate the slightly excess length.

Third-generation and later lay barges are indeed highly sophisticated systems, enabling pipelines to be laid in more severe sea states, up to H_s of 5–6 m, and in deep water, up to 600 m and potentially more. Among the most advanced are SAIPEM's *Castoro Sei*, which successfully laid the lines from Tunis to Sicily, and the *Semac*, now renamed the *Bar 420*, which laid the 36-in. FLAGS lines in the North Sea in record time. See Figures 12.2.12, 12.2.13, and 12.2.14.

Third-generation lay barges operate as follows:

1. A stable platform is provided, generally being a semisubmersible but in a few cases a very long (over 200-m) shipshape vessel.
2. Dynamic positioning is employed for lateral control of the barge, using two anchor lines only for the axial pull ahead to offset the pipeline tension.
3. The stinger is now fixed to the stern of the barge and cantilevered out behind in a long curve.
4. The pipeline is laid down the centerline, not down the side.
5. Higher tension is provided.
6. Advanced welding systems are employed to speed the welding process.
7. The pipe lengths are double- or even triple-jointed on board prior to being placed in the laying line.

As a last step in the advance to a fourth-generation system, the line will be laid in a vertical catenary (or J-curve) rather than horizontally with overbend. This catenary method of laying utilizes a hinged ramp or mast, tilted slightly off-vertical, on which triple-jointed pipe sections are placed. The pipe then descends through a slot or moon pool to bend slowly to the horizontal as it nears the seafloor. The axial tension is applied parallel to the ramp, that is, almost vertically. This system enables pipelines to be laid in very deep water. See Figure 12.2.15.

The method has been known for many years, having first been tested by the French in 100 m of water in the Bay of Biscay about 1970. However, it has become practicable only with the development of fast, automatic means of welding, since all weld-

Figure 12.2.12 Third-generation pipelayer *Semac* built by Exxon–Shell and now owned by Brown and Root.

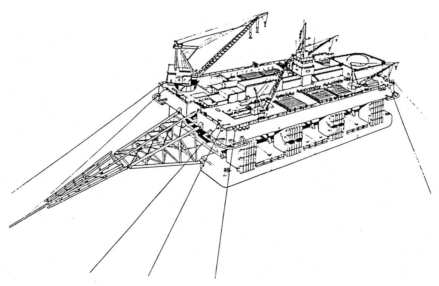

Figure 12.2.13 Semisubmersible lay barge *Semac I*.

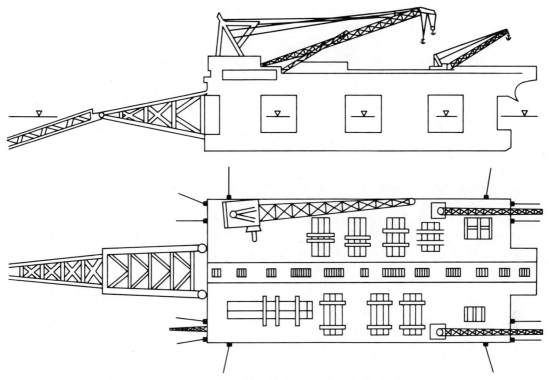

Figure 12.2.14 *Semac I* (bar 420), third-generation pipelaying barge.

ing must be carried out at one station. Electron-beam welding, high-frequency induction welding, and friction welding are some of the systems being perfected for this application, but other systems are also promising. Flash-butt welding, for example, has a typical cycle of 3 minutes.

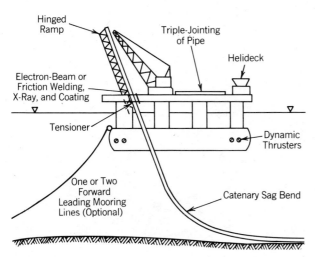

Figure 12.2.15 Catenary or "J-curve" method of pipeline installation.

The axial tension is now largely determined by the weight of pipe hanging below the lay barge. This reduces the forward-leading tension requirements to the point where this thrust can now be applied by dynamic thrusters, thus eliminating all mooring lines. This in turn eliminates the lateral acceleration forces acting on the barge and stinger as a whole. The work now can proceed through relatively severe sea states, the limiting operation being that of pipe delivery and transfer.

Semisubmersible pipe delivery barges and advanced systems for transfer at sea by a yard-and-stay system are in turn under study as a means of extending this last constraint.

12.3 Reel Barge

A significant innovation, originally directed to the installation of small-diameter flow lines but subsequently extended to pipelines 12 in. (300 mm) and even 16 in. (400 mm) in diameter, is the concept of winding a long length of line on a huge reel and then laying it in a manner similar to an underwater cable.

The first reel barges had a horizontal reel on which the line was spooled. This meant that the line was laid off one side of the barge, making it difficult to move the barge ahead on line. A subsequent "second-generation" reel barge, the *Apache*, has a large, vertically mounted reel. See Figure 12.3.1.

A line designed for laying by reel barge can have no concrete weight coating but must have thick enough pipe walls to give negative buoyancy even when empty. This of course is relatively economical for smaller-diameter pipe. The steel quality must be such that it can undergo bending beyond yield during winding, and again during unwinding and straightening. The coating must also be able to be bent without cracking or loss of adhesion; epoxy coatings have been developed which will undergo this bending without damage.

The basic procedure is as follows. The line is made up in long lengths at a shore base. The reel barge moors at the dock and pulls the line onto the reel through a spiral J-tube which bends the pipe beyond yield to the proper curvature. The tube and the spiral are designed so that the pipe bends without significant ovaling and without buckling.

Then the reel barge goes out to location. Startup generally occurs at the platform, where the end of the pipeline is pulled off the reel, through a straightener and tensioner, over a short ramp or stinger, down to a J-tube at the base of the platform, and up to the deck.

The reel barge then lays away from the platform. The straightener is a spiral with an overcorrecting bulge that brings the pipe back to a straight configuration. This develops significant frictional resistance, which in many cases may be all the tension needed for the laying of the line. Additional tension can be supplied by the powered reel or by a conventional tracked tensioner.

The reel barge now lays out the entire line, letting the end down onto the seafloor by means of a line from a constant-tension winch. The end is buoyed to facilitate recovery for welding to the next reel length.

Reel barges can be fitted so that one reel is being wound up at the shore base while one reel is being laid.

In many cases, the reel has enough capacity to lay a full-length flow line. As the diameter of line increases, the storage length of course decreases. The number of turns that can be placed on a drum is a function of pipe diameter, wall thickness, and tension, so as to prevent crushing of the pipe. The *Apache*, which is currently the largest reel barge, has the following capacities:

Pipe Diameter (in.)	Length on One Reel [ft. (m)]
8.725	360,000 (110,000)
12.75	140,000 (43,000)
16.00	92,000 (30,000)
24.00	24,000 (7,300)

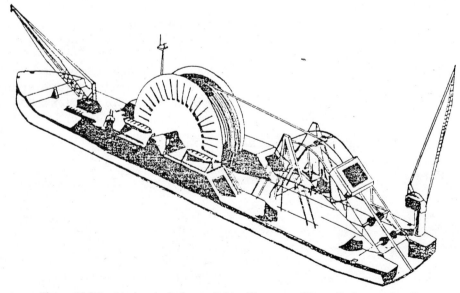

Figure 12.3.1 *Apache*, pipelaying reel ship. (Courtesy of Santa Fe International.)

The reel barge generally is moved by tug power alone, so that the layout can be accomplished expeditiously. Thus the reel barge is able to complete the laying of a length within a relatively short weather window, often of only a few hours' duration, and hence avoids the problems of storm waves and high winds.

12.4 Bottom-Pull Method

The bottom-pull method has been developed and extensively used to install pipelines through the coastal zone, so as to extend out to loading terminals in deep water. It has been further developed in recent years as a means of installing relatively short lines even in deep offshore areas.

Initial discussion will be directed to those lines which extend from shore out a distance of several thousand meters.

The program is as follows:

1. The pipeline is assembled on shore in parallel segments of 200–300 m in length.
2. A launching ramp with roller supports is constructed, leading out through the inner surf zone.
3. The inner surf zone may be protected by a sheet pile cofferdam so that a trench will stay open.
4. The first 200- to 300-m length of pipe is made up on the launching ramp, with joints welded and coated.
5. Since the ramp is inclined, the pipe is restrained from longitudinal movement by a holdback winch at the landward end.
6. The seaward end is fitted with a nose section, consisting of pig storage for one or two pigs, a positively buoyant nose, and a swivel. In some cases a sheave may be attached seaward of the swivel, with supports or a buoyant tank to keep the sheave from flipping over during the pull.
7. A pulling barge is anchored offshore, on line, at a distance of 1000 m or so.
8. On board a very large winch is installed, one- or two-drum, having high pulling capacity, for example, 300,000-lb line pull on a full drum. See Figure 12.4.1.

This winch is connected by wire lines around equalizing sheaves to two bow anchor lines, with large anchors set well out to sea.

9. When all is ready and the weather forecast is favorable, the first section of line is pulled out through the surf zone. When its landward end reaches the beach, pulling stops and the pipeline is stopped off. The next 200- to 300-m length of pipeline is rolled sideways into the launching ramp and the joint welded and coated. The next pull is made.
10. Now the barge itself must move seaward. Its anchors are reset. A third section is placed on the ramp, welded, and pulled. The pulling force is that needed to overcome friction.

Friction on the launching ramp can be reduced by the use of rollers or small rail cars to support the pipe. The pipe here is in the air, thus having its full weight exerted on the ramp. Movement seaward can be helped by the use of side boom cats or by an assisting caterpillar tread tensioner being used in reverse to push the pipe out. As noted earlier, initial sections of pipe may require restraint by use of a holdback winch.

Once underwater, the empty line has only its buoyant weight. This must be slightly negative. This results in friction on the seafloor. It is this friction which the pulling barge must overcome.

Friction coefficients have been measured in the range of 0.4–0.5 for the dynamic, moving condition, but rise to 0.6–0.8 when the pull is stopped to weld on a new section. Conservative values up to 1.0 are often used in planning since if the line cannot be moved, it will be a total loss.

The pipeline needs enough net weight to be sta-

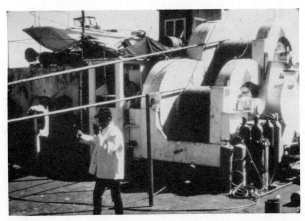

Figure 12.4.1 Installing pipeline in Quintero Bay, Chile, by bottom-pull method.

354 Installation of Steel Submarine Pipelines

ble on the seafloor and not move laterally. The amount depends on the surf, current, and seafloor conditions, but typical values for coastal lines range from 15–50 lb/ft (0.26–0.74 KN/m).

It is the total friction force developed when the line is fully laid that limits the length that can be pulled by this method. If we assume a net weight of 30 lb/ft, a friction factor of 1.0, and a winch having 300,000 lb of pull on full drum, then the maximum length that can be pulled with a single line is 10,000 ft. This can be slightly extended by making short pulls at the end so as to keep the winch drum half-full, since the winch can apply more force under this condition.

By using two parts of pulling line and a sheave at the nose, the potential overall length of line can be doubled. However, the risk of jamming of the line in the sheave makes this solution acceptable only if the required force cannot reasonably be provided with a single line.

The bottom-pull method is extremely sensitive to weight and displacement tolerances, since the net weight, a small value, is the difference between two large numbers. Therefore, great care has to be taken to control and monitor the actual values.

The principal potential variances are:

1. Steel pipe wall thickness (often 3–5 percent over)
2. Steel pipe diameter
3. Concrete weight coating thickness
4. Unit weight of concrete
5. Water absorption into concrete during pull

The weight coating is often applied in such a way that the ends are much thicker than the midsection. This needs to be accounted for. In some cases it can be compensated by an under-tolerance in applying the field coating at the joint.

The effect of these tolerances will be illustrated using the example in Table 12.4.1.

If the winch capacity, even using short pulls at the end and a two-part line, is only 700,000 lb, this means the pipeline cannot be completed and at the best will require work doubling the installation cost; at the worst, the line will have to be abandoned.

In order to measure and check tolerances, the following procedure has been found effective.

Three random but typically appearing pipe sections of 40-ft nominal length are selected for weighing. They are placed in seawater for 24 hours and then lifted out and accurately weighed. Their steel pipe wall thickness is calipered and the diameter measured. The circumference of the coated sections is also measured at three points along the length.

Then all subsequent pipe sections are measured for steel pipe wall thickness and for diameter and measured with a tape for circumference of the coated section.

The above will enable net weights to be calculated within 2–5 percent, if the pipes are relatively uniform.

Once the pipeline has been pulled, it is flooded for stability. A test plug is fitted on the inner end to permit hydrostatic testing. After testing, one pig is then activated by compressed air to empty the line.

The second pig is there for potential problems such as buckling. Buckling of a pulled pipe does not occur in the vertical plane, as when laying from a lay barge, but in the horizontal plane, usually due to longshore currents.

When even longer lengths of pipeline are to be pulled, three options are available:

1. Increase the winch capacity. Conventional winches have upper limits, so a cable grip jack device with spooling drum may be used, increasing single-line pulling force to 1,000,000 lb. Wire rope of 1,000,000-lb working strength would be excessively large and moreover might have excessive bottom friction itself. On a

TABLE 12.4.1 Effect of tolerances on installation of submarine pipeline by Bottom-Pull method.

	Nominal	Actual
Air weight	1,100 lb/ft	1,140 lb/ft
Displaced water weight	1,070 lb/ft	1,090 lb/ft
Net weight	30 lb/ft	50 lb/ft
Total force for 20,000 ft, using a starting friction value of 1.0	600,000 lb	1,000,000 lb

project across Spencer Gulf in Australia, the contractor used high-strength 10-in. pipe, empty, as a pulling line, and a cable grip jack, in this case fitted with pipe grips, as the pulling winch.

2. Pull one line out its maximum distance. Pull a second line out beyond it, so that the inner end of the second is at the outer end of the first. Make a connection.

   This procedure was employed for an offshore terminal off Antigua. The ends of the two lines were brought as close together as possible, a "pup" section fabricated from template measurements, and flanged connections used.

3. Decrease net weight and apply very careful monitoring. On a long pull across the Bay of Trieste, Yugoslavia, each length of pipe was weighed in water in a special tank.

The net weight can also be reduced, and hence the required pulling force lowered, if floats are attached to the line. Oil drums have often been used in the past but prove rather crude and unreliable. Polyurethane floats can be accurately designed and attached by straps.

This increases the risk of problems during installation, since the floats add significantly to the drag force from waves and current. On occasion they have been torn loose: if numerous floats are torn off, the line may become too heavy to move, and thus end up as a catastrophic loss.

Assuming satisfactory installation, the floats must later be cut off. While divers were used in the past, mechanical equipment has been designed to travel along the completed pipeline severing the straps. ROV's have also been used.

Despite the potential problems, floats are a viable and accepted solution for heavy pipelines. See Figure 12.4.2.

The reason a swivel is installed at the nose of a pulled line is to prevent the natural twisting of the wire line under tension from imparting twist to the pipeline. The nose is made buoyant and sometimes shaped like a sled to prevent it from digging in as it is pulled. A pendant and buoy is often fitted to the nose to enable its progress to be observed visually and to facilitate recovery in the event of problems.

Often a flanged elbow is incorporated in the nose piece to facilitate connection to a riser or hose.

Figure 12.4.2 The bottom-pull method, used for the initial section of the North Rankin A pipeline, utilizes temporary buoyancy floats to control submerged weight and hence reduce frictional resistance.

A pulled pipeline normally will follow the path of the pulling force, so that it is usually possible to pull around a curve. However, on a hard sand bottom this is not always true, and in such a case the line may drag sideways. One solution is to make the line heavier, that is, increase the net weight to give it more stability. In other cases, lines to anchors have been rigged so that periodically the curve is pulled back to position. Solutions such as this, or trying to pull around a pile, often result in buckling in the horizontal plane.

A better solution is to spread crushed rock on the seafloor at the zone where the bend must take place in order to get more lateral stability.

Another solution is to secure several shots of chain inside the pipe, held at the location of the bend or in the heavy surf zone by a line running back to the holdback winch's second drum. This way, the extra weight of the chain stays at the critical location while the line is pulled past.

In one unfortunate contract, the owner furnished coated steel line which was almost in equilibrium as to buoyancy, having only a few pounds of negative net weight. The contractor, faced with an oncoming storm, decided to go ahead with the pull and try to get the line in place and flooded before the storm hit. As indicated in Chapter 2, long-period swells run out ahead of a storm. Therefore, as the pull was in progress the swells came in at about a 45° angle. While they refracted around to the normal in shallow water, there was still a net volume of water to be displaced to the south, resulting in a strong wave-induced longshore current. This bowed the pipeline out until it buckled. The contractor aggravated the situation by attaching a line at the buckle leading to a tractor on shore; this attempt to pull sidewise broke the pipe and the entire line had to be abandoned.

On the next try, the contractor used an ingenious trick. He filled the line with rock salt to give it weight, pulled the line out properly, then washed the salt out.

One more tale of disaster will be given to illustrate the interaction of the hydrodynamic, weight–buoyancy, and structural aspects. This line was in the Bay of Fundy, with 10-m tides twice a day. The concrete coating was reinforced only with a very light mesh resembling chicken wire. The launching ramp terminated at high tide; the line was to be pulled across the long tidal flats at high tide and out to an offloading buoy.

As the pulling operation commenced, the offshore anchors of the barge slipped. By the time these were reset, the tide was falling, exposing the line on the mud flat. Now as they pulled, they had the increased friction of the pipe's air weight burying the line in the mud flat and developing excessive friction. They then held on until the tide came in and released the pipe from the mud, but now the winds and sea were kicking up and the line bowed laterally. This caused the concrete weight coating to crack, the light wire mesh to break, and the line to float. Eventually the line broke and ended up on the beach, where it had to be abandoned due to multiple kinks and buckles.

Unfortunately, variations on this theme have occurred elsewhere in the world; for example, one of the first submarine lines to Kharg Island, Iran, also reportedly ended up as "spaghetti" on the beach. Recently, a number of cases of "floatup" of pipelines have occurred in the North Sea, apparently due to breaking off of large segments of concrete coating due to wave-generated pore pressures within cracks and delaminations and to flexing of the line under vortex-shedding movements.

The lessons are clear. To all the recommendations regarding weight and buoyancy control must be added the need to ensure adequate reinforcement in the concrete weight coating.

Several clever schemes have been developed to extend the use of the bottom-pull method. One of these involves the prelaying of a wire line cable on the seafloor, with a series of heavy anchors at the outer end, so that the line's full strength can be developed. Then a special pulling head is attached which incorporates hydraulically actuated cable grip hoists. One arrangement has these jacks inside the nose piece, feeding the line through a watertight sleeve. This enables the hydraulic lines to be fed through the pipe. In this scheme, the pipeline "swallows" the cable as it pulls itself out along the line. In this system, the weight of the pipeline is increased by the weight of the line.

In another arrangement the cable grip hoists are located outside the nose and the pipeline crawls over the cable, pulling itself forward. This system, developed by Harold Anderson, maintains the weight of the pipeline, and hence the frictional resistance, at a constant value per unit length.

Another system, developed by Dr. William Buehring, also is based on the use of a prelaid cable. In this scheme, after the cable is laid, a drilling vessel takes in on the bight of the line by direct pull, thus obtaining the mechanical advantage from the geometry as well as the guidance from the prelaid cable.

Use of one of the above systems of pulling along a prelaid cable enables the line to be laid in a moderately sinuous pattern, and thus it can be laid around outcrops and steep changes in bathymetry.

The bottom-pull method has been successfully extended to the installation of relatively short deepwater installations such as interconnecting lines between platforms and flow lines. The pulling force is usually that of a large tug and hence is limited to the bollard pull which the tug can exert, with the maximum force being in the range 50–100 tons. Being laid in deeper water, out of the surf zone, the net weight on the bottom can be reduced to a bare minimum, say 10–15 lb/ft.

In 1983, a 2.4-mile-long bundle consisting of a 12-in.-diameter oil line and a 4-in.-diameter fuel gas line was pulled off from Ninety-Mile Beach in Bass Straits, Australia, and towed 68 miles to connect Fortesque and Halibut platforms. The launch from the beach, which included the jointing of sections, took 21 hours, the bottom tow 33 hours. To reduce friction force and to prevent digging in, 1600-ft-long sections at each end were buoyed with pontoons so as to raise above the seabed with slight positive buoyancy. This enabled the new bundle to be pulled over an existing line. As the tow approached the platforms, the end sections were flexed laterally, using winches on the platforms, until mating fittings on the ends of the pipeline were mated with receiving fittings on risers from the pipeline. ROVs were then used to disconnect the 3-in. tow cable and all 88 pontoons.

A similar method was used in 1977 to install the 36-in. connection line, 2200 m long, between the Statfjord A and B platforms in the North Sea, laying it in a trench that had been previously dug with a plow.

Care has to be exercised with this method to make course changes very gradually and to avoid rock outcrop areas, since if the coating is abraded or spalled, the delicate weight–buoyancy balance can be upset with disastrous results. One of the earlier attempts to use this method involved a relatively long bottom tow with severe course changes to avoid known minefields in the North Sea. The progressive damage to the coating from the sharp bends led to eventual loss of the line.

12.5 Controlled Above-Bottom Pull

Continued efforts to develop a reliable method for transport of lines over the seafloor have led to the development of a number of ingenious methods. See Figure 12.5.1. One of these is the "controlled above-bottom-pull" method, in which the line itself is designed for slight positive buoyancy. Short lengths of chain are attached at frequent intervals to give the overall combination a negative buoyancy. Thus it is the end of the chain which drags on the seafloor, not the pipe.

The chains automatically control the net underwater weight of the combined system; if the pipeline tends to rise, it lifts more chain off the bottom; if the pipeline tends to sag down, more chain rides on the bottom, reducing the downward pull on the line.

The friction force is now determined by the weight of the "tails" of the chains which drag on the seafloor. The length of tail is in turn determined by the variations in bathymetry over a short distance and the safety required to offset tolerances in weight and buoyancy. While these can be calculated ahead of time, once the pipeline has been launched in relatively shallow water it can be inspected by diver and adjustments in chain length made before pulling out to deep water.

The attachments of the chain to the pipeline must be properly detailed so as to prevent chafing and abrasion. A weak link may be installed so as to ensure that should a chain snag on an underwater obstruction, it will break free before buckling or crimping the pipe.

The controlled bottom-pull method has proven cost-effective in installations in the North Sea and offshore California, especially for lines of limited length, of the order of 3000 m. During tow, at 3–4 knots, the pipeline may "fly" at a depth of 30–40 m below the surface. As it approaches location and the tow is slowed, it gradually lowers to just above the seafloor, with the chains dragging the bottom. Then it is pulled into final position and ballasted down onto the seafloor.

A larger diameter pipeline may be employed as a carrier line for several smaller flow lines and cables bundled inside. Obviously, it is essential that the carrier pipe maintain its watertight integrity during the tow. Internal pressurization has been utilized to overcome any leakage. Unfortunately, in at least two recent cases, the combined line has

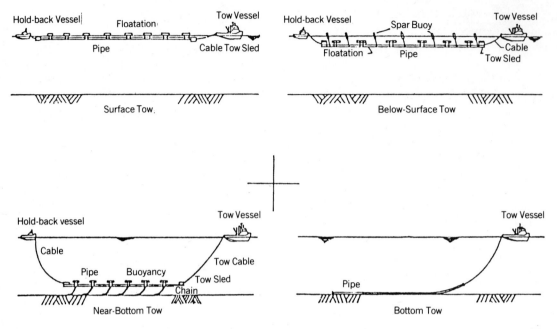

Figure 12.5.1 Flow line–pipeline tow installation methods.

prematurely flooded and sunk, damaging the flow lines being carried inside the carrier pipe. This would indicate that consideration needs to be given to methods of "damage control" of in-leakage due to possible overstress during tow. Appropriate steps might include use of foam, compressed air, multiple pipes, or subdivision.

12.6 Flotation

The idea of moving long lengths of pipeline while floating and then progressively sinking the line to the seafloor at the site has attracted many contractors over the years. It appears relatively simple; the line may be given the net positive buoyancy by attachment of floats, which will later be cut free.

To keep the pipeline in line and prevent buckling due to waves, wind, and so on, one boat tows while a second boat astern acts as a drag. This keeps the line under tension.

In some cases, it may even be appropriate to reduce the weight coating to where the empty pipeline has inherent net positive buoyancy; in this case, the line, once ballasted full and placed in service on the seafloor, can never be emptied.

Unfortunately, this method of flotation has a number of serious drawbacks, which can only be overcome by thorough engineering; even then there may be excessive risk. The first problem is that of waves acting on the floating line, causing it to "snake" in response to the short-crested waves. This may damage the coating and cause it to fall off; in turn, the weight balance and stability in service are affected. Even with the line in tension, over several kilometers the lateral and vertical forces will alternate over many thousands of cycles and can eventually lead to coating damage.

Such a line must obviously be towed out in calm weather. It is very susceptible to even small storms such as squalls. A number of offshore lines have broken up and been lost in this way.

The second problem has to do with the attachment of temporary floats. For a moderately long line, there will be hundreds of these. Under the wave action, some attachments may fatigue and fail, thus leading to local areas where the line takes on an increasing sag.

The third problem to be overcome is that of ballasting down to the seafloor. If floats are cut loose at one end, that end will bend downward sharply and may buckle. The same or worse can happen if attempts are made to introduce water ballast into the line. It will run to one end or a low point; this will cause the rest of the ballast to run there, and the line will take a sharp bend. Therefore, it may be necessary to leave every second or third float attached until that section of line is on the seafloor. Floats must be checked for collapse under hydrostatic pressure.

The French have developed a well-engineered system of air-filled rubber (neoprene) bags, which have overcome these problems quite effectively. This method is described in Section 12.9.

Successful installations of relatively short lines in calm water have been made by the flotation method, although most of these were in protected or semiprotected waters.

The essential point is that the entire sequence must be thoroughly engineered to ensure success. Adequate redundancy must be provided to ensure that the loss of any few buoys does not lead to progressive failure.

To facilitate connections at the platforms, as the line nears its final location, a line from a J-tube on the platform is affixed to the pipeline end. The buoys at that end are progressively released, while a winch on the platform pulls the end down, either to mate with the J-tube or run on up into the J-tube to deck level.

12.7 Controlled Underwater Flotation (Controlled Subsurface Pull)

The controlled underwater flotation method has been developed to overcome some of the deficiencies of the surface flotation method described in Section 12.6. In this method, the pipeline, having slight net negative buoyancy, is towed at a depth of 5 m or so below the surface, where it is much less affected by local waves and not at all by wind.

Support for the line is by hinged or articulated spar buoys attached at frequent intervals. These provide a relatively constant upward force, one that is "soft," that is, not very responsive to the changes in sea level due to waves. With their small waterline plane, they do not respond significantly to wind-driven waves, and the frequency of response of the system becomes very long (over 1 minute). This system therefore virtually eliminates the first major problem of flotation.

The line is kept under tension in the same manner, with one boat towing, another acting as a stern drag.

Upon arrival at the site, the line must be lowered to the seafloor. If the spar buoys are articulated, it is practicable to remove progressively the top sections of the spars, thus allowing the line to sink under slight negative buoyancy. Alternatively, weights can be progressively attached, which offset the spars' buoyancy and again allow the line to bend downward in a gradual curve. Of course, when the line is on the seafloor, any remaining spars are removed by a diver or ROV equipped with cutters.

This method of controlled subsurface pull has been successfully used for flow lines and platform interconnection lines in the North Sea.

The controlled subsurface-pull and controlled above-bottom-pull methods have been combined by C. G. Doris to constitute what they have named as the "guide rope method" for placing flow line bundles. The bundle assembly includes both floats and chains. When the pipeline is empty, it floats 30 m below the surface: when filled, it floats 10 m above the bottom. Thus, it can "fly" over obstacles, rock outcrops, escarpments, and other pipelines.

One such bundle of 4- and 2-in. pipes was towed 13 miles just below the surface, and then sunk to the above-bottom mode and the ends pulled into subsea manifold connections, using a guide funnel and pull-in lines through sheaves.

Thus, this method appears to have great potential in the expanding area of subsea completions.

For installing pipelines between the underwater manifold and the Cormorant A platform in the North Sea, an 8-in. oil line plus 2-3-in. TFL well test service lines were made up inside insulated sleeves which in turn were placed within two carrier pipes, one 26-in. line and one 24-in. line. The bundle was then towed in two 3.3-km lengths, kept at mid-depth by the chains for a distance of 490 km using one tug of 75-ton maximum bollard pull pulling and another of 35-ton maximum bollard pull restraining, that is, keeping the line in tension. The average pulling forces were 50-ton pull, and 12-ton restraint, and the average speed 5.5 knots. Depth below surface was 30–100 m.

The carrier pipes were pressurized with nitrogen gas to 15 atm to prevent water in-leakage. A full contingency plan was developed to cope with possible accidents or difficulties, such as the snagging of a chain on an obstruction. The position of the pipes as they approached the field installations was monitored by acoustic transponders and side-scan sonar.

12.8 J-Tube Method: Single- and Double-Pull

Curved J-tubes can be pre-installed on a jacket. After the platform is in place, a messenger line (also pre-installed in the J-tube) is connected to a long pulling line from shore or a barge.

The pipeline is made up vertically on the platform, using steel quality and coating similar to that used for reel barge operations (Section 12.4). This practice of successive vertical jointing is called *stovepiping*.

The barge or shore winches now pull the end of the line down the J-tube, around the bend, and out over the seafloor as a bottom pull. Thus the length of line which can be pulled is limited by bottom friction.

One way to extend this length is to do so with a double pull. The first length of line, having an inner diameter larger than the outer diameter of the final line, is pulled as far as it will go. Then the final line is pulled inside the first.

Since the friction factor of steel on steel is about 0.1 and can be reduced by lubricants, the second line can be pulled well beyond the end of the first, with the same available force.

With high-pressure lines, this method requires the provision of a substantial amount of excess pipe. With lower-pressure lines, where flanged couplings may be accepted, the only cost penalty is the larger size of the first length of line and the cost of the underwater connection.

Of course, a fully welded connection can be made by using a habitat, as described earlier.

12.9 S-Curve

The S-curve system, referred to earlier in Section 12.6, employs inflated bags of neoprene or rubber which exert a buoyant effect on the line. As bags are pulled below water, they partially (and eventually fully) collapse, thus reducing their uplift force.

The French experiment referred to earlier was used successfully to make a test installation of a small-diameter steel line in water 2500 m deep in the Mediterranean. The short length of line was towed to the work boat and fed up over the bow to the deck where the bags were attached and then out over the stern. The sinking was initiated by forcing the end under water; from then on the line automatically took its S-shape and so went to the seafloor without buckling. It was later successfully retrieved by lifting up on the end, after which the bags progressively expanded and brought the pipe to the surface in a gradual curve.

In a prototype installation, the makeup of the pipe and the attachment of the bags would probably be on a lay barge.

12.10 Bundled Pipes

The simultaneous installation of two or more pipes in a bundle is feasible by many of the above methods. The reason is usually, but not always, to handle several products.

The main requirement is to ensure that the attachments are sufficiently rugged to take the stresses imposed during laying without failure and without damaging protective coatings.

The use of two or more lines in a bundle enlarges the opportunities for the construction engineer to control weights and buoyancies to suit his needs. For example, if an oil line and gas line are pulled together, the oil line, in service, may provide the net negative weight to stabilize the system, even though the gas line may be near neutral buoyancy.

Similarly, inclusion of a second or third line in a bundle may reduce the net weight during installation yet increase stability in place after it is flooded. In this case, the additional line becomes an expensive but effective means of stabilization of the system.

12.11 Laying under Ice

To install pipelines below the ice, several variations on the bottom-pull method have been developed. These have assumed that the work would be done in winter and that the ice would be "fast ice," with little movement over the period involved in the pipelaying operations.

Initially it is necessary to run a messenger line. Holes can be cut in the ice at intervals along the line and acoustic transponders installed at each. An ROV can then be lowered through one hole and programmed to lay out a messenger line to each hole in succession.

The messenger line is then used to pull a wire rope line (the pull line). The pipe is then made up on a launching ways and pulled into place by the conventional bottom-pull method.

At the shore approaches, the prior installation of a larger-diameter pipe through which the principal pipeline is to be pulled may be required for permafrost insulation. This casing can also be used during construction in the manner described in Sections 12.4 and 12.8 as a means of reducing friction and separating the near-shore excavation and construction from the main pull.

Installation of pipelines in the Arctic is described in more detail in Chapter 20.

12.12 Burial of Pipelines

The burial of pipelines is often required in order to provide protection to the pipeline against repetitive pounding under wave action and the impact of dropped anchors and trawl boards and to prevent loss of fishing gear by bottom fishermen. Burial of the pipe also permits the pipe to be designed with less net weight (less coating) which in turn reduces the bending stresses during pipe laying.

The "pounding" referred to above is especially serious in the surf zone, as well as in shallow water where vortex shedding by wave-induced currents can cause alternate raising and lowering of sections of the line, leading to fatigue. Concrete coating can be ruptured and break off, allowing the line to rise. The same phenomenon can occur due to high currents alone, in locations such as Cook Inlet.

In the inner portion of the surf zone, the damage may be aggravated by direct wave impact and by abrasion from moving sand and gravel.

Burial can be accomplished by laying the line in a predredged trench or by subsequent trenching after the line is laid. A similar protection may be given to a surface-laid line by covering it with rock.

Where underwater sand dunes are migrating, as in the southern portion of the English Channel, then predredging by a trailer suction hopper dredge has proven practicable. The line is predredged to a stable elevation, and then the pipeline is laid.

Through the surf zone, a variety of solutions are employed. At beaches where the surf and longshore currents are relatively mild, a hydraulic dredge may be employed to overdredge a channel through the beach. The line can then be pulled ashore from the lay barge, allowing the sands to naturally backfill the trench. However, where the beach is subjected to heavy pounding from storm surf, over a period of time, the iterative raising of the pore pressures in the sand may jack the pipe up and out to exposure. Thus the pipe must be sufficiently heavy in this zone so that in service it remains stable. This may require extra jacketing, the use of the double-pipe concept, pipeline anchors, or high-density backfill over filter fabric.

Instances of such raising and exposure are reported from such widespread areas as Cook Inlet, Alaska, Ninety-Mile Beach on Bass Strait, Australia, and the Strait of Magellan.

Another method of course is the use of a sheet pile cofferdam through the surf zone, which keeps the trench open while the line is pulled through it.

Finally, a tunnel or tube can be preconstructed through the surf zone. This can be concrete or steel pipe prelaid in a cofferdam, a directionally drilled hole with a casing of larger-diameter pipe, or a precast concrete tube constructed "in the wet."

Significant progress has been made in recent years with the extension of directional drilling techniques to the installation of pipelines (and cables) through the surf zone. The drill is set up on the shore, as a slant rig. It then drills a shallow, curved hole out under the beach zone, "daylighting" in 10–15 m of water. Such holes have been successfully drilled up to 40 in. in diameter and 2000 m in length.

Meanwhile a length of pipeline has been pulled out or laid out on the seafloor, beyond the exit hole.

A reamer is now attached to the drill head, and then the pipeline is attached with a swivel. The drill now pulls back to shore, reaming the hole and pulling the pipeline in behind it.

Present technology enables this method to be considered for diameters up to 36 in. and even 48 in. and for lengths up to 6000 ft. This technique is best suited for use in stiff silts and clays and weak rock, especially near the seafloor exit where the depth of overburden is reduced.

A large precast concrete tunnel or tube was constructed at Cove Point, Maryland, through which an LNG line was run. Precast tube segments were installed on the west coast of Norway, on a very exposed rocky coast. The Statpipe Gas Line was then pulled through the tunnel.

Sheet pile cofferdams have been very widely used. The special considerations affecting their construction are:

1. A trestle must be constructed for access. Usually steel pipe piles are used.

2. The steel sheet piles must, in general, penetrate below the final excavated level of the trench.
3. Bracing must be provided to resist wave and current forces.
4. The bracing and sheet piles must be connected so as to resist alternating forces without suffering fatigue. Welding in such an environment is not very satisfactory, hence high-strength bolting appears most suitable.
5. Sand will tend to build up on one side and erode from the other due to the interference with normal littoral drift. Scour on the downstream side, especially at the outer nose, may erode the sand and undermine the sheet piles.
6. Removal of the trestle and sheet piles may loosen the soils to the point where they will subsequently erode more readily. Cutting off the piles at the sand line may be more appropriate.
7. The outer end of the cofferdam, being normally open, will allow a great deal of wave energy to be focused into the trench. This may or may not be serious depending on the soils, waves, and trench profile. If necessary, the outer end can be closed temporarily with sheet piles, and then opened to allow the line to be pulled through.

For burial of pipelines in deeper water, trenching has most often been carried out by a jet sled, designed to be guided by the pipe and to excavate the soil beneath it so that the line will sink below the seafloor. The jet sled may be designed to run on the pipe, using rubber wheels. The machine must be designed to ensure that its tires cannot damage the coating. See Figure 12.12.1.

In one case in southern California, repeated running of the jet sled over the line did break up the coating to the extent that the line had to be replaced.

Other jet sleds are designed to skid, crawl, or run on the adjoining seabed, being centered and guided by the pipe but not supported by it. A prob-

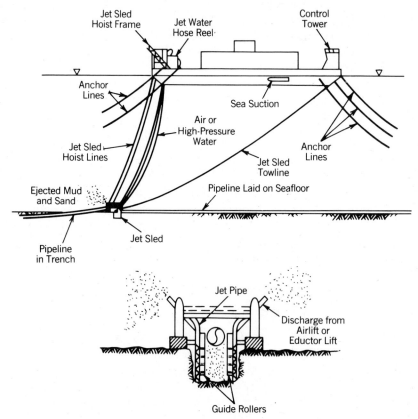

Figure 12.12.1 Jet sled operation for pipe burial. (Adapted from *Offshore Platforms and Pipelining*, Petroleum Publishing Co., Dallas, Texas, 1976.

lem here arises, of course, if the trench side slopes become too flat due to encountering loose sediments.

Excavation under the pipe may be accomplished by a combination of jetting, airlift or eductor removal, or mechanical cutting. See Figure 12.12.2.

The emphasis is on powerful equipment. See Figures 12.12.3 and 12.12.4. Ideally, it should be able to cut to the required depth in one pass. Multiple passes not only are costly, almost in proportion to the number required, but may become less effective due to the increasing depth and progressive infill of the trench.

A steel pipeline has significant bending rigidity and strength. Hence it will not move down into a dredged trench unless the dredged length is sufficient to cause it to deflect to the bottom. Therefore, the trench when cut must stay open long enough to enable the line to feed itself to the bottom of the trench.

Fortunately, this is usually not a problem in most deep-water pipeline installations, since bottom currents and sediment infill are usually limited over the short-period and relatively calm seas in which the operation will be carried out.

Power requirements for jet sleds and trenching machines are high. As much as 32,000 HP has been used to power the jets and eductors of a large pipeline burial system used to trench boulder clay in the North Sea. For the Santa Fe pipeline bury barge *Creek,* eight engines drive jet pumps to produce 16,000 gpm (76 m^3/sec) at 2500-psi (17-MPa) pressure.

Eductors are more efficient than airlifts in the removal of material.

In most cases, once pipelines have been trenched and lowered to their designed elevation, natural sediment transport has been counted on to fill the trench. If a pipeline is to be backfilled by dumping or placing sand, care has to be taken that the flowing sand, which is temporarily a high-density fluid, does not raise the line out of the trench. This has occurred on both small and large pipelines, to the great embarassment of all concerned.

Another method of pipe burial involves the principle of liquefaction. By introducing water and air under the pipeline along a length, the sand is "fluidized," allowing the pipeline to sink of its own weight. Obviously, this works best in easily liquefied materials such as fine sands and silts. Vibration applied to the pipe—as, for example, from inside—aids the process. This method has so far been used,

Figure 12.12.2 Deep-water jet sled employs a two-stage seawater eductor system to excavate trench for pipeline burial.

to the author's knowledge, only for relatively short lengths of line, as through a beach zone.

Ever more sophisticated trenching and burial equipment has been developed, such as mechanical trenching machines. Like the jet sleds, these are guided by the prelaid pipeline, but they take their support from sleds on tracks at the sides. Rotating trenchers excavate beneath the pipe and throw the material to the side.

The newest form of trenching is that of plowing. A monstrous plow is pulled along the seafloor, stabilized against tipping by widespread outrigger sleds or, more recently, rotating shares (wheels). The plow digs the trench, forcing the excavated material up on the sides. See Figure 12.12.5.

This development was pioneered by R. J. Brown and has proven very successful in the firm clays of the North Sea. The plows have been designed for the soils expected to be encountered, heavy in North Sea clays, lighter in recent sediments in the Beaufort Sea. In the Bass Strait, Australia, an 80-ton plow was used to post-trench the line, with the

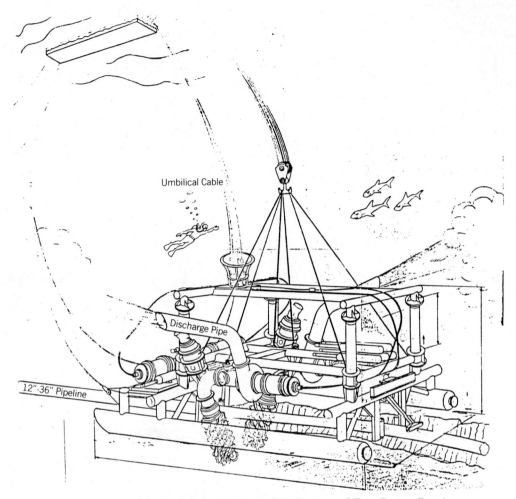

Figure 12.12.3 Jet sled for pipeline burial. (Courtesy of Toyo Pumps Corp.)

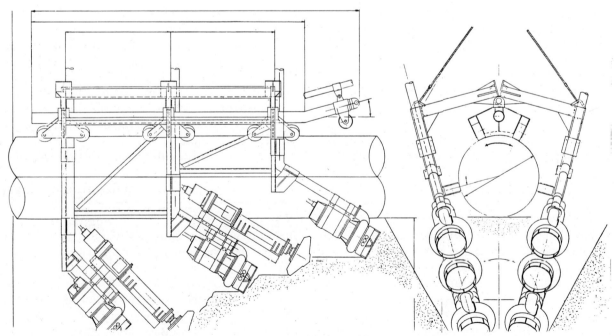

Figure 12.12.4 Jet sled device for pipeline burial uses multiple submersible pumps with agitators. (Courtesy of Toyo Pumps Corp.)

12.12 Burial of Pipelines

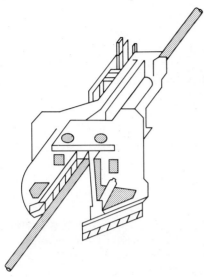

Figure 12.12.5 Pipeline trenching plow for Gullfaks–Statfjord C connection. (Courtesy of R. J. Brown Associates.)

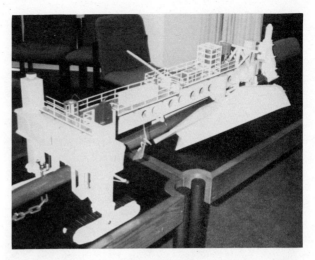

Figure 12.12.6 Model of plow used to trench and bury pipeline from North Rankin A platform, on Northwest Shelf of Australia. (Courtesy of Woodside Petroleum.)

plow designed to ride around the pipeline without touching it except at curves. This pipe dug a furrow up to 1.2 m deep in sand and partially cemented sandstone.

Towing forces generally are one to two times the weight of the plow. Traction is provided by a large, dynamically positioned towboat for relatively light plows or by an offshore derrick barge for heavy plows. It might be practicable for the plow to be rigged with a large hydraulic linear jack so as to pull itself along a prelaid wire rope or even an uncoated pipeline.

One of the most spectacular uses of the plow to date was on the Northwest Shelf of Australia, where an enlarged version of the plow weighing 380 tons trenched a 1- to 2.3-m-deep trench for 118 km of 46-in. O.D. line in only 1 month. See Figures 12.12.6, 12.12.7, and 12.12.8. The plow had to dig through limestone and cap rock, requiring up to 460 tons pull, whereas in softer materials, sand plus silt and clay, 250-ton pulling force was sufficient. Plowing rates reported were 15–45 m/min in sand, 10–20 m/min in sand over rock, and 5–10 m/min in limestone.

Difficulties arose principally in soft material, where the plow dug itself in too deeply.

The plow was pulled by a large offshore pipelaying barge, using a chain-pulling line and developing its reaction force from the barge's anchor lines.

The sequence of startup was as follows: The two mating (positioning) cones were lowered over the pipeline, being spaced apart 40 m by a strut. Divers guided them, so as to seat over the pipe on the seafloor. The plow was now floated to position over the pipe and ballasted down. Removable female sleeves fitted over the preplaced mating cones. With the plow now accurately in place, the pipe was pulled up into roller guides at each end. The plowshares were then lowered hydraulically so as to meet underneath the pipe, where they are clamped together.

The forward part of the plow rides on two outrigger crawlers, which are Caterpillar D-9 tractor underbodies.

The derrick barge has hydraulic controls which enable it to raise or lower the plow in relation to the crawlers, thus regulating the depth of cut.

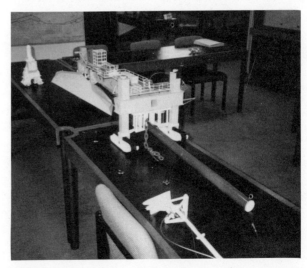

Figure 12.12.7 View of model of plow from other side.

Figure 12.12.8 Recovery of plow after successful trenching of 118 km of pipeline on Northwest Shelf, Australia. (Courtesy of Woodside Petroleum.)

The plow concept uses a long beam so as to automatically even out irregularities on the seafloor. R. J. Brown has suggested that use of computerized controls, reacting to leading sensors, will be developed in the future so as to handle even rough seafloor profiles.

It is believed that a similar system could be employed to scrape the spoil piles back into the trench should early backfill be required.

Plowing appears especially attractive in Arctic soils where lines will have to be trenched deeply (up to 3 m) in order to protect them from the scour of sea ice pressure ridge keels.

For the connection line between the Gullfaks A and Statfjord C platforms, it is anticipated that the hard boulder–clay will require several passes of the plow. Hydraulically operated grader blades are being designed so as to push the spoil banks aside, leaving a flat level surface for the following passes of the plow.

Most recently, on the Heimdahl pipeline in the North Sea, the 115-km pipeline was post-trenched by plow in only 11.5 days. The plow weighed 145 tons. It was deployed onto the pipeline in 150-m water depth in 19 hours. Recovery after completion required 13 hours.

The most recently developed plows have multi-pass capability, with small shares that ride the previously plowed pipe under sensor control so as to clear the pipe for the principal plowing action.

When cap rock or rock outcroppings must be trenched, it is normal first to break them up with explosives. These can be shaped charges laid on the seafloor or drilled in, using high-pressure jet drills or percussion drills. With cap rock it is important not to drill through the hard overlying layers, since then the explosion will take place under the cap, resulting in it being broken into large slabs only, which are extremely difficult to excavate.

Through overconsolidated silts and in permafrost, high-pressure jet drilling has generally been found more effective than rotary or pneumatic drilling.

Rock breakers (huge chisels repeatedly raised and dropped, or driven with an impact hammer) can also be used effectively for cap rock, breaking it downward into the softer soils below.

Covering of a pipeline with rock has been carried out very effectively by Dutch engineers, using a converted trailer suction hopper dredge, dynamically positioned, fitted with an inclined ladder and conveyor belts, discharging the rock down the ladder so as to encapsulate the exposed line. See Figure 12.12.9. Use of the ladder ensured accurate deposition and minimized the impact of falling rock. More recently, this same contractor has proposed the use of a flexible tremie tube, of steel and polypropylene, hanging vertically under the rock-dumping vessel, thus capable of controlling the deposition of rock in depths over 100 m.

Sarmac of Italy has developed flexible mattresses of rock-filled fabric which can be placed over submarine pipelines to protect them. They were utilized by Snam–Progetti to cover the gas pipelines from Algeria to Sicily, at depths of 500–600 m. They were lowered to the seafloor, using a structural steel frame that was able to be automatically released once the mattress had been placed on the line. These mattresses are also being used to protect a pipeline at a location where it is crossed by another line.

Flexible (articulated) mats of concrete blocks were used to cover portions of the gas pipeline on the Northwest Shelf of Austrailia, where it was laid on a bare rock seafloor.

Pipeline anchors have been used to hold down the pipe at beach crossings and in areas of high bottom currents, especially when the line lies on a hard, bare seafloor. These are usually screw anchors, which are drilled into the soil and attached to an inverted U-clamp which fits over the pipe. While these anchors can be installed by divers, this

Figure 12.12.9 Various methods of placing rock to protect submarine pipelines against erosion, wave-induced movements, and damage from dropped objects and fishing trawl boards. (Courtesy of ACZ Marine Contractors.)

is slow and expensive. Systems have therefore been developed for installing them from a barge directly.

For crossing beach areas where high currents and waves have created a deep layer of cobbles, the protection of the pipeline poses significant difficulties. Such a situation occurred on a recent gas line in the Strait of Magellan and previously occurred in Cook Inlet, Alaska. Anchors appear impracticable, and burial extremely difficult to implement, because of the difficulties of trenching. High-density riprap (iron ore) has been suggested as a covering protection.

Such a potentially hazardous zone led the Norwegian company Statoil to build the previously mentioned precast, prestressed concrete tunnel through this coastal zone. Other possible solutions include double-jacketed pipe with the annulus filled with high-density grout, and covering the exposed line with articulated concrete mats.

12.13 Support of Pipelines

When laying pipelines across an uneven seafloor of hard material, such as an area of rock outcrops, it may become necessary to provide supports to prevent excessive sag moments in the pipeline span.

In shallow water, sandbags and grout-filled bags have been stacked in by divers. Burlap bags, half-filled with fresh concrete mix of low slump, are best, because the grout exuding through the burlap mesh will knit the adjoining bags together.

In deeper and more exposed waters, neoprene and flexible fabric bags can be placed by divers and pumped full of grout.

For the 520-mile Statpipe gas-gathering system, the crossing of the Norwegian Trench, with its steep rock escarpments, results in free spans up to 100 m. In addition to grout bags, steel support frames were designed to give intermediate support to the pipe.

Eternal Father, strong to save
Whose hand doth still the mighty wave
Who bids the restless ocean deep
Its own appointed limits keep;
O hear us when we cry to Thee
For those in peril on the Sea.

NAVY HYMN

13

Nonsteel Pipelines and Cables

Submarine pipelines have been constructed of a wide variety of materials designed to suit the specific environment, operational, and cost criteria involved. In Chapter 12, steel pipelines were addressed, since these constitute the bulk of deepwater pipelines for the transport of oil and gas. Their selection is based both on the need to accommodate high internal pressures in service without leakage and the bending and combined stresses developed during installation.

Other materials widely used for submarine pipelines are polyethylene, fiber-reinforced glass, concrete, and flexible composites of steel and neoprene. Cables for submarine transmission of electric power are another type of flexible pipeline, involving somewhat similar offshore construction techniques.

13.1 Polyethylene Lines

Polyethylene can be used to produce a highly flexible and collapsible line which is resistant to chemical attack and has low friction. It has therefore been utilized for test installations for ocean thermal energy conversion (OTEC) plants, so far involving relatively small diameters (1.5 m and less). This material has a density less than seawater and hence has to be weighted or anchored down so as to stay properly submerged.

One installation for an OTEC test plant off the island of Hawaii utilized polyethylene pipe made up in long floating strings in quiet water and towed to the site, where anchors with short Kevlar lines were used to progressively submerge it to where it floated just above the rough and steeply sloping rock floor. As the weights were attached, the polyethylene line progressively collapsed (buckled) but regained its circular shape as it went underwater. So the buckle traveled along the line as it was laid, without permanent damage.

Reinforced saddles of polyethylene were fitted at regular intervals; here is where the mooring lines were attached, so as to hold the line near the seafloor in an inverted catenary.

Stress concentrations at moorings and fittings must be avoided, as polyethylene is subject to internal fatigue under sustained high stress.

An advantage of the polyethylene line over a steel line is that no ferrous ions are picked up by the seawater; thus it is suitable for use in supplying aquaculture facilities.

Another, larger polyethylene line was successfully suspended in a vertical hanging position under an OTEC test facility to a depth of several thousand feet. Later, when this test had been successfully completed, the line had been upended to horizontal and was being towed to a new test site. Reportedly, the heavy end was supported by wire lines; during the tow and under the wave action, these chafed and the concentrated stresses on the polyethylene led to a local failure and loss of the line. (It was later salvaged.)

13.2 Fiber-Reinforced Glass Pipes

Fiber-reinforced glass (FRG) has been used for a number of sewer outfalls, as well as for slurry transport of salt. The material is highly resistant to chemical attack. It can be coated so as to protect it from UV degradation.

It is light in weight, hence even relatively large-diameter (2-m) lines can be readily handled in long lengths, set by a derrick barge, and joined or, alternatively, pulled out from the beach. Because it is of low density, added weight may be necessary to enable it to be properly seated on the seafloor. As it is placed, precast concrete saddles are usually lowered over the line to hold it in position under wave action.

Because of its light weight, FRG lines are hard to lay and hold in place in the surf zone. Weight can temporarily be added internally—for example, chain wrapped with canvas and/or fiber rope so as not to abrade the interior. Even sandbags have been used.

As with polyethylene lines, details of saddle bearings must be carefully developed and accurately constructed so as to prevent local wear and point bearing. FRG is susceptible to abrasion and to internal fatigue under sustained stress concentrations.

13.3 Flexible Pipelines and Risers

In recent years, the technology for the design and manufacture of steel-reinforced plastic hoses has advanced significantly so that reliable lines for crude oil and petroleum product service are now available up to at least 16-in. diameter. These lines consist of several layers of neoprene double-wrapped with helixes of steel in a manner resembling that of armored power cables. Being flexible, they can be transported on reels, especially lines of a size used in flow lines. Larger sizes may be faked out on the deck of a barge.

These pipelines have to be led off the barge over a curved cradle, not unlike a small stinger. Rollers will prevent abrasion. Tension must be applied to control the lowering.

Since such lines are usually laid with their end open to the sea so as to avoid buckling collapse due to hydrostatic pressure, the net weight of the line may approach that of a steel line. Such flexible lines are usually so short that they are laid in one day or less; hence weather conditions may be selected to minimize the required tension.

In very deep water or where the tension value exceeds the allowable value for that line, a wire rope may be married to it so as to give it support and to take the tension force.

Great care must be used to avoid cutting or abrading such lines. Nylon or other fabric must be used for the slings with which the pipe is handled.

Coflexip has developed an integrated system using a large-diameter wheel trencher on bottom crawlers to enable the simultaneous trenching and laying of the line. See Figure 13.3.1. Such a system was employed for laying flow lines and cables in the coral seabed of the Zakum Field, in Abu Dhabi.

Coflexip has developed a method to install its flexible pipe in 450 m of water in the Montanazo Field off Barcelona, Spain. It will utilize a very sophisticated positioning system, with a reference array of 30 subsea transponders, and an ROV in monitoring pipe touchdown.

13.4 Outfall and Intake Lines

A great many lines of reinforced concrete have been laid through surf zones and on out to 60–70 m of water depth to serve as sewer outfalls, circulating water discharge lines and intakes for cooling water, and so forth.

These lines have varied from about 1–6 m in

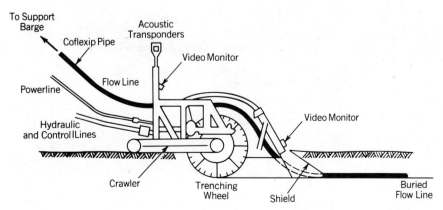

Figure 13.3.1 Trenching and laying machine for buried flow lines. (Courtesy of Coflexip.)

diameter. Most commonly, they have been of the bell-and-spigot type, with each segment (4 to 8 m long) being laid as a single piece. Lines have been laid in a trench and then backfilled or laid on the seafloor and covered with rock.

The laying of concrete pipe segments in the open ocean and their successful jointing requires that a procedure be developed for control of the segment as its support is transferred to the seafloor in order to eliminate the heave and surge motions transmitted from the surface. Two solutions are used. In one a jack-up construction rig is used. The rig positions itself and then raises on its legs. A crane must lift a pipe segment from the transport barge to the jack-up.

Alternatively, several segments may be made up near shore, in a string, temporarily tied together by external cables and floated out to location. Flotation is provided by external tanks, by inflatable bulkheads at the ends of the pipe string, or by an internal sausage-shaped pontoon.

A main problem is that of transfer from the floating mode to the jack-up, especially if relatively heavy seas are running. The pipeline string must be fed in between the legs of the jack-up, using extreme care not to impact them, because of the danger of buckling a leg. The pipe sections are then lowered to the seafloor and entered into the previously laid bells, as described in the following section.

Another problem with the use of a jack-up rig is the need to move continually, that is, only one string of 15 to 30 m in length can be laid from one setup. At each position, the rig must be jacked up and later down and hence is sensitive to the sea state at these critical times.

The other, more widely used method is to use a cradle, often colloquially called a "horse." The cradle is set on the deck of the derrick barge. The pipe segment, usually just one length of 4 to 8 m, is lifted from the transport barge and set in the cradle, where it is temporarily secured. Then the cradle is lowered overboard and seated on the seafloor, approximately in position, just ahead of the previously laid segment. See Figures 13.4.1, 13.4.2, and 13.4.3.

Divers now descend wire lines to an observation and control point at the joint. They first check the bell of the previously laid pipe to ensure it is free of debris. Then, operating hydraulic rams, they enter the spigot end of the new pipe into the bell end of the previously laid section.

The joint is now checked all around for compliance with the prescribed width of gap or opening (usually about 30 mm) and an external sand seal run around the gap. A rubber air or water hose of proper diameter, previously fitted around the new pipe, can be rolled into the gap rather readily and makes an effective sand seal.

The pipe is now checked for correct alignment,

Figure 13.4.1 Installing concrete outfall sewer pipe on San Francisco Ocean Outfall Project across San Andreas Fault and on out 4 miles into the sea. (Courtesy of Morrison–Knudsen Inc. and Ocean Outfall Constructors.)

13.4 Outfall and Intake Lines

Figure 13.4.2 Laying ocean outfall concrete pipe, using cradle. (Courtesy of Morrison–Knudsen Inc. and Ocean Outfall Constructors.)

both vertical and horizontal, and crushed rock fed in underneath to support it.

The reaction to the positioning forces is supplied by the cradle resting on the seafloor or trench bottom. The rock referred to above can be carried down with the cradle in hoppers alongside the pipe.

Using jets or vibrators, the rock is caused to flow in so as to completely fill under the pipe.

Now the divers attach a test hose and pressurize the zone within the double O-ring of the joint. See Figure 13.4.4.

The last operation before raising the cradle is for the diver to check by hand that there is no rock trapped in the bell at the leading edge of the pipe.

In some cases, in order to provide additional thrust for seating the pipe, a wire line is run through the center of the pipe to a winch on shore. This wire line is extended as each pipe section is laid by shackling on another piece of wire rope. A C-hook at the far end then fits over the bell; as the line is taken up, it forces the spigot into the previous bell. Alternatively, hydrostatic thrust may be used.

In the past, concrete or weighted-timber sleepers have been used to support the leading edge of each pipe section. Now their use is largely discontinued, due to the stress concentrations on the pipe. Gravel-filled bags, preplaced by diver, are sometimes used as a "sleeper." They are less rigid than a concrete sleeper, but their use requires that the subsequent bedding rock be well compacted under the remaining length of pipe.

For smaller-diameter outfalls and intakes, a steel–concrete cylinder pipe is often employed. It is lighter in weight and has greater beam strength, enabling longer lengths to be laid at a time.

When laying pipelines through heavy-surf zones, a sheet pile cofferdam is used, as noted earlier, to still the water and to keep the trench free from sand until the line can be laid. In some cases, where sands

Figure 13.4.3 Concrete pipe is suspended in cradle preparatory to lowering into sea. (Courtesy of Morrison–Knudsen Inc. and Ocean Beach Outfall Constructors.)

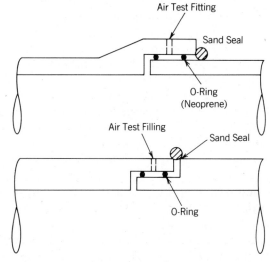

Figure 13.4.4 Sealing of joints of concrete pipe segments.

overlie rock and the trench must be excavated into the rock, it is impracticable to install sheet piles.

One system used successfully—for example, at Durban, South Africa—is to first install larger-diameter pipe instead of sheet piles as a temporary protection for the permanent line. A trestle was constructed alongside the line of the trench by cantilevering out one bent at a time, using steel pipe piles drilled and socketed into the rock. Then drill carriages were cantilevered over the side to run down casing and drill into the rock. Powder was loaded and the rock shattered by blasting.

Now one section of 20 m or so was excavated to grade, using a crane on the trestle with clamshell bucket. As soon as the 20-m section had been dug to grade, a pipe length was entered. This pipe had an inner diameter some 500 mm greater than the outer diameter of the permanent line. The leading end of each pipe had a temporary bulkhead to keep out sand.

The following day, another 10 m was excavated, the temporary bulkhead removed, and a new pipe segment laid.

On stormy days, work was suspended. Laying in such a manner inevitably involved tolerances—hence the allowance of 500 mm in diameter so as to accommodate laying inaccuracies.

When the pipe segments reached the outer end of the surf zone, the permanent line was pulled through from shore.

The annulus was then pumped full of concrete, thus giving additional weight for stability in the surf zone. In other cases, a sand slurry fill might be preferable.

Concrete pipe lines have been pulled out to sea, much as described for steel pipelines. The pipeline lengths are made up on shore and connected together by posttensioning tendons run through the pipe segments. Joint flexibility is provided either by bell-and-spigot joints, with O-rings, or by neoprene gaskets between the pipe ends. In the last case, special means must be employed to ensure long-term corrosion protection for the tendons. A number of concrete pipelines similar to the above have been installed in New Zealand using this pulling technique.

Some outfall and intake lines require support, because of either soft soils or erratic seafloor. In some such cases, longer-length structural box culverts are prefabricated, using reinforced and precast concrete. They may be designed to span 30–40 m or more between supports.

In soft ground, piles may be driven as a support. Both steel pipe and prestressed concrete piles have been used by attaching brackets which support the cap yet permit adjustment in position as the cap is set. In shallower water, timber piles have been so used, being driven below water with a follower to an exact grade. A prefabricated steel or concrete cap is then placed over these piles and secured to them. Alternatively, the cap may be set first and used as a template for the piles. See Figure 13.4.5.

The caps may have bearing pads preattached, along with guides to enable accurate seating of the pipeline (tunnel) units.

In the case where the seafloor consists of irregular rock outcrop, steel forms tailored to the bottom may be set and filled with tremie concrete. Divers are used to make the measurements and to seal the joint between the forms and rock.

At the outer end of the pipelines, intake or discharge structures are usually required. These may be specially fabricated pipe segments, with diffuser ports, or even caissons designed to minimize the water velocity and maximize dispersion.

The latter segments, often weighing many hundreds of tons, are often installed by dividing them into precast concrete segments match-cast so as to fit each other or with joints subsequently injected with grout or epoxy so as to be tight.

Suitable guides and mating cones are provided to fit segments in proper position relative to the adjoining segments.

Large caissons may be floated out and seated on the seafloor, much as described in Chapters 10 and 11. Because of the relatively small plan dimensions of such structures, and hence low waterplane moment of inertia, special care must be given to ensuring stability at all stages of installation.

A few large-diameter steel pipelines have also been used, fabricated in segments and placed in relatively long lengths—e.g., 15 m (50 ft)—with connecting joints, either O-rings or flanged and bolted. These lines have proven to present significant difficulties for the constructor due to distortion in fabrication and handling, making it difficult to fit up the joints. Their light weight gives similar problems to those encountered when placing fiberglass lines in the sea and may require temporary ballasting, either internally or by means of external saddles, so as to prevent displacement by current and waves.

Use of a well-designed cradle that supports the pipe so that it will take the same shape as the pre-

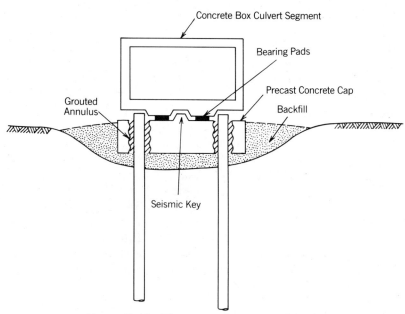

Figure 13.4.5 Pile-supported concrete culvert.

viously set section appears to be indicated by this experience.

13.5 Cable Laying

The first and most dramatic application of ocean engineering and construction was the bold laying of communication cables across the oceans, uninhibited by the depths and, in most cases, able to span the irregularities of the seafloor by intentionally laying in a loose snake-line pattern as opposed to a straight line. These early cables were not buried.

Almost equally remarkable was the development of the ability to locate, grapple for, and pick up severed ends, splice with waterproof splices, and re-lay.

In recent years, the need has developed to lay power cables across open water areas. The currently most advanced application is the construction of four power cables across the English Channel, trenched in the limestone seafloor. Proposed but not yet built are the power lines across the Straits of Belle Isle, between Newfoundland and Labrador, an irregular rock seafloor gouged and polished by the keels of icebergs. Another proposed deep power cable would link the principal Hawaiian Islands.

Key elements in the technological development of subsea cables have been the perfection of armored cable using double-wound helixes of steel over the insulated copper or aluminum cable (see Fig. 13.5.1) and the development of rock-trenching machines capable of cutting a narrow trench up to 1 m deep in hard rock while laying a guide cable in the bottom of the trench.

These cutting wheels, fitted with diamond or carbide teeth, must be supported by outrigger sleds or tracked carriages which run on the seafloor. Therefore, the seafloor must have been first cleaned of surficial sediments and debris by one or more of the dredging or jetting schemes discussed in the section on dredging, for example, the use of a trailer suction hopper dredge.

The cutter–trencher, essentially a wheel, cuts a trench 200–300 mm wide and a meter or so deep.

Figure 13.5.1 Submarine power cable. (Courtesy of Kerite Co.)

As it moves on, it lays a wire rope line into the bottom of the cut trench. This remains there so as to later serve as a guide for the power cable.

When the trench is complete, the power cable is laid. See Figures 13.5.2 and 13.5.3. High-pressure water jets operate from a sled which is guided by the prelaid cable. These jets clean out any debris. The sled device resembles a pipeline trenching sled and may incorporate eductors or airlifts in addition to the jets. See Figure 13.5.4.

The power cable is then fed into the trench, using the previously laid wire line as a guide.

The power cable differs from the pipeline in its much greater flexibility, which enables it to be reeled onto a drum of moderate radius and subsequently unreeled without damage. However, as power cables grow in size and armoring, their flexibility decreases, so that they require a laying barge not too dissimilar from a reel barge for pipe laying.

A review of the above procedure will highlight several critical features. First, the seafloor must be either naturally level or artificially leveled to the extent that the trencher can cut a relatively uniform depth. This may require blasting of some outcrops and ledges, using shaped charges or even drilling. The problem with explosives is that the remaining rock is fractured, which can cause the trenching cutter to bind or jam.

In many seafloor soils, a heavy plow, such as that described in Section 12.12, may prove suitable for direct trenching or for pre-levelling to enable the wheel to work.

The methods needed to clean and level the seafloor prior to trenching will vary widely, depending on the character, depth, and hardness of the seafloor surficial materials. In some cases, the trailer suction hopper dredge will prove most efficient. In others, a sled device, similar to the larger pipeline plows, may be applicable, dragging boulders and loose rock aside.

If blasting becomes necessary at discontinuities in profile such as ledges, then minimal charges should be used in order to avoid deep fracturing.

The current power cable crossings of the English Channel consist of four cables, with two being constructed by the British using the methods described above. The seafloor is being cleaned by a trailer suction hopper dredge. In some areas, bucket dredges are being used to clear the route of boulders. In areas of irregular rock, explosives (shaped charges) have been employed. An underwater crawler has been used to survey the route prior to

Figure 13.5.2 Cable-laying barge. (Courtesy of Volker–Stevin Offshore.)

13.5 Cable Laying

Figure 13.5.3 Laying and burying of submarine cables. (Courtesy of Volker–Stevin Offshore.)

Figure 13.5.4 Cable-laying and trenching sled. (Courtesy of Toyo Pumps.)

trenching. The trench cutter is a self-propelled RTM III trencher, having a 4.5-m-diameter drum with 180 cutting picks (teeth). It cuts a 600-mm-wide slot to a depth of 1.5 m in the limestone and sandstone seafloor. The trenchers can climb a 1.5-m ledge or ridge.

In one 600-m section, the sandstone boulders were dredged out by drag line. Unfortunately, prior to the arrival of the trencher, some of the boulders had fallen back in due to the high currents and side slope erosion. It would appear prudent when predredging such an area either to clear a wide swath or to temporarily backfill the cleared trench with sand. Then a trailer suction hopper dredge could clean it out just ahead of the trencher. After the trencher passes, a second temporary backfill may be necessary to keep the boulders out until the cable layer arrives.

After the trencher has cut one slot across the channel, the cable layer starts across, using a sled equipped with high-pressure water jets, enabling the sled to follow the prelaid wire line and thus install the cable at the bottom of the trench.

The French, on the other hand, are using a combined trencher-cable layer, which lays the cable in the bottom of the trench as it is cut. The trencher is preceded by a wheeled submarine. Overall navigation guidance is by the SYLEDIS positioning system.

The sea was rough and stormy
The tempest howled and wailed
And the sea-fog, like a ghost
Haunted that dreary coast,
But onward still I sailed.

LONGFELLOW, "THE DISCOVERY OF THE NORTH CAPE"

14

Topside Installation

14.1 General

In recent years, almost all of the topside facilities have been first fabricated into modules and then transported by barge and set on the platform by an offshore derrick barge. The capacity of offshore derrick barges has steadily grown to where 1000- to 1200-ton modules are commonplace and individual lifts of 4000 tons and more have been made. The current record is the deck of the Esmond platform, at 5400 tons.

The purpose of using even larger modules is to enable more of the fitup and testing to be done at the shore site. This not only has the advantage of enabling the work to be done under optimal conditions, but disperses the work so that it can be accomplished concurrently with other modules and other structural work.

14.2 Module Erection

The modules are set onto the module support frame, which is a skeletonized deck structure. See Figure 14.2.1. Some will be set onto skid beams and skidded and jacked to final position; others may be set directly.

The modules must be structurally adequate in themselves both for the temporary loads imposed during transport and installation and for the permanent loads due to the operations and environment.

The structural support frame for each module must first support the vessels, piping, and so forth within it and then transfer the forces developed by dead and live loads and environmental loads to other modules or the module support frame.

Lifting of such extreme loads must follow the general principles of heavy offshore lifts outlined in Section 6.3 and in addition must be thoroughly engineered for all stages of the operation. Picking points and padeyes must transfer the forces to the slings. The slings, with their angles in three-dimensional space, must in turn transfer the loads to the hook. Where more than one crane will be involved in the lift, the interaction of loads between the cranes must be considered, including the effect of tolerances in boom position, sea-induced motion, and the change in the derrick barges' waterplanes as the load comes onto them.

Picking loads are dynamic; adequate allowance must be made for dynamic amplification in lifting force, as well as in lateral swing. This latter can be greatly reduced by power-controlled tag lines.

Low-temperature effects, possibly causing embrittlement under impact loads, need to be addressed and suitable steels and welding procedures adopted.

Many modern heavy lifts of modules are assisted by on-board computers monitoring the loads, the radii, and the position of the booms. (See also Section 6.3.)

Modules are usually loaded onto a barge at a shipyard or shore base by skidding out, much as a jacket is loaded out. Dimensions are smaller and total weight much less, but loads may be more concentrated.

The modules must then be properly tied down for sea. See Section 9.2.

Engineered slings are pre-attached to each module so that all that remains to be done as the lift commences is to raise each sling up over the hook by means of the crane's whip line. Meanwhile, the tie downs are cut loose.

When sea conditions appear favorable, the module is lifted clear of the barge, slowly rotated to position, and set in its place. See Figure 14.2.2. Auxiliary means such as powered tag lines on

14.2 Module Erection

Figure 14.2.1 Setting module into module frame, Statfjord B platform, Norway.

last may be transferred to the stern as soon as the load touches down. The object is to prevent the load from being inadvertently lifted back up as a subsequent wave raises the derrick barge's stern before there is enough slack in the falls.

Up on deck, each module is then welded to the module support frame.

A typical series of modules will include the following:

Utilities modules
Control room module
Quarters modules
Helideck
Wellhead module
Separation module
Dehydration module
Pig-launching module
Generator module
Switchgear module
Metering module
Bulk storage modules
Pedestal cranes
Drilling modules
Drilling derrick
Flare stack
Casing and drill string laydown racks

Figure 14.2.2 Semisubmersible heavy-lift crane barge lifting 2000-ton module.

the deck of the platform, tapered guides, and fenders are used to help seat the module in correct position.

There is a need to move smoothly and quickly to a setdown so as not to expose the system to higher waves or low-cycle fatigue. It is desirable to incorporate tolerance in initial positioning into the structural design. Once the unit is set, jacks can then move it to final exact position. Jacking points must be provided in the module frame.

The problem of overhaul during setdown—that is, of getting rid of a load from the hook when there are 24 or more parts of line in the hoist blocks—is a difficult one. A free overhaul clutch for the crane hoist is the main solution. In some cases, bal-

14.3 Hook-up

The hook-up of these modules and their subsequent testing is highly demanding in terms of manpower and support. It delays the start of production of oil or gas and hence adversely affects cash flow.

In recent years, the complexities of hook-up have led to overruns in cost and time of hook-up of 100 percent or more. To reduce these, the first step is to use larger and fewer modules, that is, more self-contained modules.

A second step is to space the modules apart by 1 m or so so as to allow a crawl space for access for interconnection. A third step is to use flexible connections to the extent permissible for the high operating pressures in the pipeline connections.

Careful control in tolerances of all interconnecting points at the time of module fabrication is essential. Templates may be used to ensure compatibility.

If the hookup work is supported by a semisubmersible derrick or "floatel," a suitable gangplank or walkway is required. This must have rollers to accommodate surge of the semisubmersible. It must be supported so that in no way can it fall even if the barge drags an anchor or parts a mooring line. Sophisticated articulation is often provided so as to compensate for much of the barge's movements.

Fire protection during hook-up is critical and must take precedence over actual work. This requires the early installation of a fire pump casing, the submersible pump, and headers around the platform decks. Until the platform system is fully established, fire hoses must be led over from the tending semisubmersible or derrick barge. A fire alarm system must be hooked up.

Life safety must be ensured; this means that life jackets and safety lines must be employed initially, that lifesaving capsules must be placed with the initial modules, and that a patrol boat must be on duty to pick up a man in case he falls overboard.

Other services and systems on board the platform must be activated at an early date. These include the generators, both primary and emergency, with their diesel fuel supply tank, and lighting systems for night work. A freshwater system must be established for potable water supply and washdown. Compressed-air systems are required both for instrument air and for utilities and tools.

Radio communication to shore and boats must be established, as well as public address systems throughout the platform. A smoke-detection system is needed in the quarters modules, as well as sprinkler and fire alarm systems. At the helideck, a foam extinguisher system must be installed. The rope landing nets must be installed and the landing lights activated.

Ventilation systems must be activated in the quarters and then within the facilities area.

Because, during the hook-up stage, welding is the principal item of work, welding generators must be installed and cables led around the deck. Heated welding rod storage needs to be provided.

In addition to the permanent platform cranes and hoists, temporary hoists and powered winches will be required during hookup.

Temporary shops and offices are needed. An X-ray lab is required. There must be an electrical shop, an instrument shop, and a general tool shop and a warehouse for bolts, pipe flanges, and the like. A paint shop is required, with its separate fire-extinguishing system.

Finally a hospital room must be available for emergency treatment of injured personnel.

It is obvious from this long summary listing that each module should be equipped as far as possible with the items needed to complete its hook-up. Beyond that, very thorough and detailed planning is needed by engineers, craft supervisors, and their foremen to insure that all needed supplies and materials arrive with the modules so as not to require separate lifts.

However, these separate items cannot be just loosely stored in the module but must be properly boxed and secured so that they cannot be displaced during the lift. Further, their weight needs to be computed and added to the calculated lift weight.

The module lift weight then becomes the sum of the equipment, piping, cables, and so on, the module frame, the lifting gear, including slings, and the tools and supplies stored on board.

Some rather spectacular lifts of modules have been made by the use of two or even three crane barges working in concert. The Statfjord A quarters modules were too high (40 m) to enable them to be lifted by a single crane. Their weight, about 1000 tons, was not too unusual, but the height and profile required that three crane barges be used. These were moored together with all deck winch controls and dynamic-positioning thruster controls at one location. The three barges picked up the module at the dock, transported it to the concrete gravity platform which was then moored in the fjord, and repositioned the barges while carrying the load so they could set it up on skid beams.

At the final lifting site, the barges were moored to the structure. Because of the short length of lines, nylon rope was used so as to have some elastic stretch to accommodate surge.

The pick was engineered with extreme care, since the two smaller crane barges could only lift 300 tons each.

On a subsequent platform, similarly high and heavy modules were set by a semisubmersible crane barge. Since this work was carried out in a fjord, the semisubmersible was not selected for minimal response to seas but rather for its extreme height when deballasted so as to ride on its pontoons. Now it was able to lift the module over the deck structure and to set the module directly in place.

14.4 Giant Modules and Transfer of Complete Deck

In a recent, very heavy offshore lift (the deck for platform Esmond), a large semisubmersible derrick barge used both of its cranes in concert, holding one dead astern while rotating the other as necessary to position the load on the platform. The module set a record for a single lift: 5400 metric tons.

For such operations, freedom from long-period swells is necessary. Work is carefully scheduled for the weather window and the derrick barge may sit several days awaiting suitable seas in which to make the lift. Of course, the large semisubmersible barge does not react to short-period wind waves of moderate height, but it is sensitive to longer-period wave energy such as that from remote storms.

Several offshore constructors have recently announced plans to build extremely large semisubmersible derrick barges having capacities of 8000–10,000 tons. Such large modules would presumably be constructed either on a quay wall, whence they would be skidded onto a barge for transport, or else on temporary pillars, much as has been done for the Statfjord B and C decks. A large barge would be floated in under the module, which would then be lifted off by deballasting. More recently, transporters, either crawler- or wheel-supported, are run under the module, then jacked up to raise the module off its supports. The transporter is then towed across a dock or quay onto the barge. See Figure 14.4.1.

With all such monstrous deck modules, the deflections due to dead load during picking need to be carefully computed and special means taken to accommodate the dimensional changes in support locations that occur as the module is picked and then set.

Appendix 5 of this book reproduces the DNV *Rules for Heavy Lifting Operations*.

The ultimate in heavy lifts of modules is that of a complete deck set as a single unit. In this concept, the deck is transported on a barge which enters between the platform legs. Then by careful deballasting and jacking, the deck is lowered so that its mating cones engage the top of the jacket legs. The barge further deballasts until free from the deck and then is pulled clear. This "hi-deck" principle

Figure 14.4.1 Loadout of 4400-ton module using wheel-supported transporters. (Courtesy of London Offshore Consultants.)

was successfully used to set the complete deck on the Maureen platform in the North Sea.

The detailed engineering for such a feat, involving the transfer at sea of 20,000 tons or more, must consider all the matters addressed in Sections 10.2.9 and 10.2.11, where the transfer of a complete deck to a concrete platform in protected waters was described. These include the differing deflections of the deck due to changing support conditions and the consequent changes in dimension at the stabbing points as the load is finally transferred to the jacket.

Differential thermal expansion must be considered. Finally, means must be developed to equalize the load between the four or more legs after the deck support has been transferred.

The Hutton deck as well as the Maureen deck were transferred as completed units. During mating of the Hutton deck, relative movement of the two structures was limited in design to 200 mm, but in the actuality only 60 mm relative displacement was experienced.

Three separate shock absorber systems were installed to absorb any impact as the hydraulic catch probes engaged the cones in the deck. One system was incorporated in the cones. A second consisted of 1.5-m pillars of polyurethane in telescoping steel casings. The third was by jacks, suitably softened by connecting the hydraulic tanks to nitrogen-filled bladders.

Feasibility studies have been carried out for Arctic offshore platforms in which giant modules 7000–10,000 tons in weight would be completed and hooked up in shipyards in temperate climates. These would then be transported to the Arctic and offloaded by skidding across onto the platform or island.

For such an offloading operation, the transport barge may be grounded at the edge of the platform, being ballasted down to seating on a prepared sand berm. This berm would have been previously screeded level, using a fixed or floating screeding frame.

The advantage of seating on the berm is that there will be no change in barge deck elevation as the module's load is transferred onto the platform or island. The disadvantage is the difficulty and cost of screeding a level berm underwater. In several past instances in the open waters of the Beaufort Sea, such screeding has been able to achieve tolerances of only about 300 mm, far too great to prevent damage to the barge.

Alternatively, the barge may be landed on pads and sandfill injected underneath as a slurry. This has proven successful in the seating of Dome Petroleum's SSDC-1.

The other method available is to keep the transport barge afloat during offloading, using carefully controlled selective ballasting to maintain the barge at a constant elevation and trim. Some of the newest jacket launch barges have this capability.

The total weight of such a giant module (e.g., 10,000 tons) is obviously not a problem since jackets weighing three times as much have been onloaded. The new factor is the concentration of loads, both on the barge and on the skidways on the platform.

Barge decks will probably have to be specially reinforced with underdeck trussing in both transverse and longitudinal directions. Skidways on a sand or gravel island will probably need pile supports. Skidways on a structural platform will have to be properly designed for each position of movement as well as for the final position of the module.

Frictional loads can be reduced by the use of Teflon-coated roller pads.

Crawler- or wheel-supported transporters can also be used to move a large module off onto the island, much as is currently practiced at Prudhoe Bay West Dock.

Once the module is in correct position, it can be jacked up to final location and permanently supported.

Similarly, in some cases the monster module will have been transported on the barge while at a low elevation so as to enhance stability while crossing the open sea. At the site, with the barge either grounded or moored alongside the platform, the module will be jacked up to a higher elevation and then skidded off.

When raising such heavy loads by jacks, cribbing should be kept close behind, with perhaps a maximum gap of 15–20 mm, so that in the event of a jack failure the impact will be minimal. Lateral support must also be provided to ensure against sidewise movement under sudden change in barge heel or trim or a local support failure.

Or where the Northern Ocean, in vast whirls,
Boils round the naked melancholy isles
Of farthest Thule, and th'Atlantic surge
Pours in among the stormy Hebrides.

JAMES THOMSON, "IN AUTUMN"

15

Underwater Repairs

15.1 General

Offshore structures are exposed not only to the extreme conditions of the environment such as wave slam, ice impact, and fatigue, but also to accidental events such as boat impact and objects dropped off the platform. The list of such accidental events that have occurred over the past several decades is myriad. It includes ramming by a supply boat that went full ahead rather than full astern, impact from the reinforced corner of a cargo barge, and impact by a derrick barge whose mooring lines had parted. The dropped-objects category includes a number of pedestal cranes pulled off their supports when they attempted to follow the movements of a supply boat and thus exceeded their allowable radii, drill collars, casing, a mud pump, and pile hammer. Anchors have been dragged across a pipeline. Leaks into underwater compartments have developed through corroded and ruptured piping.

In the environmental category, horizontal bracings near the waterline have been excited by vortex action and subjected to vertical cycling beyond that for which they were designed, with consequent failure by fatigue. Defective welds and heat-affected zones have led to crack development, and its subsequent propagation has been accelerated by corrosion in the crack. A platform may also be damaged by operational failure, which leads to flooding or overloads. Scour may undermine the legs and lead to excessive lateral response. Finally, corrosion may occur which weakens the structure beyond allowable limits.

Thus repairs become necessary. Obviously, each such undertaking is highly case-specific and needs to be engineered appropriately to the particular needs of the case.

One fundamental principle is that the carrying out of the repair must not increase the risk of failure. This may require auxiliary strengthening before the damaged member is cut out for repair. It may mean that repairs must be delayed until a more favorable season, when the environmental loads are reduced. It may mean that limits are placed on operations until the repairs are completed. Repairs may have to be carried out step by step.

A second principle is that the repaired or reconstructed element must not adversely affect the performance of the structure. As an example, if the reconstructed member will be significantly stiffer than the original member, a dynamic re-analysis of the entire platform may be required.

In order to reduce the time of repairs to a minimum, not only to limit the costs but especially to limit the time when the platform is in a weakened condition, the repair procedures must be planned in extreme detail and all necessary tools, rigging, fittings, and so on provided as one package.

For complex repairs, a rehearsal with the crew may be advisable so as to reduce the problems of communication during the actual repair.

Many repairs are located near the sea–air interface, and hence the repair work will have to be carried out under conditions of high currents as well as wave turbulence due to incident, refracted, and reflected waves. Some local protection may be able to be provided by the derrick barge acting as a floating breakwater.

A major problem with large offshore structures is that of identifying locations so that divers, submersibles, and so forth may readily return to specific spots. The pre-installation of large, highly visible (yellow or orange) numerals on platform legs and walls is now routine practice. However, for a specific repair, additional local markers must be installed. Attachment of a wire guide line as a first

382 Underwater Repairs

step will facilitate descent and location and help the diver to maintain position even in a strong current.

Cleaning of marine growth is another operation which must be carried out at an early stage. High-pressure waterjets are the most effective means, although in order to thoroughly inspect or commence repair of a crack in a weld, for example, supplemental wire brushing may be necessary.

Inspection can be carried out by divers, a diving bell, or an ROV with video. The latter two are replacing much of the diver work in order to reduce costs and increase safety. Especially when it is necessary to inspect inside the jacket frame, the ROV will eliminate the potential for divers to be trapped.

When divers must go into a frame or under a structure, at least two divers must work as a team, with one tending the lines for the other. As a first operation, a guide line can be carried in and attached, enabling the diver to easily retrace his path.

The Underwater Engineering Group of the United Kingdom is actively developing case studies of repairs and is carrying out research and development into new, more effective methods.

When corrosion is a cause of damage, the repairs must include a determination of the cause. Steps, such as cathodic protection, must be instituted to prevent on-going corrosion.

15.2 Repairs to Steel Jacket–Type Structures

When diagonal or horizontal braces have fractured or been badly distorted, the damaged section can be cut out. By use of a template and careful measurement, a "pup" is now cut to exact length and end profile, with the ends beveled for full-penetration welding. External clamps are used to hold the pup in position. Now a habitat is placed around the brace and dewatered, and the full-penetration welds are made. See Figure 15.2.1.

Alternatively, underwater "wet" welding may be employed, provided tests show that satisfactory quality can be attained.

Another solution to this problem is to cut out the damaged portion of the brace and then slip in a longer section of slightly smaller diameter but thicker-walled pipe to which packers and grout fittings have been attached. By the use of cementing techniques similar to pile-to-jacket sleeve connections, the strength of the brace may be restored. The packers are inflated at each end and the grout injected and vented into a standpipe at the high end to ensure that all water has been ejected. See Figure 15.2.2.

The third scheme uses both an internal and external sleeve. Only the external sleeve is structural. The internal sleeve is merely an internal form to facilitate grouting. See Figure 15.2.3.

Special grouts are used, having expansive and high-bond capabilities in order to develop the transfer in as short a length of overlap as possible. The provisions of standard API RP2A concerning grout transfer from pile to sleeve are followed as a guide. Multiple weld beads or shear keys are used to enable grout bond-shear transfer in a short length. Epoxy injection is an alternate method.

Another scheme involves the insert of a heavy internal structural member of smaller diameter than the original tubular. This new section can be a heavy-walled tubular with shear rings welded on the ends or even a rolled shape, for example, an H-pile. At the lower end, a pig or packer is attached. An external sleeve is clamped over the gap, using two halves, with flanged and gasketed joints. Then the brace is pumped full of concrete. The concrete is actually a fine concrete, using cement plus sand, with the sand graded up to about 6-mm maximum size. See Figure 15.2.4.

Hydraulic expanders are in the process of development, primarily for use in expanding piles against jacket sleeves so as to transfer the load by a combination of direct shear on the corrugated surfaces and friction, thus supplementing or replacing load transfer by grout. This "swaging" process appears especially useful for repair of damaged tubular bracing.

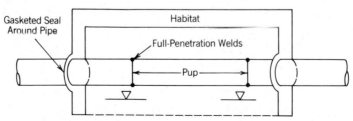

Figure 15.2.1 Repairs to damaged braces: scheme 1.

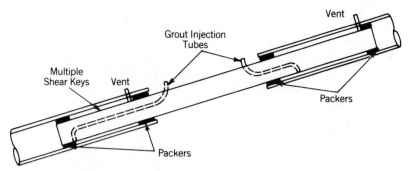

Figure 15.2.2 Repairs to damaged braces: scheme 2.

It is of course to be noted that all these schemes are but variants on one scheme with effort directed toward simplicity and reliability.

Other schemes have been used, especially where axial loads are light, involving clamped external sleeves. These may be in two halves and use high-strength bolts to draw them in tight against the arms of the original brace. Load transfer is by friction of steel on steel.

Recently Chevron reported the discovery of a serious fracture in one of the underwater tubular braces of the Ninian Southern platform. The fracture was discovered by an ROV during routine inspection. The fracture actually was a complete severance of the tubular and is believed to have been due to a faulty weld on a fabrication window used to gain access for a backup weld at a node.

It is interesting to note how many serious fractures in tubulars have occurred not at the critical nodes but rather at temporary closures or attachments: the Alexander Kjelland tragedy was due to a faulty weld in attaching a sonic transducer.

Repairs were carried out by Chevron at a depth of 43 m. The tubular was 1200 mm in diameter, fabricated of 30-mm plate. Because the brace would have to take cyclic tension, the decision was made to use a welded repair. It was feared that a bolted and grouted connection might not have the necessary fatigue resistance under the required stress ranges and number of cycles involved.

Oceaneering Inc. built a specially designed underwater habitat in which a team of saturation divers would work. The fractured faces were removed and sent for a metallurgical examination. The two ends of the braces were now prepared for the weld. Then a lead-filled template was fitted to each end of the tubular to get an exact impression of the exposed ends. From these impressions, transition pieces of 500-mm plate were fabricated and then welded in place, leaving a gap between them so as to enable access for a backup weld on the insides.

Special collars had been welded around the transition pieces. High-strength bolts were now tensioned to draw the collars together with 400 tons of tension force.

A 50-mm "pup" (a closure length or spool) was now fitted into the gap and welded from the outside. The cooling of the weld added to the pretension in the brace, so that the final stress was 600 tons of tension, thus restoring the state of stress that was originally in the member under static load conditions. See Figure 15.2.5.

All welds were inspected and the platform was recertified.

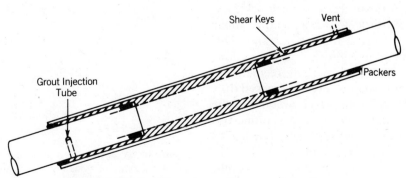

Figure 15.2.3 Repairs to damaged braces: scheme 3.

384 *Underwater Repairs*

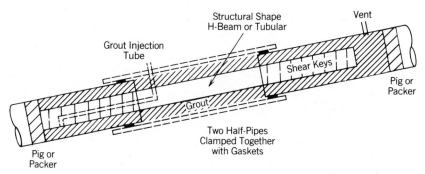

Figure 15.2.4 Repairs to damaged braces: scheme 4.

Cracks in welds may be treated in a number of ways, depending on their size and extent, the service required, and the cause of the fracture.

In any event, the external surface must be ground smooth. After inspection, if the crack is in the non-severe category, small holes may be drilled at each end to act as crack-propagation arrestors. The holes and crack may now be filled with epoxy, using crack-injection techniques.

More serious cracks must be gouged out and re-welded, using either a habitat or wet-welding techniques.

Cracked nodes have been repaired by clamping on an external nodal sleeve made up of as many sections as required and injecting the space with epoxy.

In some minor cases of dented jacket bracing or legs, it may be sufficient to fill the leg with grout so as to inhibit compressive buckling.

15.3 Repairs to Concrete Offshore Structures

As with steel structures, concrete structures may be damaged by impact, collision, and dropped objects. They may also be damaged by over- or under-pressurization beyond that for which they were designed, resulting in flexural and shear cracks. Excessive foundation settlement may lead to shear cracks. Repeated cycling at high stress ranges, especially if they extend into tension, may cause low-cycle fatigue of the concrete or reinforcement.

The Beryl A platform was severely cracked during construction by ballasting which overlooked the temporary differential in heads on the numerous small interstitial cells, called star cells. This led to excessive tension plus shear, resulting in through-wall cracks and displacements of the star cell walls.

Repairs were carried out expeditiously and adequately using the following techniques.

1. Removal of cracked and crushed concrete by chipping by hand or with a very small air hammer so as not to damage further the adjacent concrete.

2. Where major displacement of the walls had occurred (up to 100 mm radially), the addition of an inner reinforced concrete wall about 200 mm thick, tied to the existing wall with drilled-in and grouted studs and expansion bolts.

3. Use of grout-intruded aggregate as a means of correcting wide cracks and strengthening walls.

4. Use of cement grout injection in moderate-width cracks.

5. Epoxy injection of all cracks, carried out by sealing the cracks externally and then injecting a fluid epoxy until material of a similar consistency flowed out the next port above. These cracks were successfully injected against an ex-

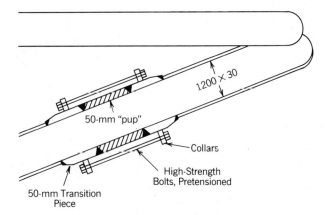

Figure 15.5.5 Schematic representation of underwater repair to tubular brace of Ninian central platform.

ternal water pressure that ranged up to a 40-m head.
6. Shotcrete application to build up the shell walls for compressive resistance.

Numerous cores plus acoustic testing verified the soundness of the repaired structure, and it was therefore certified for installation. It was installed successfully, despite having to sustain differential heads up to 100 m. The repairs have shown no problem in 10 years of service.

When the Statfjord A platform was ballasted down to mount the deck, a small leak was noted in a juncture wall with another cell. Attempts to epoxy-inject this crack were less successful than with Beryl A. In that instance, the cracks were though the wall and hence crossed by reinforcing steel. In the case of Statfjord A, the crack was laminar and had essentially no reinforcing across it. Even though the injection pressure was limited, the act of injection did propagate the crack. A decision was made to limit injections and to fill the adjoining star cells with a long-term corrosion-inhibiting gel.

It would appear from the experience at Statfjord A plus that on subsequent long-span bridges, that stitch bolts should be drilled in and grouted before attempting to epoxy-inject a laminar (in-plane) crack.

One of the shafts of a North Sea concrete platform was impacted by an anchor-handling barge whose heavily reinforced corner caused significant cracking at and just below the waterline. Some spray leakage occurred. Repairs were successfully carried out by use of an external work caisson lowered on the exterior of the shaft, clamped to it, and dewatered. The damaged concrete was then cut out and additional reinforcing bars installed. New concrete was then placed as grout-intruded aggregate, using a colloidal mixing process to increase the fluidity of the grout. Epoxy was injected into the construction joint to fill any shrinkage cracks; very little "take" was noted.

This procedure would seem to be applicable to a severely damaged peripheral wall of an Arctic structure impacted by ice or an iceberg. Where lesser displacements are noted, cracks could be epoxy-injected without the need for an external caisson. The concrete must, however, be adequately warm to enable the epoxy to cure. An insulation blanket on the outside plus prolonged warming on the inside might be required in some cases.

The loss of underpressure in a concrete oil storage platform may lead to cracking. Essentially this is an overpressure situation in which the oil inside, being less dense than the water with which it is automatically equalized at the bottom, exerts an upward pressure on the roof of the storage caisson, causing multiple horizontal cracks in the cell walls just below their juncture with the roof.

These can be repaired by epoxy injection, applied by divers from the exterior periphery of the base caisson. Access to the star cells may be very difficult, and, depending on the circumstances, other means of crack stopping may be preferred. For example, depending on a detailed evaluation of the design, it may be feasible to fill the star cells with a stabilized or gelled drilling fluid, which will prevent leakage and future corrosion.

Repair procedures for potential major damage to concrete structures have been developed and tested by Italian engineers in connection with the approval of the Genoa floating concrete drydock. These included the case of severed prestressing tendons, for which the repair procedure consisted of cutting out the damaged concrete, splicing the severed tendons, and restoring the prestressing force by internal jacks.

More recently, Taylor–Woodrow demonstrated a means of repairing severed tendons underwater. The damaged concrete was removed by high-pressure water jets operated by divers. The damaged and distorted reinforcing bars were cut out by an oxy-arc burner. Broken prestressing tendons were then coupled to prestressing bars. A prefabricated reinforcing grid was then placed and tied to the existing steel. Prefabricated formwork was secured and gasketed to the existing structure. Then a high-strength concrete mix was pumped in from the bottom. After curing the concrete, the formwork was stripped and the bars stressed and grouted. Tests showed that the structural strength was fully restored.

External prestressing has been successfully used to repair damaged bridge girders and would appear practicable for use in underwater structural repairs as well. Holes drilled into structural elements could be used for the end anchors. Polyethylene-encased prestressing strands would have long-term durability.

Leakage has occurred along nests of small-diameter grout piping. Cement grout injection in stages, using techniques developed in tunneling, has proven successful.

15.4 Repairs to Foundations

The foundations of offshore structures are an integral part of the system that resist the environmental forces and support the operating loads. Hence they must be maintained in their design condition.

Scour around jacket-type structures is of two types. One is an areal scour, which lowers the seafloor in a dishlike shape. The remedy here is to dump or chute in rock which will have two characteristics:

1. It is small enough so that it will not work its way down into the soil.
2. It is large enough so that it will not scour under currents and wave-induced forces.

A permeable material is required in order to prevent trapping wave-induced pore pressures which can reach several tons per square meter.

The above requirements can be met with a two-layered system, one of small rock or gravel, the other of large rock, but this is very difficult and costly to place around and between the legs of a jacket at great depths. A single blended placement of rock is less efficient, requiring more quantity of material but enabling placement in one operation.

Another effective scheme for scour prevention is to replace the small rock by a filter fabric and then cover it with larger rock. The filter fabric can best be assembled by two layers, sewn together. One layer is a fine mesh to prevent sand migration; the other is a coarse mesh of heavy polypropylene, reinforced with stainless steel wire.

Mattresses can be made up of reinforced filter fabric with concrete blocks attached. They can then be lowered into place, using a steel frame. Mattresses filled with fine and coarse gravel have also been used. Still another system employs an integral filter fabric to which multiple closely spaced neoprene bags are affixed. After placement, the bags are pumped full of grout.

The other type of scour is a localized erosion around individual legs. This is often accompanied by gaps forming around the head of the pile as it deflects under lateral loads. Here the use of an initial layer of small rock (e.g., pea gravel) can be very effective by working down into the gap. Larger rock on top will then protect the smaller rock from erosion.

A gas blowout under a platform can lead not only to cratering but also to a general loss of support for the piles.

To restore the capacity and safety of the foundation usually involves a major effort. First, the crater can be filled with small rock or coarse sand placed through a tremie tube. Then both the newly placed material and the adjoining sediments must be consolidated, since the entire zone will probably have been significantly loosened by the escaping gas. The consolidation has to be approached with great care, since the resultant settlement of the upper soils may lead to distortion of the structure. For that reason, densification by shock (dynamic compaction) or vibration may not be suitable. Placement of an extensive blanket of rock may be safer, with careful monitoring of both soils and structure at frequent intervals.

When, due to accident or blowout, the pile capacity has been reduced, insert piles can be installed through the primary piles and joined to them by grouting. These insert piles are usually extended beyond the tips of the primary piles so as to obtain additional bearing and friction load transfer for axial loads. Insert piles, properly grouted, will also stiffen the piles and hence improve their capacity to resist lateral loads.

As an alternative to a drilled and grouted insert pile, a belled footing may be constructed using a belling tool. This solution may be preferable in stratified soils or where skin friction transfer is uncertain.

Because the overhang of the deck will probably limit the accessibility, requiring the insert pile to be assembled in short lengths and preventing the use of a hammer, a riser can be installed and reverse-circulation drilling employed. Once the hole is drilled, the insert pile can be lowered to position, using either welded joints or mechanical connectors. Then grout can be injected into the annulus, bonding the insert pile to the soil and to the primary pile.

The above assumes that geotechnical investigations show adequate soil at the greater depths. See Chapter 16 for more details on drilled and grouted insert piles and belled piles.

On two of the early concrete structures in the North Sea, after the platform was installed, drilling was commenced in order to install the conductors. As the drills penetrated the temporary closures of the concrete base slab, with the water level in the drilling shaft being lower than sea level, substantial quantities of foundation sand rushed up into the drilling shaft, leaving a void of several hundred cubic meters' size below. Some "piping" even led in from the outside perimeter due to the difference in heads.

To repair and fill this void, an underbase grout mix having a low heat of hydration, was flowed in under the slab, working in several stages so as to ensure complete fill. The zone on the periphery which had been disturbed by the "piping" was filled with small rock. Subsequent behavior of the platform, over a period of nine years, has been entirely satisfactory.

To prevent this adverse phenomenon on future platforms, water levels in the drilling shaft are required to be equalized, especially during conductor installation. Steel skirts, which isolate the drilling compartments, are built into the base slab design.

15.5 Fire Damage

Fire is of course one of the most dreaded events on an offshore platform, and extensive measures are taken to prevent it. In some more recent structures, fire-protective insulation is placed around principal structural members of the module support frame, and tubular members of this frame are sometimes filled with water.

Of course extensive spray, Halon, and deluge fire systems are also provided on deck and in utility shafts.

When structural members have been subject to overheating and distortion due to fire, a complete evaluation and analysis must be made. While in extreme cases a member may have to be cut out and replaced, in others it may be practicable to reinforce it internally or externally so as to restore its capacity.

Fire in one of the shafts of a drilling platform, in addition to damaging the exposed steel members inside, may cause spalling of the inside of the shaft walls. In most cases, heavy sandblasting, then placement of wire mesh properly anchored to the existing concrete, followed by shotcrete, will prove adequate.

15.6 Pipeline Repairs

In Chapter 12, repairs to pipelines damaged during installation were described. These included the use of a hyperbaric chamber lowered down over the line, enabling cutout and replacement in the dry.

Wet welding as a means of repair is still under development; satisfactory welds have been made in many cases, but their overall reliability is still not fully proven. Friction welding appears to be one especially promising method, but requires that the new section of pipe be rotated; this is not always practicable.

In relatively shallow water, pipelines may be raised to the surface for repair; the methods and precautions required are discussed in Chapter 12.

Pipelines in service are often damaged by dropped or dragged anchors, sometimes by commercial or naval shipping, but most often from derrick barges and workboats working in the same area. Trawl boards from fishing vessels may damage the coating or even the pipe, although usually it is the lines to the board which break, resulting in its loss and a claim by the fishing boat.

The damage to pipelines may range from concrete weight coating being broken off, to dents, to holing, and even to the line being ruptured and the ends dragged apart. Leakage will usually be detected by the very sensitive pressure differentials employed in monitoring pipeline operation or by sheens of oil appearing on the surface. Exact locations can usually be found by acoustic means, by hydrocarbon sensors, or by internal instrumented pigs.

Side-scan sonar can be employed to determine gross positions of the line and segments. Then divers, manned submersibles, or ROVs equipped with video can be employed to give detailed information.

The coatings must first be removed from the damaged area, using a high-pressure water jet, an underwater concrete saw or grinder, and cutters for the mesh. Once cleaned, the damaged area must be carefully profiled and a determination made as to "cutout." Sometimes the damaged section can be cut out and an external sleeve slipped over the gap. In other cases, a pup will be more suitable.

When the rupture is a case of cracking or splitting, external split sleeves may be adequate. Tightly torqued together and with seal welding at the ends, sometimes augmented by epoxy injection, the line may be restored to full service. See Figure 15.6.1.

Using a sleeve secured to the pipeline with thrust screws, wet-welding techniques were employed to repair a 36-in. submarine line in the Mediterranean. See Figure 15.6.2. A number of other systems ap-

388 Underwater Repairs

Figure 15.6.1 Rupture in underwater pipeline at junction with lateral, repaired by external sleeves. (Courtesy of Plidco.)

plicable to shallow water have been described in Chapter 12. These include the use of a hyperbaric chamber ("dry habitat") pressurized so as to permit welds to be made in the dry. Special gasses must be used so that the quality of the weld is not affected adversely. See Figure 15.6.3.

Continued advances in the development of hyperbaric chambers now extends their effective depth to about 300 m.

A number of systems have been developed in recent years to enable pipeline repairs to be made at greater depths and, in many cases, without the need to waterflood the line. In one system, the line is depressurized and a small hole is drilled just beyond each end of the damaged section. Inflatable plugs are inserted through the holes and inflated. The gap is now tested to verify that the seals are tight; then the damaged section is cut out. A new pup is inserted, using divers trained for hyperbaric welding. Once completed, the plugs are removed by pigging.

SNAM is developing a system known as SAS for deep-water submarine pipeline repairs. It consists of five modules: a thruster module which contains power, sensing, and transportation components; a dredge module; a pipe preparation module; a spool-cutter module; and a pantographer module. This last measures the exact length and configuration for the new spool piece.

Various types of mechanical connectors are under development. These include cold-forging tools

Figure 15.6.2 Using a sleeve secured to the pipeline with thrust screws, wet-welding techniques were employed to repair a 36-in. submarine line in the Mediterranean. (Courtesy of Plidco.)

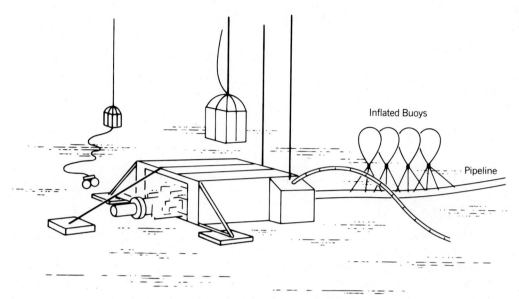

Figure 15.6.3 Pipe repair operation (Comex) for deep-water hyperbaric operations.

such as those being developed by Cameron Iron Works and Big Inch Marine.

It appears that adequate means of pipeline repairs at depths up to 1000 m will be available in the near future.

Break, break, break,
On thy cold grey stones, O sea!
And I would that my tongue could utter
The thoughts that arise in me.

TENNYSON, "BREAK, BREAK, BREAK"

16

Strengthening Existing Platforms

16.1 General

There appears to be a growing need for the strengthening of existing platforms in order to accomplish one or more of the following objectives:

1. Carry increased loads from additional equipment needed to support gas and/or water injection and other secondary recovery operations
2. Extend the life of an older platform
3. Upgrade the platform to withstand greater environmental forces
4. Improve the structural system to overcome deficiencies that have been discovered since initial installation
5. Upgrade an exploratory drilling structure to a drilling and production platform

Four categories of strengthening have been recognized:

1. Increasing the strength and rigidity of individual structural members or assemblies
2. Increasing pile capacity to withstand greater axial loads
3. Increasing structure–soil interactive capacity to resist lateral loads
4. Increasing shear and bearing capacity of gravity-based platforms

16.2 Strengthening Existing Members or Assemblies

As noted earlier in Chapter 15 on repairs, horizontal and diagonal bracing near and just below the waterline is often subjected to vortex shedding from the orbital velocity of the waves, leading to a failure from fatigue.

One means of overcoming this problem has been to install underwater K-braces between a lower node and the midspan of the member in question.

Tubular members, with split half-rings at the ends are carefully templated and fabricated to the as-built dimensions. The new member is then placed, and the other half of the rings or clamps installed and drawn up with torqued high-strength bolts. Epoxy grout is then injected.

The installation of additional bracing such as this obviously changes the response of the structural members, as well as introducing new loads into the nodes. A reanalysis of the platform is therefore required before such strengthening is carried out.

Another and simpler method to prevent fatigue due to vortex shedding is to install spoilers on the brace in question. These can be clamped on. While they slightly increase the load on that brace, the effect is usually limited to the member in question, with minimal effect on the remainder of the structure.

On the Ninian central platform, a number of deck chord members were reinforced to carry heavier axial and bearing loads by filling them with concrete. In order to achieve complete filling, the grout-intruded aggregate method was used; that is, a grout pipe was placed inside the tubular, and then the tubular was filled with coarse aggregate. A sand–cement grout of high fluidity was then pumped through the pipe, which was gradually withdrawn as the voids between the aggregate particles were filled.

It would of course also be possible to insert a steel shape into the tubular before placing the aggregate. The steel shape should be configured so as to provide for proper flow of the grout and escape of entrapped air and bleed water; for horizontal members, a heavy-walled tubular will probably prove best.

Brown and Root have developed a method by which bracing members of existing platforms can be strengthened by procedures similar to that described above, with the additional inclusion of an ungrouted sleeve through which a posttensioning tendon is run. The tendon is then anchored and stressed, using a hydraulic jack specially adapted for underwater use. Then the tendon is grouted. See Figure 16.2.1.

This method is especially useful when it is desired to increase the axial tensile capacity of that member. Depending on details of the node, the tendon may be run through the nodes and stressed externally to them.

When the upper chord of deck trusses or the upper flange of deck girders is required to carry heavier bearing loads from the modules, heavy bearing plates can be fitted on, with stiffeners to spread the concentrated load to the tubulars. In some cases, it may be desirable to fill the tubular member with concrete at the location of the bearings so as to prevent local deformations and to distribute the concentrated loads.

The nodes at the intersection of the legs and braces of jackets may be given increased resistance to buckling and ovaling by filling with fine concrete (typically 10 mm. aggregate and silica fume-cement mix). Where a pin pile is located within the jacket leg, the annular space may be grouted.

Another method used to strengthen both nodes and braces is to clamp on an external sleeve, in two halves. These can be squeezed around the brace by the torquing of high-strength bolts, and the annular space is injected by epoxy or cement grout.

16.3 Increasing Capacity of Existing Piles for Axial Loads

Increasing the capacity of existing piles to sustain increased axial loads is a problem which has been encountered a number of times in cases where calcareous sands have been encountered and where a reevaluation of their performance under dynamic loads has led to the decision to increase their capacity. It also occurs where a platform must carry heavier loads than those for which it was originally designed.

The most straightforward decision is usually to install insert piles. Since the overhang of the platform usually limits the clear working space to 10 m or less, the insert piles will have to be installed by drilling rather than driving.

A work deck is installed, upon which a short drill rig can operate. Then a riser pipe must be run down to the top of the existing primary pile and sealed to it, sufficiently tight to enable an 8- to 10-m positive head to be maintained within the riser. Flange or coupling joints are used for the riser.

Then the drill string is assembled and a hole drilled for the insert pile down to the required depth.

An insert pile must now be made up. Welded connections are standard, but a great many are required due to the short length of each segment that can be fitted in under the deck. The welding has to be carried out under conditions of spray and wind; hence, protection is required. Nondestructive testing, usually X-ray, takes time and height. The alternative is to use a high-strength tapered-screw mechanical coupling. Such fittings should be qualified by tests under dynamic and cyclic loading. Based on the state of the art as of this writing, the screwed connections would appear to be reliable and much more practicable.

With the insert pile, a grout pipe is run. An inflatable packer is provided at the tip to prevent grout from filling the insert pile. Alternatively, a float shoe may be installed. Cement grout, sand-cement grout, or fine concrete is then placed, until instrumentation shows that the annulus is full.

Shear transfer to the insert pile can be enhanced by welding on shear rings. To prevent bleed water collecting under the ring and thus reducing direct load transfer, it is recommended that the undersurface of shear rings be beveled. The shear ring may also be spiraled up the insert pile. See Figure 16.3.1.

Means must be provided to prevent the insert pile from floating in the denser grout since the insert pile itself will usually not be filled with concrete, at least not simultaneously. It may be wedged to lock it in place or weighted by filling with barite drilling mud.

To save the time and cost of coming out with the drill string and then setting the insert pile, it may sometimes be expedient and adequate just to grout the drill string in place in the drilled hole using an expendable bit. This was done, for example, on several piles for the strengthening of the Kingfish platforms in the Bass Strait, Australia.

Another solution is to utilize the end-bearing ca-

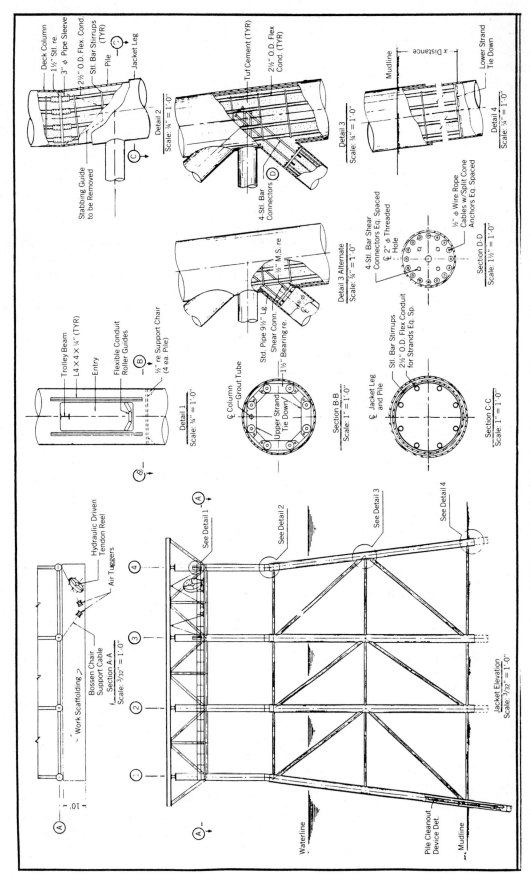

Figure 16.2.1 Platform strengthening. (Courtesy of Brown and Root, Inc.)

16.3 Increasing Capacity of Existing Piles for Axial Loads

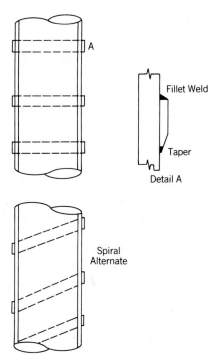

Figure 16.3.1 Shear keys on insert piles.

and hopefully bind with the soils. Special admixtures are available to help penetrate sandy soils.

The problems are many, however. The sands, being saturated, are relatively impermeable, since any grout flowing into them must push water out. Grout always follows the path of least resistance, trying to find an exit to the sea. Thus while on paper we can draw symmetrical grouted bulbs, in practice they are rarely achieved. If too high a pressure is developed, the grout may fracture the formation and escape into the sea.

Other types of grouts, such as polymers and chemical grouts, penetrate dense soils more readily but also suffer from the problems of inconsistent performance. Shell Chemical Co. has developed a low viscosity epoxy material, Eposand, which shows promise for penetrating sands of low permeability.

Two- or three-stage grouting may help to achieve an effective bulb. Use of packers above and below the grout injection points has been proposed by Halliburton as a means of controlling the grouting.

The ballasting or weighting of a pile to give greater uplift capacity is similar to the weighting of pacity of an existing pile, since this will have often been disregarded in the initial calculations. To do this, the material inside the existing pile must first be removed.

On the same Kingfish platform referred to above, elongated holes were cut into each jacket leg, above water, for access to the pin pile. A concentric group of pipes were made up, using segments only 4 m long. See Figure 16.3.2.

These sections were progressively made up and lowered down the existing pile. Obstructions such as neoprene pile closures had to be fished out. Using the jet plus airlift, the plug was cleaned to within a few meters of the bottom. Then a sand–cement grout was injected so as to form a 10-m-long plug. The pipes were now withdrawn and an illmenite (iron ore) slurry fed into the pile to give it added weight to resist uplift. To reduce heat of hydration of the rather massive plug of concrete, and hence prevent disruptive cracking, blast furnace slag-cement or cement-pozzolan mixes may be used.

Pressure grouting around the tip and walls of an existing pile have been often proposed. Holes can be perforated in the walls of the pile by controlled explosive casing-perforator devices. Grout of very high fluidity and low surface tension can then be flowed into the pile, to exit from the ports

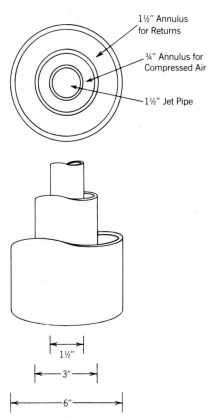

Figure 16.3.2 Multiple pipe segments for cleaning out plug inside existing pile.

a table leg. Earlier, use of illmenite slurry was suggested. Barites added to drilling mud are expensive but effective and can utilize the drilling and mud facilities already on site. Finally, magnetite iron or even lead shot graded for maximum density can be used.

Depending on the character of the soil, drilled and grouted insert piles may not be suitable or may be excessive in length. Where good bearing strata exist near the tip of the existing pile, a belled pile footing may be constructed in much the same manner as the drilled and grouted insert pile.

A riser is run, and then the drill string. After drilling to the desired depth (T.D.) the belling tool is operated, gradually belling out a truncated cone. Reverse circulation is used, with airlift assist. The air injection should be well above the bell, up in the primary pile, so as not to reduce fluid density in the bell and cause caving.

When the belling is completed, spot mudding may be used, if necessary, to hold the bell open, although the internal head in the riser, being 5–10m higher than sea level, will help by causing all flow to be outward. Either a bentonite mud preconverted to calcium bentonite or a polymer mud should be used so as to be compatible with the cement.

Next the bell is concreted. A concrete containing small aggregate, up to 10 mm maximum, is preferred to straight cement grout because of higher tensile strength, lower heat of hydration and less tendency for a brittle mode of failure.

In any case, the cementitious materials should have low heat of hydration: a 70:30 blast furnace slag–portland cement mix is suggested. Alternatively, a cement with 50 percent pozzolan replacement may be used. The mix may be pre-cooled.

Either tremie or pumping may be employed, provided pipe sizes, concrete mixes, and procedures are selected to prevent excessive rates of flow. For depths up to at least 250 m, and probably greater, the tremie method appears to give better control, especially if the formation surrounding the bell is sensitive to hydraulic fracture.

Shear transfer from the insert pile to the bell has to take place over a relatively short length, and hence special shear transfer devices are usually required, such as reinforcing bars grouped in bundles or shear keys on both inside and outside of the insert pile. End bearing plates may also be used, providing direct load transfer from the insert pile to the bell.

A float shoe or end plate may be provided at the tip of the pile to prevent concrete from filling the insert pile. The insert pile must be weighted or secured to the main pile to prevent it from floating.

A secondary system may be necessary for grouting the annulus between insert and primary pile, in order to ensure complete filling of this critical zone.

16.4 Increasing Lateral Capacity of Piles and Structure in Their Interaction with the Near-Surface Soil

In many soils, the cyclic lateral deflections of the structure cause a gap around the pile near its head, thereby increasing the amplitude of lateral displacements under storm waves and progressively weakening the resistance.

The placement of small rock (e.g., pea gravel) on the surface in order to feed down into any gap and plug it has previously been noted.

On a more extensive scale, a thick layer of graded rock placed around the jacket legs may help to confine the soil beneath and prevent its local liquefaction under the pumping action of the pile movement.

The mud mats, originally intended for temporary use during installation only, offer potential means of enhancing lateral load transfer to the soil. Provided they are kept in full contact with the soil and prevented from undermining by scour, they often can transfer considerable lateral and vertical loads in many types of soil.

To ensure full contact with the soil, underbase grout injection can be employed. First a boundary of small rock or filter fabric with heavier rock on top is established, and then holes are drilled in the mud mat for injection of a thixotropic grout—that is, one containing a thixotropic admixture such as bentonite or Methocel—so that it will not just flow through the rock and out to sea.

Then the antiscour blanket of graded rock or filter fabric plus rock or articulated concrete blocks is completed, extending well out from the mud mat periphery so as to prevent loss of fines from under the mud mat by pumping action.

The rock should be brought up well over the mud mat itself to ensure continuity.

Another means of enhancing lateral resistance, which may be used in combination with one of the above, is grout injection around the piles or through the mud mat, into the soils beneath, especially if these soils are sandy. This grouting is different from that used as underbase fill in that the grout selected must be very fluid and penetrating. Although subject to the erratic flow patterns described earlier, in these shallow and limited zones enough injection points can be used to reasonably cover the entire zone.

For a small platform in the Bombay High Field of India, where excessive lateral displacements were occurring due to crushing of calcareous sands just below the surface, insert piles were installed and grouted. This significantly stiffened the piles, reducing strains.

16.5 Strengthening of Concrete and Steel Gravity-Based Platforms

The typical gravity-based platform depends for its stability on the near-surface soils. Where these are known to be weak prior to initial installation, various methods of improving the capacity of the existing soils may be implemented. These are described in detail in Chapter 7 and include dredging, dredging and refill, embankments, wick drains, freezing, and sand piles. In this section, the strengthening of the soils *after* installation will be addressed.

Where sliding shear resistance needs to be improved, the placement of an embankment around the periphery can lengthen and strengthen the potential shear surface. To prevent shear failure upward through the embankment, it will preferably be constructed of fractured rock so as to interlock. When gravel or sand are used, a thicker layer will be required.

Depending on the seafloor soils, it may be necessary to first place a filter fabric or a layer of smaller rock. Depending on the depth and the wave and current environment, it may be necessary to cover the embankment with heavier rock that will remain stable. Such an external embankment or berm, placed after installation, will give direct passive resistance to the structure as well as lengthen the shear path in the native soils below and, depending on the soils and the time, increase the shear strength of the native soils around the perimeter.

External berms, with either a graded rock filter or filter fabric, are an excellent way to prevent liquefaction and pumping of sand from under the edges of the structure.

In Arctic and sub-Arctic areas, a berm may also serve to intercept a deep-keeled ice feature and ground it, slowing or stopping it by the passive resistance of the material along the sides and ahead of the keel.

Placement of such an external berm or embankment is a relatively cost-effective way for upgrading an exploratory drilling platform to a production platform, since the difference in ice features that must be withstood is predominantly (but not entirely) reflected in their draft and hence keel depth.

To enhance the shear resistance of the soils underlying an existing platform, shallow wells may be drilled into permeable strata and drained into the interior of the platform, thus consolidating them over a period of time. In the meantime, these drains are effective in preventing pore pressure buildup and the possibility of liquefaction. Since this may cause settlement of the platform, the effect of downdrag on the conductors must be considered.

To increase the dead weight of a gravity-based structure—as, for example, when it is desired to increase frictional resistance to sliding—additional ballast may be added.

Illmenite and similarly fine-ground iron ores have been placed as a slurry and then decanted. These can achieve specific gravities of 3 and even 4, which gives a net gain of 2–3 tons/m^3 (20–30 kN/m^3) when placed in a cell which previously was filled with salt-water ballast only. Native sand can similarly be used, but the net gain is limited to about 0.9 ton/m^3 (9 kN/m^3).

Another method of increasing the effective weight of a platform having continuous peripheral skirts is to drain the underbase area into the structure, controlling it so as to create a net negative underpressure, but limiting it to well below that which will lead to piping of the sea in under the skirts. Note that this will simultaneously strengthen the soils and reduce any tendency toward liquefaction. In sandy or silty soils, it may be desirable to lay out an impermeable apron around the periphery, properly weighted down with rock, to ensure that piping will not occur during a storm.

Weight can also be added on top of the roof of the base caisson in the form of concrete placed underwater. There is usually reserve capacity for such heavier external loads, since the design of the roofs is usually determined by the installation condition. Such additional concrete on the roof may also improve the resistance to the impact of dropped objects.

Experience appears to indicate that collision from boats and barges is a more common event than initially thought. In some older platforms, it may be desired to increase the impact resistance of the shaft walls near the waterline by increasing the thickness of the wall. Work will normally be done from the inside, by lowering the water level inside below the zone in question.

Shear dowels may be drilled and grouted in. The existing concrete surface should be heavily sandblasted. Additional circumferential and vertical reinforcing steel is secured to the dowels. As the new concrete is placed, a bonding epoxy may be progressively sprayed on the existing wall, just above the level of the fresh concrete.

To enhance the bending capacity of a shaft, the wall of the shaft may be thickened through the region of high moments. In addition to the steps outlined above, vertical ducts are placed in the new concrete, through which prestressing tendons are then installed, stressed, and grouted. If the upgrading is for the purpose of enhancing the ultimate strength only, then unstressed steel bars may be used. These measures may cause a drastic change in the platform's and deck's response, since both the stiffness and ultimate strength are being increased. Hence this work should only be undertaken after a thorough dynamic analysis of the revised structure.

Occasionally it becomes necessary to construct a new penetration through a shaft which is below water level so that there is a head of water on the outside. Work will usually be scheduled for the summer weather window.

The shaft is dewatered inside to below the location and staging placed. On the exterior, a steel caisson with gaskets is affixed to the structure wall by drilled-in anchors. These can be supplemented by taut wraps of wire rope taken up by turnbuckles. The drilled anchors should not extend as deep as the existing vertical posttensioning ducts. See Figure 16.5.1.

Now, working on the inside, the location of the ducts is carefully determined. As-built drawings will give the approximate location, but this should be verified by careful, slow drilling of a small hole, say 10 mm in diameter, since often minor displacement of ducts will have occurred during slip-forming and concreting.

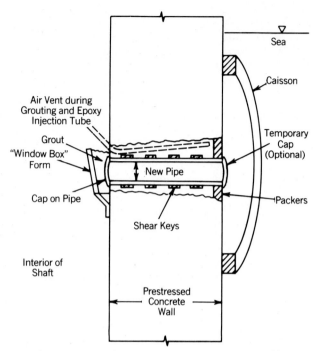

Figure 16.5.1 Installing new pipe penetration through existing shaft wall.

Next a central hole is drilled through so as to drain the water from the space between the external caisson and the wall; this will squeeze the caisson tighter against the wall.

Now, using the central hole as a pilot, the hole for the new pipe sleeve is drilled through the wall. It will have a diameter 10–40 mm larger than the outer diameter of the pipe sleeve which is to be installed.

While it is desirable from all counts to miss the conventional reinforcing steel insofar as practicable, the critical thing is to miss the posttensioning ducts. An occasional cut-reinforcing bar can normally be accepted.

The hole is now enlarged on the outer side so as to be tapered. The walls are roughened. The pipe sleeve, with shear rings attached, is fitted with an external packer or expanding ring. The pipe sleeve is inserted, the packer or ring activated, and the sleeve fixed tight in the center of the tapered hole.

Two small-diameter (3-mm) plastic tubes are fitted, one just below the pipe sleeve, the other just under the top of the hole. A nonshrink cement grout is now pumped in. Using a window box on the inside, a slight positive head of grout is maintained until initial set. Now epoxy is injected through the two small tubes, which are slowly withdrawn. The intent is to fill any small bleed water gap under the pipe or top of the hole. The new pipe is capped on the inside.

Next, a valve in the caisson is opened, so that full water pressure acts on the new penetration. If all is satisfactory, with no leakage, the caisson is removed. A diver can then place an underwater-setting epoxy sealant around the external edges of both pipe and hole, where the packers were.

If leakage is noted, small holes are drilled along the leaking seam and epoxy injected. This may be done even against an external water head.

Gaze down at the black satin waves,
Stars reflected in tracks of silver light.
Look, see, where the waters are foaming,
Sparkling sea-jewels are tossed into sight.

AUTHOR, "SHADOWS OF GOLD"

17
Salvage and Removal

17.1 General

Current regulations of many countries require the removal of offshore platforms and other structures when they have finished serving their purpose and are no longer in use. In most cases, the requirement is that they be removed to a point 2–5 m below the mudline.

Increasingly, the constructor is being required to develop his full procedure for eventual removal at the time of final design and to set it forth in a manual for approval by the authorities.

In most early cases, there has been little prior planning for removal and hence no fittings or details have been built into the structure to facilitate removal. Even where these have been built in, over the years these may have become fouled by corrosion and marine growth and are no longer operable.

Indeed, in many cases, miscellaneous construction during installation or during service may have blocked easy access to carry out the work. Pile-stabbing guides and lifting cones are some examples; drill cuttings and antiscour riprap may be others.

A number of studies are currently underway in both Europe and the United States to re-appraise the regulations and to re-evaluate the need for removal to below the mudline. In some cases, a platform may have continued utility as a lighthouse, radar station, or offshore scientific or educational laboratory or be suitable for conversion to a platform for generating power by wind, waves, or ocean thermal gradient.

Offshore platforms become natural habitats for marine life, furnishing protection and breeding places for a wide variety of organisms. A study of Rincon Island off the coast of California has been very revealing in this regard. Prior to construction of this offshore drilling island, the area had been well documented as a "marine desert" with its featureless bathymetry swept clean and bare by waves and currents. After construction of the island, with its slopes covered by riprap and concrete armor units, a host of organisms, over 2500 varieties, has found shelter, so that the island is now an official Marine Preserve.

It is well known that the best sport fishing occurs around offshore steel jacket platforms; indeed, some smaller jackets which were abandoned have been purposefully reinstalled as artificial reefs. Finally, some structures may be reused as breakwaters or the support for breakwaters.

Disposal of offshore structures must comply with the IMCO international agreement on ocean dumping and with national regulations. Obviously, this means that disposal will not be permitted in waters where present or future navigation might be impeded or where they would interfere with bottom trawl fishing.

Disposal onshore is of course made complex and costly by the draft of these huge platforms and the limited number of areas where structures might be beached for cutting up and disposal.

Three types of structures will be investigated to develop general principles and possible solutions for removal. Obviously any particular platform will have to be addressed in specific detail as to requirements, methods, and control of operations.

Perhaps even more than in initial installation, there are risks involved in removal, risks that during salvage an accident will occur or the structure become unstable, presenting the constructor with a more difficult and costly or even a nearly impossible operation. Therefore, just as in installation, risks must be enumerated and evaluated and contingency plans prepared.

17.2 Piled Structures (Terminals, Trestles, Shallow-Water Platforms)

The first step is to remove all superstructure facilities and equipment. Care must be taken when burning abandoned pipelines and so forth to ensure that they are gas-free. A decision must be made as to whether to work from on top of the structure or afloat. Working on top means a smaller crew, few limitations due to weather or sea state, and inexpensive equipment. On the other hand, lift capacities (weight and radius) are severely limited.

Working afloat gives more maneuverability, the opportunity to make heavy lifts of large sections, and immediate availability of auxiliary and supporting facilities. Daily costs are high, and the rig will be subject to the effect of the seas.

In practice, a combination of the two methods will often be used. Removal of the deck structure, either before or during pile removal, is generally straightforward but requires a careful study of material handling—what will be the sequence and the route for off-delivery of salvaged elements.

Piles may be cut with casing cutters, working from a drill rig. The drilling equipment may be supported on the pile itself, from a derrick boom, or from the mast of a drill rig on deck. The casing cutter uses an expanding bit.

Explosives may also be used to cut off piles. These are shaped charges, lowered down the pile. Positive means must be taken to ensure that they do not get tilted or jammed in the pile; otherwise the resultant distorted pile may be much more difficult to cut by any means.

A third method is to dredge alongside the row of piles, so that they are exposed to the required depth, and then to burn them off using divers and a jet lance.

In some cases, piles may simply be cut free from the structure and pulled. In sands, a large vibratory hammer with a steady pull may be effective, especially if combined with a jetting operation. Impact extractors are available, but generally they are of a size suitable only for relatively small piles such as those used in harbor construction.

In silts and fine sands, overfilling of a pipe pile with water increases the pore pressures and decreases the effective lateral pressure and skin friction. The water may be filled up to the top of the pile. It may then feed to the soil through the open tip. A free jet run to the tip may help to break up a densified and relatively impermeable plug. The pile may first be perforated by controlled explosives so as to facilitate the flow of water into the soil. If the situation permits, a jet may be run down alongside the pile on the outside.

Jacking, especially in conjunction with jetting, has been used. The main problem is finding a support for the jack reaction. Usually this must be the adjacent structure, which then must be checked to ensure adequate capacity.

To break piles loose for subsequent removal by one of the above methods, the pile may initially be driven down a half-meter or so to break any setup adhesion that may have developed. This is especially effective in clay soils. Then extraction may take place.

An ingenious method was developed for removal of the 2-m-diameter × 80-m-long dolphin and temporary trestle piles installed to support the construction of the Oosterschelde Storm Surge Barrier in the Netherlands. The piles were capped with a steel dome welded to the top of the pile. Water was then pumped in and the pressure raised until the pile jacked itself out. When high pressures were required, there was danger of piping developing around the tip of the pile. Therefore, a blinding course of very fine sand was placed, which was of low permeability, hence reducing pressure as the water penetrated below the tip. This blinding course was typically a 1-m deep fill in the pile and often had to be reinstalled several times as the pile moved up.

Air pressure is not used in such cases as it could conceivably lead to catastrophic explosion of the pile cap. However, if the pile is first water-filled, then air can be used to apply the pressure to the water; the volume is now so small as to reduce the potential hazard.

If a pile has been frozen into permafrost, it may best be freed by first breaking up the plug with a high-pressure jet (even if frozen), removing the plug by airlift, and then injecting steam into the water in the pile until adfreeze is broken.

17.3 Offshore Drilling and Production Platforms (Jackets with Skirt Piles)

Jackets with skirt piles are the structures currently of greatest interest. Many of the platforms in the Gulf of Mexico have been in place over 25 years;

the reservoirs are essentially exhausted, and the regulations currently in effect require removal to below the mudline.

The costs of removal will be high. This by itself makes the concept of secondary recovery and continued production attractive.

Following is a scenario for removal which could be applicable to such a platform.

1. Cement and plug wells. Purge and disconnect all risers and flow lines.
2. Remove all deck equipment and facilities. Remove module support structure above cellar deck.
3. Set drill rig and crane on deck.
4. Cut and remove conductors and attached casing.
5. Using drill string, with or without a casing, enter each skirt pile. Cut pile below mudline with a casing cutter–type tool. Repeat for all piles. Structure is now resting on mud mats and partially on cut piles where they bear.
6. Jet around mud mats and lower bracing. Wash off any buildup of soils on mud mats or bracing.
7. Attach temporary buoyancy tanks, either near the waterline or at the top of the skirt piles, using an internal expanding plug.

   If we assume a 100-m water depth and a 5-m spacing of skirt piles, a temporary buoyancy tank slightly less than 5 m in diameter × 100 m high will have a net buoyancy of about 500 tons (5 MN). If we attach four such tanks at each leg (or one quad unit at each leg), then we have 80 MN or 8000 tons of uplift available, about the weight of the typical jacket plus the portions of the grouted piles in the sleeves.

   Once each tank is connected, it can be secured against the adjacent jacket leg.

   Each such temporary tank will have been floated out, upended by ballasting, moved to position at the leg, and ballasted down to mate with the pile using a mating cone. Now the gripping device will be lowered through the central sleeve and hydraulically activated to grip the pile.

   The tanks must be so located and sized that when the jacket finally is fully afloat, it will assume an acceptable attitude and draft.

8. When all temporary tanks are in position, the crane and drill rig are removed. The cellar deck is removed.
9. Make fast towlines from tug so as to exert a forward pull.
10. Using compressed air, deballast all temporary buoyancy tanks. Provide positive vents at lower end of tank so that overpressurization cannot occur.

    The structure should now float free.

    British Standard BS 6235 warns that dewatering by air injection requires attention to detail and control systems similar to submarine practice, since the consequences of failure of the system, particularly with large volumes of compressed air, may be disastrous. These risks include free-surface effects, rapid expansion of the air volume as the structure rises, and, in the event of a crack, explosive propagation and rupture.

11. If the temporary tanks will not give enough buoyancy, the jacket itself may be given some buoyancy by capping the top of the legs, plugging all vents in braces (by diver), and injecting compressed air.

    Note that 100-m depth requires 10 atm or 150 psi air pressure.

12. Tow structure in vertical mode to disposal site.
13. Cut all connections of tanks by explosive cutters, actuated acoustically. Tanks will now float free, ready for recovery, while jacket sinks to the seafloor.

The tanks assumed above will have a total weight of about 20 × 100 or 2000 tons, which at current (1986) fabrication prices will be about U.S. $300,000–400,000. To these must be added the 20- to 500-ton hydraulic grippers designed to expand and lock against the walls of the pile. No attempt is made to salvage these since they will often become jammed in the pile.

An alternate to the above, if permitted by the authorities, is to attach fewer buoyancy tanks, on one side only. Then a pull by lines from the derrick barge can tip the platform over and allow it to rest on its side on the seafloor. Explosive charges can be detonated to crumple the legs on one side.

The lines must be capable of free overhaul so that they will pay out freely as the jacket falls.

In lieu of the derrick pull, two or more linear hydraulic jacks, remotely operated, can be installed on deck with lines leading to preplaced anchors. In this case the jacks will be expendable.

Once the jacket is lying on its side on the seafloor, with only a moderate net weight (moderate negative buoyancy), then it can be dragged on the seafloor to its final disposal site, just as a pipeline or plow is dragged.

Many of the shallow-water platforms have jackets that weigh only a few hundred tons. Therefore, after step 5 the derrick barge can rig slings to the jacket and pick the jacket up off the seafloor. Tag lines are attached to snug the jacket in against the stern of the barge, and the jacket is transported to shallow water or to shore for disposal.

If desired, the jacket can be placed on a barge for transport provided that suitable cribbing is placed to distribute the load on the barge deck so that the protruding stubs of piles will not punch through the deck.

17.4 Gravity-Base Platforms

As noted earlier, gravity-base platforms have primarily been built of reinforced and prestressed concrete, although several are of steel and some recently constructed GBS platforms are of hybrid steel–concrete construction.

These platforms are characterized by large base caissons, the latter having originally provided flotation during transport and installation. A major concern in removal is that of excessive breakout force due to soil shear strength and cohesion and adhesion of soil with the base as well as with skirts and dowels.

The basic concept in salvage and removal is to refloat the structure and then tow it to the disposal site. A typical procedure might be as follows (see Fig. 17.4.1):

1. Cement and plug wells. Purge and disconnect connecting flow lines and risers. Cut and re-

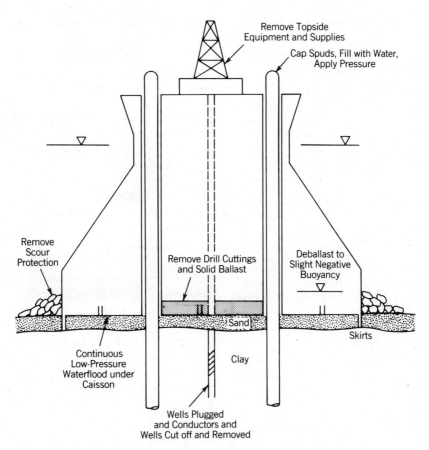

Figure 17.4.1 Removal of gravity-base platform.

move conductors and attached well casing. Plug conductor holes in base with concrete plugs. Cut piping connections loose and plug penetrations by means of concreting (grouting) or steel caps.
2. Remove deck modules and deck structure.
3. Salvage any desired equipment from utility and riser shaft.
4. Remove exterior ballast walls near base (cut posttensioning ties), allowing ballast to spill out over seafloor. Plug any openings which may have been formed or cut for piping penetrations and so forth.
5. Remove solid ballast in interior that may have been placed after installation on site.

    This ballast will probably have been either sand or slurried iron ore. This can be removed by airlift or eductor or specialized equipment such as Marconaflo pumps, which have jets incorporated to agitate the material and slurry it, facilitating its removal by airlift or eductor.

    The extent of removal depends on the computation of weights and availability of access to compartments. It would of course be desirable to install access sleeves or manholes in the platform at the time of original construction so as to facilitate this operation.
6. At this stage, the ballast compartments are fully flooded.

    Using pipes leading to the underbase, inject water underneath at a low, steadily maintained pressure, slightly above ambient at the base elevation.

    The pressure must be low enough that it cannot cause piping under the skirts. Once piping occurs, little additional benefit can be attained by underbase waterflood.

    Maintain pressure for up to 24 hours.
7. Deballast caisson to a slight positive buoyancy. If structure does not break free, deballast one side more than the other so as to tip caisson off. Once caisson breaks free at one edge, water will enter and break all suction.

    Limit deballasting to the point where, if structure breaks free, it will not rise above the level at which it is still fully stable.

    This can be a very critical stage because some excess positive buoyancy must be provided to extract the dowels. Once the structure breaks free, it will rise until equilibrium is reached. A special check must be made that at this stage the structure is still stable.
8. If, for any reason, piping does take place under one of the outer skirts, suspend the operation and inject grout into the local area affected so as to prevent loss of waterflood overpressure. Allow the grout to harden, and then resume operations.
9. Tow structure to disposal site.

When an exploratory drilling platform of the GBS type is being refloated for relocation, then normally the deck and equipment will not need to be removed. Thus steps 2 and 3 are not applicable in this case.

Compressed air is generally not desirable and may be dangerous. First, if used under the caisson to help overcome suction, the high-pressure air will tend to escape laterally, leading to piping. Secondly, if compressed air is used internally to expel the water, as the structure rises, the external head decreases. The air then expands more. Thus the structure tends to rise much further and faster than planned.

Thirdly, this expanding bubble of air creates a free-surface effect, traveling to the high side, where it exerts even more upward force, developing an overturning moment.

In ship salvage, the use of compressed air has led to disastrous results. Excess air is pumped in in order to overcome the suction. The ship breaks loose, and rises up. The air bubble expands. The ship accelerates. It rises to the surface, but now the air bubble is all on one side and the ship turns over, the air escapes, and the ship plunges back to the bottom!

In the case of the jacket, where air is used to expel water in steps 8 and 9, the air bubble is confined in vertical pipes or tanks, so stability is not a problem. Total volume was limited by the vents in the temporary tanks, so the structure could not rise excessively high after breaking loose.

It is obvious that extremely careful calculations are needed, taking into account not only the original structure's weights, displaced volumes, and so on, but also changes that have occurred since, such as:

1. Marine growth
2. Drill cuttings stored inside

3. Weight of underbase grout which sticks to base
4. Setup of soil on skirts and dowels, increasing extraction force
5. Ballast or dropped material on caisson roof

In addition, the structural adequacy of various critical members must be looked at carefully, since the forces exerted by buoyancy are extremely great. Corrosion or physical damage may have occurred over the intervening years.

17.5 New Developments in Salvage Techniques

New techniques are rapidly being developed for salvage, many of which are applicable to the removal of offshore platforms. Guidance can be obtained from ship salvage experts.

One of these new techniques is the use of foam instead of compressed air. Foam has the advantage of displacing a fixed volume of water, with no change as the structure rises off the seafloor. Polyurethane foams are relatively low in cost but have depth limitations of about 100 m. Syntactic foams can be designed for performance at great depths, but become increasingly expensive.

Expandable neoprene balls can be used to seal tubular members, facilitating emptying or injection of foam.

Inflatable neoprene buoys are available which can be attached at many locations so as to reduce the net underwater weight. An attempt was made to salvage the damaged Frigg DPI jacket this way, attaching inflatable buoys. On launching, the temporary buoyancy tanks in the pile guides of the jacket had collapsed, causing it to sink to the North Sea seafloor. Unfortunately, during salvage operations, a summer storm tore many of the buoys loose. Working conditions at the site were very difficult, so the salvage effort was abandoned. The jacket was finally half-floated, half-dragged to a deep-water disposal site.

Who can say of the sea that it is old?
Distilled by the sun, kneaded by the moon,
It is renewed in a year, in a day, or in an hour.

THOMAS HARDY, *The Return of the Native*

18

Constructibility

18.1 General

The construction of offshore platforms has been a heavily cyclic industry, responding almost frantically to the discovery of new oil provinces such as the North Sea or significant changes in price level, such as that which followed the OPEC oil embargo of 1973–74. During intervening periods, as the market stabilized, the industry matured and became more orderly construction-wise and cost-wise. By the time the next boom arrived, technology had also changed, requiring a learning curve before the industry as a whole was geared to the new demands.

The result has been very significant cost overruns and schedule delays in the early periods of each new development, followed by a gradual steadying as estimates rose to meet actual data and competitive forces brought costs down closer to their targets.

This has been demonstrated in the North Sea. In the late 1970s cost overruns of almost 100 percent occurred. By the mid-1980s a stable industry was turning out both steel and concrete platforms within the estimated costs and time schedules. Currently, this same cycle appears to be running its course in the Arctic and the deep water oil provinces.

The cost of structures has a significant influence on the viability of the offshore development, largely because it is an early capital expenditure. Similarly, the time lag between structure expenditure and oil income has tended to increase rather dramatically as the projects have moved to environmentally more demanding areas—deeper water, more remote locations—and with greater ecological, social, and political constraints. This is especially true in the Arctic, where the cost of exploratory drilling platforms is very high and environmental restrictions plus winter ice limit the number of holes that can be drilled each year.

Pipeline construction offshore and onshore is extremely capital-intensive and requires a large continuous throughput to justify the expenditure. In many locations (e.g., the Arctic) the line must be deeply buried; in others it is required to be covered with rock. Indeed, the total capital investments are approaching the level where many projects which are technically feasible may remain economically infeasible.

Engineers and constructors involved in the design and construction of structures can influence the cost of their portion of the development by sound design, construction planning, and construction management. These latter two aspects are germane to the subject matter of this book and can be encompassed by the single term *constructibility*.

In an offshore production platform, while the structure may run several hundred million dollars, it still typically represents only 20 percent of the total cost of development. The other aspects are the production and processing facilities, the drilling, and the transport facilities. Long trunk pipelines may easily match the cost of the offshore platforms and their associated items. In the Arctic the proportions may be somewhat altered, with the platform becoming relatively more important because of the ice loads which it must resist.

Nevertheless, the offshore structures per se are not usually the major cost of field development. However, they are a significant part of that total cost, and they are usually on or close to the critical path in schedule. They comprise one area where sound innovative design, competent construction planning, and competent construction management can achieve meaningful savings.

Constructibility encompasses the concept development, the integration of design with construction, the selection of construction methods, facilities, and stages, the procurement and assembly of materials and fabricated components, the organization and supervision of the work, and the training of workers. It includes analysis and planning, quality control and assurance, safety engineering, cost estimating, and budget control.

It also includes an item of special concern to offshore structures: weight control. It addresses personnel and material transport and access, craneage and lift planning, and the furnishing of lighting and utilities. Constructibility employs work simplifications and standardization techniques in order to overcome the difficulties inherent in complex and sophisticated construction in an offshore environment. Finally, its scope includes deployment, installation, and subsequent removal, relocation, or salvage.

In the following sections, specific aspects of constructibility will be addressed.

18.2 Construction Stages

An offshore structure goes through a series of very distinct stages as it moves from fabrication to offloading (or floatout), to completion afloat, to transport, to installation, and to module erection and hookup. The stages pertinent to each of the various types of structures have been described in some detail in the previous chapters. In constructibility planning, it is essential to formally set these stages forth by title, description, and schematic drawing. See Figures 18.2.1 and 18.2.2.

Obviously the first cut will deal with major stages of construction. Each of these major stages can then be subdivided into the detailed stages required.

The stages should be further portrayed by a series of appropriate drawings or sketches; isometric drawings have been found extremely useful. The drawings should be essentially outline in character, with key items pertinent to that stage shown in heavy lines. The purpose is to eliminate aspects not essential to that stage so that the key elements can be clearly recognized. Thus, while they are based on engineering design drawings, they differ from them in emphasis, clarity, and use. Computer-aided design (CAD) is especially effective in enabling three-dimensional portrayal of the successive stages. See Figure 18.2.3.

Experience in the preparation of such descrip-

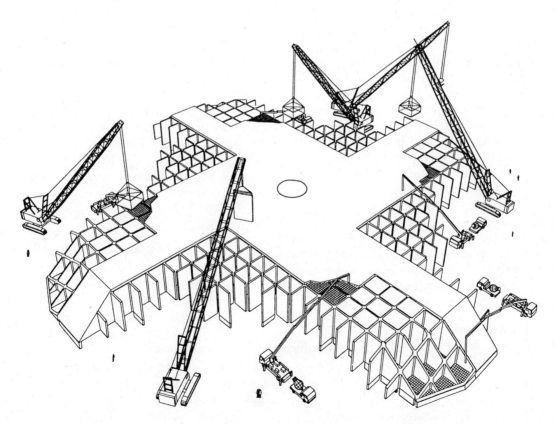

Figure 18.2.1 Construction stage planning diagrams. (Courtesy of Kajima Corp.)

406 Constructibility

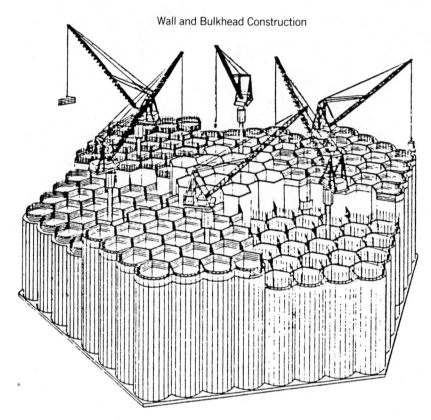

Figure 18.2.2 CAD applied to constructibility planning. (Courtesy of Kajima Corp.)

tions and drawings has shown that serious errors have occurred due to jumping past intermediate stages which have been incorrectly assumed to be unimportant or self-evident. The entire purpose of constructibility planning is negated when this happens, because it is just these skipped stages that so often turn out to be critical.

Once the constructor is satisfied that all the stages have been set forth, then engineering evaluations can be made of each such stage to ensure proper structural, geotechnical, mechanical, and hydrodynamic performance. As was noted in the chapters on steel and concrete structures and embankments, many elements are subjected to higher forces and stresses during these construction stages than under the design environmental loads. Examples are:

Steel piles during driving
Pipeline bending and radial compression during installation
Legs and bracing of steel jackets during launching
Cell walls of gravity-based structures during deck mating

For many of the stages, the key issue will involve the interaction of two or more disciplines. For example, ballasting by means of mechanical systems is intimately related to the structural capacity under differential heads, the stability performance afloat, and the instrumentation with its real-time readout.

Key considerations which have been inadequately addressed in the planning of previous structures include:

1. Draft, with relation to available water depth
2. Stability during all stages of installation
3. Hydrodynamic response of structure during tow, especially acceleration forces acting on mechanical installations
4. Effect of pressure and temperature changes on function of instrumentation, valves, and minicomputers
5. Initial contact with seafloor and the interactive effects of trapped water trying to escape
6. Breakout forces from the seafloor and resultant draft and stability when relocating or removing the structure

7. Snap loads of mooring lines due to stored energy from long-period excursions; use of fairleads and sheaves of too small diameter
8. Effect of shallow water and minimal underkeel clearance on wave characteristics, structure or vessel response, squat, yaw, and seafloor scour
9. Control of draft and stability in event of ruptured ballast line, jamming of valve, or carrying away of bulkhead, allowing internal flooding
10. Human error in ballasting control—adoption of controls, training, and system isolation as needed to prevent catastrophe
11. Arrangement of lines and umbilical control cables to prevent fouling during critical operations
12. Inadequate weight and tolerance control during fabrication
13. Inadequate consideration of tolerance in differential heads of ballast water in compartments

The division of the project into stages and the subdivision of each stage into actual steps is a procedure by which the most efficient method can be selected for each step. Sound judgment and experience will tend to integrate closely related steps within each stage. However, the limitation of such an approach is that the "forest may be obscured by the trees." Therefore, a conscious overall evaluation must also be made from the holistic point of view to ensure coordination and integration of all steps and stages. In the hands of an experienced constructor, such an overview may result in incisive decisions as to the program and direction of the work.

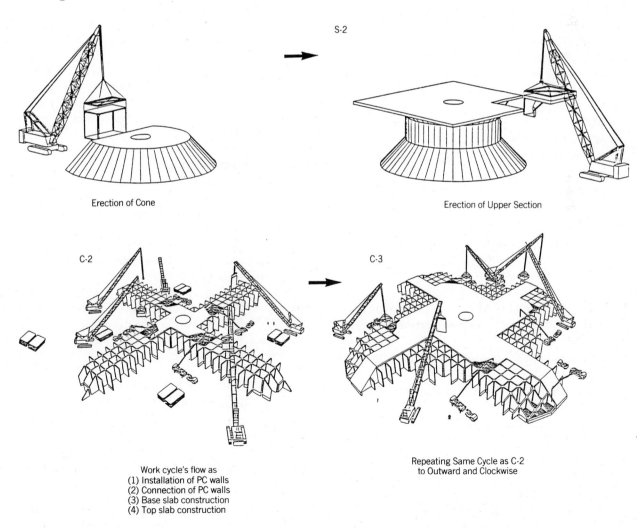

Figure 18.2.3 Stages of construction portrayed by CAD schematics. (Courtesy of Kajima Corp.)

In offshore construction, however, with its revolutionary developments in equipment, tools, and instrumentation, with its new structures and systems and environments, specific experience may not exist. Instead of relying solely on intuition, therefore, the conscious use of constructibility planning and evaluation of stages should lead to a more rational and effective program.

18.3 Principles of Construction

Some of the principles which can be beneficially applied to reduce the time and cost of construction are:

1. Subdivision into as large components and modules as it is possible to fabricate and assemble.
2. Concurrent fabrication of major components in the most favorable location and under the most favorable conditions applicable to each component.
3. Planning the flow of components to their assembly site.
4. Providing adequate facilities and equipment for assembly. This may include such items as synchrolifts, heavy-lift cranes, both land-based and barge-mounted, drydocks, basins, and so forth.
5. Simplification of configurations.
6. Standardization of details, grades, and sizes insofar as practicable.
7. Avoidance of excessively tight tolerances; provision for flexibility and adjustment in connections, especially in mechanical system piping.
8. Selection of structural systems that will utilize skills and trades on a relatively continuous and uniform basis.
9. Avoidance of intermittent peaks in the demand for the labor force; selection of construction methods that involve relatively uniform demand.
10. Avoidance of procedures that are overly sensitive to weather conditions; arranging shop prefabrication and painting of elements which are very sensitive.
11. Mechanical systems to be incorporated in or on the structure should be modularized into the largest possible components, even if this requires additional structural support or interruption of the construction of the structure proper.
12. Construction methods should be selected which are appropriate to the specific structure. Avoid fixation on only one method—e.g. concrete pumping, slip-forming, welding, barge launching. Be versatile in choice of methods.

18.4 Facilities and Methods for Initial Fabrication and Launching

In almost all cases, the early stages of construction are carried out at a shore base. This base may be purpose-built for this one project or may be a relatively permanent facility.

The area for such a facility must be adequate to accommodate not only the structure and/or components themselves but also storage of materials, access roads, support buildings, and infrastructure facilities.

Offshore structures are typically large in scope and will require a large number of men over a substantial period. Therefore, it will usually prove economical to expend the effort and money to build first-class facilities—that is, with proper surfacing, roads, structures, utilities, and so forth—to enable men and equipment to work efficiently.

The work will almost always go on around the clock; therefore, adequate lighting is required. The work will almost always continue even in inclement weather; therefore, adequate enclosures must be provided, as appropriate to the work, especially for welding and painting, with adequate change rooms for the workers.

By definition, the structure will move from the onshore yard to afloat, either self-floating or on a barge. This requires bulkheads, dredging, and dolphins adequate to ensure the safe transfer of the structure to the water-borne mode.

A number of ingenious methods have been developed to facilitate this movement from onshore to offshore. Some of these are briefly described below.

18.4.1 Construction in a Basin, with a Temporary Closure of Earthfill, Steel Sheet Piles, or Caisson Gates (See Figure 18.4.1.1).

Relatively permanent basins usually use caisson gates as the closure, since they enable rapid removal

18.4 Facilities and Methods for Initial Fabrication and Launching 409

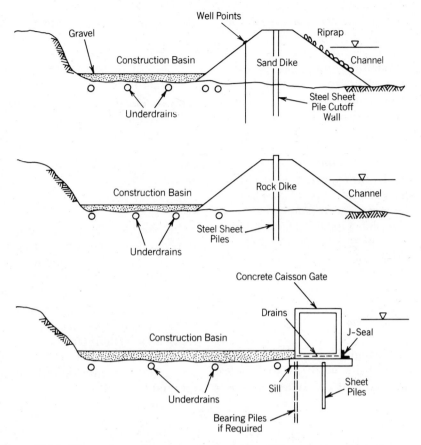

Figure 18.4.1 Three schemes for construction basins for offshore structure fabrication.

and reinstallation. This system has been used for the construction of the offshore terminal structures in Queensland, Australia, for the steel jackets for the North Sea at Nigg Bay, Scotland, and for the offshore platforms at Loch Kishorn, Scotland.

Steel sheet piles in a rock dike have been used for the Condeep production at Stavanger, Norway.

Sand dikes with a steel sheet pile cut off and well points were used to construct the 66 piers of the Oosterschelde Storm Surge Barrier in The Netherlands.

18.4.2 Launching From a Ways

Very large and heavy structures have been launched from building ways—for example, tankers and subaqueous tube segments. Side launching usually results in much lower structural stresses. However, it is essential that the launch be uniform and that one end not hang up or lag behind the other.

End-0 launching produces high bending moments as the stern is picked up by the buoyancy of the water. The bow meantime is transferring very concentrated loads onto the ways, and in turn the bow itself is experiencing very heavy concentrated forces.

Jackets are usually launched end-0 because of their tapered configuration. However, the Lena guyed tower, with its rectangular cross-section, was successfully side-launched, and as noted in Chapter 9, recent Japanese studies indicate that it may be applicable for tapered configurations as well.

The prestressed concrete floating phosphate plant Rogamex was constructed on ways in Singapore and side-launched.

18.4.3 Sand Jacking

In the as yet rarely used method of sand jacking, a basin is excavated by dredging, keeping the basin full of water. The basin is then filled with sand up to a working grade. A temporary rock surfacing is placed.

410   Constructibility

The structure is now constructed at normal yard grade, with full access.

When it comes time to launch, the sand is excavated by suction from under the structure, using jetting as necessary to promote horizontal flow of the sand so that relatively uniform load distribution occurs. Stresses in the structure are continuously monitored, as are excavated depths along the structure's sides. Appropriate adjustments are made in the sand removal operations.

When fully excavated, the structure floats free and is towed out. The sand fill can now be replaced.

This method eliminates the problems involved in dewatering a basin while enabling all work to be carried out at yard grade. See Figure 18.4.3.1.

Jetting and eductor piping may be preinstalled in the sand fill to facilitate dredging and promote flow.

18.4.4 Rolling In

Large-diameter piles, cylinders, and tubes may be launched by rolling down a ways. As with side-launching, it is essential that the cylinder move down parallel to the shore and that one end does not hang up.

18.4.5 Jacking Down

The advent of modern hydraulic lift systems enables modules, for example, to be skidded out over the barge slip on girders and then lowered down onto barges. Such a facility is especially well fitted for repetitive loadout operations.

18.4.6 Barge Launching

Many sizable offshore structures have been constructed in segments on a large barge or in a floating drydock. The segment is then launched (or floated off).

This system is especially well adapted to the launching of a subsea template. The template may be assembled on the barge. Alternatively, it may be fabricated onshore and skidded onto the barge for transport and launching.

During the launching, the barge is usually submerged by flooding. In many cases the main body

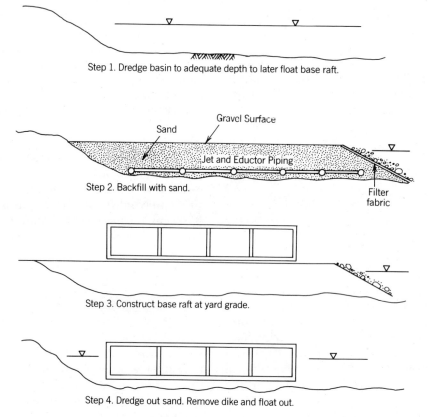

Figure 18.4.3.1 Sand-jacking method of construction and launching.

of the barge is completely submerged so that the structure can float directly off. Stability during and after the submergence and launching becomes a major concern.

As the deck of the barge goes underwater, the waterplane area is now reduced to that provided by the structure on deck. At this critical stage, the center of buoyancy is essentially the geometric center of the barge. The center of gravity of the combined system (barge plus ballast plus structure) typically is still quite high, so the righting moment furnished by the waterplane is very important. It no longer is that of the barge but now is limited to that of the structure. The free-surface effects of the water ballast used for submergence must also be taken into account. To overcome these, it is usual practice to have some compartments topped up, others empty, and only a few with a free surface. The structural effects of such unequal loading must then be considered.

At deeper submergence the structure starts to lift off. Now its waterplane no longer assists the barge in maintaining stability. Accidents have occurred in which the barge rotated uncontrollably during this stage.

To provide stability control, the barge is usually fitted with columns at one or both ends, which give enough waterplane moment of inertia to provide stability.

These columns also enable the draft of the barge to be controlled. See Figure 18.4.6.1.

The barge is of course subjected to an external hydrostatic head in excess of that normal to conventional barges. Obviously a specially designed barge is called for, or else a standard barge must be modified by internal strengthening as necessary and by sealing of the vents and other deck fittings. Heavy-duty submersible barges used for the ocean tow of jack-up rigs, dredges, and so forth may be available for the construction.

Use of compressed air inside the barge to offset the hydrostatic head is only practicable provided that sufficient internal compartmentation exists to avoid excessive free-surface effects, and that the sides and deck(s) are strong enough to resist the air pressure.

18.5 Assembly and Jointing Afloat

Jointing afloat of large structural units is now a well-established art. The Hondo steel jacket, over 280 m in total length, was constructed in two segments,

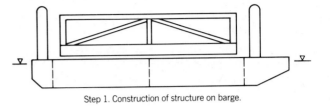
Step 1. Construction of structure on barge.

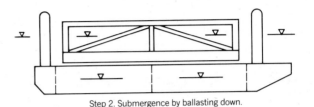
Step 2. Submergence by ballasting down.

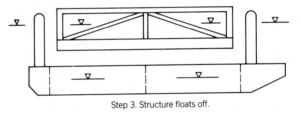
Step 3. Structure floats off.

Figure 18.4.6.1 Barge launching of subsea template.

which were joined while afloat in partially protected waters. Mating cones, hydraulic ram locking devices, and internal welding were used to provide full structural continuity.

A large concrete floating drydock was constructed in northern Spain by first fabricating barge-size segments on a barge, launching them, and then joining them by prestressing and tremie concrete.

The hulls of large tankers have been constructed in two halves and then launched and joined together while afloat. A temporary cofferdam was used to enable dewatering. Heavy rods and bolts held the sections in alignment for welding.

Subaqueous tunnel sections and large outfall and intake tunnel sections have been joined underwater by various combinations of bolting, prestressing, and underwater concreting and grouting.

The Valdez Floating Container Terminal was built in two 100-m-long sections and then towed 1000 miles to the site and joined by concreting and prestressing the joint.

The Hood Canal Floating Bridge was similarly constructed of large concrete sections which were joined at the site by grouting and prestressing.

In general, the principles applied were the following:

1. Use of large mating cones and sockets for intial positioning
2. Some means of temporarily fixing the sections relative to each other with an essentially rigid connection so as to prevent differential movement in any of the six degrees of freedom
3. Sealing the joint zone so as to make it watertight
4. Dewatering the joint zone
5. Permanent jointing by bolting, welding, or prestressing plus concreting, grouting, or epoxy injection
6. Permanent sealing against water in-leakage

18.6 Material Selection and Procedures

The design will of course have determined the specifications for the materials based on their performance in service. Constructibility considerations will now go further, as the constructor addresses the practicability of building the structure to meet the specifications.

With steels, for example, welding procedures and materials are intimately related to the ambient temperature at which the work will be carried out. The constructor has an opportunity to optimize these by one or more of the following steps.

1. The constructor may elect to carry out the majority of the welding within protected, heated, and dry enclosures.
2. The constructor may elect to use preheat and/or postweld treatment to attain the required results.
3. The constructor may elect to purchase specially processed steels which are less sensitive to the conditions.

With concrete structures, the constructor has even more alternatives from which to select the optimum combination. He may increase the cement content in order to gain workability and early strength. He may use a superplasticizer admixture to improve workability and strength and lessen the need for vibration. He may add air entrainment to improve workability and prevent segregation. He may include condensed silica fumes in the mix to increase strength.

The timing and sequence of addition of the various components of a concrete mix have a decided effect on its properties. For example, air entrainment should usually be added at the end of the mixing cycle.

Aggregate selection and gradation may be modified. Surface characteristics, absorption, strength, and thermal properties are all-important parameters.

To control heat of hydration and thermal gradients (and hence cracking), aggregates may be precooled, ice may be used instead of water, and cement type may be changed. Pozzolan or blast furnace slag may be used to replace a portion of the cement.

The method of delivery and placement of the concrete affects its quality. Pumping of concrete, for example, compresses the air entrainment bubbles. It forces water into absorptive aggregates, thus gradually stiffening the concrete mix.

The curing of fresh concrete and the insulation provided to the forms and freshly exposed surfaces are of great importance in preventing shrinkage and thermal cracking and ensuring durable concrete.

With embankment materials, the in-place density and side slopes are very sensitive to the gradation (fines) and the method of deposition. Overflowing of dredged materials may be effective in reducing fines. Sometimes materials may be blended from two or more sources in order to obtain an optimum mix.

Soil and rock materials may be deposited underwater in a variety of methods: dumped as a mass, placed through a tremie tube, discharged hydraulically at the surface, or discharged at the seafloor through a specially designed separator.

The constructor may have to decide between greater care (and cost) in placement and a supplemental operation of densification by vibration.

18.7 Construction Procedures

Within the context of each construction stage, suitable procedures have to be developed so as to meet the following criteria:

1. Compliance with the specifications and drawings
2. Assurance of meeting quality requirements
3. Ability to meet schedule requirements

18.7 Construction Procedures

4. Adaptability to equipment, facilities, and skills available
5. Economy in overall performance: lowest possible costs consistent with items 1, 2, and 3
6. Minimum risk of accident or delay

Each major operation within each stage is analyzed as to the most efficient way of construction.

Since the two largest expenditures within the control of the constructor are the fabrication of the structure and the hookup of mechanical facilities, the principle attention insofar as efficiency is concerned is directed to these two phases. However, the phases involving heavy lifting, loadout, launching, delivery, and site installation, while not heavily labor-intensive, are controlling from a technical and equipment viewpoint, so attention must be directed to them also to ensure technical performance and safety.

Thus the focus in the procedures for the different stages of the work differs, in one case being directed to efficiency, in the other to equipment selection and technical performance.

Evaluating the procedures and selecting methods is essentially a series of sub-optimizations. The constructor temporarily isolates each, placing boundaries at each end of the stage, and develops the most efficient methods for that stage.

Due to the immense amount of work in fabrication, whether steel or concrete, the approach should follow the same logic and patterns as those used in the Japanese shipbuilding industry. There the work is broken down into as many sub-units as practicable. Each is then fabricated in the most favorable attitude (often upside down) and under the most favorable conditions. Advantage is then taken of the great advances in transport and lifting gear to move large components to the assembly site.

Since offshore structures are assembled on or near the water, this opens the opportunity for wide dispersion of the fabrication site for the components and their subsequent transport by water.

The assembly proper can then be afloat in a sheltered location, in a drydock, graving dock, or basin, or on land at a launching facility. Use is made of heavy lift transporters of several thousand tons' capacity, sheer-legs crane barges up to 3000 tons' capacity, of syncrolifts capable of handling 50,000-ton components, and overhead gantry (bridge) cranes of 600 or more tons' capacity.

Large crawler cranes and crane barges can be used together, in parallel, to raise the complete sides of steel jackets or to lift huge modules. Obviously, very close coordination and control will be required. The planning must consider the changing distribution of loads and radii as the lift takes place.

Final assembly is facilitated by having detailed the fitup so that the connecting pieces are automatically guided to exact location.

Obviously, accuracy is essential. The detailed engineering must consider the effects of thermal differences and of distortions due to dead weight in each of the different attitudes.

With regard to the fabrication of steel tubular components, the decision must be made as to where to place the junctures, whether at nodes or in mid-length. Since the nodes are three-dimensional, fit up is usually much more difficult there. If the juncture is in the mid-length of a tubular, the nodes can be first erected to their correct position, the tubular "pup" cut to exact length as measured in the field, and the girth weld readily made. However, this then involves two more joints. Another system is to precut and contour one end, allowing the other to run long. After the first has been welded in place, the other end is field-cut to length.

The most modern yards now have computer-controlled cutting and beveling of the members, which ensures exact fit at the nodes.

With regard to a jacket or large module frame, how should subassemblies be selected? Should the jacket be split into its several panels for component fabrication, as is extensively practiced at the McDermott and Brown and Root yards on the Gulf of Mexico, or should it be split into three-dimensional space frames, as used by NKK for the North Rankin platform?

With concrete structures, several decisions have to be made. Will all elements be cast in place or will some or any be precast? For the Ninian central platform, several hundred concrete shell units were precast in southern England and then transported to northwestern Scotland and erected by sheer-legs crane barge. Precast cellular internals were combined with cast-in-place external walls in the Super CIDS platform. See Figures 18.7.1, 18.7.2, and 18.7.3.

While precasting offers opportunities for dispersion and sub-optimization, it also requires consideration of lifting capabilities and joint details.

The next decision is in regard to cast-in-place concrete. Should slip-forming be used or panel

Figure 18.7.1 Concrete mid-body of Global Marine's Super CIDS "GBS-1" being constructed in NKK graving dock at Tsu, Japan.

forms? Slip forms have been very successfully used on the Condeeps, for example, but require a large surge in manpower during short periods of time. Will the climbing rods and yokes interfere excessively with embedment and reinforcement placement? Panel forms enable the reinforcing installation to be carried out at different locations from the concrete placement, facilitating dispersion and equalization of manpower requirements, but requiring more construction joints.

A third major decision is the method to be adopted for concrete delivery. Will it be by pumping, as used on the Seatank and Doris structures, or by bucket, as used on the Condeeps?

Most of the manpower requirements on a concrete offshore structure relate to the reinforcing

Figure 18.7.2 Details of construction of prestressed concrete caisson for Super CIDS.

Figure 18.7.3 Concrete mid-body of Super CIDS is floated out of construction dock for mating with steel base and deck sections. (Courtesy of Global Marine Development Corp.)

installation. Should the bars be handled individually, as is practiced on slip-forming operations, or should preassembled cages be used? How should splices be made: by lap, weld, or mechanical connectors? Will color coding of reinforcing and prepackaging of the reinforcing bars for individual zones save time for the placing crew?

In particular, should the stirrups, of which so many are required in the typical offshore structure, be bent bars, closed, prewelded loops, or mechanically headed bars?

A very important role can be played in decision making for critical fabrication operations by use of mockups. Full-size sections of the structure, whether intersecting tubular nodes or the juncture of concrete shells, are selected. They are then fully fabricated, with all inserts, posttensioning ducts, reinforcing steel, stiffener plates, and so on. This mockup enables the visualization of the interaction of the many details and the practicability of welding, steel placement, and concreting.

Such mockups have invariably proven their worth, especially if carried out by some of the individuals who will be responsible for their subsequent construction in the field.

Returning to the sub-optimization process, many operations during the fabrication and erection will have to be carried out at high elevations, 50–100 m above the base and land. Since the workers will require staging, can this be pre-attached before erection? See Figure 18.7.4.

If precast concrete or steel components are to be erected, what can be done to facilitate their initial setting quickly and accurately? On the Ninian central platform, bearing plates with screw adjusting nuts were set in the previous concrete pour. A survey crew then surveyed each plate accurately, adjusting it to the proper level and scribing the exact location on the bearing plate where the bearing of the shell should sit.

Before each shell element (200–300 tons) was lifted, tag lines were attached. The workers on top could then catch the tag lines and guide the shell unit to the proper location without waiting for further survey checks.

The mockups also serve a valuable purpose in training the workers, especially if they are shown the results of their work. For example, if placing concrete among congested reinforcement results in honeycomb and rock pockets, they will visually see the need for vibration. If welding studs with excessive heat results in plate warping, they will understand the reason for the apparently excessive shifting of location and imposed time lags.

One lesson from Japanese shipbuilding practice that has been demonstrated repeatedly in their construction of offshore equipment and structures (see Figs. 18.7.5 and 18.7.6) is that teams of workers, comprised of several trades and assigned to a specific task group with the objective of completing all work within a specific zone, are much more productive than a highly centralized organization by

416 *Constructibility*

Figure 18.7.4 Prefabricated staging was attached to these precast segments before erection—Ninian central platform.

trades, with each trade then responsible for all work within its classification throughout the structure. Their analysis of worker productivity has shown that the most important elements are:

1. Good access and adequate workroom
2. Favorable position for working
3. Ability to pace one's own work without excessive dependence on the progress of fellow workers
4. Immediate availability, close to hand, of tools and materials
5. Clearly defined work program and procedures
6. Identification of the individual worker as part of a team

The adoptions of the "zone" concept or organization, as opposed to the "trade" concept, is a return to decentralization as opposed to centraliza-

Figure 18.7.5 Japanese system of modular construction was utilized in fabrication of sub-base of Super CIDS Arctic exploratory platform.

Figure 18.7.6 Steel caisson Molikpaq nears completion of fabrication in Japan. It will later be towed to Beaufort Sea of Canada where it will be seated on the seafloor and filled with sand. (Courtesy of Gulf Canada.)

tion. Since basic skills and techniques continue to be implemented by trade, the new approach resembles the matrix system of organization which has been adopted by some large engineering organizations.

The task group teams will of course be reassigned and reconstituted as the needs demand. Even where specialist subcontractors are involved (e.g., for posttensioning), the task group organization appears to give greater overall efficiency and reliability.

Construction procedures offshore are planned with primary consideration being given to the sea states and weather conditions under which they will be performed.

Chen and Rawstron in their paper "Systems Approach to Offshore Construction Project Planning and Scheduling," *Marine Technology Journal*, October 1983, make use of advanced simulation techniques in the planning of offshore construction operations. Limiting sea states are determined for various operations such as module lifting, pile driving, pipe laying, and saturation diving, and the effects of vessel motions on operations are evaluated.

From such analytical techniques, the duration of operations, adequacy of equipment, sequence of work, and risk of delay or cost overrun can be evaluated.

Construction offshore must similarly be planned stage-by-stage in order to ensure the most efficient operation. This is most effectively done by a series of sketches, showing the arrangements of the equipment, structure, and support vessels in plan, along with the location of anchors and mooring buoys and the lead of mooring lines.

It is essential that each substage be shown so that, as equipment moves, the new leads of lines and the new locations for support craft are clearly apparent. Crane radii can then be plotted.

Isometric or vertical elevations will ensure that there will be no interference between boom and structure during critical lifts and that tag lines can control the lift.

18.8 Access

A much-overlooked aspect of constructibility is that of providing access for personnel and equipment to the areas where they must work. Workers need safe and convenient access. Studies have shown that up to 50 percent of a worker's work day is associated with moving. It is inefficient and expensive to have workers climb ladders, thread or force their way through congested reinforcing steel, climb across scaffolds, and walk planks. Proper and safe access needs to be engineered. See Figure 18.8.1.

Figure 18.8.1 Heavily congested work site offshore, with inadequate access, reduces efficiency.

An offshore structure during fabrication is usually a repetition of many similar cells or frames. Statfjord B Condeep, for example, had 90 identical skirts and 24 almost identical cells. It is very easy for workers to get temporarily lost. Markers easily read at night or in rain are needed, identifying locations both in plan and elevation.

Similarly crane location, reach, and swing need to be carefully laid out to ensure that the boom will not hit the side of the structure as it reaches out to set the load and that loads can be set within the allowable radius as well as picked. This requires a three-dimensional study.

Internal communication must be planned for general supervision and for guidance of lifts, control of concrete slump, and so forth.

Lighting for night work must consider shadows cast by the structure and cranes.

18.9 Tolerances

Offshore structures are not only among the largest structures built by man but they are structures which must be moved, floated, and rotated. They are subject to a wide variety of external loads during construction. They must then interconnect with other systems. Tolerances therefore become of far more than normal importance.

Weight control is critical. Steel jackets typically are launched and are then supposed to return dynamically to float on an even trim with only 1 m or so of freeboard.

Concrete structures, when ballasted down to receive the deck, have typically only one percent or less of reserve displacement at this stage. Weight control procedures are therefore instituted and an organization set up to control the weights during the entire construction period.

For steel structures, weight variances and items are those such as:

1. Variations in thickness (steel plates usually run to the plus side)
2. Variations in diameter
3. Stiffener plates
4. Lifting attachments
5. Weld material (usually overruns)
6. Erection bolts
7. Slings
8. Closure plates
9. Scaffolding
10. Instrumentation
11. Grout piping and vent piping
12. Coatings and paint
13. Anodes and anode supports

For concrete structures, weight variances and items typically include:

1. Variations in wall thickness (usually over)
2. Variation in geometry
3. Reinforcing bar length (usually over)
4. Spacer bars, chairs, and supports
5. Embedments
6. Posttensioning anchorages
7. Ducts (empty at some stages)
8. Unit weight of concrete
9. Water absorption into concrete
10. Scaffolding
11. Bearing plates
12. Ballast quantitites
13. Ballast unit weight in place (density, water, etc.)

The other factor contrasting with weight control is geometry control, which affects both buoyancy and structural performance.

For a cylindrical structure, whether a pipeline or a concrete structure, the tolerance controls may include:

1. Out-of-roundness
    a. Two diameters at 90°
    b. Best-fit circle
    c. True circle
    d. Local variations from circle
2. Diameter
3. Wall thickness
4. Displacement of centerline
    a. From true position
    b. From relative position

Geometry control is important to ensure against buckling under external hydrostatic head and for subsequent fitting of other components—for example, embedments, piping, and precast units.

Pipeline segments are typically difficult to accurately evaluate for displacement and hence buoyancy, because of variations in coating thickness along the length, especially bulging of the coating at the end of each double-jointed segment.

With underwater embankments, elevation and slope tolerances must be realistic and related to the prevalent sea state under which the work will be undertaken, the equipment available, and the surveying facilities installed. For some of the earlier steel caissons (hulls) installed in the Alaskan and Canadian Beaufort Seas, very tight tolerances were required for the screeded surface on which they were to be seated in order to prevent local damage to the bottom of the hulls. Adoption of more readily achievable tolerances at an early stage might have then fed back into the structural design of the caisson's bottom, with a significant reduction in site work under a critically limited schedule and extremely costly conditions.

18.10 Survey Control

Survey control is of course intimately associated with geometry control but extends beyond it to guide the fabrication and erection process before and during construction.

Where components must fit to others, it is often difficult to establish the proper reference line. One must be selected—for example, a line connecting the center of two best-fit circles. The other points must be properly related in all three planes.

Templates will often be found to be the best method of transferring complex interactive dimensions.

Match-casting of precast concrete members has been highly successful in assuring later fit. It was utilized effectively for the breakwater segments of the Ninian central platform. Care has to be taken in such match-casting to avoid distortions due to thermal effects, for example, warping during steam-curing.

Similar match-fitting and templating can be used with steel fabrications, again recognizing potential distortions due to welding.

Templating has been effectively used by the Japanese module fabricators to ensure proper fit between adjacent modules and thus facilitate connection.

Proper survey control procedures must also be set up for erection of space frames, of which the steel jacket is the most common example. Distances are large, 50 m or so, and points are high in the air and hence of limited accessibility. The sun's heat may cause significant elongation of upper members during the afternoon, while lower members on the ground are partially restrained by friction as well as perhaps seeing lesser temperature rise due to shade. Deadweight deflections may account for even greater distortions, since the jacket is usually fabricated in a different attitude from that in which it is installed. Diagonal measurements often provide the best check.

When structures are afloat, there is always difficulty in establishing reference lines, especially the vertical lines. Lasers can be rigidly mounted at the base, accurately set normal to the base. They can then project a normal line, called "vertical," even if the structure is slightly listed due to ballasting or deadweight.

Even a relatively rigid structure such as a concrete gravity-base structure undergoes significant deflections during construction due to deadweight and ballasting eccentricities. Thus the shafts may deflect outward during ballasting down to receive the deck, the ballasting typically producing a hogging moment and deflection in the structure as a whole. See Figure 18.10.1.

Survey methods for final site location are discussed in Section 6.7.

The key point of this section is to emphasize the role that survey control plays in the planning of construction.

18.11 Quality Control and Assurance

The establishment of a quality control (QC) manual and a quality assurance (QA) program is an essential aspect of constructibility.

The first task is to set up what the requirements are. They of course include those specified by the designer. If the designer has used only general requirements, such as "compliance with ASTM Specification A————," it is necessary to determine which elements of that specification are important and, further, which will be determined or measured in the construction process.

To these the constructor must add those requirements necessary to enable him to carry out his work in accordance with the materials selected and procedures adopted—for example, preheat temperature, humidity control for painting, and moisture control for above-water embankments or early strength for concrete.

In establishing these lists, every effort should be made to reduce the number to the bare essential minimum. The nuclear reactor syndrome that "everything that can be measured must be measured" must be avoided. Paperwork must not become more important than the structure.

The quality assurance program should then provide for the identification and recording of the critical items which may be important for future reference. QA should not be used as a whip by which to ensure that the inspectors are doing their job.

Where defects can be immediately corrected, they should be.

Agreement should be reached before construction starts as to what records are to be recorded and which data (e.g., radiographs) are to be kept. Those which are so kept must be properly identified and stored.

Only that which is essential for the proper performance of the structure should be tested and inspected. Only that number of tests should be made which are necessary to ensure maintenance of quality on a statistically defensible basis.

The reason for the above exhortations on limiting inspection and tests is that experience has shown that we have usually collected far more data than we can reduce, evaluate, and use and that by so doing we do not sufficiently emphasize the key properties which are truly important for performance.

Examples of mistaken programs are the taking and testing of an excessive number of cylinders for concrete compressive strength, where fewer cylinders supplemented by Schmidt hammer testing might be more appropriate, and excessive reliance on X-ray for welding under conditions where cold lap may be the more likely defect.

Figure 18.10.1 Two mammoth sheer-legs crane barges support structural steel mud base of Super CIDS as concrete mid-body section is ballasted down to mate. (Courtesy of Global Marine Development.)

18.12 Safety

The engineering of a safety plan for the large offshore project requires careful job-specific study by the construction and engineering personnel responsible for executing the project. They should develop a manual to apply to their project in which various safety risks are identified and appropriate preventive or mitigating measures adopted.

Of the many safety precautions, procedures, and equipment required by various regulatory agencies, which are important to implement on this job? Which are irrelevant or nonapplicable? Which may be detrimental and hence require a special exemption?

A general law of one of the Australian states required that all man-hoists be powered down. This was written for building work on shore. Its mistaken application offshore can be very dangerous, since in the transfer of workers by cage or Billy Pugh net, the ability to throw the clutch out of gear and to freely overhaul is essential for safe transfer between heaving vessels and platforms.

Is additional scaffolding required? Will lifelines and snap-on belts be of value on some high work? Should safety nets be provided, and if so, is their purpose to save a worker who falls or the protection of those working underneath or both? The design of the net and its supports should then be appropriate to the purpose.

What provision is made for workers to stand aside while loads of steel or concrete are being lowered? Will walkways or recesses be usable at such times? Red plastic caps on projecting reinforcing bars will prevent deep scratches, protect eyes, and prevent puncture wounds. They should be attached before the bars are delivered.

What about man overboard? Arctic waters are at $-2°$ C: a human can live only a few minutes in such water. At such temperatures, boats will not always start instantly. Should engines be left running and the boat be manned at all times? A continuously manned and operated lifeboat is now required around the platforms being constructed afloat in Norwegian waters.

Even if a man can swim, and even if he has a life jacket on, what does he grab hold of in a choppy sea and strong surface current? Fiber lines floating in the water can be trailed out from the structure. They should be well marked by buoys to prevent fouling boats' propellers.

Are there searchlights that can be used to illuminate the man in the water?

Fire is the scourge of the sea and especially so in the Arctic and sub-Arctic when piping and valves become frozen and intakes clogged with frazil ice. What secondary means are available for fighting fire?

If a worker is injured, what means are available for evacuating him from a congested location inside the structure to a shoreside hospital? As the structure nears completion, there will probably be excellent facilities, but in the early stages, temporary means must be planned.

Finally, regarding training: Major emergencies such as fire, collision, explosion, or imminent overturning require the coordinated action of several hundred workers, many of whom are not offshore-oriented and offshore-trained. Evacuation may need to be carried out under conditions of darkness, wet decks, loss of power, high winds, and a stormy sea. The diverse groups of workers aboard need to be organized into crews, and the crews need instruction and rehearsal. This matter is especially difficult where numerous specialist subcontractor personnel (e.g., X-ray technicians) are on board on a temporary basis and hence unfamiliar with the organization and the vessel or platform.

18.13 Control of Construction: Feedback and Modification

An offshore structure is a major undertaking on two fronts, because of (1) the effect of the sheer size, complexity, and interdisciplinary aspects, and (2) the dynamic movement, transport, rotations, and so on, carried out on a grand scale, sometimes involving over half a million tons and a structure the size of our largest high-rise buildings.

Construction management will have carefully planned each operation. Now as the work goes on, how is the success or lack of it monitored? What warning signals will be sent, and how will they be recognized in time for corrective action?

Referring first to the productivity of fabrication and erection, careful monitoring can be carried out on the basis of schedule, unit costs or percentage of completion, man-hour or crew-day requirements, all compared with budgeted costs and time. It is not enough to try to control by flagging exceptions; the 10 percent overrun or underrun may apply to an insignificant item or one which will soon be competed and hence beyond correction.

Rather, the major components of the work need

to be identified: schedules and budgets assigned, with consideration of the learning curve, and the special conditions. These key items are then closely monitored, usually on a crew-day basis.

Constructibility planning must of course include an interface with the critical path scheduling. The critical path method (CPM) is a valuable technique for evaluating and controlling the various operations. The growing use of microcomputers in the field enhances the ability to identify critical elements of progress early so as to enable appropriate action to be taken.

Critical path schedules are of course constantly updated. While most attention goes quite naturally to the items that lag, consideration must also be given to the opportunities that present themselves when work goes faster than scheduled. The Statfjord C platform was ahead of schedule on several early items; others were then accelerated to enable the completed structure to be placed several months ahead of target.

The second type of construction control relates to technically critical operations. What early indications will there be if serious engineering problems are imminent? Prior study has to be given to this matter for each critical operation. The instrumentation can be installed and observation schedule and procedures established so as to ensure that timely warning is received.

Examples of early feedback are unexplained discrepancies between weight control, ballast control, and observed draft. Another is a trim or list that is inexplicable or beyond predictions. Rupture of erection bolts may indicate excessive built-in stresses. Cracking of welds may be due to poor welding or to excessive stress. Are ground settlements occuring under the jacket as it is erected? Frequent levels should be run.

In the upending process, is the attitude matching that predicted on the basis of ballasting calculations? If not, watertight closures may have ruptured, or piles that were carried may have broken loose. From detailed consideration of each major observation, the needed data, their timeliness, and their relevance can be determined.

Experience on major projects onshore and off shore where serious accidents have occurred has shown in hindsight that warning phenomena had often been observed but had been disregarded because of overconfidence that the engineering and construction control was infallible.

18.14 Contingency Planning

"Murphy's law" postulates that "what can go wrong, will go wrong, and at the worst possible time." For each detailed planning phase, a list of credible potential accidents and errors needs to be listed, including especially those due to human error. Human errors become more likely and more serious under the adverse conditions under which personnel must work. See Figure 18.14.1. Each of these potential accidents and errors is then examined in detail. What can be done to prevent them? The preventive step may be physical (structural or mechanical), or it may be the assignment of a specially trained worker.

Examples are manifold. To offset a stripped valve stem or jammed gate, valve position indicators may be installed, with a remote readout at the con-

Figure 18.14.1 Construction operations must continue even under adverse sea conditions.

trol station, so as to verify that the valve really is open or closed. Valves may be arranged in series, with a space between, so as to provide a backup in case some foreign object gets in. External screens may be provided over intakes. To prevent snagging and ripping off from a boat line, guards may be installed over the screens.

The above series of steps is now standard in the Norwegian North Sea, ever since a wire line got sucked into the Frigg ballasting line and kept two valves from closing. Fortunately, this occurred near the end of the installation and did not result in serious damage to the structure, but it could have been catastrophic.

We learn primarily from past mistakes, so the advice of experienced personnel is invaluable in preparing and reviewing contingency lists. However, the initial listing can be quite fully prepared even by a less experienced engineer, by extrapolating from more conventional problems on land and in surface vessels and by addressing his imagination to the situation. However, some of the contingent accidents cannot readily be prevented. For these, backup equipment should be provided. Special care must be taken to prevent progressive collapse—for example, when a crane boom collapses and falls across the falsework, which allows structural elements to drop, holing the bottom.

Nevertheless, some contingent events will be judged so serious as to require a major change in construction procedure, even at an increase in cost or time.

18.15 Manuals

From the previous sections of this chapter, manuals are now prepared covering each major stage and each important or critical component of the construction process. A list of such manuals could include the following

1. For a steel jacket–pin pile structure:
   a. Welding procedures
   b. Node fabrication
   c. Erection of jacket legs
   d. Survey control
   e. J-tube installation
   f. Loadout
   g. Towing to site
   h. Launching
   i. Up-ending
   j. Positioning and landing
   k. Pile installation
   l. Grouting of piles to sleeves
   m. Conductor installation
   n. Deck girder erection
   o. Module erection
   p. Scour protection
   q. Riser installation
   r. Instrumentation
   s. Salvage and removal
2. For a typical concrete offshore platform:
   a. Skirt installation
   b. Base raft construction
   c. Air cushion
   d. Dock flooding
   e. Floatout
   f. Mooring at deep-water site
   g. Construction afloat
   h. Ballast control
   i. Weight control
   j. Geometry control
   k. Towing to mating site
   l. Mooring at deck-mating site
   m. Deck supports (for deck fabrication)
   n. Deck girder erection
   o. Module erection
   p. Deck loadout
   q. Deck transport
   r. Emergency mooring of deck
   s. Deck mating
   t. Deck outfitting
   u. Towing to site
   v. Installation at site
   w. Penetration phase
   x. Underbase grouting
   y. Scour protection
   z. Conductor installation
   aa. Riser pull-in
   bb. Instrumentation
   cc. Inclining test
   dd. Salvage and removal

424 *Constructibility*

Obviously not every structure needs all the above manuals. Many of the items listed may be small enough in scope for a particular structure that they can be combined.

As with the earlier division into stages, the important thing is not to overlook or gloss over a substage, for in accordance with a corollary to Murphy's law, this will turn out to be the critical one.

The preparation of each of these manuals requires the participation of all involved parties, including contractors and subcontractors, and all disciplines. Thus it turns out to be an effective means of communication and of making each group aware of the others' needs and concerns at that stage.

A draft of the manual is then circulated for review to management, design engineering, field construction supervisors, consultants, key subcontractors, and insurance surveyors. They are asked to review in detail and comment. Not only do constructive suggestions for improvement arise, but this review makes each party even more fully aware of the operation and enables each to focus on critical aspects:

1. The first section in each manual defines the scope of work to be covered and lists the other manuals which interface.
2. The next section includes the relevant drawings and specifications.
3. A few specially prepared summary drawings are included, relevant to the work covered in that manual.
4. Sources of material, as it will arrive, are identified.
5. Equipment available is identified.
6. Relevant weather and sea data are set forth.
7. The many substages of procedure are listed, with sketches of each such substage followed by calculated weights, ballast quantities, draft, freeboard, and so on, as may be applicable. Important tolerances are listed.
8. Quality control requirements are set forth.
9. The survey and measurement program is described, along with acceptable tolerances and corrective methods.
10. Special safety requirements are set forth.
11. A contingency plan is attached.

In fact, as applicable, each of the previous sections of this chapter form the basis for a summary section in the manual.

It is important that these manuals be issued in time for adequate review and revision if needed. Similarly, it is important for the reviewers to do their work promptly, allowing time for needed revision and recirculation.

18.16 On-Site Instruction Sheets

While during construction applicable drawings from the design and those showing temporary construction will be in the construction site office, as will the manuals, these are hardly suitable for use out on the platform structure. Complementary sets of construction drawings are therefore prepared, using the above documents as a source. One set is prepared for each substage or major operation. Unlike design drawings, these drawings show only those elements which are essential for that construction phase. Isometric drawings may be used for certain steps. The tolerances applicable to each step are clearly shown. Critical requirements from the specifications and instructions are noted with arrows pointing to the affected location. Each step is shown on a single drawing. Serious errors have arisen when two or more steps are combined "to save paper." Auxiliary gear and equipment are listed: slings, guides, tag lines, jacks, and so on. A bill of all auxiliary gear is tabulated so that its availability can be checked. This bill includes safety equipment. These drawings are then issued on waterproof paper for actual use in the field by the construction personnel. These drawings are especially useful in the rigging, lifting, and launching operations. For these, successive drawings can show the different positions of the load, the booms, and the lines as the load or structure is transferred from one location to the next.

In the walls of a concrete offshore platform, multiple embedments are required—internally, to support utility shaft decks, conductor guides, piping, and hangars, and externally, to provide for riser attachments and anodes. To ensure proper location of these during slip-forming, one contractor has prepared charts giving the embedment requirements and their locations for each "slice" of each shaft or cell. The slices are 1 m high; thus several hundred such drawings are required.

In addition to the embedments, the prestressing

duct requirements, reinforcing steel, and mechanical installations are also detailed for each slice.

A new $1 million articulated pipeline "stinger" was being connected to the stern of a pipelaying barge in the Bass Strait, Australia. The connection was detailed with 60 high-strength bolts. These bolts were of course shown on the drawing, and the specifications for the bolts themselves were in an accompanying manual. Far down in the manual was a note on how to torque the bolts, but this section never reached the field superintendant or crew. Having no instructions, they did not torque the bolts at all. The stinger was attached, work started, and the connection failed in fatigue within two days, dropping the stinger in a crumpled heap and buckling the pipeline.

Subsequent similar matters have been handled by sending assembly drawings to the field with the torquing instructions and other critical requirements clearly noted right on the drawings, not buried in an accompanying specification.

18.17 Risk and Reliability Evaluation

Risks associated with the various construction stages and procedures can be identified and a qualitative evaluation, at least, made of their reliability and safety involved. The word *qualitative* seems appropriate, even where some effort is made at quantification, because each operation has many unique aspects and because the data base is generally inadequate.

Risks which have been identified on previous structures include:

1. Delay in materials, fabrication, hookup, testing, and approvals
2. Excess hydrostatic heads acting on compartments or through piping, ducts, etc.
3. Loss of compressed air
4. Flooding due to external damage, piping failure, valve failure, plug or bulkhead rupture
5. Overtopping due to waves
6. Free surface water from spray, rain, leaking manholes
7. Structural cracking due to differential settlement or ballasting errors
8. Mooring line failure during storm
9. Anchor dragging
10. Fire and explosion
11. Storms—wind, waves, and high currents
12. Tsunamis
13. Dynamic amplification of motion
14. Acceleration forces on deck equipment
15. Failure of tie-downs
16. Shifting of load
17. Tug breakdown
18. Broken towline
19. Ice jamming of towline
20. Ice jamming under and around structure
21. Excessive yaw and sway
22. Grounding
23. Tug stopped—structure overruns tug
24. Loss of stability during final placement
25. Lateral "skidding" due to trapped water underneath (water cushion)
26. Loss of reference markers
27. Malfunction of instrumentation
28. Seafloor irregularities, hard spots, boulders, etc. previously unidentified
29. Excessively stiff soil
30. Excessively soft soils
31. Storm or fog during installation
32. Piles failing to develop resistance
33. Piles showing excessive resistance above design tip elevation
34. Excessive scour during installation
35. Inability to break suction effect during removal
36. Launched structure failing to float at proper draft or proper attitude, i.e., in list or trim
37. Structural damage on launching
38. Lines fouled on projecting fittings

The above list is obviously incomplete and includes both major and minor items. One of the frustrating aspects of offshore construction is that minor risks, when they occur, often combine to create major problems.

When the initial Seatank underwater storage tank was installed as a demonstration project in the Bay of Biscay in 1970, if failed catastrophically due to a long series of events:

1. As the structure was submerged near its waterline, the hydrodynamic effect of the waves

caused it to "hang up," requiring additional ballasting.
2. This delayed the operation, and then the weather worsened.
3. The tugs had difficulty positioning themselves.
4. The added ballast to compensate for item 1 led to "plunging" later, i.e., too rapid sinking when the structure was more deeply submerged.
5. This caused the boat lines to foul on manholes, ripping at least one open.
6. The combination of items 4 and 5 caused the structure to plunge to greater depths before compressed air could be injected for internal pressure compensation.
7. The structure imploded.

The subsequent successful installation of some 20 large concrete platforms has shown that preventive measures can be taken and that they can prove fully adequate. They include limiting reliance on compressed air to augmentation of safety factor, with the structure being designed to withstand external hydrostatic heads even after the loss of air. Positioning tugs are fitted with bow thrusters to enable them to position themselves in wind and waves. Manholes are recessed and fittings are protected by guards against snagging by towlines. Columns, shafts, or temporary tanks are installed to prevent sudden loss of stability and draft control.

Careful evaluation of risks and reliability is essential to the selection of the appropriate method. Frequently, the results can be very positive; a procedure which appears excessively dangerous, such as launching a 30,000-ton jacket the size of a highrise building, mounting a complex deck weighing 20,000 tons on a preset jacket in the North Sea, or mating an articulated loading column with its base while both are floating at an inclination, can, with thorough engineering, be made into a sound and reliable undertaking. Conversely, a relatively "simple" operation such as setting a module on the deck of a platform may be excessively hazardous if it is treated superficially and carelessly—if, for example, inadequate attention is given to padeyes and sling leg orientation.

Risk and reliability evaluation is obviously closely related to contingency planning. The latter, however, is intended to establish specific procedures to prevent or mitigate risks after the overall plan has been established. Conversely, risk and reliability evaluation is intended to serve as a broad guide and overview to ensure that sub-optimization techniques have not led to adoption of excessively risky procedures and that areas of high risk will be re-investigated so as to reduce their probability of realization and mitigate the consequences.

The architects of these clipper ships were like poets who transmute nature's message into song, obeying what wind and wave had taught them, to create the noblest of all sailing vessels and the most beautiful creations of man, a perfect balance of spars and sails to the curving lines of the black hull, and this harmony of mass, form, and color was practiced to the music of dancing waves and of brave winds whistling in the rigging.... For a few brief years they flashed their splendor around the world, then disappeared.

SAMUEL ELIOT MORISON,
The Oxford History of the American People

19

Construction in the Deep Sea

19.1 General

The deep sea is one of the two new frontiers of offshore activity, the other being the Arctic. At what depth does the term *deep sea* apply? This is a highly arbitrary limit, which changes as developments take place. A decade ago, 100 m was considered deep and 200 m was defined as the arbitrary limit of the continental shelf. As of this writing, conventional platform construction practice is being extended to over 400 m. Guyed towers, articulated buoyant towers, and tension leg platforms have been designed for 500 m and more. Subsea systems for production of oil in the deeper waters, out to 600 m and more, have been tested. Pipelines have been successfully installed in 600 m between Tunisia and Sicily. Shell Oil is constructing a fixed steel platform in 410-m (1350-ft) water depth in Green Canyon, off Louisiana, and other companies are studying guyed towers and tension leg platforms for depths up to 500 m in the same area. Military activities, such as the deployment of acoustic sensors for recovery of armament and equipment, have reportedly been carried out at depths up to 6000 m.

Test facilities for ocean thermal energy conversion (OTEC) have included pipelines installed to a depth of 600 m. Tests of manganese nodule mining equipment have been conducted at a depth of 2000 m.

For the purposes of this chapter, the *deep sea* will be defined as those depths at which manned intervention appears to be no longer economically practicable (i.e., over 300 m) and where hydrostatic pressures dominate design and construction, so that new equipment, systems, and procedures become necessary.

The deep-sea frontier is rapidly emerging as an area of great interest. Exploratory drilling for oil has already been carried out at depths of over 2000 m. There is informed speculation that the sedimentary fans at the base of the continental slopes may have high potential for petroleum reservoirs. Chevron has announced plans for a subsea completion in 2500 ft (800 m) in the Gulf of Valencia off Spain. The Deep Sea Drilling Project included the successful drilling and reentry of a hole at a depth of 6000 m.

Potential exploitation of the polysulfide mineral deposits from mid-ocean rifts will require specialized dredging operations, with equipment and materials capable of operating in hot brine. Manganese nodules are concentrated on plateaus and basins lying at 2000–4000 m depth, requiring efficient dredging systems capable of operating at such depths.

OTEC systems are generally based on the utilization of the cold water from 1000 m depth. The floating structures for this concept may require mooring in 4000 m.

The deployment of sensor devices with cable moorings and of large surface and subsurface buoys is currently being carried out throughout almost the entire range of ocean depths.

Cables and pipelines are being studied for crossings of straits where the water depths range from 300 to 600 m. A bridge across the Straits of Gilbralter, now under pre-feasibility study, will require piers in from 300 to 500 m of depth.

Studies and tests have been carried out by the U.S. Navy for mass concrete placements in the deep ocean.

Scientific exploration such as that proposed to identify the existence of neutrinos will require extensive deployment of cables, sensors, and moorings at extreme water depths. The Deep Undersea Muon and Neutrino Detection (DUMAND) Project will involve the placement of an array of sensors 250 × 250 × 500 m in plan at a depth of 4500 m.

19.2 Considerations and Phenomena Related to Deep-Sea Operations

Depth effects which are of concern to the constructor include the following:

1. Extreme hydrostatic pressures.
2. Change in gross structural dimensions and hence volume and buoyancy due to high external pressures.
3. Density changes in liquids, including seawater, due to high pressure and low temperature.
4. Reduction in volume of solids due to bulk modulus effects (usually important only for low-modulus materials such as polyurethane foam).
5. Absorption of water into concrete and other solids.
6. Absorption of gases into solids.
7. Miscibility of water and other fluids.
8. Change in resistance of materials due to high triaxial stress states.
9. Density and other currents. At depths of 1000 m the currents may be of the following order:
Density currents: 0.2–0.5 knots
Internal wave-generated currents: up to 0.6 knots
Tsunami currents: up to 0.6 knots
Currents may produce vortex shedding and thus require the installation of "spoilers." They may also produce dynamic responses in long risers and strumming vibrations in long cables.
10. Internal waves.
11. Density layers (stratification) of sea.
12. Leaks in seals of hydraulic systems, electrical connectors, etc. due to high pressure.
13. Difficulty of control as a result of time lag in response of hydraulic systems due to the long length of lines.
14. Static and dynamic strains (stretch) in cables, casing, rods, etc. due to long length.
15. Remote sensing and control requirements for positioning, orientation, guidance, etc.
16. Interaction of pressure and temperature on highly compressed gases, with rise in temperature when pressurized and sudden drop (even below freezing) when pressure is released.

19.3 Techniques for Deep-Sea Construction

The constructor has available a number of techniques for meeting the special needs of the deep sea.

1. Electronic and acoustic sensing devices have been developed which enable remarkably accurate measurement and control of orientation and positions, both true and relative. These include gyros, inertial guidance, photographic and acoustic imaging, video, and sonic devices.
2. Many of these devices can be effectively deployed from ROVs using fiber optics for transmission of information. They can also be deployed on the structure itself. A number of these were successfully employed on the Cognac platform during its installation. Suitable systems have been field-proven in deep exploratory drilling operations, including reentry operations of the *Glomar Challenger* at 20,000 ft (6000 m).
3. Dynamic positioning, both at the surface of the sea, where propellor-driven thrusters are usually employed, and at depth, where jet thrusters are more applicable. These can be computer-controlled to maintain positions as determined by input data from satellites on the surface and acoustic transponders in the sea and on the seafloor.
4. Use of deaerated seawater as a hydraulic fluid.
5. Use of low-density fluids which still possess low compressibility and hence permit balancing of fluid pressures. These include gasoline, propane, oil, and solvents. Several solvents are available which are safe to handle, have minimal miscibility with water, and have specific gravities in the range of 0.55–0.60. After use, these solvents can be displaced by seawater and recovered to a tanker.
6. Syntactic foams (closed-cell), possessing low density but capable of resisting hydrostatic pressures up to 6000 m and more.

7. High-density materials for weight control. These include barite-weighted drilling mud and iron ore slurry.
8. Development of neutral buoyancy materials of high strength, such as fiber ropes and polyethylene pipe.
9. Use of drill casing and drill string for lowering of heavy objects. These can then also be used for transmission of fluids. Casing maintained empty may partially offset its dead weight by buoyancy.
10. Development of supporting techniques such as arc-flame cutting. Recent studies have been made of the effectiveness of underwater arcs and flames at great depths. Underwater flames involve an arc to ignite the preheating flames, the use of premixed flames fueled by hydrogen or methane, and an oxygen jet to burn away the preheated material. Underwater flames seem to have no inherent depth limitations and in fact may perform even better in deep water than in shallow water. Underwater arcs are more adversely affected by the pressure, water chemistry, and heat-sink. It appears possible to strike and maintain an arc, but further research and development will be needed to ensure efficient operations at great depths.
11. Development of self-contained power sources such as nickel-hydrogen, silver-hydrogen, and lithium thionyl chloride batteries, which are pressure-compensated.
12. An abrasive water jet cutting system has been developed by Ocean Systems Engineering, Inc., to cut pipe or piles in deep water.

19.4 Properties of Materials for Use in the Deep Sea

The high hydrostatic pressures experienced with increasing depth cause the properties of many materials, such as fluids, to change from those associated with more normal, near-surface operations. In this new environment, these properties become of importance for construction operations. Table 19.4.1 gives the properties of seawater at various depths.

Propane, crude oil, gasoline, diesel oil, and solvents are fluids possessing buoyancy in seawater, thus capable of reducing the effective weight of large structures during installation. Their properties at various depths are given in Tables 19.4.2 through 19.4.6. Attention is directed to the fact that these fluids have different degrees of miscibility with seawater. Table 19.4.7 gives the properties of syntactic foam. Both "syntactic" and "syntectic" are used in the technical literature, but syntactic is adopted in this book.

Heavy fluids and solids are often required in or-

TABLE 19.4.1 Properties of Seawater at Various Depths

Depth	Specific Gravity	Unit Weight	Water Temperature	Pressure	
English System					
(ft)		(lb/ft^3)	(°F)	(ksf)	(psi)
3000	1.030	64.25	37	193	1340
6000	1.034	64.5	34	386	2680
9000	1.038	64.75	34	581	4040
12000	1.042	65.0	33	776	5400
SI System					
(m)		(kN/m^3)	(°C)	(MN/m^2)	(MPa)
1000	1.030	10.09	3	10.09	10.09
2000	1.035	10.13	1	20.32	20.32
3000	1.039	10.18	1	30.50	30.50
4000	1.043	10.22	1	40.72	40.72

TABLE 19.4.2 Properties of Propane at Various Depths

Depth	Water Temperature	Pressure	Unit Volume	Density
\multicolumn{5}{c}{English System}				
(ft)	(°F)	(psi)	(ft^3/lb)	(lb/ft^3)
0	60	100	0.0320	31.0
1000	43	444	0.0306	32.7
2000	39	888	0.0300	33.3
3000	37	1332	0.0298	33.5
4000	35	1776	0.0296	33.8
5000	34	2220	0.0294	34.0
6000	33	2664	0.0291	34.4
\multicolumn{5}{c}{SI System}				
(m)	(°C)	(MPa)	(m^3/kN)	(kN/m^3)
0	16	0.71	0.206	4.860
500	5	5.20	0.195	5.130
1000	3	10.40	0.192	5.220
1500	2	15.60	0.190	5.260
2000	1	20.80	0.189	5.300
2500	1	26.00	0.188	5.340
3000	1	31.20	0.186	5.400

Note: Propane is miscible with seawater.

TABLE 19.4.3 Properties of Crude Oil

API gravity	Density lb/ft^3	Density KN/m^3
20°	58.2	9.14
35°	53	8.30
42°	50.8	7.96

$$\text{API gravity} = \left(\frac{141.5}{\text{Sp. Gr. at } 70°\text{F}} - 131.5\right)$$

Example: $20 = \left(\dfrac{141.5}{\text{S.G.}} - 131.5\right)$

S.G. = 0.935

$0.935 \times 62.4 = 58.2$ lb/ft^3

$58.2 \times \dfrac{157}{1000} = 9.14$ KN/m^3

TABLE 19.4.4 Properties of Gasoline

API gravity = 67.5° at 1 atm at 60°F
  Density = 44.4 lb/ft^3 = 7.0 kN/m^3
At 3000 ft and 33°F (1°C):
  Density = 46.1 lb/ft^3 = 7.25 kN/m^3
Other grades show density at 3000 ft of 47.7 lb/ft^3 = 7.5 kN/m^3

Note: Gasoline is slightly miscible with seawater.

TABLE 19.4.5 Properties of Diesel Oil

API gravity = 40°
  Density = 51.4 lb/ft^3 = 8.06 kN/m^3

TABLE 19.4.6 Properties of Solvent Mixture: Heptane Plus Hexane

Heptane: 62° API vapor pressure at 100°F (38°C) = 1.6 psi (11 kPa)

Hexane: 75.2° API vapor pressure at 100°F (38°C) = 5 psi (36 kPa)

Mixing 60% hexane plus 40% heptane gives API 70° at 68°F (20°C); nominal internal pressure about 5 psi (36 kPa)

Assumed increase in density with depth and lower temperature is similar to that of gasoline—that is 4%, giving a unit weight of 45.5 lb/ft^3 at 3000 ft or 7.45 kN/m^3 at 1000-m depth.

Note: Heptane and hexane have low miscibility with seawater.

TABLE 19.4.7 Properties of Syntactic Foams

Lightweight Syntactic Foam

Density = 35 lb/ft^3 = 5.5 kN/m^3
  Strength to withstand 2000 ft (600 m) of external hydrostatic head

High-Strength Syntactic Foam

Density = 42 lb/ft^3 = 6.6 kN/m^3
  Strength to withstand 20,000 ft (6000 m) of external hydrostatic head

TABLE 19.4.8 Properties of Weighted Drilling Muds

Density			Mix Proportions per Barrel of 42 gal			
(lb/gal)	(lb/ft^3)	(kN/m^3)	Bentonite	Barite	Lignosulfonate	Caustic
16	120	18.9	8.5	13	4	1
18	135	21.2	8.5	523	4	1
20	150	23.6	8.5	634	4	1

For heavier densities, add other finely ground materials:

Material	Specific Gravity
Barite	4.2–4.3
Galena	6.5–6.7
Iron oxide	4.9–5.3
Iron particles	7.8
Lead powder	11.4

der to ballast structures down against the buoyancy of the seawater. Table 19.4.8 gives the properties of weighted drilling muds, while Table 19.4.9 gives the properties of bulk solids.

To lower and position deep sea structures, wire rope, chain, drill pipe, and drill casing have been used. The properties of these are listed in Tables 19.4.10 through 19.4.12.

Fiber ropes offer the advantage of relatively high strength with near-neutral buoyancy. Properties of several of the more commonly used rope materials are listed in Table 19.4.13. Kevlar (aramid fiber rope) offers the strength of steel wire line at only one-fifth of the weight in air and only one-twentieth of the weight in water. It is only one-eightieth the weight of equivalent chain in water. Mooring and hoisting lines are available up to 1,200,000 lb (550 tons) breaking strength. Hence they are admirably suited for deep-water application. Figures 19.4.1 and 19.4.2 give the effective load carrying capacity of wire rope and drill pipe at various water depths.

19.5 Construction in the Deep Sea

Construction concepts and methods are rapidly developing for operations in the deep sea. Among the most promising structural concept is the tension-leg platform, developed as a drilling and production platform for use in water depths as great as 2000 to 3000 m. See Section 11.5. The guyed tower is another candidate, for up to a maximum depth of 600 to perhaps 800 m. See Section 11.4.

The buoyant articulated deep water tower developed by C. G. Doris has potential application to the 1000 m. depth. See Section 11.3 and Figure 19.5.1.

The positioning of equipment and of structures

TABLE 19.4.9 Properties of Bulk Solids for Weighting and Ballasting Purposes

Material	Density			Bulk Weight in Air		Bulk Weight in Seawater	
	Solid Specific Gravity	Solid State					
		(lb/ft^3)	(kN/m^3)	(lb/ft^3)	(kN/m^3)	lb/ft^3	kN/m^3
Siliceous or limestone sand	2.64	165	26	105–115	17–18	41–51	7–8
Iron sands (oxides or sulfides)[a]	4.8–5	300–312	48–49	195–220	31–35	131–156	21–25

[a]Iron sulfides may have corrosive effects on steel and concrete reinforcement.

TABLE 19.4.10 Weight of Chain and Wire Rope

Size	Weight in Air (lb/ft)	Weight in Air (kN/m)	Weight in Seawater lb/ft	Weight in Seawater kN/m	Proof Test (lb)	Proof Test (MN)
3¼" chain	105	1.53	91	1.34	804,000	3.65
4" chain	152	2.22	132	1.93	1,200,000	5.45
5" chain	232	3.40	202	2.96	—	—
6" chain	323	4.70	281	4.11	—	—
4" wire rope	29.6	0.43	25.8	0.38	—	—

in deep water presents new problems. Spread moorings can be used but only by reducing the weight of the mooring lines by means of buoyancy. Low density lines such as KEVLAR can be used, or floats may be attached to conventional wire rope lines. Alternatively, casing can be used by filling it with syntactic foam or a low density fluid.

As structures are submerged, they can presumably be positioned by dynamic thrusters, locked in by on-board computer through acoustic transponders to surface vessels and thence to satellites. Alternatively, they can use preset seafloor transponders to maintain relative position.

Submerged buoyant structures can be kept afloat at prescribed elevations off the seafloor by the use of weighted tethers. If they rise, they pick up more tether weight (for example, chain) and hence return to their original elevation. Objects lowered on rope or casing must consider the dynamic response of the lowering vessel as it responds to the waves, as well as the inertial effects of the object, with its added mass of water that must also be accelerated. To overcome this, giant heave compensators were devised for the *Glomar Explorer* and were used to overcome roll, pitch, and heave effects.

Macoperi has announced plans to build a giant derrick barge with two 6000-ton cranes capable of lifting 12,000 tons. It will include the ability to lower

TABLE 19.4.11 Weight of Drill Pipe

Size	Weight in Air (lb/ft)	Weight in Air (kN/m)	Weight in Seawater lb/ft	Weight in Seawater kN/m³
3½"	15.50	0.23	13.5	0.20
4"	15.70	0.23	13.7	0.20
4½"	20	0.29	17.4	0.25
5"	19.5	0.285	17.0	0.25
6⅝"	31.9	0.465	27.8	0.405
8⅝"	40	0.58	34.9	0.504

Drill Pipe Is Available in the Following Grades

Grade	Yield Strength (kips/in.²)	Yield Strength (MPa)	Ultimate Strength (kips/in.²)	Ultimate Strength (MPa)	Ultimate Elongation (%)
D	55	400	95	680	18
E	75	530	100	710	18
G	105	750	120	860	15
S	135	960	150	1070	—

TABLE 19.4.12 Properties of Drill Casing

Size—diameter (in.)	Weight in Air (lb/ft)	Weight in Air (kN/m)	Weight in Seawater (open ended) lb/ft	Weight in Seawater (open ended) kN/m	Buoyant Weight of Closed Casing (lb/ft)	Buoyant Weight of Closed Casing (kN/m)
6⅝	32	0.47	28	0.41	−16.6	−0.24
8⅝	49	0.72	43	0.63	−23.0	−0.34
10¾	55.5	0.81	48	0.70	−15.0	−0.22
13⅜	85	1.24	74	1.08	−22.5	−0.33
18⅝	96.5	1.41	84	1.23	+25.5	+0.37
24½	113	1.65	98	1.44	+97	+1.42

Note 1: Minus sign (−) indicates weight exceeds buoyant force; plus sign (+) indicates buoyant force exceeds weight. These net values assume air-filled casing.

Note 2: If the 24½" casing is filled with a buoyant fluid such as heptane plus hexane, the net weight in water becomes slightly negative.

Weight of pipe in air	−113 lb/ft
Weight of fluid in air	−135 lb/ft
Displacement	+210 lb/ft
Net weight in water	− 38 lb/ft = −0.5 kN/m

TABLE 19.4.13 Properties of Typical Fiber Ropes

Size		Nystron Braid[a] (Nylon/Polyester)			Nylon/Multifilament Polypropylene Rope[b]			Polyester Rope[c]			Kevlar[d]		
Diameter (in.)	Circumference (in.)	Weight per 100 ft (lb)	Minimum Breaking Strength (lb)	Elastic Elongation at 25% of Breaking Strength (%)	Weight per 100 ft (lb)	Minimum Breaking Strength (lb)	Elastic Elongation at 25% of Breaking Strength (%)	Weight per 100 ft (lb)	Minimum Breaking Strength (lb)	Elastic Elongation at 25% of Breaking Strength (%)	Weight per 100 ft (lb)	Minimum Breaking Strength (lb)	Elastic Elongation at 25% of Breaking Strength (%)
2	6	114	121,000	7	93	88,400	7	124	105,400	3	132	172,000	1
2¼	7										180	224,000	1
3	9	268	272,000	7	210	193,000	7	294	236,300	3	—	—	—
4	12	470	460,000	7	371	329,000	7	515	399,500	3	—	—	—
5	15	719	683,000	7	590	505,000	7	788	593,300	3	—	—	—
6	18	988	921,000	7	836	698,000	7	985	731,850	3	—	—	—
7	21	1,348	1,233,000	7	1,080	884,000	7	1,478	1,071,850	3	—	—	—

[a] E at 25% B.S. = 140 000 psi.
[b] This rope has neutral buoyancy! E at 25% B.S. = 100 000 psi.
[c] E at 25% B.S. = 280 000 ps.
[d] E at 25% B.S. = 1,400,000 psi.

Source: From *Samson Rope Manual* No. 2.77, 2nd edition, Aug 1977, Samson Ocean Systems, Inc. Boston, Mass.

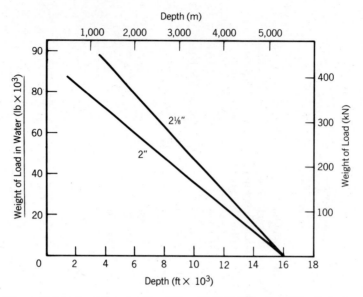

Figure 19.4.1 Load Carrying Capability of 6 × 41 Fiber Core Wire Rope.

900-ton loads to a depth of 450 m. The barge will be 190 × 87 m in deck area.

Free fall can be used to emplace anchors. The Naval Civil Engineering Laboratory at Port Hueneme, California, has developed such anchors which fall free to the seafloor and then explosively propel the anchor into the seafloor. Free-fall objects achieve a terminal velocity which depends on their drag. Free-fall deployment may also be "controlled" by installing multiple buoys. These buoys can then be progressively flooded to enable the object to descend in steps.

French engineers have developed "breather" buoys, which decrease in volume and hence buoyancy as they descend. These have been successfully used in laying a test section of pipeline in 2500-m water depth.

"Glide" can be used with submersibles and ROVs to control the rate of descent.

"Pulling down" of a buoyant structure against a

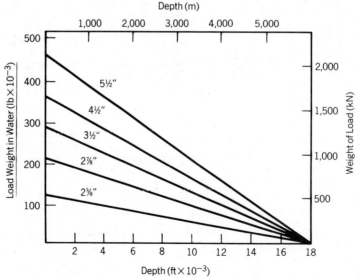

Figure 19.4.2 Load Carrying Capability of API Drill Pipe.

19.5 Construction in the Deep Sea

Figure 19.5.1 Buoyant articulated deep-water gravity tower proposed by C. G. Doris for application up to 1000-m water depth.

seafloor anchor is an effective system for decoupling the system from the surface wave effects once the structure is below the surface.

When landing on a deep seafloor, there are potential problems due to excessive penetration into the seafloor ooze. This layer of very soft material ("soup"), which may actually be in colloidal suspension, may not have been revealed in the geotechnical investigations due to lack of acoustic reflection and failure to be retained in sampling tubes.

Legs extending downward from the structure in the form of large dowels may help to stabilize the initial penetration. The legs may have steps of increased diameter, so that total penetration is limited.

The turbidity cloud caused by landing in the soft ooze must be considered, as it will impede the use of TV for positioning control. Prior steps may be taken to blanket the ooze, as outlined in Chapter 7 on seafloor modifications. The U.S. Naval Civil Engineering Laboratory at Port Hueneme, California, is continuing development of such means to suppress turbidity due to colloidal suspension of seafloor oozes.

For placement of concrete at depths ranging from hundreds to thousands of meters, at least two methods have been developed. In one, the concrete is transported in a long tube for discharge at the sea level. In the other, the concrete is pumped down a pipeline. An oversanded mix is used and the diameter of the pipeline is reduced so that the friction limits the velocity to about 3 m/sec and thus prevents segregation. Aggregates should either be coated or presaturated so as to prevent a change in character of the mix due to absorption of water by the aggregates under pressure.

Newly developed admixtures which prevent segregation may make it possible to place concrete underwater by use of buckets or other discrete devices.

Cement slurries (grout) have long been placed at great depths by the oil-drilling industry, where they have been used to cement casing strings, plug wells, and so forth.

When concrete or grout is used in large volumes, the heat of hydration must be considered, and special cementing mixtures such as blast furnace slag cement or cement plus pozzolan must be employed to reduce the heat and consequent disruption of the concrete or grout.

For breaking objects loose from the seafloor, waterflooding underneath is considered the most effective method. The pressure must be kept low enough to prevent piping and hence escape to the sea. In soils of low permeability, many hours of such flooding may be required to raise the internal pore pressure in the soils sufficiently to overcome the suction effect.

Deep-ocean dredging operations have been studied in detail for the mining of manganese nodules from the deep seafloor, and test operations have been carried out at depths up to 4000 m. The use of airlifts has been found to be an effective and efficient method. Because of the large volumetric expansion of air near to the surface, the airlift is employed to raise the material only as far as a submerged pump capsule. Conventional pumps are then used to raise the nodules the additional 100 m to the surface vessel. See Figure 19.5.2.

Seafloor soils can be consolidated by suction

436   Construction in the Deep Sea

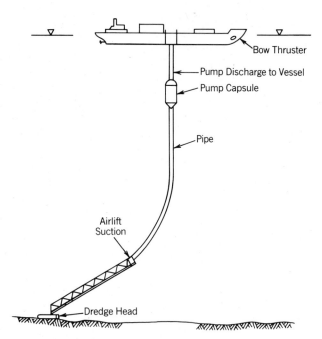

Figure 19.5.2 Deep-ocean dredging system. (Courtesy of Deepsea Ventures, Inc.)

drainage, carried out after the structure is emplaced or even before installation, by drainage from under an impervious membrane. Such a membrane might be of plastic, although it would be extremely difficult to place at depth unless attached to the base of the structure itself. However, a mixture of cement and bentonite flowed over an area of seafloor may also serve as a "membrane" to seal the soils.

For installing gravity-based structures in deep water where the seafloor is known to be irregular—as, for example, with rock outcrops—one solution is to dump rock so as to create a submerged embankment on which to seat the structure. The rock may be dumped from a bottom-dump barge in one mass so as to minimize segregation during the descent. The rock should be presaturated so as to dispel all air. After dumping, further consolidation can be obtained by dynamic compaction (i.e., the repeated dropping of a heavy ram or explosives) or by intense surface vibration. Screeding at depth is very difficult, so a preferred solution is to equip the gravity-based structure with long skirts and to equalize bearing by underbase grouting after landing.

Dumping of rock by use of a chute to depths of 300 m has been described in Section 7.7. The use of this flexible chute, suspended from a ship, provides control and prevents segregation. It may prove feasible to extend the depth range of this type of placement and thus provide a high degree of control. See Figure 19.5.3.

For structures such as subsea templates seated on the deep seafloor, it may be necessary to transfer manipulators and service modules between them and a surface vessel. Popup buoys may be attached to such a structure, to be released on acoustic signal and thus provide a guide line for subsequently lowering or guiding a manipulator or structural element to an exact mating with the previously installed element.

Tensioned guide lines of this type were extensively employed in the 1960s and 1970s at water depths up to 500 m for reentry of drilling strings into casing. They have now been largely replaced by acoustic and inertial guidance, as developed and used on the *Glomar Challenger* to reenter a casing at 6000 m.

As an illustration as to how structures may be placed at great depth, the following example is given. A large gravity anchor block or underwater oil storage tank is to be placed at 6000m depth. Procedures need to be developed. It will be recognized that these procedures are extensions of the system successfully used in lowering the bases of the Hutton TLP. See Figure 19.5.4.

The structure is constructed as a steel–concrete sandwich (hybrid) design so as to have an impervious membrane on both faces and thus eliminate the problems of absorption of fluids under pressure.

The structure is so configured that it can be completely filled with a low-density fluid such as a mixture of hexane and heptane and still remain afloat with minimal freeboard.

In shallow water, the structure is submerged to sit on the seafloor by using a barite-weighted drilling mud to fill some external ballast tanks. A semisubmersible drilling vessel is floated in over the structure and connected by drill casing. The structure is now snugged up under the semisubmersible and made fast with drill casing for tow to the deepwater site.

On arrival, the external ballast tanks are again filled with high-density fluid (barite-weighted drilling mud) so that the structure is negatively buoyant. With calm seas, the structure is released to be suspended from the drill casing. The drill casing is in turn held by linear jacks.

The structure is lowered by linear jacks. Pneumohydraulic motion compensators will be used to respond to the short-period displacements, while the

Figure 19.5.3 Placing rock at depths greater than 100 m, using a flexible chute and guided discharge. See also Figure 7.7.3. (Courtesy of ACZ Marine Contractors.)

semisubmersible vessel will respond to the longer-period motions. The most severe stresses and stress ranges will occur when the structure is passing through the shallow depths. These stresses generally reduce as the structure descends to deeper levels.

As various depth horizons are reached, additional buoyancy fluid is added (under pressure) so as to offset the compression and reduction of volume in the fluid due to the increasing hydrostatic pressure and the increased weight of drill casing.

As the structure itself reduces in volume and the low-density fluid compresses further, the high-density drilling mud (or iron sand) is discharged so as to keep the weight on the drill casing within allowable limits.

Upon touchdown, additional high-density drilling fluid or salt water is used to displace the low-density fluids in order to provide proper on-bottom stability. Since this is a gravity anchor, for which weight is necessary, high-density drilling fluid or iron ore slurry will be used to fill the internal compartments.

Other lowering systems have been successfully employed in deep water. Both subsea drilling templates and undercarriages for ocean mining have been slung in under the "mother" vessel" by the simple procedure of rigging wire rope lines, ballasting the template or carriage to slight negative buoyancy, and lowering it by means of a derrick barge or similar vessel. As it descends, it is drawn in under the "mother" vessel. The dynamic forces

438 Construction in the Deep Sea

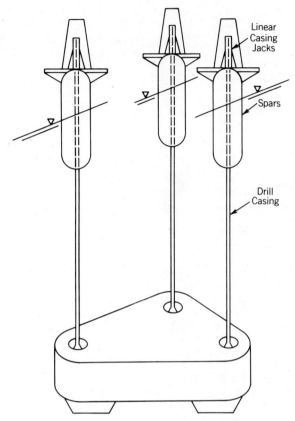

Figure 19.5.4 Lowering a heavy mass in the deep sea.

of lowering are absorbed by the wire rope lines. To increase their length and hence their stretch, they will be made up in multiple parts. Neutrally buoyant fiber lines, such as nylon, have the added advantage of lower modulus of elasticity and hence greater ability to absorb dynamic loads.

In any event, a complete dynamic analysis needs to be made of the coupled systems to ensure against harmonic resonant response at all stages of descent.

A procedure similar to the above has been proposed for assembly of the 30-m-diameter OTEC cold-water pipe. See Section 11.11.

Linear winch systems are now available with up to 2000-ton capacity, working with wire rope up to 5⅜ in. diameter or casing up to at least 10-in. diameter. The winch system is able to pass socket connectors through the jack.

The installation of pipelines at great depths has already been pioneered by the French engineers with their successful test installation of a small diameter line at 2500 m by use of the S-curve method with breather buoys. Further description of this method is given in Section 12.9. Additional systems for deep-water pipe laying are described in other sections of Chapter 12; these include the fourth-generation lay barge and the reel barge, laying by the catenary ("J-curve") method, the "J-tube" system (See Fig. 19.5.5) and the "bundled pipes" system.

External buoyancy floats of syntactic foam are attached to risers in the deep sea to render them almost neutrally buoyant and thus reduce the dead-load tension to acceptable values. Dynamic stress ranges must still be accommodated, so normally a clump weight will be affixed at the lower end to give a steady-state tension level, around which the actual stress state varies due to dynamic forces, without allowing the net tension to ever decrease to zero.

The construction and installation of subsea production systems and underwater oil storage were described in Chapter 11. These systems, along with floating production systems and tension leg platforms, are emerging as the trend in development of oil and gas in the deep sea. See Figure 19.5.6. Methods for mooring of floating systems such as spars or special-purpose vessels, are being devel-

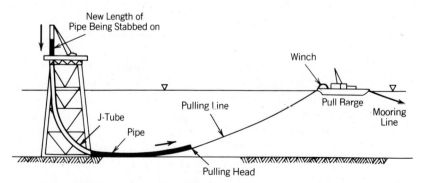

Figure 19.5.5 Reverse J-tube pipelaying method for deep water (Exxon).

19.5 Construction in the Deep Sea

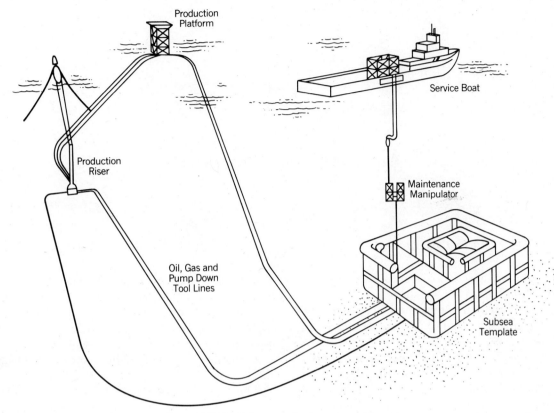

Figure 19.5.6 Subsea production system (Exxon).

oped along several lines: the use of preset spread moorings in which the deadweight is partially offset by buoyancy, either empty drill casing or syntactic foam blocks, and dynamic positioning using either diesel fuel or process gas as available.

Drill casing is especially attractive for deep moorings because of its high unit and gross strength along with the ability to fill it with either low-density fluids or syntactic foam for counteracting buoyancy.

Recent developments such as those described earlier in this chapter clearly indicate that there will be a growing need for advanced construction methods capable of operating in this highly demanding environment.

I must go down to the seas again,
  for the call of the running tide
Is a wild call and a clear call that
  cannot be denied;
And all I ask is a windy day,
  with the white clouds flying,
And the flung spray, and the blown spume,
  and the sea-gulls crying.

JOHN MASEFIELD, "SEA FEVER"

20

Arctic Marine Structures

20.1 General

The Arctic Ocean and the adjacent sub-Arctic Seas are major challenges for offshore construction. These frontier areas are dominated by perhaps the most severe environmental conditions yet addressed for offshore development.

This chapter is largely a synthesis, collecting the construction procedures, data, and guidance that relate to the Arctic Offshore. Much of this material has been previously presented in individual topic-oriented chapters. Because of the important role of the Arctic and because work in Arctic regions requires a multidisciplinary approach, it seems appropriate to gather together in one chapter the multiple aspects to which consideration must be given if structures are to be economically and safely built there.

Development of the Arctic offshore areas has been greatly accelerated by the leasing programs of Canada and the United States. Structures have been required for exploratory drilling and the first production structures are under construction. Extensive engineering has been directed to the development of new concepts, suitable for the sea ice environment. See Figures 20.1.1 and 20.1.2.

Reference should be made to the detailed description of sea ice and icebergs in Section 1.10; to the discussions of geotechnical properties of the Arctic seafloor in Sections 2.5, 2.6, 2.7, and 2.8 (overconsolidated silts, permafrost, weak Arctic silts and clays, and ice scour); and to special ecological considerations in the Arctic in Section 3.1, especially noise, open leads in ice, oil spill in ice-covered waters, and disturbance of critical migration routes.

20.2 Sea Ice and Icebergs

The Arctic Ocean proper is dominated by the polar pack, a disk of permanent ice 1500 km in diameter circulating clockwise in a gyre centered at about 80°N and 150°W. The polar pack is largely composed of multiyear ice floes interspersed with ridges and fractured by leads. The average thickness is about 4 m.

The U.S. Navy–NOAA* Joint Ice Center publishes weekly maps of ice conditions in the Arctic and sub-Arctic. In addition, a prediction of future ice conditions is published monthly. Examples of such maps for summer (August) and winter (December) are shown in Figures 20.2.1 through 20.2.3. Note in the 7 August 1984 map (Fig 20.2.1) the relatively narrow corridor of ice-free water around the north of Point Barrow. It is under these conditions that the successful tow of the Molikpaq and GBS-1 was made. This tow is described in more detail in section 20.11 of this chapter on "Deployment of Structures in the Arctic." Figures 20.2.1 through 20.2.3 show how the ice conditions in the Arctic and sub-Arctic vary with the seasons.

Around the periphery of the Arctic Ocean, in the shallow waters of the continental shelves, is the zone which varies from open water in the brief summer to annual ice in the winter. This annual ice is relatively immobile—hence its name, "fast ice."

Between the pack and the fast ice is the highly dynamic shear zone, dominated by both annual and multiyear ridges and by massive floes which break off from the pack. This is also known as the "stahmuki zone."

---

*NOAA = U.S. National Oceanographic and Atmospheric Administration.

20.2 Sea Ice and Icebergs 441

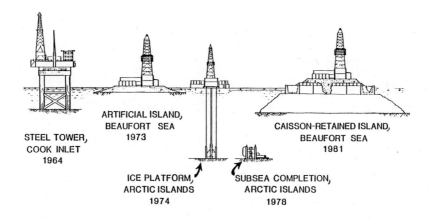

Figure 20.1.1 Arctic development milestones I.

When the moving ice encounters an obstruction, whether it be a shoal or a structure, rubble piles build up. These rubble piles may protect (buffer) the structure from further impact by floes, but conversely, they remain as massive ice features well into the following summer, thus impeding access for resupply.

When the ice sheet is driven onto a sloping shore, it tends to ride up onto and over the beach. When the sea level has been raised by a storm surge, the ride up may proceed 30–100 m inland from the normal shoreline. Rideup may similarly be a threat to artificial islands and platforms.

Seasonally, the shear zone, which is where much

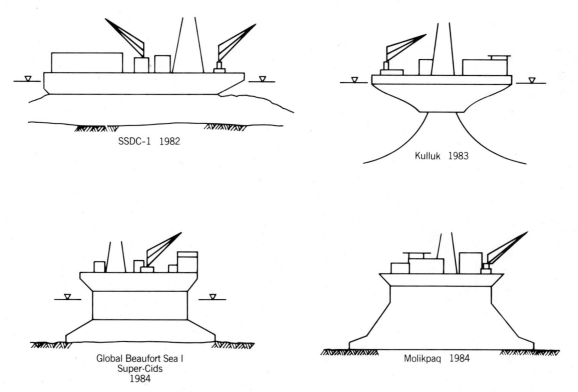

Figure 20.1.2 Arctic development milestones II.

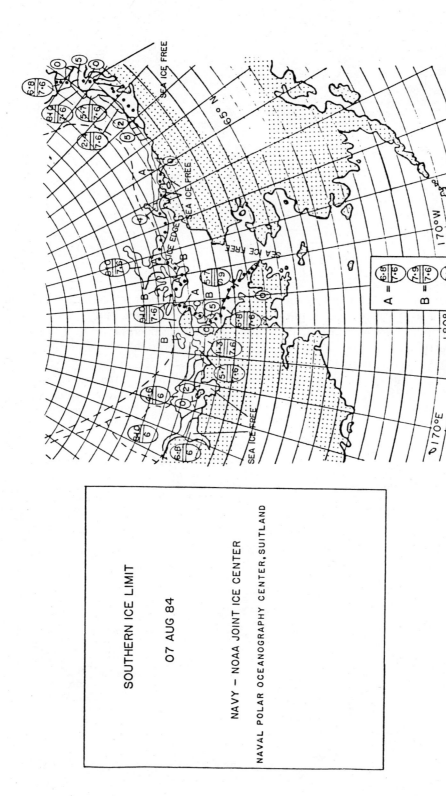

Figure 20.2.1 Sea ice conditions in western Arctic, August.

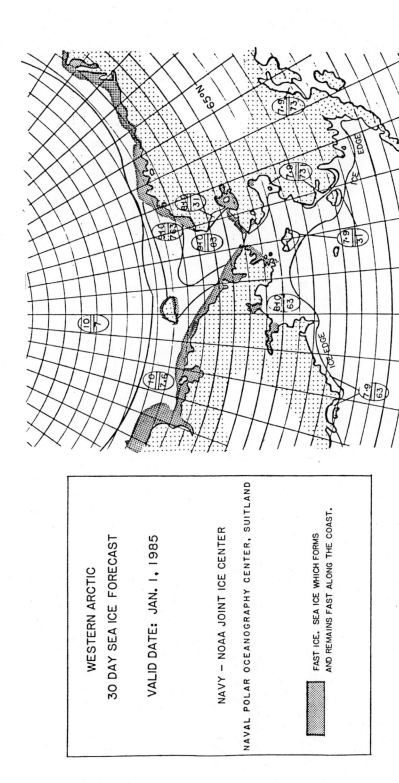

Figure 20.2.2 Forecast sea ice conditions, western Arctic, for January.

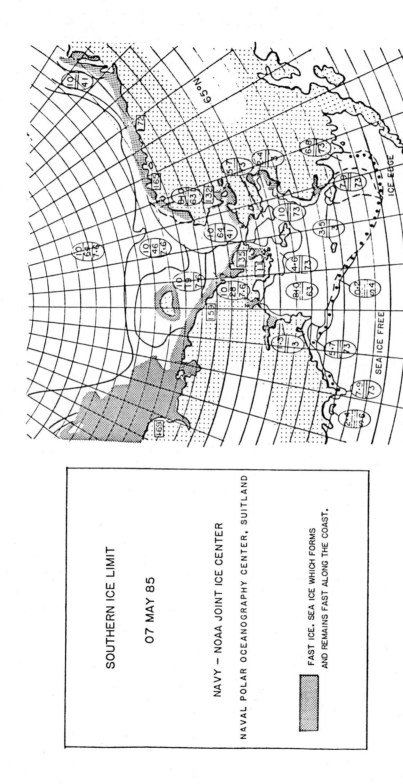

Figure 20.2.3 Sea ice conditions for western Arctic, May.

of the potential offshore petroleum development lies, typically undergoes the following sequence:

1. *July–September:* Open water, with several intense storms. One or more invasions by multiyear icefloes.
2. *September–November:* Freezeup. Thin ice forms, restricting movement of ice floes, but also restricting operations. Ice breakers and ice-strengthened vessels are needed.
3. *November–May:* Winter ice conditions. Movement of ice slow and erratic, driven by wind and currents.
4. *May–July:* Spring breakup. Dynamic movement of ice, with both annual and multiyear features.

Within the Arctic polar pack are several anomalies. One is the polyna, which is a large open-water area which forms even in the winter. The other is that of ice islands and ice island fragments. These start as huge tabular bergs which break off from a glacial ice sheet on Ellesmere Island and are caught up in the polar pack. Being of glacial origin, they are freshwater ice. As they ground in shallow water, or as thermal and impact stresses cause fractures, they break up into smaller fragments.

In the sub-Arctic areas, other ice-dominated environments exist. Between Greenland and Eastern Canada, in Baffin Bay, Davis Strait, and as far south as Newfoundland, icebergs are a primary consideration. These are largely blocky bergs, ranging in size up to 10 million tons or even more. As they melt and break up, growlers of several thousand tons or so form, and finally the bergy bits, small enough to be acclerated by storm waves yet difficult to detect either visually or by radar. In the winter these bergs are locked in the annual sea ice which forms, yet movement of the agglomerated mass still continues.

The worst incidence of icebergs is in April and May, which is also the worst time for fog. Annual incidence of bergs varies by almost two orders of magnitude. Reasons for this extreme variability are not clearly understood.

In the Chukchi Sea, the open-water season is 3–4 months long. The area is subject to frequent multiyear ice floe invasions, punctuated by intense storms.

The Bering Sea sees only annual ice. However, south of Nome, the ice is very dynamic, with sheet ice forming and then being driven south by the wind, leaving open water to form a new sheet. The sheets are rafted upon one another, and many annual ridges are formed. Ice driven up onto the shallow flats of the Yukon Delta forms monstrous rubble piles which last until late in the following summer.

Then, in the early summer, the area is subject to occasional major storms, with high waves steepened by the shallow water and significant storm surges up to 3 m. Occasionally the ice rubble mounds are lifted up so as to float off and drift northward from Norton Sound as "floe bergs."

Further south in the Bering Sea, in the Navarin Basin, the winter ice is less thick and of lower strength due to the higher water temperature. Due to the effect of storm winds, large annual ridges can still form, with consolidated zones in their mid-depth.

20.3 Atmospheric Conditions

Several of these conditions have a significant effect on construction activities in the Arctic. Low temperatures are of course a controlling phenomenon of the Arctic, reaching to $-50°C$ in winter. Summer temperatures are typically $10°C$. Water temperature is generally $-2°C$, rising in late summer to $+8°$ near shore. Low temperatures affect not only operations but may cause brittle failure of conventional steels.

Fog tends to form at the edge of the pack ice during summer. This is due to the cold air off the ice flowing over the warm open water. Since the edge of the pack is never far offshore, this means that during the construction season, visibility can be severely restricted.

Offshore eastern Canada this condition may extend well out over the open sea. This type of fog tends to hang close to the sea surface and often is limited to 10–15 m in height, above which visibility may be fine.

An even more serious impedance to operations is that of the "whiteout," when sky, land, and sea become one vast white haze, with no reference features.

Atmospheric icing of superstructures, boat masts, rigging, and crane booms can build up rapidly and create serious problems of weight. Typically it can build to 3–4 in. of solid ice or 12 in. and more of porous ice. It is most serious in the southern reaches of the sub-Arctic where the air has more moisture. API Bulletin 2N warns that accretion of ice on the superstructure of vessels or structures can cause local overstressing or reduction of overall stability by increasing topside weight and the exposed areas to wind. Atmospheric icing buildup can be reduced by using tubulars instead of shapes and by means of low-friction coatings such as polyethylene.

Radio communication can be adversely affected by the Aurora Borealis *("northern lights")* and associated electromagnetic disturbances. These are usually most severe near the time of the equinoxes (March 15 and September 15).

Magnetic compasses are essentially useless since the north magnetic pole is on Baffin Island. However, satellite navigation is available. Loran C coverage is poor in the Arctic. The Global Positioning System, which become operational for commercial purposes by 1986, gives excellent navigational control.

Windstorms are often relatively local in extent and very intense, reaching 30 m/sec velocity or occasionally even more, but with a usual duration of only 6–12 hours. They are difficult to forecast and hence arrive with only a few hours' warning or no warning at all. When these winds occur at times of low temperature, wind chill effects can be extreme, reaching the equivalent of $-70°C$. See Figure 20.3.1.

20.4 Arctic Seafloor and Geotechnics

The seafloor soils of the Arctic are among the most complex and difficult in the world. Along the shores of the Arctic proper, the shelf is very shallow for a long distance out, up to 60 km and more, at which point the depth may be only 100 m. On this shelf are recent Holocene deposits of silt (from such glacier-fed streams as the Mackenzie), degrading to clayey silts and silty clays the farther the location from the river mouth.

These weak recent sediments may be from 2 to 20 m in thickness. Near the edge of the shelf, extensive slumping has been identified, even though the slope is very flat. Below these recent sediments, in many cases overconsolidated silts are encountered up to 10 m or more in thickness. These represent an extremely difficult problem for the constructor. They are so dense that it is almost impossible to drive a pile into them and extremely difficult to excavate or cut. Yet if they are broken up, the silt goes into suspension as an almost colloidal material.

The most effective means of penetrating them is with a high-pressure water jet. Mechanical drills and cutters will work but experience excessive abrasion unless augmented by a jet.

These overconsolidated silts are in turn often underlain by dense to very dense sands. Subsea

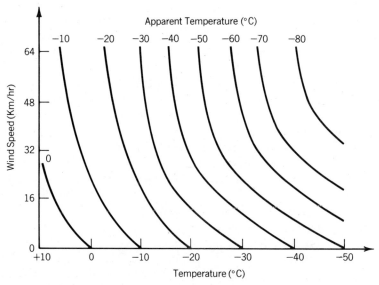

Figure 20.3.1 Apparent temperature due to wind chill.

permafrost may be encountered, even well out into the shear zone. In the permafrost zone, the upper levels are usually partially ice-bonded, with lenses of ice, whereas at greater depths the sands are fully bonded.

Between the top of the sands and the silt deposits described above often lies a thin stratum, only 1–2 m thick, where the silt is extremely weak, almost fluid. Many explanations have been proposed for this phenomenon; one is that the water from the melting permafrost and associated gases have permeated up through the partially bonded sands to be trapped under the impermeable silts that overlie, thus breaking down the silt structure and increasing the pore pressures.

Shallow methane gas has been encountered at a number of locations. Care must be taken in drilling and pile driving to avoid accidents due to small explosions.

Occasional ice lenses have also been reported in the silt deposits, well above the body of subsea permafrost.

Another anomaly is the reported presence of boulders on the surface of the seafloor, out near the edge of the shelf. These are reportedly of relatively small size (cobbles to small boulders) and hence may not be of significant concern.

The surface of the Arctic shelf, out to a depth of 50 m or so, has been repeatedly plowed by the keels of ice ridges. Furrows are generally 1–2 m deep, 10–20 m wide, with ridges forced up on the sides. Old furrows have refilled with loose sediments. The scour patterns are varied and crossed as the direction of ice movement has changed. It is postulated that the entire shelf, out to 50- or 60-m depth, has been continuously scoured over recent geologic times. See Figure 20.4.1.

Another scour phenomenon is that of strudl-scour. In the early spring, when the rivers thaw, large quantities of water flow out over the shore fast ice. Eventually they find a weak point and break through to the sea below. The velocities are high enough to scour a small crater, 20–40 m in diameter, 10 m or more in depth.

Pingos, the hillocks that arise on the coastal plains due to frost heave, are also occasionally encountered underwater, as relics from the period when the sea surface was lowered. Pingos can have a base diameter of 100 m and a height of 50 m. Forward-looking sonar should be employed when navigating in Arctic shelf waters. Relic pingos have left pock marks up to 100 m in diameter and 10 m deep on the floor of the North Sea and other sub-Arctic areas. These pock marks are believed to have been formed when the ice core melted, allowing the pingo to collapse.

Offshore Labrador and eastern Canada, glaciers have scoured the seafloor clean in many shallow areas. Scour marks from iceberg keels are evident at such locations as the Strait of Belle Isle. Deep scours at depths of 300 m or more have been found. These are perhaps similar to those found in the Antarctic out to a depth of 500 m, which are believed due to glacial tongues which have extended out into the sea in relatively recent times.

In Norton Sound, extensive sand deposits occur, loose near the surface, often densified by wave action at depths of several meters. Overlying these sands are 1–2 m of very loose recent deposits from the Yukon and Kuskokwim rivers. These loose silty sands are very susceptible to movement by currents and waves. The underlying sands are often gas-charged with methane gas.

The Navarin Basin is composed of deep deposits of extremely weak silty clay deltaic deposits, susceptible to slumping even on relatively shallow slopes.

In the southern areas of the sub-Arctic, that is, offshore Newfoundland and in the North Aleutian Basin of the Bering Sea, very dense sands are encountered, the densification apparently being due to storm wave action.

Another geotechnical phenomenon, but of apparent more concern to drillers than constructors, is that of clathrates. These methane hydrates exist as stable solids under proper conditions of temperature and pressure, but expand to 500 times greater volume as methane gas, once the limits of temperature and pressure are exceeded. Clathrates are believed to exist only at depths of several hundred meters or so. However, gas from decomposed clathrates may be one cause of the very weak silt stratum at the interface with the sands described earlier.

The sudden release of methane gas from clathrates during drilling may adversely affect foundation soils and may cause severe downdrag on piles and conductors. Drilling techniques have been developed to prevent this phenomenon from occurring, for example, by allowing early release of pressure before casing is installed.

The Arctic is generally an area of low seismicity.

448 Arctic Marine Structures

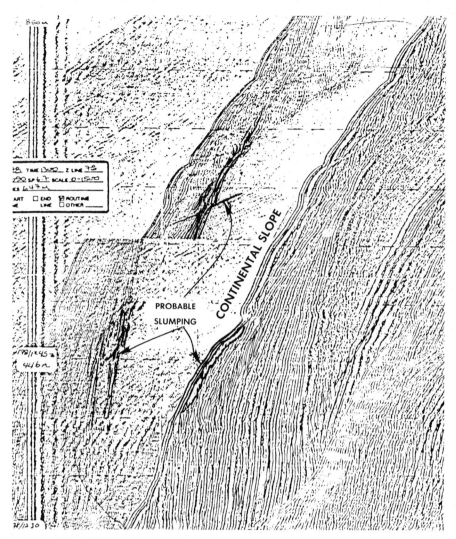

Figure 20.4.1 Acoustic imaging of Arctic seafloor, Alaskan coastal shelf.

However, there are local areas of seismic activity such as near Camden Bay on the North Slope of Alaska and a major area in the northeastern Beaufort Sea, just west of the Canadian Arctic Islands. Earthquakes in this last area may affect structures and facilities in the eastern Beaufort Sea as far south as Tuktoyaktuk, thus requiring consideration of such phenomena as liquefaction of silts and sands.

In the southern Bering Sea, earthquakes must again be considered due to the proximity of the active plate margin just south of the Aleutian chain.

For the constructor, operations in these seismic environments may involve more strict requirements for gradation of imported sands and for their densification and consolidation.

The shoreline of the Arctic coasts and barrier islands is especially sensitive to physical and thermal changes. The permafrost line is near to the surface, actually only 30 cm or so below the onshore tundra. It is this frozen soil which protects the coast, often forming a low bluff under summer wave action, with a thin layer of gravel on the beach proper. Excavations and other constructional activity can upset this equilibrium, leading to large-scale and progressive erosion of the fine sediments.

The barrier islands of the Beaufort Sea are under a more or less continuous process of erosion and deposition. Along the North Slope of Alaska, the currents are predominantly from east to west, driven by the Beaufort gyre. Hence, the islands are eroding on the east and accreting on the western sides. Man-built structures, islands, seawalls, and the like interrupt this process locally.

20.5 Oceanographic

The frequent storms in the summer affect the open-water zone, which has very limited fetch in the north–south direction but much greater fetch in the east–west direction. Typically the waves will be of only 2 m significant height, but when the ice pack has been driven northward, so as to leave 200 km of open water or so, higher waves are possible.

The wave period will be similarly affected, being usually 4–5 seconds but increasing to 6 or 7 seconds when the open-water fetch is extensive. Waves will tend to be quite steep due to the shallow water in the zones of most current construction interest.

Around structures and especially embankments, refraction patterns and directional spread from the rapidly moving storms may create confused and turbulent seas.

The high winds pick up the top of the breaking waves as spray. The spray may become of significant amount, transporting several hundred tons of water onto a structure during a storm, creating problems of drainage. Since the spray can shoot up to 30 m or more around a structure, it can pose problems for helicopter rescue during a storm. During early fall, when the sea is still open but the air is below freezing, severe icing problems may ensue.

The storm durations are usually relatively short, with a total duration of 24–36 hours and a period of peak intensity of 6 hours or less.

There are several interactive phenomena which may lead to significant storm surges, creating both raised water level ($+2$ m) and depressed water level (-1 m). These differences in water level are caused by the low barometric pressure, the strong winds, and the currents. The effect of surges is abnormally great in the Arctic because of the long, shallow shelf and the narrow "river" that exists for the water to escape between the land and the pack ice.

This same combination leads to rather severe currents in the open-water zone, especially between Harrison Bay and Point Barrow, where the bathymetry and close presence of the pack lead to currents of almost constant velocity from the surface to the seafloor.

The Bering Sea is subjected to severe cyclonic storms, not unlike those of the North Sea, especially in its southern stretches. As the storm-driven waves reach the shallow waters of the Yukon delta and Norton Sound, they are depth-limited, becoming very short and steep and piling up several meters of storm surge.

20.6 Ecological Considerations

The Arctic has its own unique ecology, with the preponderance of activity taking place on the coastal plains and in the adjacent open water. Large colonies of geese and other shore birds nest along the coast. Within the leads and open water, photoplankton bloom in the summer flourishes, leading to an ecological chain that culminates in the bowhead and other whales, seals, and the polar bear.

Native Inuits are also part of the ecology. Over the past 12,000 years, they have developed a unique culture that has enabled survival despite the harsh environment.

Construction activities and development obviously have an impact upon this ecology. Some impacts appear positive. The indigenous population has better food, education, health, and communications than it has ever enjoyed. Many of the larger fish, birds, and mammals seem to be doing well, with little adverse effects on their numbers, although they have obviously had to adapt to the changes. The caribou, for example, have learned to cross the Alyeska pipeline, and the shore birds and bears pay little attention to the helicopters.

However, other concerns raised by environmental protectionists are real and have resulted in rules which can and must be followed. Grayling and other fish migrate very close to the shoreline. Jetties and causeways may adversely affect them. Loud noises are believed to adversely affect the bowhead whales, at least if they are in the immediate vicinity, although there is a great difference of opinion as to distance and degree. Low-flying aircraft, high-powered boats, and the dredging of gravel through a pipeline are believed to affect nearby whales.

Dragline dredging of gravel from rivers and river deltas must be carried out in furrows parallel to the flow, not cross-channel, in order not to interfere with fish swimming upstream to breed.

Great concern has been raised about oil spills, especially in the spring, when there is substantial coverage of broken ice. Industry in both Alaska and Canada has therefore formed special oil cleanup task forces so as to be able to respond quickly in case of a spill.

450 *Arctic Marine Structures*

The effects of oil in Arctic waters is a subject of much debate. In the winter, the rough underside of the ice, especially the keels of ridges, will contain the oil. The cold water causes the heavier fractions to congeal and drop to the seafloor.

In the open-water season, the problem is similar to that in temperate zones. Fortunately, most bird activity is inland on the multiple shallow freshwater lakes and tundra rather than on the sea proper.

The spring breakup is the time of greatest concern, with current efforts by industry concentrating on development of a capability to contain and clean up any spill, even in the broken ice.

A constructor will therefore be under special requirements to so conduct his operations as to prevent disruption and adverse impact upon the ecology. This may limit certain activities that would otherwise seem appropriate. These limitations are regional and national in character, with some activities permitted in Canadian waters which are prohibited in Alaskan waters and vice versa. For example, many activities will be prohibited during the fall migration of the bowhead whale in September and October, which are the same periods when the constructor is trying to finish his work for the season.

Two elements of the ecology are aggressive to humans, the polar bear and the mosquito. The polar bear considers humans his natural prey. He is quick and agile on the ice and in and under the water. Isolated survey and geotechnical crews must be accompanied by an indigenous hunter ("bear watch"). Care must be taken in approaching an abandoned camp or facility. Bears are attracted to human activities and are very curious: They have been known to climb aboard vessels and rigs which were locked in the ice.

The mosquito is less deadly, carrying no serious diseases, but nevertheless emerging in summer in such great numbers as to impede activity. Fortunately, its realm rarely extends more than a mile offshore.

20.7 Logistics and Operations

The Arctic is like the desert, with great distances between the small habitations, a generally flat and featureless terrain, low precipitation, great extremes of temperature, a small indigenous population, and very little infrastructure to support activities. Aggravating the problems is the short construction season.

Point Barrow, at the northern tip of Alaska, is a critical logistics pivot around which all floating structures, barges, and vessels must transit. Point Barrow itself is some 2200 miles from the Pacific Northwest of the United States, almost 3000 miles from Japan and Korea. Vessels and floating structures must cross the North Pacific just to reach the passes through the Aleutian Chain and then proceed through the shallow but sometimes rough waters of the Bering and Chukchi seas to reach Barrow. See Figure 20.7.1.

At Barrow, the ice usually recedes about the first of August. Thus passage is delayed, even though further east the Beaufort Sea may have had partial open water in late June or early July.

The polar pack never recedes far from Point Barrow, thus restricting the channel to a few miles. Unfortunately, these few miles have limited water depth (about 10 m) extending out some 7 miles. The channel typically remains open until about September 15, when freezeup starts.

The other route to the Beaufort Sea is by barge down the Mackenzie River, which generally breaks up in mid-June, thus allowing access 6 weeks to 2 months earlier than Barrow, but of course restricted as to size of cargo.

The Alyeska Pipeline Road is available for commodity shipment through most of the year, with limitations in late spring due to frost heave. Air service is generally good to the Beaufort Sea, with adequate fields and communication. Planes have landed successfully on ice airfields. Airdrops have been used for many commodities—even timber piles, for example.

While other areas of the Arctic and sub-Arctic may have somewhat less demanding logistical requirements, none are easy nor inexpensive. Staffing points are required, support bases must be developed, and winter layover areas must be established.

Shore bases must have a year-round water supply, year-round fire protection, provision for disposal of garbage so as not to attract bears, and provision for sewage disposal. This last will generally be chemical, since the permafrost is near the surface. Sudden winds can be very serious, especially if they collapse or tear off a roof in winter, when the outside temperature is $-40°C$.

Because of the conditions described above, aircraft and special types of vehicles have been exten-

Figure 20.7.1 Caissons for caisson-retained island Tarsiut were delivered on large submersible barge to the Canadian Arctic.

sively used in construction. Several thousand rock gabions were transported by helicopter to provide additional protection against wave action for Tarsiut Island. Icebreaker supply boats have been used to haul construction materials and equipment. Fixed-wing aircraft up to the C-5 Hercules, have made airdrops on the ice. An air-supported vehicle of the hovercraft type has proven its ability not only to ride over both ice and water but incidentally to use its air to break thin ice ahead of itself, allowing barges and boats to operate as late as November 1.

Helicopters have been used to tow loaded hovercraft and ice sleds carrying cargo. An Archimedes-screw-propelled vehicle has proven able to cross both open water and ice in essentially all seasons. For under-ice surveys, ROVs appear to offer the best potential, although under-ice diving has been extensively employed in the Arctic Islands of Canada.

An ice-strengthened cargo barge pushed by two pusher tugs has proven effective in opening the channel around Point Barrow in thin ice.

Meanwhile, advanced icebreakers for industrial and commercial support are being built and deployed by Canada, Finland and the USSR. The U.S. Coast Guard icebreakers are unfortunately designed for multipurpose activities, making them essentially unavailable for Arctic marine construction. The offshore petroleum industry has developed designs for icebreakers and ice-strengthened supply boats ready for procurement as the need arises.

Construction operations in the Arctic must be planned in extreme detail. Literally each tool and each material item needs to be considered as to where and when it is needed, how it gets there, and how it will be used in the extreme cold, wind chill, and darkness. Equipment needs to be specifically designed for the Arctic. Ordinary greases freeze up; silicon grease must be used instead.

20.8 Earthwork in the Arctic Offshore

Most of the structures constructed to the date of this writing have been in the shallow waters of the Mackenzie delta and adjacent Canadian Beaufort Sea and in Prudhoe Bay. The Canadian structures have been largely built of sand, dredged and placed by hydraulic dredges when a sand source could be found nearby, or with trailing suction hopper dredges when the source was distant. In some cases, the source has been 50–80 km away from the site. The deepest water in which such an island has been constructed to date is 62 ft (19 m) at Issungnak, which involved almost 5 million m³ of sand. See Figure 20.8.1.

Deposition has been largely by hydraulic or barge discharge at the surface, resulting in very flat side slopes, as flat as 15:1. Use of special discharge tremie pipes, with devices at the tip to slow the velocity, have resulted in denser placement and steeper slopes, about 5:1 to 6:1.

It is of course not only a matter of placement technique but also of gradation which determines the slope and density. Some islands have experienced progressive liquefaction when they were built up too rapidly with material that did not achieve the required density. In one case, multiple slope failures were triggered one after the other and the embankment spread out so far that it had to be abandoned.

Slope protection is required as the island emerges from the water. The sudden storms of the open-water season can quickly erode thousands and even tens of thousands of cubic meters of sand.

One solution that has been adopted—with only moderate success, however—has been to build the island up to just below sea level, hold until late in the season when most of the storm activity is over, and then rapidly complete the above-water portion.

The short duration of the open-water season requires a large dredging capacity and continuous operation. Trailer suction hopper dredges with ice-strengthened hulls have been able to work up to November 1 in the eastern Beaufort Sea.

One potential problem for above-water embankments is the inclusion of ice fragments in the fill. These may thaw in the following summer or when hot oil is produced, leading to sudden slumping. This has occurred in similar construction ashore.

For the Alaskan Beaufort, gravel has been available from the river deltas of the Sagavanirktok, Kuparuk, and Colville rivers. This gravel has often been transported by ice roads on the fast ice during January, February and March for direct dumping through the ice. Side slopes of 3:1 are achieved. For islands further offshore, such as Mukluk, the gravel was stockpiled in the winter and then hauled by a fleet of barges for placement the following summer. When scraped or pushed off barges, gravel achieves slopes of about 5:1. See Figure 20.8.2.

Slope protection has been largely accomplished by the use of plastic bags, filled with sand, placed in one or two courses over a heavy filter fabric. The bags are UV-stabilized, 2–4 yd³ in size and double-strength. These have worked well for exploratory islands but suffer tearing and dislocation from the ice during ice movement and breakup, requiring substantial annual maintenance. See Figures 20.8.3 and 20.8.4.

Articulated concrete mats, attached to or laid over filter fabric, have performed well in tests and appear to be more permanent and require less maintenance. As a result, they have been adopted for use on North Star Island and Endicott Island, two of the first production islands to be built in the offshore Arctic. See Figure 20.8.5.

Freezing of embankments, both above and below water, has attracted many engineers. Natural freezeback occurs in above-water and near-surface soils. Artificial freezing has been proposed to stabilize and strengthen underwater embankments.

However, there are several difficulties. First, although the natural air temperature is low, the heat of fusion is the same whether in the Arctic or in a temperate zone. Secondly and more seriously, there is little experience with saline soils. Such as there is indicates that brine channels are formed, with the freeze front progressively driving increasing brine concentrations ahead so as to form unfrozen brine lenses and hence potential failure planes.

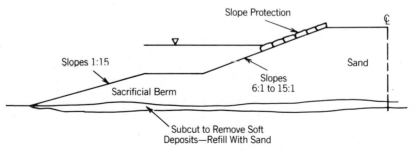

Figure 20.8.1 Sand island.

20.8 Earthwork in the Arctic Offshore

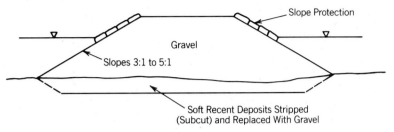

Figure 20.8.2 Gravel island.

A construction dike formed around the site for the Prudhoe Bay Salt Water Treatment Plant (for the Waterflood Facility) was designed on the basis of natural freezeback. Portions of this failed, however, apparently due to unfrozen lenses of high-salinity water.

Freeze pipes can be installed from the embankment or structure in drilled or jetted-in pipes or laid as mats between layers of embankment.

To be effective in increasing shear and bearing capacity for Arctic offshore structures, the soil temperature should probably be lowered to at least $-10°C$.

Frost heave during freezing must be considered. The effects can probably be minimized by following a progressive pattern so as to drive the expansion out from under the structure.

Freezing takes time, usually several months on a large installation such as those contemplated for offshore structures. Once frozen, the condition can be rather easily maintained by either active brine circulation at low volumes or by passive means such as the thermopiles used on the Alyeska pipeline in Alaska.

Frozen soils sustain short-term loads well but have high creep under sustained loads such as may occur under winter ice movements.

The soft, silty clay soils of the Arctic seafloor place severe requirements on concrete and steel structures designed for use in the shear zone. Therefore, a great deal of attention has been devoted in recent years to the development of effective and economical methods of improving the soils. Since the silty clay and clayey silt sediments are anisotropic in character, with greater horizontal than vertical permeability, the concept of consolidation

Figure 20.8.3 Mukluk Island, a gravel island with extensive slope protection, located in the Beaufort Sea, Alaska. (Courtesy of Sohio Petroleum.)

Figure 20.8.4 Construction of Seal Island, Beaufort Sea, Alaska. (Courtesy of Shell Oil.)

through drainage and surcharge appears attractive. To accomplish this requires the placement of an impervious membrane, the installation of wick or sand drains, and the provision of pumping capability. Practical considerations indicate that often the best way to carry out the consolidation is after the structure is installed, using the base of the structure plus its skirts as the membrane and draining it to the interior.

It is also considered practicable to install the vertical drains prior to arrival of the structure, provided an initial sand or sand and gravel embankment has been placed over the impermeable surficial soils of the seafloor.

Figure 20.8.5 Articulated concrete mats installed on Beaufort Sea production island. (Courtesy of Tekmarine, Inc.)

20.8 Earthwork in the Arctic Offshore

Another scheme is based on the installation of sand piles prior to arrival of the structure, designed to enhance the bearing and shear resistance of the soil initially, to be supplemented by drainage and consolidation later. Such underwater sand piles could be installed using a 1-m-diameter steel pipe mandrel with a hinged or expendable plug in its tip. This would be driven to the required depth by an impact or vibratory hammer. During driving, the mandrel would be filled with sand by means of a hopper. The top of the pipe would then be closed and low air pressure applied to keep the sand in place in the hole while the mandrel is withdrawn. Only 10–20 psi of air pressure is required.

Dredging of the softer materials and their subsequent backfilling with imported sand is a highly practicable and effective solution if the depth of the soft material is limited to a few meters and if the soft material will stand on reasonable side slopes during the dredging. Both these requirements are met in many areas of the eastern Beaufort Sea. The advantage of such a system is that it is positive, that it is not based on time-dependent consolidation, and that it is not affected significantly by the characteristics of the weak soils that will be removed.

Surcharging around the perimeter of a structure may be an effective means of enhancing the resistance to sliding by prolonging the shear path. This surcharge can best be applied after the structure has been installed and thus can use the structure as a reference marker for control of the dumping operations.

Another means proposed has been that of penetration, in which the structure is equipped with long skirts so as to penetrate through the soft soils. This scheme has been proposed for the Troll Field in the Norwegian North Sea.

In the soft soils of the Arctic and sub-Arctic such as the Navarin Basin, the penetration may be aided by removal of the soft soils from within the skirts, for example, by jet and airlift or by jet and eductor. See Figure 20.8.6.

The eductor is essentially independent of column height and hence more effective for discharge over the base of the structure, whereas the airlift is more

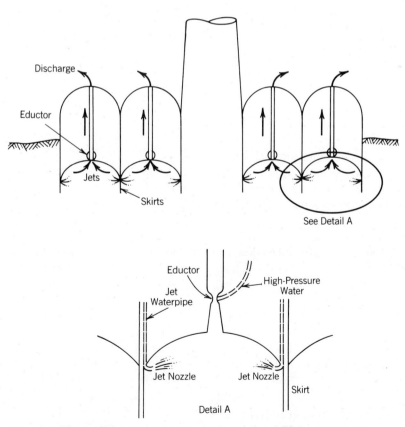

Figure 20.8.6 Use of jets and eductors (or airlifts) to remove soft material from within skirt compartment.

effective with a high column and discharge at sea level.

Finally, it is practicable to support a caisson structure by means of piles. Sleeves would be provided within the structure or as skirts attached to the structure. Piles would be driven through the sleeves, to be later connected by grout. The key problems are the length of sleeve necessary to develop the pile capacity, which is difficult to provide within the depth constraints typical of Arctic coastal shelf, and the problem of obtaining adequate grout strength under the low temperatures of the Arctic seawater. High-alumina cement has been used effectively for the cementing of oil wells and for the grouting of piles on the Sag River Bridge.

20.9 Ice Structures

It is only natural that early attention was given to the use of ice as a structural material for temporary structures in the Arctic. In World War I, a considerable amount of research was directed toward a mixture of ice and sawdust, known as *Pi-Crete*, which it was hoped would be suitable for use in constructing temporary floating airfields so as to enable the planes of that day to transit the Atlantic. In more recent times ice has been utilized for ice roads out over the fast ice and as temporary drilling platforms in the fast ice wedged in the channels between the Arctic Islands.

The ice roads are constructed by progressive flooding, allowing each layer to freeze. Dikes are built of snow and broken ice rubble. The water is obtained from holes drilled through the ice. Such roads have carried heavy trucks and have been used for the delivery of gravel to offshore sites. The gravel is then dumped through slots cut in the ice at the site.

The movement of trucks, especially when in convoy at fairly high speeds, can produce resonant undulations in the flexible floating strip and lead to failure. Spacing and speed need to be controlled so as to prevent this.

API Bulletin 2N, in the section "Design of Ice Roads," states:

Attention is directed to the problems of edge loading, and of cracks opening due to tidal, thermal, wind and other forces.

Thermal cracks generally appear in ice roads following construction. Such cracks are particularly noticeable following snow clearing and large changes in temperature. Wet cracks may heal—dry cracks often have to be repaired with slush or water.

A particularly dangerous situation can occur if two wet cracks join to form a wedge.

Attention is directed to the spacing of vehicles traveling in the same and opposing directions, to dynamic amplification due to waves created by the vehicle, and to fatigue, caused by heavily-loaded trucks traveling at close intervals.

The life of drilling pads built up of layers of ice by progressive flooding and freezing has been extended into the early summer by the use of polyurethane insulating blankets.

Considerable experimentation has been carried out by Union Oil, Exxon, Sohio, and Amoco in developing artificial ice islands for use as year-round drilling islands in the shallower waters of the Beaufort Sea. Water has been sprayed out over the ice using the large circular spray systems more usually employed in farm irrigation. This allows almost instant freezing and thus a continuous operation.

A major problem with these islands occurs in spring, as the opening water thaws and erodes under the perimeter ice edges, causing the resultant overhangs to break off. Other problems encountered are propagating cracks due to thermal stresses.

API Bulletin 2N, in sec. 9.3, "Maintenance of Frozen Structures," addresses the need to monitor the behavior of frozen soils and embankments:

Cracks in the frozen earth structure may be repaired by filling them with water.

An adequate internal temperature monitoring system is required.

Maintenance of desired internal temperatures may be achieved by:
  a. Balancing natural cooling and heating
  b. Selective insulation, such as installing a cover in the spring, and removing it in the fall
  c. Convective refrigeration, such as "freeze piles"
  d. Active refrigeration

Artificial ice rubble is being generated around more conventional drilling structures of steel and concrete in order to provide a cushion against the impact of large ice floes. This has been done in several ways:

1. Construct an underwater berm or embankment of sand or gravel on which the ice features will

ground, thus progressively forming an encasing rubble pile.

2. Accelerating this natural action by mechanically adding ice so as to weight down the sheet adjacent to the structure until it grounds. This can be done by icebreakers pushing the ice ahead during the early winter or by dozers working on the ice that is at least temporarily fast to the structure. Cranes have been used to lift and place blocks of ice cut from the adjacent sheet.

3. Spraying water out from monitors mounted on the structure. The water freezes as it falls through the air and quickly generates massive rubble. This system was developed by Sohio and Exxon and is being used to provide a rubble encasement around the Super CIDS exploratory drilling structure at a location offshore Cape Halkett, Alaska, in a water depth of 16 m. Amoco has successfully constructed an exploratory drilling platform of ice in 7 meters of water.

4. Considerable experimentation has been carried out by Sohio on the use of steel dolphins (pyramidal frames) seated on the seafloor, designed to intercept the moving ice and cause it to raft, thereby forming a rubble pile.

The use of ice for temporary structures will undoubtedly be extended as further work is carried out in the Arctic. Its limitations are due to the fact that ice is a solid at a temperature near to its melting point, that it is subject to brittle fracture, and that thermal strains and water erosion can lead to early and sometimes sudden failure.

20.10 Steel and Concrete Structures for the Arctic

A wide variety of structures of steel and concrete have been designed for the Arctic and sub-Arctic, designed to resist the high lateral forces from the ice and to transmit these forces down to the foundation soils.

20.10.1 Steel Tower Platforms

Tower-type structures have been used for many years in Cook Inlet, Alaska. These steel structures are constructed like jackets, being built in a fabricating yard, launched and delivered as a self-floating structure, upended to set on the seafloor, with steel piles driven or drilled and grouted, either through the large legs or as skirt piles; see Figure 20.10.1. These structures are suited for moderate-depth waters and moderate ice conditions such as those of the southern Bering Sea.

The legs of the jacket must be spread widely apart to prevent the ice sheet from arching between them. Through the ice zone, the legs will require special reinforcement, e.g., sandwich steel-concrete construction, in order to resist ice impact.

20.10.2 Caisson-Retained Islands

Caisson-retained islands have been developed as an extension of the surface-piercing island concept described earlier. These perimeter caissons provide protection through the air–water interface, preventing wave erosion during construction, reducing fill quantities significantly, and facilitating access to the structure. They reduce the overall construction time by permitting concurrent construction operations. See Figures 20.10.2.1, 20.10.2.2, and 20.10.2.3.

The caissons themselves may be of steel or con-

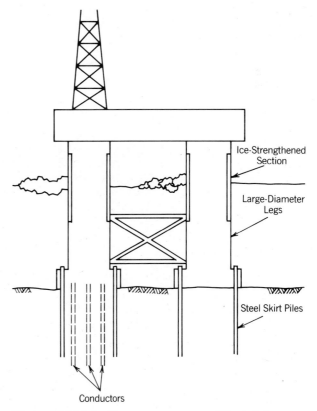

Figure 20.10.1.1 Tower-type structure (Cook Inlet-type, modified) for moderate sea ice conditions.

458 Arctic Marine Structures

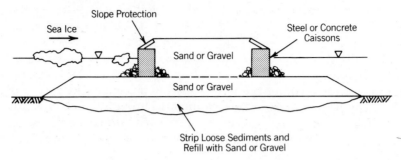

Figure 20.10.2.1 Caisson-retained island.

crete, filled with sand. Joints are provided between the adjoining caissons.

A concrete caisson-retained structure was built for Tarsiut Island and a steel caisson–retained structure is being employed by Esso, both in the Canadian Beaufort Sea.

20.10.3 Gravity-Base Caissons

Gravity-base caissons are large structures of concrete and/or steel founded either on the existing seafloor or on a prepared underwater embankment. Dome's SSDC-1 was such a structure, constructed from a section of a VLCC hull reinforced internally by concrete and steel. Global Marine's Super CIDS (GBS-1) is a gravity-based caisson constructed of three sections, the base and deck being steel, the midsection being of prestressed lightweight concrete, with the three sections joined together to act monolithically. Gulf Canada's Molikpaq (MAC) is a gravity-based caisson of steel, filled with sand when on site.

Many concepts have been developed, but they all tend toward axisymmetrical structures, circular or polygonal in plan, with sloping or vertical walls; see Figures 20.10.3.1, 20.10.3.2, and 20.10.3.3. The cone shown in Figure 20.10.3.3 is designed to fail the ice in bending (flexure), whereas the near-ver-

Figure 20.10.2.2 Four caissons installed for Tarsiut Caisson Retained Island, Beaufort Sea, Canada. (Courtesy of Gulf Canada and Dome Petroleum).

Figure 20.10.2.3 Completing the filling operations at Tarsiut Caisson Retained Island. (Courtesy of Gulf Canada.)

Figure 20.10.3.1 BWACS concept for exploratory drilling platform for shallow Arctic waters. (Courtesy of Zapata Drilling and Brian Watts Assoc.)

460 *Arctic Marine Structures*

Figure 20.10.3.2 SAMS concept for exploratory drilling platform in weak soils of Arctic Ocean (Courtesy of Sohio Petroleum, ABAM, PMB Systems Inc., and Ben C. Gerwick Inc.)

tical-sided caissons shown in Figures 20.10.3.1 and 20.10.3.2 are designed to fail the ice in crushing and shear.

Additional resistance against sliding in weak soils can be developed by the use of piles or spuds, such as those designed for the Sohio SAMS. See Figure 20.10.3.4.

20.10.4 Jack-up Structures

Considerable conceptual design effort has been given to the development of jack-up rigs of various types designed to float into location on a large hull, which will then be seated on the soil, raising its deck above the sea and ice. The narrow column then

Figure 20.10.3.3 ACES concrete offshore platform for use in offshore areas of Arctic Ocean of Alaska and Canada. (Courtesy of Brian Watt Assoc.)

20.10 Steel and Concrete Structures for the Arctic

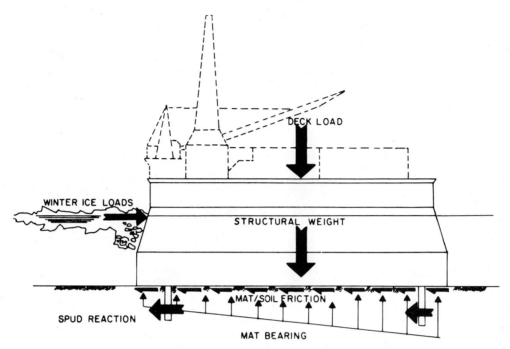

Figure 20.10.3.4 Schematic presentation of shear transfer of lateral ice pressure to foundation soils by use of spuds. (SAMS concept: Sohio.)

presents a minimum face to the ice and is intended to reduce ice forces acting on the structures. See Figure 20.10.4.1.

While the concept may eventually be developed into reality, none has yet been built because of concerns over whether the extreme ice loads, such as those from a multiyear floe or short heavy ridge, will really be reduced by the narrow shaft and because of potential problems of dynamic amplification under continuous ice crushing, which has led to the failure of several lighthouses in the Baltic Sea.

20.10.5 Deep-Water Structures

Bottom-founded structures for deeper water are being conceived as cones, stepped pyramids, or monopods in order to reduce quantities of material, limit the weight on the seafloor, and reduce the ice forces. Their construction and installation would be similar to those for the North Sea gravity-based platforms; see Figures 20.10.5.1 and 20.10.5.2. Installation of these structures requires very careful calculations of stability as the structure is ballasted down, due to the decreasing waterplane. Means of countering this, such as the use of temporary buoyancy tanks, are described later in this chapter, in Section 20.12, "Installation at Site."

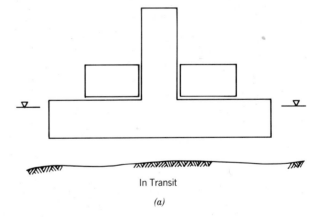

In Transit

(a)

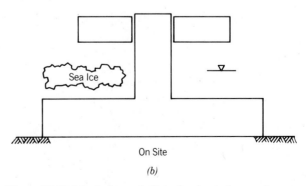

On Site

(b)

Figure 20.10.4.1 Jack-up platform for Arctic (concept).

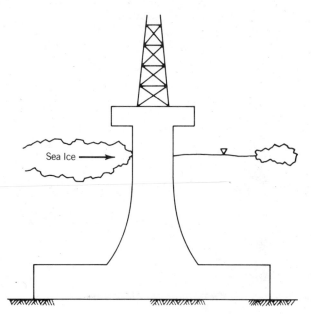

Figure 20.10.5.1 Monotower concept.

20.10.6 Floating Structures

For deeper waters, floating structures have been developed to be permanently or temporarily moored to high-capacity anchors. These structures generally resemble the caisson structures and are built of steel, concrete, or a combination of the two.

The Kulluk exploratory drilling platform, designed to work in moderate sea ice, is an inverted cone, so that the ice is broken and deflected downward. This structure is built of steel; an earlier design of this concept based on concrete was developed by the author; see Figure 20.10.6.1. The Kulluk has successfully extended its drilling season to the first part of December.

A spar-type structure in concrete was designed for Exxon, to break the ice upward or downward depending on the draft; see Figure 20.10.6.2. This structure was designed for sub-Arctic areas, with unconsolidated ridges up to 10–15 m.

For less severe ice conditions, shipshape hulls with turret moorings have been proposed. These are designed to weathervane with the ice movement. Thrusters may be installed to optimize this rotation and to take care of adverse combinations of wind, current, and ice. The bow will usually be wider than the body so as to facilitate clearance of the broken ice. Sides of the hull may be inclined inward so as to break the ice downward.

The major problem for all floating structures is the design of high-capacity moorings and anchors, since the forces in heavy ice are almost an order of magnitude higher than those experienced from wave forces in such locations as the North Sea.

20.10.7 Well Protectors

Once the structure is properly moored on location, a glory hole is usually excavated in the seafloor, enabling well control gear to be placed below the

Figure 20.10.5.2 Stepped-pyramid concept for gravity-base structures designed to resist iceberg impact.

20.11 Deployment of Structures in the Arctic

Figure 20.10.6.1 Kulluk floating drilling platform being deployed in Canadian Beaufort Sea. (Courtesy of Gulf Canada.)

level of potential ice scour. This has been accomplished by very powerful jets and by the use of a vertical dredge head. Glory holes are obviously best suited to floating drill vessels.

An alternative to the glory hole, better suited to seafloor-founded installations, is to sink a well protector caisson (of perhaps 10-m diameter) down into the seafloor using a combination of jets and the weight of the shell along with airlift or eductor excavation. Such well protector caissons can be either of steel or concrete. A double steel shell can be easily transported to the site, hung under the drilling derrick, and pumped full of concrete which will set at low temperatures. Then the weight of the concrete is available to aid the penetration. Another alternative is the use of precast concrete segments, posttensioned together vertically with prestressing bars.

20.11 Deployment of Structures in the Arctic

The various caisson-type structures are typically planned for construction in the warm-water ports of the temperate regions for delivery afloat to the Arctic. While the caissons for the Tarsiut Island Caisson Retained Island were delivered on a large submersible barge, most have been and will continue to be delivered as self-floating structures, complete and monolithic, fully outfitted, so as to minimize the subsequent labor in the Arctic. See Figure 20.11.1.

The tow across the North Pacific, whether from Japan–Korea or from the Pacific Coast of North America, is so long as to have a high probability of encountering a major summer storm enroute, with its relative long-period waves typical of the Pacific. It is therefore necessary to thoroughly investigate

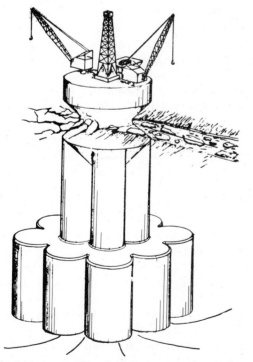

Figure 20.10.6.2 Floating spar platform for use in sub-Arctic waters is designed to break ice both downward and upward. (Courtesy of Exxon Production Research.)

Figure 20.11.1 Global Marine's Super CIDS GBS-1 under tow from Japan to the Arctic.

the dynamic response of the structure in such sea states. While recent advances in hydrodynamic analysis such as strip theory make it possible to analyze the response of such a structure, it is difficult to include damping aspects without a physical model test. Fortunately, in many cases these will show that the theoretical amplifications are significantly damped, so that the structure's response remains within acceptable limits.

Special attention has to be directed to cones and similar structures where the wave runup over the sloping surface may lead to erratic lurching.

In a major storm, the tugs may have to cast the structure free. The route should therefore be selected so as to keep adequate sea room. After the storm the tugs can pick up the tow once again. Free-floating hawsers (Kevlar or nylon) attached to spare pendants will facilitate the pickup.

Most tows will be required to be manned, in order to operate pumps if necessary. Full provision must be made for the safety of the men on board, in compliance with the regulations of the relevant agencies, for example, the U.S. and the Canadian Coast Guards.

Because of the length of the tow, provision may have to be made for refueling of the tugs en route.

The size and number of tugs will vary with the structure's size, displacement, and so forth, but most frequently more than one tug of high horsepower (greater than 16,000 HP) will be required. Arrangements for multiple tugs will be similar to those in the North Sea. Because of the longer length of Pacific waves, towlines will probably be longer. Axisymmetrical structures such as those used in the Arctic have little directional stability, tending to yaw excessively. Skegs will usually be found effective.

Upon arrival at Point Barrow, tows will stand by awaiting the opening of a suitable channel through the ice. The depth at Point Barrow inside the shoal 7 miles offshore is a maximum of 10–11 m. Unfortunately, the pack ice usually hugs this shoal until September, leading to a low probability of successful passage around the seaward side of the shoal. When several class 4 to class 6 icebreakers become available, then deeper-draft tows will be limited only by the shallow waters of the Bering Sea and draft at the installation site.

The towing of large offshore platform structures in the shallow waters of the Arctic under conditions of broken ice, up to $3/10$ and $4/10$ coverage, presents many new problems. However, it has been satisfactorily accomplished on several major wide-beam structures, the Prudhoe Bay Seawater Treatment Facility, the Kulluk, the SSDC-1, the Molikpaq, and the Super CIDS (GBS-1) being among the important achievements.

The Molikpaq and GBS-1 were towed in convoy, led by a class 4 icebreaker vessel. Ice conditions became extremely heavy as the convoy approached Harrison Bay. Towing with one boat on a diagonal proved most effective. Towline snagging on ice resulted in severe impact forces on the towline, attachments, and boat. The tow speed averaged 1.3 knots over the almost 500 miles of distance from Point Barrow to Herschel Island. For the GBS-1, the tow from Point Barrow to Point Halkett took 4 days. Drafts of the two structures were 9–10 m.

From Point Barrow eastward through the Beaufort Sea the water depth is very limited. The underkeel clearance is minimal, so most ice must clear around the sides. If the structure is polygonal, towing with one point forward is best so that the broken ice will be forced to the side, preventing buildup of a rubble pile ahead.

Due to shallow water in the Beaufort Sea, sway, yaw, and squat are potential problems. In actual experience, low speed, the use of several boats, and the confinement of the ice have tended to minimize these adverse effects. Towing skegs may be effective in limiting the sway and yaw.

The towlines for crossing the Pacific will normally have been secured below the waterline, at the center of rotation, in order to keep a favorable trim, which is usually slightly down by the stern.

For the Point Barrow passage, trim will need to be equalized in order to minimize draft. If the towlines are below water, they may tend to snag under ice blocks due to the irregular speed of the tugs in the ice. If attached above water, they may create an unfavorable trim down by the head. A vertically oriented bridle may be the optimum solution.

On the Molikpaq tow, the initial passage was made by pulling with two icebreaker tugs. As one tug encountered heavy ice, it would slow. Then as the ice broke and the tug surged forward, the towline would snap up, causing severe impact on the fittings of both tug and tow and threatening to break the towline. API Bulletin 2N, sec. 7.3, notes:

The effects of ice accretion on mooring lines, both above and below water, should be considered. Mooring line buoyancy and drag forces, and consequently tension forces, may be affected by icing.

This is an area where nylon or Kevlar towlines may be more suitable than wire lines. In the case of the Molikpaq, they switched to using one icebreaker to break ice and other to tow; this improved the situation.

Crowley Maritime has used pusher tugs effectively to push an ice-strengthened barge so as to break a channel in thin ice. Consideration should be given to installing pusher notches and fittings on the stern of the structure itself and to using it as an icebreaker, since most structures have the necessary strength to resist the impact of ice.

If in the future, deep-water structures will have to be towed through or deployed in pack ice, then the structure may well be configured *at the towing draft waterline* so as to efficiently break ice and to deflect the broken ice to the side so that it will not jam the passage. Past experience (e.g., the *Manhattan* voyage) has shown that the ice pack can be opened to pass a 30-m-wide ship. In the future, the structures may be 150–200 m in beam.

Where skirts are installed under the structure, an air cushion may be used to reduce draft during the Point Barrow passage. Since with partial ice coverage there is a low sea state, the "plug" can typically be reduced to about 0.5 m.

20.12 Installation at Site

Caisson structures for the Arctic will usually be installed in a manner similar to that employed in the North Sea. Upon arrival at the site, the tugs will fan out in a star formation, controlling lateral position and orientation as the structure is ballasted down to the seafloor. Positioning is controlled by a combination of satellite navigation, medium-range electronic-positioning systems such as SYLEDIS, and seafloor transponders.

Some of the conical structures and monopods proposed for the Arctic encounter stability problems at this stage. There are two solutions.

In waters of limited depth, a conical structure may be tipped down, keeping waterplane stability until one side has touched down and then using the seafloor for stability as the structure is ballasted down to its permanent attitude. Such a system has been used in shallow waters in the Gulf of Mexico and is presumed suitable for exploratory drilling platforms in the Arctic. However, it is highly questionable for permanent production platforms because of the angle to which the deck and consequently the processing equipment must be tilted and because of potential disturbance to the foundation.

The other method is to install temporary buoyant columns which give righting moment stability to the conical structure during the critical submergence phase. After installation, these temporary columns are removed.

For all gravity-base structures, as they near the seafloor, a cushion of water will be trapped underneath which must be vented internally or displaced laterally. Dowels or spuds can be dropped so as to engage the bottom and hold the structure in position. These spuds will have to penetrate the soil; in some cases they may have to be locked to the structure in order to use its weight to force penetration. The rate of descent can be reduced so as to allow the water to escape without creating piping channels. Slow descent and adequate venting of the trapped water will also prevent lateral "skidding" of the structure.

The structure will be seated down upon a seafloor which overall is usually level, but which will have been plowed by ice scour and thus have numerous local deviations in profile and strength, with ridges and furrows, so as to create very high local pressures acting on the base. Figures 20.12.1 and 20.12.2 show Global Marine's GBS-1 (Super-CIDS) seated on location in the Beaufort Sea.

In order to provide a uniform foundation for a gravity-based structure, attempts have been made to pre-place sandfill and then screed it off as a level base to receive the structure. This was successfully carried out underwater for the Prudhoe Bay Waterflood Project Seawater Treatment Facility for a barge-mounted facility 610 × 150 ft in plan; however, this was done in shallow-water Prudhoe Bay within a diked-off area where the sea was essentially still.

Out in the open sea, attempts so far to screed from floating equipment have achieved only a ±30-cm tolerance, despite the use of heave compensators. Recent reports indicate that for the 1984 installation of Molikpaq, a tolerance of 15 cm was achieved by use of a computer-controlled drag arm from the dredge. For several of these structures, the constructors resorted to underbase filling with a sand slurry flowed in through piping installed within the structure. The sand is first mixed into a slurry, then pumped with a very low head, approximately that of gravity, to flow out under the base of the structure. This is similar to underbase grouting as practiced under the North Sea structures, but has the advantage of being a permeable, cohesionless foundation, able to adjust to settlements, to be refilled if necessary; and to facilitate subsequent refloating.

When using sand underbase fill, the structure must first be seated, either on three or four pads that have been pre-leveled or on mud mat–type feet extending down from the structure. For the SSDC-1, the pads were concrete slabs, with a polyurethane mattress affixed, which provided support under the low bearing pressures during founding but crushed when the structure was fully ballasted.

For permanent structures, a low-strength grout can be pumped in underneath instead of a sand slurry. However, for a permanent production island, where thaw subsidence is expected to cause a dished settlement under the center of the structure, sand may be selected because of its ease of periodic refill.

If grout is to be used, return valves should be used which enable grout lines to be flushed with water after initial grouting so that they can be reused later. The grout mix chosen must be capable of hydrating and gaining strength under temperatures of 0°C to −2°C.

Skirts are generally required to seal around the perimeter of the caisson to prevent scour, enhance shear transfer to the soil, and contain sand or grout underbase fill. Upon landing, the skirts must be forced to penetrate by the weight of the structure. In the shallow waters typical of the Arctic, the structure weight may be limited. Hence maximum ballasting weight will often be required.

Some exploratory and probably all production structures will be ballasted with sand slurry so as to provide the needed weight for stability under maximum ice loads. This slurry will be pumped in through pre-installed pipes and discharged into the various ballast tanks.

After the structure is seated, scour protection may be required. This can best be done if filter fabric has been pre-attached to the sides of the structure just above the base. After seating, the fabric can be laid out laterally and rock or articulated concrete mats laid. The two concepts can be merged into one, with the articulated mats being pre-attached and transported by the structure until founding, after which they are laid out laterally.

If bearing piles or spud piles are to be installed after seating, their driving procedures will depend on the soil conditions which are anticipated. A structure developed for Sohio known as SAMS (Sohio Arctic Mobil Structure—not yet constructed as of 1986) incorporated steel spuds 84 in.

Figure 20.12.1 GBS-1, the Global Marine Super CIDS exploratory drilling platform, during summer open-water season in Alaskan Beaufort Sea.

20.12 Installation at Site

Figure 20.12.2 GBS-1 (Super CIDS) on location in Alaskan Beaufort Sea. Note creation of artificial rubble pile by spray. (Courtesy of Global Marine Development Co.)

(2.12 m) in diameter, with walls up to 3 in. (75 mm) thick. After landing, the spuds were to be driven or jacked through weak overlying soils into stronger soils below. See Figures 20.12.3, 20.12.4, and 20.12.5.

In overconsolidated silts, high-pressure jetting within the pile was felt to be essential to break up the plug. Concurrent driving with an impact or vibratory hammer should then achieve penetration.

Similarly, in fully ice-bonded sands (permafrost), high-pressure jetting is believed to be the most effective means of breaking down the material in order to achieve penetration of piles.

For subsequent removal of piles driven into permafrost, any freezeback can be destroyed by injecting steam into the water of the core of the pile. Raising the height of water in the pile to above sea level will help to break the bond of the soils by in-

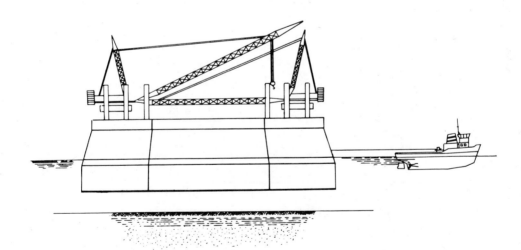

(1) DEPLOYMENT
1. STRUCTURE TOWED TO SITE WITH SPUDS PREINSTALLED IN SLEEVES
2. SPUDS HELD IN PLACE BY SLIPS & WELDED RIBS
3. ALL EQUIPMENT ON BOARD
4. SLEEVES CLOSED WITH RUBBER CLOSURES

Figure 20.12.3 Sohio's mobile Arctic exploratory drilling platform SAMS being deployed to site of installation. Courtesy of PMB Systems Engineering Inc., ABAM Engineers, and Ben C. Gerwick, Inc.

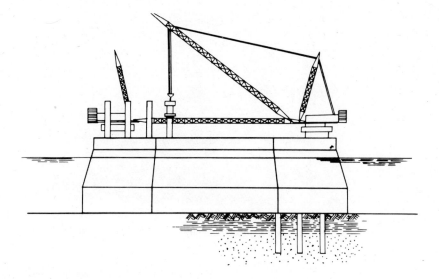

② INSTALLATION
1. SPUDS VIBRATED DOWN (SHOWN) OR JETTED TO GRADE.
2. TOP OF SPUDS DRIVEN BELOW TOP OF DECK.
3. SPUDS SHIMMED AND BOLTED OR WELDED IN PLACE.

Figure 20.12.4 Installing spud piles for SAMS exploratory drilling platform. (Courtesy of Sohio Petroleum.)

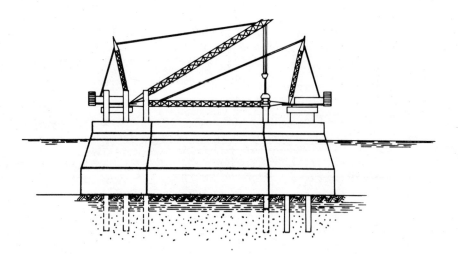

③ REMOVAL
1. SPUDS REMOVED BY JACKING (SHOWN), VIBRATORY HAMMER, JETTING OR DEBALLASTING CAISSON.
2. SPUDS EITHER HELD BY SLIPS (SHOWN) OR LIFTED CLEAR AND PLACED ON DECK.
3. ENTIRE SPUD REMOVED OR TWO-PIECE SPUD USED AND BOTTOM PIECE ABANDONED (OPTIONAL).

Figure 20.12.5 Removal process for extraction of spuds—Sohio's SAMS.

creasing the effective pore pressure in the surrounding soils.

The pile may then be removed by vibratory pulling or by jacking. Alternatively, the pile may be capped and internal water pressure applied. As noted earlier, this system was successfully used to remove the 2-m-diameter, 100-m-long steel cylinder piles installed as temporary dolphins for the Oosterschelde Storm Surge Barrier in the Netherlands. To prevent the water from escaping under the tip of the pile, a plug of very fine sand was placed, which allowed the permeation of water but resisted piping, thus allowing the development of a hydrostatic head against the cap that was sufficient to remove the pile.

Finally, to break the structure free for relocation or salvage, sand ballast is removed by slurry pump and the structure deballasted to near-neutral buoyancy. Sustained low-pressure water flooding under the structure will be found very effective. Too high a pressure may cause piping and prevent subsequent maintenance of a low overpressure. When the waterflooding has equalized underbase pore pressures, one end of the structure is deballasted further so as to lift off, and then the other end is raised. In 1985, the GBS-1 (Super CIDS) was rather easily broken loose from its clay foundation in the Beaufort Sea by underbase water flooding. Thus it appears that the suction effect can be overcome by this method.

As plans are being made to extend exploratory drilling into deeper waters, consideration is being given to the use of a subbase structure on which an exploratory drilling caisson could be set. Both steel and concrete subbases have been proposed. Confidence in their use is based on the successful mating of the Super CIDS concrete ice-belt caisson to the steel "mud base." While this was carried out in fully protected waters in Japan, it appears practicable even in exposed locations of the Beaufort Sea under favorable weather conditions.

Since the subbase will be completely submerged, its submergence to sit on the seafloor may involve instabilities. Use of several temporary buoyancy tanks, attached at the corners, will enable complete control as well as serving as guides for the mating. The next problem is that of the bearing of the caisson, 100–150 m in diameter, on the subbase. The Super CIDS used a special rubber asphalt mix to equalize the bearing. A subbase has been designed by Dome Petroleum which would employ a compressible foam interface.

During construction of offshore structures in the Arctic, consideration must be given to the possibility of damage or destruction of the partially completed structure due to waves or sea ice. This is especially critical if the structure must be left uncompleted during the winter. Caisson-retained islands, using relatively small caisson elements, are especially subject to such piecemeal destruction.

Caisson-retained islands have, however, been adopted as a viable concept because they solve the problem of construction of an embankment through the sea–air interface, where so much damage has occurred to embankments. These caissons are seated on an underwater embankment close to each other. Their articulation enables them to accomodate minor differences in the embankment surface.

The installation and removal of the Tarsiut Island caissons are very instructive for future Arctic construction. The four caissons, constructed of prestressed and reinforced lightweight concrete, were constructed in Vancouver and transported to the Beaufort Sea by a very large submersible carrier barge, which submerged to load the caissons and then deballasted to lighter draft for the tow. Near Herschel Island, the barge was holed by ice. The barge was therefore ballasted down and the caissons were floated off and towed the remaining distance as self-floaters.

The sand island embankment had been constructed to elevation -6 m. Seats for the caissons had been screeded level to a tolerance of about ± 15 cm. Due to an error in global surveying, the caissons were not set down on their pre-screeded seats but some 20 m off location. This resulted in some severe bending in at least one of the caissons and consequent flexural cracking. More serious was the fact that it—and to a lesser degree the others—did not fully bed in the sand.

This survey error is not unique in marine construction and was due to the relocation of the survey tower during sand placement. It does point out the desirability of providing backup surveys and gross visual markers that will enable the working tugs and crews to guide the caissons to position. Spar buoys, Shelton articulated buoys, or driven pile markers should be used on future installations of this nature.

The caissons were skillfully positioned relative to each other, using winches mounted on the caissons and mooring lines to anchors and, later, to previously set caissons. Care was taken to keep all mooring lines as long as possible so as to absorb surge energy in the stretch of the lines. Tolerances achieved were extremely good, validating this system for positioning. Steel gates were then used to close the openings; however, they were backfilled with sand instead of the gravel originally planned. See Figure 20.12.6.

During a summer storm, the waves were focused into the re-entrant angle at the joints and caused piping of the sand from under the caissons, allowing the gates to be dislodged. Under continued battering and with the loss of the supporting sand behind, the gates failed inward, and 16,000 m³ of sand were washed out. The several lessons learned point to the following:

1. The need to prevent piping and erosion underneath the caissons, especially near their ends.
2. The need to structurally secure the closures against wave and ice forces, including cyclic loading.
3. The need to eliminate reentrant angles in the perimeter so as to avoid concentration of wave energy.

This author believes that a preferable method of closure is by an inner and outer arc of steel sheet piles, connecting with interlocks in the caissons; see Figure 20.12.7. Another potential solution is the use of rock-filled gabions.

Other problems encountered at Tarsiut Island were wave energy focusing due to the shallow draft of the berm, wave runup, mach-stem effects of high wave buildup as the waves ran along the sides of the caissons, pyramidal waves at the lee corner which had been originally designated as a supply and personnel evacuation location, and spray. The spray and runup deposited large quantities of water on the island, which caused serious local erosion as it drained. On the other hand, the berm did cause the early ice to ground and form a protective rubble pile.

The island was completed the first season by heroic construction efforts extending into November and carried on under very adverse circumstances. That winter, the height of the peripheral walls was raised by installing gabions (rock-filled nets of steel), which provided good protection and drainage. Unfortunately, they were not tied together, so that in a subsequent summer storm the waves dislodged some of the top gabions. Subsequently they were all linked together.

The island not only survived, but fulfilled its purpose well. When removal was required, the frozen sandfill was excavated by backhoe, the caissons were washed clean, and upon dewatering, they floated free, with almost negligible leakage despite the external structural cracking which had occurred during initial setdown in the erroneous location.

The experiences of recent years at Tarsiut Island and elsewhere have shown that embankments and caissons must be designed and constructed so as to be stable during the construction period, under summer storms and under a summer pack ice incursion.

The Esso Canada caisson-retained island (Fig 20.12.8) uses multiple steel caissons which, although articulated, are stressed together so as to transfer shear and axial forces and thus act as a flexible but monolithic whole. At any such joint, there are potentially six degrees of freedom and six degrees of mis-match which must be accommodated. The design of the joint and connection details must accommodate the tolerances in all degrees, enabling the joint connection to be constructed rapidly yet still provide the structural and sealing functions.

The Esso caisson was assembled afloat, towed to location, and seated on a prepared berm. It was then filled with sand which froze during the winter, forming a highly stable seawall.

20.13 Ice Condition Surveys and Ice Management

Area-wide ice conditions can best be obtained by satellite. Unfortunately, satellites as of 1986 can obtain images only when there are no clouds. The

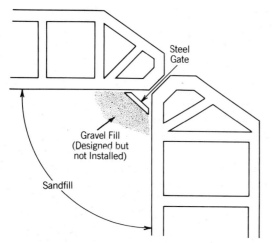

Figure 20.12.6 Details of closure between caissons at Tarsiut.

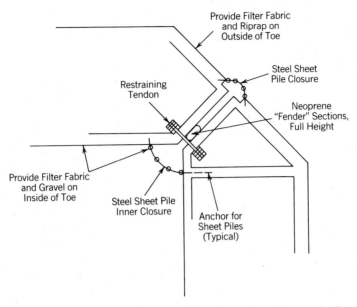

Figure 20.12.7 Proposed joint detail for caissons of a caisson-retained island.

nature of the ice edge during critical periods is to generate fog. Hence the present satellite photos are of limited value. When the cloud cover permits, however, passive microwave sensing enables the determination of the ice extent and concentration. Multiyear ice and ice islands can be distinguished from first-year ice.

In 1987 and 1989, European and Japanese polar-orbiting satellites are scheduled. These will use synthetic aperture radar (SAR) enabling them to get all-weather images.

Within local regions, the individual companies developing the offshore oil leases fly fixed-wing aircraft with side-looking radar (SLAR) or SAR.

Progress is being made in using laser and radar methods to obtain estimates of ice thickness. Multiyear ice can be distinguished, to some degree at least, from first-year ice. Structure-mounted lasers

Figure 20.12.8 Esso Canada's steel ring caisson-retained island consists of articulated steel boxes which are filled with sand when on location.

now make it possible to obtain indications of sail heights of approaching ridges, which give a rough measure of the overall ridge size. Underwater sonar measurements of approaching ridge keels is being developed.

Under-ice surveys on a large-scale statistical basis can be obtained by submarine with upward-looking sonar. ROVs using fiberoptics are capable of monitoring close-in underwater ice profiles in the winter, when the ice pack is slow-moving.

Active ice management has been attempted both in the deep Arctic and in iceberg areas.

API Bulletin 2N, sec. 9, "Active Ice Management," recommends the following active defense mechanisms:

1. A moat around the structure
2. A narrow slot
3. Remove ice rubble
4. Heating the surface of the structure
5. Installing an air bubbler system to cause water circulation
6. Use of air cushion systems to aid in breaking ice
7. Mechanically cutting or grinding the ice
8. Use of insulating mats or snow to minimize ice thickness

With regard to item 3, it is more common in recent practice to purposefully form rubble piles as an aid to reduce ice impact and cushion local forces. If the rubble pile can be grounded, then it will also transmit some of the global ice force into the seafloor directly; see Figure 20.13.1. Use of spray to generate artificial rubble has also been applied on the GBS-1 (Super CIDS) in Harrison Bay. See Figure 20.13.2.

In addition, Canmar Drilling and Gulf Canada have been using icebreaker vessels as an effective means of breaking up the oncoming ice floes. At present, they have only been able to break up the first-year ice, so as to extend the use of floating drilling vessels to about November 30.

In the deep Arctic, icebreakers can circulate upstream of the structure, breaking the floes. This has become proven technology in support of the floating drill vessel *Kulluk* as well as with the more conventional drill ships.

In fast-ice regions, where the movement is only a few tens of meters per winter, slots can be cut by powered saw and the resultant blocks lifted out

Figure 20.13.1 Tarsiut Caisson Retained Island is encased in rubble during mid-winter. Darker area at top is Arctic ice pack. (Courtesy of Dome Petroleum.)

Figure 20.13.2 GBS-1 (Super CIDS) carries out drilling operations in Harrison Bay, offshore Alaskan North Slope. (Courtesy of Global Marine Development.)

by crane. Continuous-saw-cutting machinery has been proposed, as has the use of high-pressure water jets.

Off the eastern seaboard of Canada, the towing of icebergs with masses up to 1,000,000 tons away from encounter with floating drilling vessels is a well-established practice. The tugs either "lasso" the berg by carrying a floating nylon or Kevlar line around it or use explosively driven embedment anchors. The tugs can practically exert only 100–200 tons of pull in total, so their main action is to divert or deflect the berg from a collision course. Considerable effort has gone into computer programs to help predict the movement of close-in bergs both prior to hook-on and after the tugs exert their thrust.

Major problems are the presence of fog around the berg, the poor reflectivity of radar from the ice, the unseen underwater extensions of the berg, and finally, the tendency for pyramidal bergs to roll over suddenly, endangering the tugs.

Explosives as a means of fracturing bergs and multiyear ice have been tried with little or no success due to the great absorptivity of the ice.

A potential for future development is the application of small amounts of methyl alcohol to reduce the surface tension of the ice and initiate early fracturing.

20.14 Durability

The Arctic exposure is unique for both steel and concrete. For steel, low temperatures raise the problem of brittle fracture and low energy absorption, not only of the steel plates but also of the welds. Major advances have been made in overcoming these problems, so that today reliable low temperature–rated steels and welding procedures are commercially available.

External steel plates are subjected to abrasion from the ice and corrosion from the salt water–air environments. The water has a high percentage of dissolved oxygen due to its low temperatures. Abrasion removes the products of corrosion, exposing fresh surfaces so that new corrosion can commence. Thus, the rate of corrosion–abrasion reaches as much as 0.3 mm/year, despite the low temperature.

Both dense epoxy and dense polyurethane coatings are available on the market which give excellent protection to the steel surface as well as reducing friction and adfreeze bond. These coatings require touchup every year or two. A small portable cofferdam can be provided to enable dewatering of local areas for touchup. For internal steel compartments, sacrificial anodes appear appropriate.

For concrete structures, the concrete will, of course, have been designed to be of very high quality, with low permeability and optimal entrained air. Both normal hard rock and high-grade lightweight aggregates will be used.

The major sources of problems for concrete structures are thermal cracking, freeze–thaw disintegration of the concrete, and abrasion by ice. Thermal cracking during fabrication is due to restrained cooling after the temperature rise due to heat of hydration. Proper mix design, selection of cement, use of pozzolans and so forth, and insulation of the forms can eliminate or minimize such cracking.

Proper detailing of reinforcement on the exposed face can serve to control cracks due to thermal strains both in fabrication and in service.

Freeze–thaw disintegration of external concrete surfaces can be prevented by use of air entrainment of the proper amount and pore spacing (this latter is the more important), by using a dense impermeable mix, and by using aggregate of low water absorption (less than 6 percent).

A special problem arises when perimeter compartments are filled with seawater above the external sea level. Water penetration into the concrete, combined with very low air temperatures, can create a freeze-front inside the concrete wall, leading eventually to delamination. The walls in this zone should therefore be coated with an internal membrane.

Abrasion by ice appears to be a complex interaction of frictional wear and adfreeze plucking. Use of a very dense concrete, such as that obtained by adding condensed silica fume to the mix, appears to give satisfactory results, based on both laboratory and field exposure tests.

Corrosion of reinforcement should not occur with a dense concrete such as that obtained by the combined addition of silica fume and superplasticizer (high-range water reducer). If the deck of the structure is concrete and if salts such as calcium chloride will be used by the operating personnel to prevent atmospheric and spray icing, then the decks should be well sloped for drainage and the top layers of reinforcing steel should be epoxy-coated.

20.15 Constructibility

Constructibility planning as set forth in Chapter 18 is even more important in the Arctic than elsewhere due to the very limited open-water season, the extreme logistical difficulties, and the large capital investment.

Contingency plans must be prepared for the cases of late opening of the ice for passage around Point Barrow and for sudden summer storms and even summer pack ice invasions during the construction period. Summer ice incursions are of special concern in the Beaufort and Chukchi seas. Boats may have their propellors damaged by ice. Critical equipment may not start in extremely cold weather. Wind-blown spray may add many hundreds of tons of water onto an island under construction.

API Bulletin 2N, sec. 7.5, states:

In areas subject to heavy sea ice, bad weather and ice conditions may mean delays in completing the tow to the final location. Possible temporary mooring sites should be selected along the towing route for refuge in case of such delays. Exceptionally poor ice conditions or weather conditions may cause sufficient delay to prevent installing the structure during the scheduled summer construction season. For this reason, it may be necessary to overwinter at a temporary location.

Safety of personnel must be given major consideration. A human can survive only a few minutes in water at $-2°C$.

Provisions must be made for firefighting in below-freezing weather. Intakes for water must not clog with frazil ice or broken ice. Provisions must be made for snow clearance, for although the amount of snow is small, the high winds can cause substantial drifts around structures.

In the sub-Arctic regions, atmospheric icing may make crane booms unusable and endanger the stability of boats. Decks and walkways can become iced. Measures of preventing or removing ice need to be planned. API Bulletin 2N, sec. 8.1.2, "Construction Conditions during the Arctic Summer," states:

Construction planning in offshore areas subject to ice incursion should allow for this contingency by providing proper equipment and personnel training. Contingency plans should include provision for ice surveillance and forecasting, active and/or passive defense systems, separation of vessels, and ice strengthening of vessels, e.g., satellite or radar, etc.

Section 8.1.4b, "Fog," continues:

Construction plans should account for the effect of fog on logistics and other visibility-dependent operations.

Section 8.1.4c, "Break-up and Freeze-up," states:

Construction plans should account for the effects of ice movement and logistics interruptions associated with break-up and freeze-up.

With regard to winter construction, API-2N in Section 8.2 and 8.3 calls attention to the need for preconstruction reconnaissance, with especial reference to leads and cracks in the ice, and to the need for lighted signs and roadway delineation markers. They should also indicate the distance and direction to a safe refuge or aid, in event of an accident or unforeseen event. Survival shacks, marked by a light, and survival drums should be placed at adequate intervals along the road.

20.16 Pipeline Installation

In areas of the Arctic where 40–60 days of open water can be assumed, a conventional lay barge could be employed. The shallow water and mild wave climate eliminate many of the problems normally associated with deep-water pipe laying. However, special welding procedures may be needed because of the low ambient temperatures in early fall.

Pipelines in the Arctic will generally have to be trenched 3–6 m deep in order to protect them against ice scour, at least out to a water depth of 50 m.

The fastest means for trenching appears to be by the use of a heavy-duty plow. It may be desirable to equip the plow with high-pressure water jets to break down the overconsolidated silt and to break down any permafrost or ice lenses encountered in the near-shore areas.

Pipelines in other areas can be pulled, working even under the ice. The pulling line must be laid out on the seafloor, possibly using a submersible or ROV to lay a messenger line. In fast ice in winter, holes may be cut at intervals and the ice used as a platform for feeding in and pulling of pipe. Winches may be installed on the ice at 1-km intervals and side-boom cats used to control overbend radii. See Figure 20.16.1.

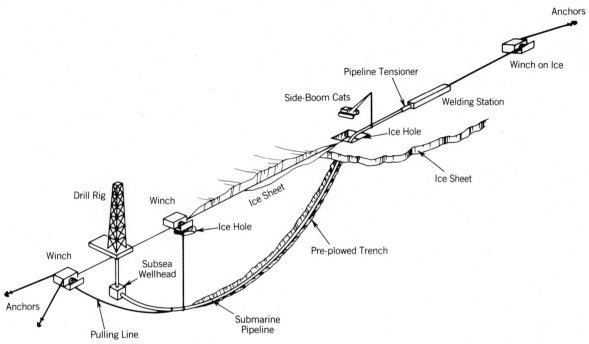

Figure 20.16.1 Panarctic's test installation of subice flow line (1979). (Courtesy of R. J. Brown and Associates.)

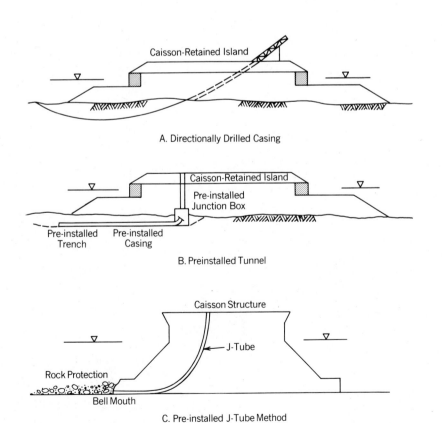

Figure 20.16.2 Concepts for pipeline tie-ins to Arctic offshore structures.

The pipeline tie-in to a structure requires careful detailing. Some structures can best be fitted with J-tubes. In other cases, a curved casing can be directionally drilled into the seafloor and used as a pull-in tube for the pipeline. In sands, a slant casing may be driven and a pre-bent elbow installed. This necessitates an underwater joint at the toe of the caisson. Such an underwater joint can be made by use of a habitat. Alternatively, the initial connection may be made by a gasketed flange, with a worker descending into the unwatered line from the platform to make the final weld. See Figure 20.16.2.

In sands, local stabilization may be required around the connection zone, using chemical grouts suitable for the low temperature. Another solution is to build in a pipe tunnel at the base of the structure. After the pipeline has been pulled in, the tunnel is dewatered to permit welding of the line.

At the shore end, the pipeline will enter a zone of permafrost. The protection of the shore from thermokarst erosion due to thawing of the frozen soil and subsequent wave erosion is an extremely important design matter. The constructor can expect that measures such as double casing with refrigerant or special insulation or a rock-filled causeway will be required.

And now there came both mist and snow,
And it grew wondrous cold:
And ice mast-high, came floating by,
As green as emerald.

The ice was here, the ice was there,
The ice was all around,
It cracked and growled and roared and howled
Like noises in a swound.

COLERIDGE "THE RIME OF THE ANCIENT MARINER"

Epilogue

The last three decades have seen amazing advances into the sea, changing our concepts and opening a new frontier in the development of resources for use by society. Its vastness, its changeableness, its profundity hold us in awe. Only recently have humans begun to think seriously about work in the deep sea, in iceberg-prone areas, and in the Arctic.

In the 1950s, the Truman Declaration established 200 meters as the outer limit of the continental shelf, because "it was obvious that no offshore structures would ever be built in deeper water." In the late 1970s several technical articles by recognized authors stated that "it would be impossible to design a structure to resist iceberg impact." It was only about 1970 that serious consideration was given to the building of structures in the Arctic, and then only in relatively shallow depths in the fast-ice zone. It was then questioned whether a structure could ever be built to resist multiyear floes. Now that all three of these psychological barriers have been crossed, how much further will construction activity proceed in the future? Are there new barriers to cross, new challenges?

Construction methods and procedures have not yet been developed which will enable the building of permanent structures in the Arctic polar pack ice.

We have yet to construct bridges across the open ocean. The bridges across the Strait of Messina and Strait of Gibraltar are still in the conceptual stage.

The Antarctic with its potential petroleum deposits within the deep sediments of the Ross, Weddell and Bellinghausen seas remains beyond our conceptual horizon.

We have not yet developed practicable and economical means for mining of polysulfide minerals from the rifts in the ocean floor.

Although floating production facilities and subsea well completions are rapidly gaining in acceptance and reliability, construction methods for the deep sea, as outlined in Chapter 19, have only recently begun to be applied. We can explore to much greater depths than we now know how to economically develop.

Floating offshore plants for the production of power, the processing of minerals, and the production of chemicals and cryogenic substances remain as yet largely a dream of the future. Offshore airports moored in the open sea and seafloor production facilities for processing of oil are yet to move from concept to reality.

The sea remains a dynamic environment, with danger ever present, as witnessed by the tragic losses of the *Alexander Kjelland* floating hotel, the *Ocean Ranger,* and the *Glomar Java Sea.*

Constructors will be called upon to play a leading role in making construction in the oceans safe, efficient, and economical: that is the challenge ahead.

As this new frontier unfolds and as society assimilates the opportunities presented by the development of the seas, it may be possible for humankind to transcend archaic concepts that have thus far limited the growth of society. For although the inital uses of the ocean are for material resources and scientific exploration, the exploitation of this vast and wonderful region of the earth will inevitably benefit our collective culture in all its aspects; intellectually, politically, and even spiritually, as we learn to work with and become an integral part of the oceanic environment.

478 *Epilogue*

*And crown thy good with brotherhood,
From sea to shining sea.*

Katherine Lee Bates,
    "America the Beautiful"

Bibliography and Technical Literature of Special Relevance to Offshore Construction

Principal Rules, Standards, and Recommendations

American Bureau of Shipping, *Rules for Building and Classing Offshore Installations*, pt. I, "Structures," New York, 1983.

American Concrete Institute, *Concrete Structures for the Arctic*, 1986, Detroit.

American Concrete Institute, ACI 357R-84, *Recommendations for Concrete Sea Structures*, Detroit, 1985.

American Petroleum Institute, API RP2A, *Planning, Designing, and Constructing Fixed Offshore Structures*, 15th ed., Dallas, 1984.

American Petroleum Institute, API 2N, *Planning, Designing, and Constructing Fixed Offshore Structures in Ice Environments*, Dallas, 1982.

British Standards Institute, BS 6235, *Code of Practice for Fixed Offshore Structures*, London, 1982.

Bureau Veritas, *Rules and Regulations for the Construction and Classification of Offshore Platforms*, Paris, 1975 (with amendments and additions No. 1, August 1982).

Canadian Standards Institute, *Standards for Offshore Structures in Frontier and Arctic Areas*, under preparation, 1986.

Fédération Internationale de la Précontrainte, Guides to Good Practice: *Sea Operations; Grouting of Vertical Tendons;* etc. Telford, London, 1985.

Fédération Internationale de la Précontrainte, *Recommendations for the Design and Construction of Concrete Sea Structures*, 4th ed., Telford, London, 1985.

Det Norske Veritas, *Rules for the Design, Construction, and Inspection of Offshore Structures*, Oslo, 1977 (rpt. 1981).

Det Norske Veritas, *Rules for Submarine Pipelines Systems*, Oslo, 1981.

Norwegian Petroleum Directorate, *Regulations for the Structural Design of Fixed Structures on the Norwegian Continental Shelf*, Stavanger, Norway, 1977. (under revision, 1986)

OCS Platform Verification Program, U.S. Minerals Management Service, (formerly US Geological Survey Conservation Division) Washington, D.C. 1979.

U.K. Department of Energy, *Guidance on the Design and Construction of Offshore Installations*, Her Majesty's Stationery Office, London, 1985.

U.S. National Research Council, Marine Board Reports: *Offshore Structures; Environmental Exposure; Activities and Safety;* and *Arctic Offshore Operations*, National Academy of Sciences, Washington, D.C., 1980–1985.

U.S. Navy Standard Tables: *Single and Repetitive Dives, Helium-Oxygen Tables;* and *Decompression Tables for Standard, Exceptional, and Extreme Exposure*, Washington, D.C.

Journals and Magazines Addressing Current Issues in Construction of Offshore Structures

International Underwater Systems Design
Journal of the American Concrete Institute
Journal of the Society of Naval Architects and Marine Engineers
Marine Technology
Noroil
Notes of the Fédération Internationale de la Précontrainte
Ocean Industry
Offshore
Offshore Engineer
Oil and Gas Journal
Petroleum Engineer International
World Dredging and Marine Construction

Proceedings and Reports of Conferences and Symposia

American Society of Civil Engineers Arctic Conference Proceedings

Asian Institute of Technology, *Geotechnical Aspects of Offshore and Nearshore Structures* (symposium), Bangkok, 1981.

"Behavior of Sea Structures", (BOSS) Conference Proceedings.

European Offshore Petroleum Conference preprints

Exxon, *Alaskan Beaufort Sea Gravel Island Design*, (technical seminar), Anchorage/Houston 1979

480 Bibliography and Technical Literature of Special Relevance to Offshore Construction

Fédération International de la Préconstrainte, (FIP) quadrennial congresses, proceedings

Institution of Civil Engineers, *Offshore Moorings* (conference), Telford, London, 1982

Offshore Brazil, Conference Proceedings

Offshore Mechanics and Arctic Engineering (OMAE) engineering symposia

Offshore Technology Conference (OTC) Preprints

POAC conferences ("Port and Ocean Construction under Arctic Conditions"), proceedings

Royal Institution of Naval Architects, *Offshore Engineering* (symposium), 1981.

Southeast Asia Offshore Technology Conference preprints

Books and Reports

Arctic Ocean Engineering for the Twenty-first Century, Marine Technology Society, Washington, D.C., 1985.

L. F. Boswell, *Platform Superstructures*, Granada Technical Books, London, 1984.

Construction Industry Research and Information Association, Underwater Engineering Group, *The Principles of Safe Diving Practice*, London, 1984.

Construction Industry Research and Information Association (CIRIA) Reports: *Research in Offshore Structures; Concrete in the Oceans;* etc., London, 1980–85.

Construction Industry Research and Information Association, Underwater Engineering Group, UEG Reports, London.

Federal Highway Administration *Tremie Concrete for Bridge Piers and Other Massive Underwater Placements*, Report FHWA/RD-81153, NTIS, Washington, D.C., 1981.

B. C. Gerwick, Jr., *Construction of Prestressed Concrete Structures*, Wiley–Interscience, New York, 1971.

N. J. Graff, *Introduction to Offshore Structures*, Gulf Publishing, Houston, 1981.

John Herbich, *Coastal and Deep Ocean Dredging*, Gulf Publishing, Houston, 1975.

John Herbich, Schiller, Dunlap, Watanabe, *Seafloor Scour*, Marcel Dekker, New York, 1984.

John E. Kenny, *Business of Diving*, Gulf Publishing, Houston, 1972.

J. G. MacGregor, Baskin, and Ellis, "A Review of Design Guidelines for Concrete Fixed Offshore Structures: A Canadian Perspective," CFER Report 84-02, Centre for Frontier Engineering Research, Calgary, Alberta, May 1984.

B. McClelland, and M. D. Reifel, *Planning and Design of Fixed Offshore Platforms*, Van Nostrand-Reinhold, New York, 1986.

Meyers, John J., Holm, Carl H., McAllister, R. F. *Handbook of Ocean Engineering and Construction*, McGraw-Hill, New York, 1969.

Naval Civil Engineering Laboratory, Techdata Sheets, Port Hueneme, Calif.

Offshore Platforms and Pipelines, Petroleum Publishing, Tulsa, 1976.

R. Sheffield, *Floating Drilling Equipment and Its Use*, Gulf Publishing, Houston, 1980.

Howard R. Talkington, *Undersea Work Systems*, Marcel Dekker, New York, 1981.

U.S. Corps of Engineers, *Shore Protection Manual*, 4th edition, 1984, Waterways Experiment Station. US Government Printing Office, Washington, D.C.

There are hundreds of technical papers relevant to the construction of offshore structures, most of which are included in the above texts, journals, and conference or symposia proceedings. Space does not permit their individual listing here.

Appendix 1

Excerpts from Det Norske Veritas, Rules for the Design, Construction, and Inspection of Offshore Structures

E1 Hydrostatic Stability and External Equilibrium

E1.1 General

E1.1.1 Requirements relating to hydrostatic stability and external equilibrium of floating and semi-floating structures are given in the following Sections of the Rules:

Section 4, which includes general design requirements. As a principle, these also apply to external equilibrium.

Section 8, which includes traditional stability requirements for floating structures in phase 0 (intact stability and damage stability).

Section 10, which includes traditional stability requirements for floating structures in phases C, T and I.

E1.1.2 As a principle, all floating and semi-floating structures should satisfy the applicable requirements of the above mentioned sections of the Rules. However, with respect to hydrostatic stability and external equilibrium, the requirements of Sections 8(R) and 10(R) replace the loading condition "Ordinary" of 4.4.4.3(R). E1.2 below is a supplement to these two Sections. The condition "Extreme" of 4.4.4.3(R) still applies. E1.3 below proposes criteria to the applied for the ultimate limit state in this condition, as well as criteria to be applied for the progressive collapse and serviceability limit states.

E1.2 Hydrostatic Stability

E1.2.1 A standardized method for the cheking of hydrostatic stability (in accordance with Sections 8(R) and 10(R)) has been established by IMCO. The following is based upon the IMCO method. Values of shape and height coefficients for the calculation of wind heeling moments may be taken as recommended by IMCO, or as given in Appendix B.

E1.2.2 The IMCO method is directly applicable only for freefloating and conventionally anchored structures, where the forces in the anchor lines have little effect on stability and floatability. E1.2.3 through E1.2.7 give in summary the IMCO method.

For semi-floating structures, and for floating structures where the anchoring system has a significant effect on stability and floatability, a modified method is proposed in E1.2.9 below.

E1.2.3 For calculation of the wind heeling moments the lever of the wind overturning force should be taken vertically from the centre of pressure of all surfaces exposed to the wind to the centre of the lateral resistance of the underwater body of the structure.

E1.2.4 The wind heeling moment curve for ship-shaped hulls may be assumed to vary as the cosine function of vessel heel. For other structures the moment should be calculated for a sufficient number of heel angles to define the curve.

E1.2.5 Determination of wind heeling moments by wind tunnel tests on a representative model of

---

Det Norske Veritas, Oslo, 1977 (rpt. 1981). By permission of DNV.

the unit may be considered an alternative to the method given in E1.2.3 and E1.2.4. The determination of heeling moment should include lift effects at various applicable heel angles, as well as drag effects.

E1.2.6 The stability requirement described in 8.2.3.7(R) is illustrated in Figure E1.1, where the righting moment curve and the heeling moment curve are depicted in the same diagram. Between zero angle and the angle of progressive flooding, the area below the righting moment curve should be 1,4 times the area below the wind heeling moment curve. $(A + C) \geq 1.4 (B + C)$.

E1.2.7 The requirement for damage stability described in 8.2.4.10(R) is illustrated in Figure E1.2.

E1.2.8 Outside the region of damage described in 8.2.4.3(R) through 8.2.4.7(R), a compartment inside an external concrete wall of at least 30 cm thickness, with no valves etc., is not considered floodable, , even if it is located adjacent to the sea or the weather deck. (See also 8.2.4.2(R)).

E1.2.9 For semi-floating structures, and for floating structures where the anchoring system has a significant effect, the initial stability may be checked by calculating the below defined (modified) metacentric height, which should not be less than 0.3 metre:

$$h_m = \frac{I_0 - \Sigma i}{V} + \frac{M}{\gamma_w V} - e$$

I_0 = moment of inertia of waterline area

i = moment of inertia of an internal, free water surface

V = displaced volume of water

M = moment per unit tilting angle, with respect to the centre of gravity of the structure, due to forces in anchor lines, tension lines, joints or other anchoring arrangement

γ_w = unit weight of water

e = distance from centre of gravity to centre of buoyancy, taken positive if the latter point is the lower of the two.

E1.3 Limit State Requirements to External Equilibrium and Rigid Body Motions

E1.3.1 Load coefficients to be used for the various limit states are given in 4.4.4(R). Regarding dynamic analyses, refer to Appendix G as well as to 4.5(R). See also E2.

E1.3.2 Provided Sections 8(R) and 10(R), as well as E1.2, are satisfied, the ultimate limit state need only be considered in the "Extreme" loading condition.

The ultimate limit state with respect to external equilibrium and rigid body motions may be reached in different ways, depending upon type of structure etc. Also, for one and the same structure, it may be necessary to consider several ultimate limit states. Typical ultimate limit states are:

when progressive flooding starts

when a mechanical limitation of a displacement (translation or rotation) is reached

when further displacement would lead to a situation essentially different from the design assumptions. (E.g. submergence of a superstructure assumed to be above water in any of the defined loading conditions).

E1.3.3 The progressive collapse limit state corresponds to the start of progressive flooding after a damage according to 8.2.4.3(R) through

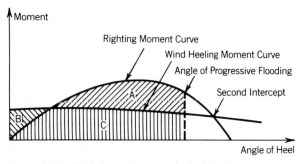

Figure E1.1 Righting moment and wind heeling curves: stability requirements.

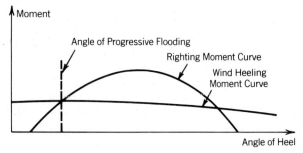

Figure E1.2 Alternate stability curves where progressive flooding occurs at low angles of heel.

8.2.4.7(R). See also E1.2.8. The structure should be assumed exposed to environmental conditions corresponding to one year return period.

E1.3.4 Serviceability limit states are normally defined by the Owner. In cases where the definitions of the serviceability limit states may affect the safety in any respect, the definitions should be subject to acceptance by DnV.

Serviceability limit states may be given in terms of acceptable accelerations, velocities or displacements under given conditions of operation.

E2 Anchoring and Mooring

E2.1 General

E2.1.1 Definitions

In E.2 the following definitions are used:

Anchoring: The attachment of a floating structure to the seabed.
Mooring: The attachment of a floating structure to another floating or fixed structure.

E2.1.2 Notation

The following symbols are used in E.2:

E = extreme (100 year) environmental load.
L = live Load.
R_k = characteristic strength (in tension) of a tension leg.
S = tension in a tension leg.
S_0 = required tension in (upper end of) a tension leg in the still water condition with zero live load L.
W_s = submerged weight of a tension leg.
δ_{mean} = mean quasistatic displacement.
δ_{motion} = oscillatory displacement amplitude.
γ_m = material coefficient.

E2.1.3 Application

E2.1.3.1 E.2 applies to any of the design phases defined in 1.1(R) during which the offshore structure is anchored or moored.

E2.2 Anchors

E2.2.1 General

E2.2.1.1 The conditions of the seabed should be taken into account in the selection of the anchor type.

E2.2.1.2 The anchor forces should be determined in accordance with E2.6.

E2.2.2 Fluke Anchors

E2.2.2.1 Ordinary fluke anchors (ship anchors) may be used provided special tests are carried out in advance to prove their suitability. Information from previous tests in similar soil conditions may replace the tests. Ordinary fluke anchors should be manufactured and proof tested according to DnV's Ship Rules. Ch. X, Sec. 9.

E2.2.2.2 For fluke anchors of special design, holding performance should be documented. The structural strength should comply with the requirements given in DnV's Rules for Offshore Structures, Sec. 6 as far as material, design, fabrication and inspection is concerned. Method of proof testing should be agreed upon in each case.

E2.2.2.3 For fluke anchors of such design that the holding capacity depends on accurate installation, documentation by underwater pictures may be required. In areas where underwater photography is not possible, the installation process should be attended by a DnV surveyor.

E2.2.3. Direct-Embedment Anchors

E2.2.3.1 Direct-embedment anchors of deep penetration and high holding power/weight ratio may be used provided the suitability of the anchors is documented in advance.

E2.3 Anchor Lines

E2.3.1 General

E2.3.1.1 Chain cables corresponding to grade NVK2 and NVK3 will be accepted. See DnV's Ship Rules, Ch. X, Sec. 8.

E2.3.1.2 Steel wire ropes are to comply with the following requirements:

The wires should be drawn galvanized or finally galvanized according to ISO-standard 2232. The nominal tensile strength should not be less than 1570 N/mm^2.

For stranded ropes the strands should be made in equal lay construction (stranded in one operation). Spiral strandroped and locked coil ropes will be accepted.

The stranded ropes should have an independent wire rope core (IWRC). The wires in steel core should be of similar tensile strength to that of the main strand.

All wire ropes for offshore use should be lubricated, with lubricants having no injurious effect on the steel wires in the rope.

Steel wire ropes for offshore use should be tested according to DnV's Ship Rules, Ch. X, Sec. 10.

E2.3.1.3 The strength of the connecting link for combined chain and wire systems should not be inferior to the strength of the anchor line.

E2.3.1.4 The approval of synthetic fibre ropes will be based on testing evidence.

E2.3.2 Catenary Lines

E2.3.2.1 For anchors not designed to carry a vertical load the length of the anchor line should be such that no vertical force will occur in any loading condition.

E2.3.3 Tension Legs

E2.3.3.1 In addition to the materials mentioned in 8.4.2.1 (R), tension legs may also be made from steel bars, prestressed concrete ties etc.

E2.3.3.2 Attention should be paid to possible abrasion where several lines are joined to a bundle as well as where the leg enters the structure.

E2.4 Auxiliary Anchoring Equipment

E2.4.1 Windlasses

E2.4.1.1 Normally, the total braking capacity of the windlass should not be less than the required strength of the anchor line.

E2.4.1.2 Cable lifters should have sufficient diameter and be so designed that unfavourable chain stresses are avoided. Cable lifters should normally be of cast steel but ferritic nodular cast iron may also be considered.

E2.4.2 Stoppers

E2.4.2.1 Chain and wire stoppers should be of a design which does not bring unfavourable stresses upon the chain or wire.

E2.4.2.2 Possible arrangement for emergency release of anchor lines will be considered in each case.

E2.4.3 Fairleads

E2.4.3.1 Fairleads fitted between the stopper and the anchor should be of the roller type and have swivel provisions.

E2.4.3.2 The fairlead diameter should be sufficiently large and the design should be such that unfavourable stresses in the anchor line are avoided.

E2.4.4 Shackles

E2.4.4.1 Shackles may be manufactured and tested according to DnV's Ship Rules. See Ch. X, Sec. 8.

E2.5 Mooring Equipment

E2.5.1 General

E2.5.1.1 Mooring equipment is subject to certification when it is essential for structural integrity, proper operation and prevention of environmental pollution.

E2.5.1.2 Standard components incorporated in the mooring equipment, such as shackles etc. may be manufactured and tested according to DnV's Ship Rules.

Possible weak links should be considered in each case.

E2.5.1.3 Yoke suspensions should comply with the requirements applicable to steel structures; see Sec. 6(R).

E2.5.2 Hawsers

E2.5.2.1 Hawsers should be of a material suitable for offshore use. Behaviour in wet condition and under cyclic load is of particular importance. Attention should be paid to possible wear of hawsers at critical points.

E2.5.3 Compensators

E2.5.3.1 Compensators based on steel springs, hydraulic/pneumatic spring systems, fibre ropes over sheaves etc. may be used.

E2.5.3.2 The compensators should be of safe design and certified materials. Possible standard components used should be manufactured and tested according to recognized codes.

E2.6 Determination of Response

E2.6.1 General

E2.6.1.1 A dynamic analysis of the system behaviour is preferable, see Appendix G, but quasistatic analysis may also be accepted upon consideration of natural frequencies of the system and its individual components.

E2.6.2 Quasistatic Analysis

E2.6.2.1 Quasistatic analysis implies that wind, current and mean wave drift forces are considered as static forces. Forces corresponding to the oscillatory wave induced motions are then added to the static forces.

E2.6.2.2 The stiffness characteristics should be determined from recognized theory.

E2.6.2.3 The anchored or moored structure will take an *equilibrium position* at which the restoring force from the anchoring or mooring system equals the sum of static forces. The distance from this position to a position corresponding to zero environmental forces is called the *mean quasistatic displacement* δ_{mean}. Due to the wave induced forces the structure will oscillate about the equilibrium position with an amplitude equal to δ_{motion}.

The total *quasistatic displacement*$_{total}$ is assumed to be the sum of the mean quasistatic displacement and the oscillatory motion

$$\delta_{total} = \delta_{mean} + \delta_{motion}$$

E2.6.2.4 If relevant, local dynamics of individual anchor and/or mooring lines should be included. The line may be excited by the time varying motions at the upper end (found from the dynamic system analysis) and by wave and current induced vortex shedding. See Appendix B.

E2.6.2.5 Based on the stiffness characteristics the local path of displacement as well as the tension in anchor and/or mooring lines and anchor forces should be documented.

E2.7 Verification of Resistance

E2.7.1 Ultimate Limit State

E2.7.1.1 The characteristic holding capacity of fluke anchors and direct embedment anchors should be taken as the conservatively assessed mean value based on information from the following sources:

Full scale tests
Model tests
Field experience
Possible theoretical calculations

All results are to comply with the actual conditions of the seabed in question.

A material coefficient (holding capacity coefficient) of $\gamma_m = 1.5$ should normally be used.

E2.7.1.2 The characteristic strength of chain cable and steel wire rope may be assumed to be the minimum breaking strength as specified by DnV's Ship Rules, Chapter X, Sections 8 and 10, or the guaranteed minimum breaking strength as provided by the manufacturer if not covered by the Rules. Normally, a material coefficient of $\gamma_m = 1.5$ should be used for chain and wire.

E2.7.1.3 The characteristic strength of synthetic fibre ropes may be assumed to be the guaranteed minimum breaking strength when new. Normally, a material coefficient of $\gamma_m = 3.5$ should be used.

E2.7.1.4 For tension legs (of tension leg platforms) the following two ultimate limit states should be considered:

1. ULS with maximum tension; see E2.7.1.5.
2. ULS with minimum tension; see E2.7.1.6 through E2.8.2.8.

These limit states should be considered in Ordinary and Extreme conditions, in accordance with 4.4.4.3 (R).

E2.7.1.5 For ULS with maximum tension 4.4.4 (R) leads to the following:

Extreme:

$$1.1 S_0 + S(1.3E) \leq \frac{R_k}{\gamma_m}$$

where

S_0 = required tension in considered leg in the still water condition with zero live load L.
E = extreme (100 year) environmental load.
$S(1.3E)$ = tension in considered leg due to 1.3E.
R_k = characteristic strength (in tension) of the considered tension leg.
γ_m = material coefficient for the tension leg, according to E2.7.1.2. For plain steel 1.3 may be used.

E2.7.1.6 For ULS with minimum tension the philosophy is that a tension (S) in the upper end of the leg of magnitude less than the submerged weight of the leg is considered as failure. In order to check this limit state one should distinguish between two cases:

(a) The pretension will not be adjusted according to variation in live load L; see E2.7.1.7.
(b) The pretension will be adjusted according to variation in live load L; see E2.7.1.8.

E2.7.1.7 For platforms with no adjustment of pretension after completed installation, the ULS with minimum tension should be checked as follows:

Ordinary:

$$0.9 S_0 - S(1.3L) - S(0.7E) \geq 1.3 W_s$$

Extreme:

$$0.9 S_0 - S(1.0L) - S(1.3E) \geq 1.0 W_s$$

where

S_0 = required tension in (upper end of) considered leg in the still water condition with zero live load L
E = extreme (100 year) environmental load
L = live load
$S(0.7E)$ = tension in considered leg due to 0.7E
$S(1.3E)$ = tension in considered leg due to 1.3E
$S(1.0L)$ = tension in considered leg due to 1.0L
$S(1.3L)$ = tension in considered leg due to 1.3L
W_s = submerged weight of considered tension leg.

E2.7.1.8 For platforms with adjustment of pretension so that the tension is independent on L, the ULS with minimum tension should be checked as follows:

Ordinary:

$$0.9 S_0 - S(0.7E) \geq 1.3 W_s$$

Extreme:

$$0.9 S_0 - S(1.3E) \geq 1.0 W_s$$

The symbols are defined E2.7.1.7.

E2.7.2 Fatigue Limit State

E2.7.2.1 For permanent anchoring and mooring systems of long design life and with serious failure consequences, fatigue data should be established for the relevant environment and a fatigue investigation carried out. The investigation should be based on the load history of the equipment.

E2.7.2.2 For chain cable and steel wire ropes fatigue data should be based on statements from manufacturers and available research results.

E2.7.2.3 For synthetic fibre ropes specific fatigue calculations are normally not required. A condition for this is that the various components will be replaced at certain intervals. A program for such replacements should be prepared in each separate case. Besides ordinary fatigue, the effect of wear, temperature-rise due to cyclic loading, long-term creep and possible other effects should be taken into account when deciding replacement intervals.

E2.7.3 Progressive Collapse Limit State

E2.7.3.1 Upon failure of one anchor line the remaining system should be able to resist the loads expected before repair without unacceptable displacements. For lines consisting of several members only failure of one member need to be considered.

E2.8 Installation

E2.8.1 General

E2.8.1.1 The position of the anchored structure should be checked with regard to permanent displacements, particularly in the first period after installation and after extreme weather conditions.

E2.8.2 Testing and Pretensioning

E2.8.2.1 After installation the anchoring arrangement should be subjected to a test load, normally corresponding to at least 50% of the minimum required line breaking strength. In case this is impractical, a reduced test load may be accepted upon consideration.

E2.8.2.2 The penetration depth of direct-embedment anchors should be verified after installation.

E2.8.3 Cross Mooring

E2.8.3.1 In case of cross mooring the anchor pattern should include catenary plans for all anchor lines. The anchor pattern and the tension in the anchor lines should be such that a clearance exists between individual lines in all conditions.

H1.9.3 Application of Structural Design Factors to a Typical Module

H1.9.3.1 Figure H1.1 gives an example of how a typical module frame can be divided into the following three areas:

Black: Padeyes and their attachments to the structures.
Shaded: Load transferring members supporting the lifting points.
White: Other members.

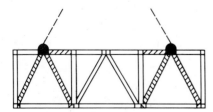

Figure H1.1 Typical module frame.

H1.9.4 Calculation of Skew Load (SKL) for Statically Indeterminate Lifts

H1.9.4.1 Direct calculations of the skew load is an alternative to selecting the SKL from H1.2.5.2.

Due to the nonlinear behaviour of the slings, the real skew load factor will normally decrease with increasing load. An allowance for this effect can be made by applying the skew load factor corresponding to 75% overload (in relation to expected maximum according to H1.3.2.2). A possible approach is described in H1.9.4.2.

H1.9.4.2 The skew load is first determined in a separate calculation. For this calculation an increased amplification factor of 1.75 times the DAF defined in H1.2.2 is applied. When the skew load factor has been determined, this skew load factor (or percentage load in each sling) is maintained for the final calculation of forces and the code check. The final calculation of forces and the code check is then carried out with the DAF defined in H1.2.2 and appropriate safety factors, etc., according to H1.4, H1.5 and H1.6.

H1.9.4.3 The assumed properties of a lifted object, slinging arrangement and in particular the length measurements of the slings contain uncertainties. It is recommended not to assume too narrow length tolerances when skew load calculations are performed. SKL below 1.1 should normally not be applied for lifting of a relatively rigid object.

Section 8 Hydrostatic Stability, Watertight Integrity and Anchoring

8.1 General

8.1.1 The general requirements concerning hydrostatic stability, watertight integrity and anchoring given in Section 8 apply to phase O for structures designed to remain permanently anchored offshore, such as conventionally anchored buoys or platforms, tension leg platforms, articulated tower structures etc. For the phases C, T and I, the requirements given in section 10 apply.

8.1.2 Anchored offshore structures are to have load line marks indicating maximum permissible draught. See 4.7.3.3.

8.2 Hydrostatic Stability

8.2.1 General

8.2.1.1 The Operations Manual is to include information, related to the stability requirements of subsection 8.2. The extent of the information is to be sufficient to give the responsible personnel the guidance necessary to ensure sufficient stability under all expected conditions.

8.2.1.2 In particular, instructions with regard to termination of the normal operating modes, and appropriate measures to be taken in deteriorating weather conditions, are to be included in the Operations Manual.

8.2.1.3 Relevant data such as curves showing draught versus displacement, centre of buoyancy, waterplane areas, centroid of waterplanes, moment of inertia of waterplanes, position of metacentre above base line, trimming and heeling unit moments, and displacement per unit immersion (tonnes/cm) are to be submitted.

Capacity data of all tanks, including position of centre of gravity and free surface correction are to be submitted.

8.2.1.4 Calculations related to the following matters are to be submitted:

  intact stability
  damage stability and floatability
  determination of weight and centre of gravity based on inclining test
  possible effects of icing on the stability.

8.2.2 Inclining Tests and Weight Determination

8.2.2.1 When the construction of the hull is completed, inclining tests are to be undertaken in order to establish the position of the centre of gravity. Further, the light weight is to be determined.

8.2.2.2 The tests are to be carried out under the surveillance of a DnV Surveyor. A report is to be submitted for acceptance.

8.2.2.3 The tests are to be carried out using the following procedure:

  The inclining angle should be of the order of 2°.
  The angles are to be measured by at least two pendulums.
  The draught is to be such that the waterline intersects the unit in a wall-sided area.
  If water is used to achieve sufficient inclining moment, the tanks involved are to be manually sounded during transfer of water. Care is to be taken to avoid trapping air in the piping system.
  The effects of external forces due to wind, waves, moorings, anchors, tugs, cranes, etc. are to be considered during the test and in the report.

8.2.3 Intact Stability Requirements

8.2.3.1 Proof of sufficient stability is to be established for all relevant loading conditions.

8.2.3.2 Conventionally moored structures are to meet the requirements of 8.2.3.3 through 8.2.3.7.

8.2.3.3 Statical stability curves are to be calculated for heeling axes parallel to the unit's two main axes in the horizontal plane. When there is reason to believe that stability about any other heeling axis may be critical, statical stability curves referred to this axis are to be calculated.

8.2.3.4 Wind heeling moment curves are normally to be calculated for wind directions parallel to the unit's two main axes in the horizontal plane. When there is reason to believe that any other wind direction is more unfavourable, wind heeling moment curves referred to this direction are to be calculated.

8.2.3.5 The wind heeling moment curves are to be calculated using a recognized method. The sustained wind speed, $V_s[_{100}]$ is to be applied, (i.e. the 100 year value).

Guidance: A method for computing the wind heeling moments is given in Appendix E.

8.2.3.6 The initial metacentric height after correction for free surface effects is not to be less than 0.3 m.

8.2.3.7 The statical stability curves (see 8.2.3.3) and the wind heeling moment curves (see 8.2.3.4 and 8.2.3.5) are to be plotted in the same diagram and shown in the Operations Manual. The areas under the statical stability curve and the wind heeling moment curve are to be calculated up to an angle of heel which is the least of:

  the angle of heel corresponding to the second intercept of the two curves,
  the angle of heel at which flooding of buoyant spaces starts.

This area for the statical stability curve is not to be less than 1.4 times the corresponding area for the wind heeling moment curve. For structures with a small waterline area a factor of 1.3 may be used instead of 1.4, upon agreement with DnV.

8.2.3.8 Tension leg and articulated tower structures, or other structures having constructional features which render the requirements of 8.2.3.2 inapplicable, will be specially considered.

8.2.3.9 For structures to which vessels may be moored, stability calculations are to take into account the simultaneous effects of:

anchor forces

mooring forces from the vessel

wind forces corresponding to the maximum wind velocity assumed to occur when the vessel is moored to the structure; see 5.2.2.4.

8.2.4 Subdivision and Damage Stability Requirements

8.2.4.1 The requirements on subdivision and damage stability given in 8.2.4 require consideration of the following two types of flooding:

(a) Accidental flooding as a result of any possible leakage.
(b) Flooding as a result of damage, e.g. collision etc. Damage areas to be considered are described in 8.2.4.3 through 8.2.4.7.

8.2.4.2 Proof of sufficient reserve buoyancy and stability is to be established for all relevant loading conditions after accidental flooding of any one compartment that can possibly be flooded, or after flooding as a result of damage.

Compartments located outside the region of damage described in 8.2.4.3 through 8.2.4.7 may be considered as not floodable, provided they are not located adjacent to the sea or the weather deck.

8.2.4.3 Damage is assumed to occur in a zone which is bounded by two horizontal planes normally positioned 5 m above and 3 m below the waterline in question.

8.2.4.4 For all types of unit, damage is assumed to occur in the shell plating of any one compartment, leaving watertight subdivision bulkheads intact. If any parts of two subdivision bulkheads are spaced less than 2.3 m apart, one of the bulkheads is to be considered as non-existent.

8.2.4.5 The horizontal penetration of damage is to be taken as 1.5 m inboard from the shell plating.

8.2.4.6 Cylinders are assumed to be penetrated radially a distance equal to 1.5 m. The circumferential extent of damage is assumed to be $\pi D/6$, where D is the diameter of the cylinder. The damage is assumed to occur at any point on the cylinder surface except at positions where it is obvious that damage cannot take place.

8.2.4.7 The vertical extent of damage is to be taken as 2.3 m in the case of vertical or horizontal members such as columns or lower hulls. For barge type units, the vertical extent of damage is to be from bottom shell to upper deck.

8.2.4.8 Realistic permeabilities of flooded spaces are to be used. For empty spaces a permeability of 0.98 is to be used, while in pump rooms, machinery spaces etc. a value of 0.85 is considered appropriate. Allowance for liquid in partially filled tanks may be made by considering the specific gravity of the liquid in relation to that of sea water.

8.2.4.9 For the initial conditions given under 8.2.4.2 the unit is assumed to be simultaneously exposed to the overturning effect of a constant, nominal wind speed of $0.5\ V_{s(100)}$ (see 8.2.3.5) acting in the direction in which the unit will heel/trim as a result of flooding of the compartment considered. The resulting still water equilibrium conditions are to comply with the requirements of 8.2.4.10.

8.2.4.10 The requirements for still water equilibrium are:

(a) The final waterline in the still water equilibrium condition after flooding, taking into account sinkage, trim and heel, is to be below the lower edge of any opening through which progressive flooding may take place. Openings, the lower edge of which is not to be submerged, include air pipes (regardless of closing appliances), ventilators, ventilation intakes and outlets, non-watertight hatches and doors. Openings such as manholes, watertight hatches, watertight doors, side scuttles of the non-opening type may be submerged.
(b) The metacentric height in the still water equilibrium condition without the effect of any wind heeling moment is to be at least 0.30 meter.

8.2.4.11 In the event that pipes, ducts, tunnels, etc. are located within the zone of damage penetration, arrangements are to be made to avoid flooding beyond the limits assumed for the various damage cases.

8.2.4.12 For the damage conditions, ballasting instructions describing proper procedures to be followed in order to reduce trim/heel, are to be included in the Operations Manual. In addition, such instructions are to be posted in the control centre of the unit in an intelligible form, preferably in visual combination with the manoeuvre arrangement of valves and pumps.

8.2.4.13 Tension leg and articulated tower structures, or other similar structures, are to remain afloat with sufficient freeboard to preclude progressive flooding with any one compartment open to the sea and simultaneously subjected to a wind force corresponding to a wind speed of $0.5V_s{(100)}$ (see 8.2.3.5).

8.2.4.14 Failure of moorings may have an adverse effect on the hydrostatic stability of the structure. It is to be verified that sufficient stability and buoyancy exist to keep the structure in an upright position after one or more of the moorings may have failed.

8.3 Watertight Integrity

8.3.1 General

8.3.1.1 The number of openings in watertight bulkheads and decks is to be kept to a minimum compatible with the design and proper working of the structure. Where penetrations of watertight decks and bulkheads are necessary for access, piping, ventilation, electrical cables, etc., arrangements are to be made to maintain the watertight integrity.

8.3.1.2 Where valves are provided at watertight boundaries to provide watertight integrity, these valves are to be capable of being operated from the bulkhead deck or weather deck, pump room or other normally manned space. Valve position indicators are to be provided at the remote control station.

8.3.2 Watertight Doors and Hatches

8.3.2.1 Watertight doors and hatches are to be remotely controlled from a central position above the bulkhead deck and are also to be operable locally from each side of the bulkhead or deck. Indicators are to be provided at the central position to indicate whether the doors are open or closed.

8.3.2.2 The requirements of 8.3.2.1 regarding remote controls may be dispensed with, provided an alarm system (e.g. light signals) is arranged showing personnel, both locally and at a central position, whether the doors and hatches in question are open or closed. A sign is to be placed on such doors and hatches to the effect that they are not to be left open.

8.4 Anchoring

8.4.1 Anchors

8.4.1.1 Anchors to be used for mooring of anchored structures are normally to be gravity anchors or pile anchors.

8.4.1.2 The design resistance of an anchor is to be determined in accordance with the requirements of Section 9. The effects of repeated loads shall be accounted for in the determination of the design resistance. If practical, verification of the design resistance of an anchor may also be based on tests.

8.4.1.3 The design of steel anchors shall satisfy the requirements of Section 6.

The design of reinforced concrete anchors shall satisfy the requirements of Section 7.

8.4.2 Anchor Lines

8.4.2.1 Anchor lines may consist of chain cable, steel wire, fibre ropes or a combination of these.

8.4.2.2 The quality of chain cables, steel wire or fibre ropes to be used for anchor lines shall satisfy a standard generally recognized for offshore application.

8.4.2.3 The following properties of anchor lines are normally to be documented:

ultimate strength
guaranteed minimum yield strength
load-elongation characteristics
fatigue resistance
durability in sea water
notch toughness

8.4.2.4 The determination of loading effects in the anchoring system is to satisfy the relevant requirements of 4.5.

8.4.2.5 The fatigue life of anchor lines is to be determined with due consideration of possible local vibrations, longitudinal and transverse, in addition to the primary loadings.

8.4.2.6 The requirements regarding safety of anchor lines will depend on the type of anchor line used. The design resistances to be used as the basis for design in ULS and FLS are to be established in accordance with the principles given in 4.4.3.2 and 4.4.4.2, and are subject to approval in each case.

8.4.2.7 Anchor lines are to be replaceable. A program for replacement of the anchor lines is to be given in the Operations Manual.

8.4.3 Mechanical joints

8.4.3.1 Criteria pertinent to the design of mechanical joints used for anchoring of structures, such as articulated towers, are subject to agreement in each case.

Section 9 Foundations

9.6.7 Jack-Up Platform Foundations

9.6.7.1 The design requirements given in Section 9 for gravity type foundations may be applied to jack-up platform foundations.

9.6.7.2 Preloading of the jack-up platform foundation in lieu of a complete foundation analysis based on comprehensive site investigations may be considered, provided that;

- the critical loading condition expressed in terms of shear stress mobilization in the foundation soil is exceeded during preloading by 30% or more
- the preload is sustained a time period estimated to be long enough to allow excess pore pressures to dissipate
- the platform elevation can be corrected for possible settlements of the foundation during operation
- it can be shown that the strength of the soil is not critically reduced due to the effects of repeated loading
- scour is considered according to 9.4.4
- the stability of the sea bottom is satisfactory.

Section 10 Marine Operations

10.1 General

10.1.1 Application

10.1.1.1 Section 10 applies to all marine operations necessary for the construction, transportation, and installation of an offshore structure, to the extent such operations may influence the safety of the structure or parts thereof.

10.1.1.2 As the requirements given in Section 10 cover various types of structure and many different types of operation, the application of the various paragraphs is to be considered in relation to the actual type of structure and to the complexity of the operations to be performed.

10.1.1.3 New concepts and solutions not adequately covered in this section will be specially considered in each case.

10.1.2 Assumptions

10.1.2.1 It is assumed that all marine operations are conducted by competent personnel and that the operation planning is based on experience and sound engineering practice.

10.1.2.2 These Rules assume that conditions during the actual operations do not depart from those assumed.

10.1.3 Surveillance

10.1.3.1 Marine operations are to be carried out in accordance with approved procedures and under the surveillance of DnV.

10.1.4 Permissible Loadings

10.1.4.1 All loadings on the structure during marine operations are to be within the limits specified in the Operations Manual. See 10.1.7.

10.1.4.2 Loads during marine operations are to be determined in accordance with the principles given in Section 5.

10.1.4.3 Loading effects are not to exceed the limits given by the requirements concerning structural resistance and serviceability. (See Sections 4, 6 and 7.)

10.1.4.4 The structural strength and behaviour of any floating unit supporting the structure are to be analysed as far as they may influence the support conditions of the structure.

10.1.5 Hydrostatic Stability

10.1.5.1 Proof of sufficient stability and reserve buoyancy is to be established for all stages of marine operations.

10.1.5.2 The following requirements are to be met:

> The metacentric height (GM) corrected for free surface and effect of possible air cushion is to be at least 1 m.
>
> The heel due to extreme wind, towing and mooring loads should not exceed 5 degrees. The wind velocity to be used in the calculations is to be selected in accordance with Section 5.
>
> The floating structure is to withstand accidental, rapid increase in loading during transfer of heavy loads, unless special precautions are taken.
>
> The structure is to remain afloat in stable equilibrium with sufficient freeboard to preclude progressive flooding with any one compartment open to the sea.

These requirements may be dispensed with in special cases provided adequate, approved precautions are taken to maintain the same degree of safety.

10.1.5.3 Inclining tests are to be performed prior to all marine operations where stability may be critical. Such inclining tests are to be performed in accordance with 8.2.2.

10.1.5.4 Structures supported on floating bases will be specially considered in each case. However, the relevant requirements of 8.2 apply.

10.1.6 Electrical and Mechanical Systems

10.1.6.1 The structure is to be equipped with all the systems necessary to maintain complete control of the structure during marine operations.

10.1.6.2 Depending on the complexity of the operation, its duration and the structure itself, a separate study may be required to determine the systems required for safe operation. The study is to include normal operations as well as emergency situations. Normally, the following systems are to be considered:

> main power supply
> emergency power supply
> electrical distribution
> machinery control systems
> valve control systems
> instrumentation systems
> bilge and ballast arrangements
> compressed air systems
> fire fighting systems
> communication systems

10.1.6.3 The systems are to be designed, built, installed and tested in accordance with relevant sections of DnV Rules for the Construction and Classification of Steel Ships or other recognized codes or standards.

10.1.6.4 All systems and equipment involved are to be tested shortly before the start of an operation. Such commissioning tests are to demonstrate that the systems are sufficiently reliable, and have the necessary capacities.

10.1.7 Documentation

10.1.7.1 For general requirements concerning documentation; see 2.3.

10.1.7.2 *Marine operations manuals:* Prior to commencement of any marine operation the details of the operation are to be fully described in operations manuals. These manuals are to cover all aspects of importance for normal operations as well as emergency situations. Generally the following aspects are to be considered:

> organization and communication
> systems and equipment involved (see 10.1.6)
> limitations imposed by environmental conditions (see 10.1.4)
> limitations imposed by structural resistance (see 10.1.4)
> limitations imposed by stability considerations (see 10.1.5)
> operational procedures.

10.1.7.3 *Marine operations records:* During a marine operation the actual procedures used and observations made are to be recorded. Where deviations from approved procedures or expected behaviour occur, the implications are to be analysed and included in the operations records together with the relevant conclusions.

10.2 Construction Afloat

10.2.1 General

10.2.1.1 Subsection 10.2 covers marine operations necessary to provide a safe and stable base for construction activities afloat. A construction base may consist of parts of the platform structure itself and/or support structures such as barges, pontoons etc.

10.2.2 Mooring Systems

10.2.2.1 The mooring system includes all arrangements necessary to keep the structure in its planned position during construction afloat. The mooring system is to be designed for all relevant loads; see section 5.

10.2.2.2 The weak link principle is to be applied when designing the mooring system; see 4.7.1.3.

10.2.3 Instrumentation Systems

10.2.3.1 For general requirements; see 10.1.6.

10.2.3.2 For proper control of the structure during construction afloat it may be necessary to make use of instrumentation to monitor:

- loads or deformations
- environmental conditions
- ballast and stability conditions
- heel, trim and draft.

10.2.3.3 Essential instruments are to be duplicated. If dependent on electric power, the standby power source is to be emergency batteries.

10.2.3.4 All instruments required by 10.2.3.2 are to be tested and calibrated to the satisfaction of DnV prior to start of operation.

10.2.4 Special Installation Equipment

10.2.4.1 Systems and equipment used during special operations such as deck mating, installation of modules etc., are to be specified. Such specifications are to be sufficiently detailed to permit complete assessment of operational feasibility and loads imposed on the structure.

10.2.4.2 To allow evaluation and approval of the special equipment, the following documentation is to be submitted to DnV:

- description of the equipment
- general lay-out drawings
- strength calculations
- material specifications
- fabrication and installation specifications.

10.3 Transportation

10.3.1 General

10.3.1.1 Subsection 10.3 covers all operations necessary to move the platform structure or major structural parts from the onshore place of fabrication/assembly to the final position. For lifting operations see 10.5.3.

10.3.1.2 The following are to be considered; the transported structure, the sea fastening arrangements, the floating units, the towing arrangements, and any special arrangements involved in the operations.

10.3.1.3 For requirements regarding instrumentation systems, see 10.1.6 and 10.2.3.

10.3.2 Transfer Operations

10.3.2.1 A transfer operation includes all the activities necessary to move a structure from one support condition to another. Such transfer operations may be performed by means of lifting, pushing, pulling or ballasting/deballasting of floating units.

10.3.2.2 If the transfer operation involves barges or other floating units, proper mooring systems are to be arranged to provide stable bases for the operation and to achieve the necessary positioning accuracy of the units.

10.3.2.3 The tolerances on supports, skidways etc., are to be specified so that the structure in no phase of the transfer operation is supported in such a way that it may be overstressed.

10.3.2.4 For barges or other floating units forming supports for the structure, a flotation study is to be carried out to verify that the support conditions are satisfactory at all stages.

10.3.3 Towing Operations

10.3.3.1 The motions and motion responses of the towed unit are to be analysed. The environmental conditions and loads are to be determined in accordance with Sections 3 and 5 respectively.

10.3.3.2 The towing arrangements and the towing force (required bollard pull of tugs) are to be sufficient to ensure proper control and speed of the towed unit in:

  adverse sea conditions
  adverse currents
  restricted waters.

10.3.3.3 The towing arrangement is to be so designed that failure will not occur in the towed unit itself (weak link principle).

10.3.3.4 The towing route is to be chosen so that adequate bottom clearance and sea room are achieved during the towing operations. Attention is to be paid to navigational accuracy, motion characteristics of the unit and possible heel/trim effects of towing forces, static wind force, ballasting etc.

10.3.3.5 Arrangements for reliable weather forecastings prior to and during the towing operations are to be provided. Weather criteria for starting operations will be evaluated in each case.

10.4 Installation

10.4.1 General

10.4.1.1 Subsection 10.4 covers the operations necessary for installing the structure at its final position. Such operations are positioning, setting, fixing etc. of the structure.

10.4.1.2 If installation operations can cause overloading of structural members or of the foundation, relevant effects of such loading are to be monitored and controlled.

10.4.1.3 Installation instrumentation: For general requirements on instrumentation systems, see 10.1.6 and 10.2.3. Instrumentation employed for control of the structure during installation may include devices for measurement of draft, penetration/settlement, inclination, ballast levels, navigational parameters and environmental conditions.

10.4.2 Launching and Upending

10.4.2.1 When a structure is to be launched from a barge, the launch barge is to be specifically equipped for this type of operation. Launchways and rocker-arm arrangements are to be considered with respect to suitability and structural strength.

10.4.2.2 The launching operation is to be planned so that loads imposed on the structure during the launch are within acceptable limits. Parameters to be specially considered are:

  freeboard of launch barge
  trim angle of launch barge
  amount of buoyancy
  position of buoyancy tanks.

10.4.2.3 The launch-prepared structure is to be designed with sufficient net buoyancy to compensate for possible inaccuracies in determination of weights and buoyancy.

10.4.2.4 It is to be verified that the structure will behave in a stable manner during the launch dive and the upending operation, and that sufficient bottom clearance is assured at all stages.

10.4.2.5 Buoyancy tanks, supports, and other intermediate equipment are to have adequate structural strength to withstand forces imposed during the launching and upending operations. Buoyancy tanks are to be closely surveyed for imperfections which could influence their structural resistance.

10.4.3 Positioning and Submergence

10.4.3.1 The structure is to be placed within the investigated area as defined in 9.2.1.3. If the sea bed has been specially prepared the maximum tolerances in positioning are dictated by the extent and the nature of the preparations.

10.4.3.2 The structure is to be lowered in a controlled manner. Due care is to be taken in assessing the inaccuracies inherent in water depth, sea bed topography, and obstructions. Sudden or large motions during touch down are to be avoided.

10.4.3.3 For requirements regarding ballasting see 10.4.4.

10.4.4 Penetration and Leveling

10.4.4.1 The ballast system is to provide ballasting rates permitting safe descent and penetration. The ballasting process is to be reversible during critical stages.

10.4.4.2 The structure is to be capable of providing sufficient ballasting capacity to overcome maximum expected penetration resistance to reach required penetration depth. As the local penetration resistance can vary across the foundation site, eccentric ballasting may be necessary to keep the platform inclination within specified limits.

10.4.4.3 The loading caused by the various ballast configurations during platform installation is to be within the limits given in 10.1.4.

10.4.5 Pile Installation

10.4.5.1 Pile installation operations are to be properly planned and executed to avoid reduction in the load carrying capacity of the various soil formations.

10.4.5.2 The piles are to be installed in a sequence providing adequate stability to the structure in all phases of installation.

10.4.5.3 Energy input to the pile and the corresponding pile set are to be recorded during pile driving operations.

10.4.5.4 Jetting is usually acceptable only inside the pile casing to a depth not affecting the soil at the pile tip. Below the tip of the pile casing only controlled drilling with carefully selected drilling fluid is to be permitted.

10.4.5.5 The placement of grout is to be carefully planned. The grout is to have pumpability and setting time consistent with the actual placement operations. The grout viscosity, density and bleeding properties are to be kept within acceptable limits.

10.4.5.6 During drilling and grouting operations fluid pressures in the drilled hole are to be within the limits set by hydraulic fracturing of the soil and by the stability of the hole itself.

10.4.6 Filling of Voids

10.4.6.1 For general requirements regarding material for filling of voids see 9.6.6.

10.4.6.2 The filling is to be done with stresses well within the limits set by considerations of structural and foundation integrity. The system used is to be able both to transport sufficient filling material to the desired location and to remove trapped water from such locations.

10.5 Construction Offshore

10.5.1 General

10.5.1.1 Subsection 10.5 covers requirements concerning the operations connected with completion of the platform structure after it is fixed to the sea bed.

10.5.2 Installation of Structural Parts

10.5.2.1 Prior to all mounting operations, such as installation of deck sections, module support structures, modueles, packages etc., geometrical tolerances on structural parts to be connected are to be established.

10.5.2.2 All placements of structural parts are to be followed by a verification that the actual support and fixation conditions are in accordance with designer's specifications.

10.5.3 Lifting

10.5.3.1 It is to be documented that structures to be lifted have structural strength adequate for the operation. Special attention is to be paid to dynamic loads.

10.5.3.2 For all lifting operations, the structural strength and general suitability of the equipment are to be considered. Such equipment may be:

cranes
crane barges
mooring system of barges
slings and shackles
spreader frames

10.5.3.3 For lifting lugs/padeyes primary structural steel is normally to be used.

10.5.3.4 Stabbing guides installed to ensure smooth placing of lifted items are to have adequate strength to withstand the impact loads likely to occur during the lifting operation. The guides are to be constructed such that the primary structure suffers no damage if the imposed loads should exceed those assumed.

10.5.4 Special Operations

10.5.4.1 Special operations include activities required to install special items on platform structures, such as conductors, scour protection, foundation drains, performance instrumentation, risers etc.

10.5.4.2 Special operations are to be considered in each case with respect to their influence on the integrity of the structure itself and its foundation.

Appendix 2

Excerpts from Fédération Internationale de la Précontrainte, Recommendations for the Design and Construction of Concrete Sea Structures, 4th ed.

4.7 General Design Considerations

4.7.1 Construction, Installation, Release and Retrieval Conditions

General. The design should take into account loading conditions on the complete or partially complete structure during construction on a slipway or in a dock, launching, completion afloat, towing or passage to site and sinking or anchoring to final position.

It should also take into account the conditions during breakout and retrieval of the structure where there is any likelihood of its being released and moved to another site during its service life. Local environmental loads, appropriate to the season where applicable, should be considered. The design for these conditions should be such that the interim and subsequent compliance of the structure with the permanent design requirements is not impaired.

Particular attention should be given to the wind, waves and currents to be anticipated during towing. These may be significantly different from those experienced at the service location.

Conditions While Structure Is Floating. The structure should be stable and have an adequate reserve of buoyancy at all stages of construction, passage to site and flotation at site. It should be capable of recovering from the angles of roll and pitch caused by the worst environmental conditions expected while afloat.

The hydrostatic and motion-induced loadings should be calculated for all stages of construction and installation of the structure.

Wave, wind and current action will induce movements in the structure while it is floating. Its periods of heave, pitch, roll, yaw, surge and sway should be calculated and their effects upon the structure, men and plant considered. Where accumulation of ice is a hazard, the effect of its mass and added resistance to wind should be included in such calculations.

All external pipework and fittings, whether temporary or permanent, should be designed and constructed to withstand anticipated environmental loads when afloat, and should be positioned or protected to prevent snagging of mooring or tow wires, retrieving gear, or likely damage from external sources, etc.

The anchorage and towing systems, points of attachment to the structure and the structure itself should be capable of absorbing snatch loads produced by wind, wave and current forces and by towing.

Installation of Structure. If the structure will be immersed during installation on site and/or during construction, the stability of the structure and the structural integrity shall be safeguarded during all phases of submersion. Particular attention should be given to

(a) uplift forces due to swells as horizontal surfaces pass below water surface level
(b) absorption of water by the structure with time, pressure and permeability producing a change in weight

---

Telford, London, 1985. By permission of Thomas Telford and Sons, Ltd.

(c) compression of air or other internal gases or fluids during descent
(d) level variations of internal ballast water
(e) hydrodynamic effects in shallow water and at time of landing.

Special attention should be given to the possibility of buckling of shell-type elements and compressive members, giving due regard to geometrical imperfections and the long-term effect on strength and stiffness.

Release and Retrieval. Where the structure is to be relocated during service, the forces imposed on the structure during release from the foundation should be determined.

Means should be provided to check or control the rapid rise of the structure once it has been released from its foundation.

In the case of gravity structures, breakout may be achieved by first reducing ballast to attain neutral buoyancy, then pumping water in below the base slab in order to form a gap.

An overpressure corresponding to the breakout force will eventually develop. Only when the breakout is complete should ballast be further reduced in order to float the structure.

Means for pumping water under the base slab should be incorporated in the initial design if retrieval or relocation are considered.

6 Construction and Installation

6.1 Construction Stages

For the type of structure covered by these Recommendations (see Section 1.1.), as much as possible of the construction work is normally performed away from the permanent site in a protected location on or near the shore. For the purpose of these Recommendations, construction is assumed to take place in stages as follows:

(a) first-stage construction in a fabrication area with the structure, initially at least, in the dry
(b) initial flotation of the partially completed structure and towing offshore; alternatively, the structure may be lifted by heavy marine lifts or floating cranes and towed offshore on barges or suspended from the floating cranes
(c) further stages of construction with the structure afloat, or temporarily grounded, in a protected location near the shore
(d) towing of the structure to its permanent location
(e) installation
(f) final construction in situ to complete the structure.

6.2 Construction

6.2.1 General

Construction methods and workmanship should follow accepted practices as described in relevant Standards, Codes and specialist literature. In general, only additional recommendations specially relevant to concrete sea structures are included here.

At no time should procedures or methods be adopted which affect the safety of the structure or lead to difficulties during later stages of construction and installation.

The design should be checked to ensure that bollards, areas of outer walls which will be pushed by tugs and parts of the structure which will be exposed to severe dynamic forces during later stages of construction are strong enough for their intended purpose.

6.2.2 Tolerances and Control of Buoyancy

Tolerances for the buoyancy and the centre of gravity of the structure should be set with due regard to the safety of the structure during all stages of construction and installation. In setting these tolerances, attention should be given to the following factors which might affect the centre of gravity and buoyancy of the structure:

(a) the unit weight of the concrete in the dry
(b) the variation with time of the absorption of water by the concrete, with due allowance for pressure gradients which could occur during all stages of construction
(c) accuracy of dimensions, in particular the thicknesses of walls and slabs
(d) control of overall configuration, particularly of

radii of curvature of cylinders and domes and the prevention of distortion during concreting
(e) the weight and weight distribution of any permanent or temporary ballast
(f) the proper functioning of the system provided to vary the ballast when floating and sinking, including the control of effective free-water planes inside the structure
(g) the vertical centre of gravity of the structure should be confirmed by an inclining test where stability may be expected to be critical.

6.2.3 Construction Joints

Construction joints should be prepared with great care wherever the structure is to remain watertight or is designed to contain oil. This applies whether the watertightness is required permanently or only temporarily as, for example, during towing and installation.

Suggested precautions to be taken when watertight construction joints are required include careful preparation of the surface, e.g. by abrasive blasting or high-pressure water jet, to remove laitance and to expose the coarse aggregate to a depth of 6 mm. An epoxide resin bonding compound sprayed on just before concreting may also be used, but in this case the surface still requires preparation. Water bars may also be used.

6.2.4 Temperature Rise in Concrete

The rise of temperature in the concrete caused by the heat of hydration of the cement, should be controlled to prevent steep temperature stress gradients which could cause cracking of the concrete (see Section 3.3.6.).

Reducing the temperature rise in the concrete may present problems, particularly with the rich mixes and thick sections demanded in many components of concrete sea structures.

Methods of control of concrete temperature include selection of coarse-ground or other cements having a lower heat of hydration, reduced rates of placing, insulation of forms and of freshly exposed concrete and use of cooling pipes. The danger of thermal shock with water curing should be borne in mind (see Section 6.2.6.).

Part of the cement may be replaced by a pozzolanic material, provided that early and sufficient testing has been conducted to prove the pozzolan's properties and its compatibility with the cement.

6.2.5 Concreting in Cold Weather

In cold weather, concreting in air temperatures below 2°C should only be carried out if special precautions are taken to protect the fresh concrete from damage by frost. The temperature of the concrete at the time of placing should be at least 5°C and the concrete should be maintained at this temperature until it has reached a strength of at least $5 N/mm^2$. Protection and insulation should be provided to the concrete where necessary.

The aggregates and water used in the mix should be free from snow, ice and frost. The temperature of the fresh concrete may be raised by heating the mixing water and/or the aggregates. Cement should never be heated. Cement should not be allowed to come into contact with water at a temperature greater than 60°C.

6.2.6 Curing of Concrete

Special attention should be paid to the curing of concrete in order to ensure maximum durability and to minimize cracking. Concrete should be cured with fresh water whenever possible to ensure that the concrete surface can be kept wet despite wind, etc. Care should be taken to avoid the rapid lowering of concrete temperatures (thermal shock) caused by applying cold water to hot concrete surfaces.

Sea water should not be used for curing reinforced or prestressed concrete although, if demanded by the construction programme, concrete may be submerged in sea water provided that it has gained sufficient strength to avoid physical damage from waves, etc.—in general 90% of the specified characteristic strength. When there is doubt about the ability to keep concrete surfaces permanently wet for the whole of the curing period, or where there is a danger of thermal shock, a heavy-duty membrane curing compound should be used.

6.2.7 Reinforcement

The reinforcement should be free from loose rust, grease, oil, deposits of salt or any other material likely to affect the durability or bond properties of the reinforcement.

The bond (development) length of bars subjected to significant impact or cyclic loading should be based on half the bond strength of that prescribed for static load unless an in-depth fatigue analysis is carried out. This applies to end anchorage lengths and to lap splices. Mechanical and welded splices may be used, provided tests show that they develop the full strength of the bars in tension, compression and fatigue.

The specified cover to the reinforcement should be maintained accurately. Special care should be taken in the cutting, bending and fixing of reinforcement to ensure that it is correctly positioned and rigidly held so as to prevent displacement during concreting.

6.2.8 Prestressing Tendons, Sheathing and Grouting

General. This section deals in the main only with those requirements of prestressed concrete which are special to sea structures. Further guidance on prestressing steels, sheathing, grouts and the procedure to be taken when storing, making up, positioning, tensioning and grouting tendons will be found in relevant Standards, Codes and specialist literature.

Tendons. All steel for prestressing tendons should be clean and free from grease, oil, deposits of salt or any other material likely to affect the durability or bond of the tendons.

During storage, prestressing tendons should be kept clear of the ground and protected from weather, moisture from the ground, sea spray and mist. No welding, flame cutting or similar operations should be carried out on or adjacent to prestressing tendons, under any circumstances where the temperature of the tendons could be raised or weld splash could fall on them.

Where protective wrappings or coatings are used on prestressing tendons, these should be chemically neutral and should not produce chemical or electrochemical corrosive attack on the tendons. Further guidance on the storage, handling and position of prestressing tendons may be found in relevant Standards, Codes and specialist literature.

Sheathing. Metal sheathing should be stored clear of the ground and protected from the weather, moisture from the ground, sea spray and mist.

All ducts should be watertight and all splices carefully taped to prevent the ingress of water, grout or concrete. During construction, the ends of ducts should be capped and sealed to prevent the entry of water and foreign material. Ducts may be protected from excessive rust by the use of vapour phase inhibitor powder.

Where ducts are to be grouted, all oil or similar material used for internal protection of the sheathing should be removed before grouting. Air vents should be provided at all crests in the duct profile. Threaded grout entries, which permit the use of a screwed connector from the grout pump, may be used with advantage.

Grouting. For long vertical tendons, the grout mixes, admixtures and grouting procedures should be checked to ensure that no water is trapped at the upper end of the tendon due to excessive bleeding or other causes. General guidance on grouting will be found in specialist literature.

6.3 Initial Flotation

Initial flotation or launching of the structure should be carried out in such a way that the structure is not subjected to excessive forces, taking into account the position of the ballast at the time of floating and that, at this stage, the structure may be incomplete and the concrete still immature. The structure should remain stable during all stages of flotation. Wherever possible, adequate damage stability and damage control should be provided, including the provision of suitable pumps and means of providing power for them.

If compressed air is introduced beneath the structure in order to reduce effective draft, the effect on any internal free-water planes should be determined and adequately controlled. Also the effect of the compressed air in changing shear forces and bending moments should be checked.

If the structure is to be lifted, the stresses in the lifting lines and the tensile and shear stresses in the concrete immediately below and surrounding the lifting points should not exceed the permissible limits. Lifting accelerations should be minimized to limit dynamic tensile stresses in the lifting lines. Care should be taken to secure the structure after initial flotation and also to ensure sufficient underkeel clearance is obtained in order to avoid accidental grounding at low tides.

For large sea structures, the normal method of initial flotation will be to let water enter the dry dock in which the first stage of the construction has taken place. Other methods which may be used include tidal launching, low-

ering by means of a floating dock or a marine lift, launching from a slipway, using a collapsing pile platform, or the removal of sand on which the construction has taken place.

6.4 Construction While Afloat or Temporarily Grounded

When further construction takes place while the structure is afloat, the structure should possess adequate stability to cope with the effects of waves, wind, currents and mooring forces. The moorings should have sufficient strength to resist all anticipated forces with an adequate factor of safety. Consideration should be given to means of maintaining damage stability and damage control at all stages of construction afloat, including the provision of suitable pumps and the means of powering them. All these matters should be given attention at the preliminary design stage. If provision of adequate damage stability is not practicable for any phase, then this should be made clear to all operating personnel. Damage control equipment and procedures should be established to minimize the risk.

A detailed plan or manual should be prepared for the ballasting of the structure during all stages of construction, and regular checks should be carried out to ensure that the buoyancy achieved agrees with the calculated values. Attention should be paid during ballasting to limitations which may exist for allowable water head on internal walls.

If the structure is temporarily grounded, the shape of the sea bed on which it is placed should be within acceptable tolerances, having regard to the strength of the concrete at the time. When grounding and subsequently lifting the structure, considerations similar to those given in Section 6.6. will apply.

6.5 Towing

6.5.1 The Structure, Superstructure, Deck and Other Equipment

Strength of the Structure. All aspects of towing should be designed and planned to ensure that the structure is not exposed to loadings greater than those for which it was designed.

Fatigue of permanent or temporary steelwork, even after the relatively few stress cycles occurring during a tow, may be a serious consideration in a corrosive environment.

Response to Motion. The motion response of the structure in all directions of freedom should be determined for the structure in the towing condition. These responses should be verified by calculation and where applicable model tests. Care should be taken in the selection of the added mass factor for use in calculations. Where the shape of the structure makes it sensitive to undesirable motions, e.g. dynamic uplift, nosediving, yaw, etc., the hydrodynamic control of the structure under tow should be adequate to minimize this effect. Checks should be made to ensure that the motions of the unit in the worst expected environmental conditions do not result in unacceptable stresses or increase in draft.

Adequate tie-downs should be provided to prevent damage to, or dislocation of, the superstructure and equipment caused by dynamic acceleration due to rolling and pitching during the tow, and also arising from inclination of the structure in a damage stability condition.

In the case of designs with base and columns, particular attention should be given to: (a) column strength, deck to column connections and other critical points of the structure; and (b) strength of internal walls when subjected to increase in hydrostatic loading from ballast water level changes for draughts at which the unit will be towed.

Stability. For undamaged stability, when towed, the structure should possess a righting energy at least 1.4 times the overturning energy induced by a 10 year, one minute sustained seasonal wind or a one-minute sustained wind that can be predicted with sufficient reliability. It should also have an adequate range of static stability taking into account anticipated motion responses, s, and an initial positive metacentric height sufficient to limit motions to those allowed for the structure.

The free surface effect of water carried as ballast should be controlled to keep it within the limits assumed in the design of the tow and determined by model tests.

If temporary buoyancy tanks are used to improve freeboard and stability, these should be securely attached to the unit, the fixings being designed to withstand all loading during towage and immersion.

Where the structure is suspended from a floating crane, a careful examination should be made of the effects of wind, waves and currents on the stability of the floating crane and its load. The dynamic tensions produced in the lifting lines by movements of the suspended loads should not exceed the safe loads on the lines. Excessive amplitudes of swing should be prevented by providing restraints on the lateral movement of the suspended structure.

For damaged stability, during floating, construction and towing stages, adequate measures should be taken to prevent the occurrence, or to limit the extent, of damage arising from likely collision. If this is not possible, the design should ensure that flooding of one compartment does not result in the capsizing or sinking of the structure, possibly after progressive collapse of internal partitions. Similarly, flooding is a danger where there are pipes or other passages through internal and external walls. Precautions should be taken to prevent or counteract the effect of damage to the cover plates, or other means provided to close such pipes or passages.

Towing Connections and Lines. An adequate number of towing connections, suitably placed, should be fitted to the structure to permit the tugs employed for the tow to be secured and released. All towing connections should have an ultimate strength not less than three times the force which would occur in the tow line, with the largest tug to be employed exerting its static bollard pull. All angles which the towline may adopt should be considered. Provision should be made to deal with emergencies caused by broken towing lines or towing pennants.

Emergency Anchors. The provision of emergency anchors and lines should be considered. These will depend on the size of the structure and the depth of water to be encountered on the tow.

Pumping. Adequate pumping capacity should be carried on the structure during the tow to control ingress of water to any compartment due to leakage, and to control the change of draught or trim should this be necessary or desirable.

6.5.2 Choice of Towing Route

General. The route to be taken during the tow should be decided by or in consultation with experts. The choice of towing route should take into account the depth of water, the strength and direction of tidal streams and currents and any navigational hazard that could affect the safe conduct of the tow.

Consideration should be given to identifying possible holding areas on the towing route, where weather and tidal factors dictate.

Depth of Water. The selected towing route should ensure that the structure is afloat at all times with an adequate underkeel clearance. Calculations of the underkeel clearance should take into account the increase in draught due to roll, pitch and heave, squat effect, towline and wind heel.

Where the soundings shown on the largest-scale navigational chart available are of questionable accuracy, e.g. where the charts are based on incomplete hydrographic surveys or there are areas subject to changes in sea-bed topography, an adequate survey of the selected route should be carried out. The planning and carrying out of surveys should be based on the accuracy and repeatability of the navigation of the survey vessel and of the towed object.

Where sand waves occur which could obstruct the tow, their shapes and locations should be determined by a sufficient number of surveys to enable their movements to be predicted. The effect of storms on sand waves should be allowed for.

It is strongly recommended that all tow routes be surveyed and the results made available in readily understood form to those responsible for the towing. The degree of survey detail will depend on the available channel width and depth and the margins required on same.

Dependence on Tides. Where safe passage over any section of the selected route is dependent on a rise of tide, published tidal information should be checked to ensure that it is truly applicable to the section considered. If the available tidal information is not completely reliable, a tidal survey should be carried out. Allowance must also be made for meteorological effects (e.g. lowering of sea level in high pressure systems, and with offshore winds).

Towing in Restricted Waters. The selected route should provide adequate sea room at the surface for the manoeuvring of tugs, and the width of channel at the sea bed should be carefully considered in relation to the width of the structure being towed and to local tidal, current and weather conditions. In restricted waters the selected navigable channel should permit the passage to be made with a minimum number of course alterations.

In restricted waters where available information

on the strength and direction of tidal streams and currents is inadequate, a survey should be made to determine such currents at the surface and, in the case of deep-draught structures, at the depth of the keel and at intermediate depths.

Abandoning Area Due to Surface Ice Danger. Consideration should be given to the evacuation from the area if threatened by ice, unless the design of the structure and towing vessels includes ice tolerance.

6.5.3 Environmental Criteria and Weather Limitations

No tow should commence other than in good weather and with a favourable weather forecast. When the tow is expected to last longer than the period covered by a reliable weather forecast, it should be designed and planned to deal safely with the most severe wind speed, wave height and current which it would be reasonable to expect during the tow. In general a ten-year statistical return period should be considered for the season of the tow.

6.5.4 Seasonal Limitations

The climate along the towing route should be investigated and towing avoided during seasons or periods when unacceptable weather is frequent or icing to an unacceptable level may be expected. If icing of the structure above water level is a possibility, measures should be prepared in advance to prevent icing reaching an extent sufficient to jeopardize the stability of the structure.

6.5.5 Towing Resistance and Horsepower Required

General. The minimum effective towing horsepower available should be at least that which will produce the required towing force at the desired speed of tow, and which will control the structure in heavy weather conditions and currents to be expected.

The minimum available towing force should be great enough to produce a towing speed which ensures that the structure can be properly controlled and will not be driven ashore or aground under the influence of adverse weather, tidal streams or ocean currents.

Where the necessary minimum horsepower can only be supplied by a combination of several tugs, consideration should be given to the maximum number of tugs which it is feasible to attach to the structure and control in a seaway.

Towing at Sea. For towing at sea, the calculated towing resistance should be based on the forces imposed on the structure by gale force winds and associated waves, together with a current of at least 1 knot. Greater current or tidal stream velocities may have to be considered in certain localities.

The calculated forces on the structure imposed by wind, waves and currents may have to be verified by model tests.

Towing in Restricted Waters. Towing in restricted waters should receive special attention. Such waters may offer shelter from the wind and the fetch may be such that waves are small, but currents, the proximity of navigational hazards, shipping density and the need to tow with shortened tow lines could necessitate a higher towing horsepower than would be required in open water. Owing to the difficulties in generating sufficient towing pull on very short tow lines attached to massive objects, it may be necessary to consider using pusher tugs. For towing in restricted waters, assistance should be obtained from a pilot having local knowledge of the waters.

6.5.6 Warnings to Shipping and Aircraft

The showing of navigation lights, the display of special signals and the use of sound signalling apparatus should conform at all times to the requirements of the *International Regulations for Preventing Collisions at Sea,* published by the Intergovernmental Maritime Consultative Organisation. Aircraft warning lights may have to be shown in accordance with the appropriate regulations.

Where the towing route passes through sea areas well used by shipping, the lead tug of the tow should transmit a warning at regular intervals to other ships giving the nature of the tow, its course, speed and position.

6.5.7 Navigation

Navigation should be such that the position of the towed structure is known at all times.

The systems and method of position fixing should be selected based on required accuracy taking into account the dimensions of the towed structure, the width and depth of the towing channel as well as the climatic conditions to be expected during the tow (including visibility).

Redundancy in each system as well as alternative method back-up systems should be provided. Particular attention should be given to the power sources to ensure uninterrupted power supply.

6.6 Installation

6.6.1 General

All aspects of the installation of the structure, including its immersion and placing on the sea bed, should be planned and carried out with the greatest care. The arrangements made for installation should ensure that the structure is placed in position within the given tolerances.

For large structures, lifting and replacing should be considered only as an emergency measure. If lifting and replacing is unavoidable, the process should be analysed carefully before it is undertaken.

Installation should be planned so as to avoid uncontrolled contact between auxiliary vessels and between these vessels and the structure. Where one structure is to be moored adjacent to another, energy-absorbing fenders may have to be provided to prevent impact damage.

Model tests should be conducted to a scale sufficiently large to establish the motions of the structure during immersion. These model tests should investigate the effects of the expected waves, currents and, if applicable, the pull from anchors and the pulling or suspension forces from auxiliary vessels.

6.6.2 Environmental Factors

Wind, Waves, and Currents. Installation should be planned to take place safely in the maximum wind speed, wave heights and currents which may reasonably be expected, taking into account the reliability of weather forecasts for the area of the site and d the expected duration of the operation. The variation of current velocity with depth should be included in the factors considered.

Effective Unit Weight of Sea Water. Variations in the effective unit weight of the sea water at the site should be investigated and taken into account when planning the immersion operation.

The effective unit weight of the sea water will vary owing to differences in salinity, temperature and silt suspension, as well as to the dynamic effects of turbulence and currents, particularly near the surface and the sea bed. These variations may have significant effects on the buoyancy and the stability of the structure during immersion.

6.6.3 Foundations

Condition of the Sea Bed. Planning of the installation should take into account the conditions of the soil at the site, including its hardness and its susceptibility to scour and suction or breakout effects.

The topography of the sea bed should be checked, attention being paid to its slope, unevenness and the occurrence of boulders. The topography should be tied in accurately to horizontal survey controls by sea-bed transponders or other means.

Planning should also take into account the possibility of erosion of the sea bed due to horizontal flow of water from beneath the structure as it nears the bottom. This flow should be kept within predetermined limits through descent speed control and/or through discharging part of the water by means of a system built into the structure.

Preparation of the Sea Bed. Preparation of the sea bed where necessary, should take place as short a time as possible before the structure is placed. Preparation may consist of dredging, levelling and trimming, provision of a screeded rock base, prior overload by surcharging or other means, such as soil densification through vibratory compaction.

Screeded rock bases should be placed with adequate control of position and tolerances. They should be protected from erosion or silt deposits during the period before the structure is placed, where this is known to be a problem.

Strengthening of Sea Bed during Installation. The sea bed may be strengthened during installation of the structure by forcing the structure to penetrate the sea bed, by consolidation through overload, by depositing tremie concrete, grout, sand, etc., or by other means.

(a) Penetration may be assisted by internal excavation using dredge cutter-heads or a jet airlift, or by external jet lubrication, vibration or cyclic overload. Where the use of jet lubrication is considered, its effect on the strength of the soil should be investigated.

(b) Overload may be provided by water ballast, sand ballast or barytes-loaded drilling mud. Reducing the water pressure beneath the structure creates apparent overload, but soil pore pressures should be controlled with extreme care to prevent the development of a 'quick' condition or 'boil' in the foundation soil.

(c) If concrete or grout is to be placed under the structure, the mix should be designed carefully to prevent segregation. The use of thixotropic admixtures may be an advantage. Placing techniques should include adequate provision for the control of pressure, expulsion of water and overflow or venting of laitance. Further details of placing concrete under water may be found in specialist literature.

(d) Reducing the water pressure beneath the foundation slab by permanent drainage increases the effective stresses in certain soils and improves the foundation stability.

Scour Protection. If scour protection is to be provided by means of stone filling, the stone should be placed in such a way as to avoid damage to the structure. After placing, the scour protection should be surveyed accurately to establish data for monitoring its performance.

Strengthening of Sea Bed after Installation. Methods for strengthening the sea bed after installation of the structure include long-term consolidation by the provision of drains. Drainage should include adequate filtration by means of porous rock or filter membranes. Adequate time should be available for the drains to produce a sufficient strengthening effect on the sea bed before extreme loading conditions are encountered. Means for drilling-in sand drains should be provided, where applicable.

6.6.4 Ballasting or Pulling Down the Structure

General. Before ballasting or pulling down is begun, the environmental conditions should be checked to make sure they are not worse than those assumed during design and planning of the operation. In addition, all systems should be checked to ensure that all aspects of the plan can be carried out as scheduled and with the required margin of safety.

Operations in unsheltered waters should depend as little as possible on the use of auxiliary floating equipment.

Once immersion has begun, it should proceed decisively. Flooding or other ballasting should begin immediately the structure is in position on the sea bed in order to provide stability against unforeseen environmental conditions as soon as possible.

Stability and Buoyancy of the Structure during Immersion. The stability of the structure at the various stages of immersion should be analysed and immersion should be planned so as to shorten the duration of critical phases.

Where the structure has a sudden reduction of the water plane, immersion should be planned to continue to a depth producing an acceptably increased stability before commencing any operation which might influence trim or heel.

Variations in buoyancy with depth and with time should be taken into account during immersion of the structure. Variations in buoyancy can be caused by absorption of water by the concrete, changes in the density of the sea water, changes in the volume of the structure, the bulk modulus of the concrete, sea water and buoyancy liquids, beaching and Venturi effects and drag due to vertical currents.

The 'beach effect' causes temporary uplift as horizontal planes on the structure submerge below the surface while a sea is running. Means should be provided for overcoming or controlling this effect and for preventing subsequent plunging as the effect disappears with depth.

Maintenance of Location. During immersion the structure may be kept on location by means of tugs and/or by using moorings.

Where moorings are used for positioning, account should be taken of the changing catenary and the possible loss of spring effect in the lines as the structure nears the sea bed. Mooring lines should be provided with adequate spring to accommodate surge loads, including those from long-period surges.

The method used to maintain location should possess adequate capacity to resist horizontal forces which may arise. Particular consideration should be given to the forces arising from the horizontal flow on 'squeezing' out of the water from beneath the structure as it nears the sea bed. These forces can be minimized by providing for the escape of the water, or by using probes (spuds) which penetrate the sea bed ahead of the main structure. If soil anchors are used, these should be designed to a holding power not exceeding 80% of the breaking load of the mooring line.

Lowering or Pulling Down. Sea-bed anchors to be used for pulling down should be designed for a working load at least equal to the pull required and in excess of the breaking load of the pulling lines. Equipment, lines and fixing points on the structure intended for lowering or pulling down the structure, should be able to withstand, with an adequate factor of safety, all forces developed in them during the lowering or pulling down process, including dynamic effects. They should be arranged, possibly by doubling up, to prevent progressive collapse of the system. Care should be taken to prevent the fouling of lines, even if winds shift to a direction normal to the seas.

Control of Water Ballast. The flooding or pumping system for controlling water ballast should possess adequate redundancy and excess capacity to deal with the possibility of leaks or inoperative valves.

All sea intake lines should be protected by spring-operated valves which close automatically in the event of loss of control air or fluid pressure. These closures should have a manual over-ride or other means of redundancy.

Compressed Air. Where compressed air or gas is used to resist hydrostatic loads, potential operational risks associated with such means should be considered, taking into account realistic variations in the pressure of the compressed air. The tolerances assumed should be consistent with the accuracy and reliability of the pressurization and control systems. Air pressures should be restricted so as not to produce a net internal overpressure.

In case the use of compressed air defines the duration of the operation, the implications of weather or tidal variations must be considered. Loss of air pressure should not lead to loss of structure.

Methods of pressurization might include the use of liquefied gases which are allowed to expand, and automatic pressurization produced by entry of water. The need for compressors for control and back-up should be considered. Procedures/equipment should prevent freezing of valves when gases are allowed to expand.

Prevention of Excessive Impacting with Sea Bed. The level and location of the structure should be monitored with extra care and accuracy as it nears the sea bed. The effects of *impacting collision* with the sea bed, suction and break-out should be taken into account and the structure should be equipped with means to control these effects.

6.7 Construction on Site

Work on site should be executed according to accepted engineering practices and as required by relevant Standards and Codes. Added emphasis should be placed on the careful planning of technical and human resources and the safety measures demanded by the marine environment. Consideration should be given to prevention of damage by collision.

R7.3 Repair of Concrete

R7.3.1 General

Methods of repair of concrete sea structures should follow generally accepted practices, the advice given in specialist literature and the recommendations of the manufacturers of the materials being used. Sea structures pose special problems of access and working conditions created by the environment. The methods chosen should enable adequate protection to be given to the work and the workmen so that a high standard of workmanship may be achieved.

Materials should be carefully chosen to be compatible with the conditions prevailing during and after their application.

The following recommendations cite standard methods and materials for repairs. It is not intended that the use of other methods and materials should be inhibited provided that they can be shown to be satisfactory.

R7.3.2 Materials

R7.3.2.1 Cement

The cement will be determined by the requirements that are put to the quality of the repair, the circumstances of the execution and the possibilities of the execution.

R7.3.2.2 Resins

Where resin materials are used, they should be of a type which is sufficiently effective in a damp or wet environment. They should be of a formulation suitable for the particular application and the manufacturer's instructions should be strictly adhered to.

Epoxide resins generally will be found to be most

suitable for the repair of sea structures. Apart from the manufacturer's instructions, general guidance on their use may be found in specialist literature.

R7.3.2.3 Aggregates

Aggregates should consist of natural sand, rock fines or dense coarse aggregate generally not exceeding 10 mm in size, but where the dimensions of the repair justify it, a larger size of aggregate may be used.

R7.3.3 Repairs to Spalled Concrete

If a concrete mix is to be used for the repair, the defective concrete should be cut out to a depth of not less than 50 mm and all loose material removed. Exposed reinforcement should be cleaned thoroughly. The concrete should preferably be damp, but surface-dry, when the patch is applied.

A thin coat of neat cement grout of epoxide-resin bonding agent should be brushed well into the prepared surface and the patching material placed immediately afterwards. Patches of this thickness should be made with concrete containing coarse aggregate and the concrete vibrated or rammed thoroughly into position. For large-scale repairs, gunite (shotcrete) may be used.

If the damaged zone is under water, the repair should be made with a shuttering, filled with a stable mortar or by a cement grout, pumped between aggregates that have been placed in advance.

Where reinforcement is exposed in the area to be repaired, it should be cleaned thoroughly and coated with an epoxide resin or other material which will isolate it from the patching concrete.

After completion, the repair should be protected from accidental damage and from waves and currents until such time as it has gained sufficient strength. Where cement mortar or concrete has been used for the repair, it should be adequately cured to prevent rapid drying out. Any covering used to protect the repair should be fixed securely in position to prevent its being displaced by winds, waves or currents.

If the thickness of the patch is less than 40 mm, a resin-based mortar should be used as the patching material.

During its period of service, a concrete sea structure may have absorbed chlorides from the environment. The patching material will be chloride-free and this difference may set up electrochemical action in the reinforcement leading to corrosion. This is the reason for providing the reinforcement with an isolating coating.

R7.3.4 Repairs to Cracks

Before a crack is repaired, its cause should be determined so that the appropriate method of repair may be chosen. The chosen method will also depend on the zone in which the crack occurs.

Where corrosion of the reinforcement has spalled the surrounding concrete, the damaged concrete should be cut away and the repair made as described in "repairs to spalled concrete". If a narrow crack has only to be sealed against the ingress of moisture and no further movement of the crack is expected, it can be reached by injection of an epoxide resin.

When continuing movement at the crack is expected, a chase should be cut along the line of the crack and this sealed with an elastic material such as a polysulphide rubber or by the insertion of a prepared neoprene or rubber bitumen sealing strip. Alternatively, a flexible cover strip may be fixed to the surface of the concrete. Where the crack is under water, an alternative method, such as injection, will have to be used.

Appendix 3

Excerpts from API RP2A, Planning, Designing, and Construction of Fixed Offshore Platforms, 15th ed.

Section 5 Installation

5.1 General

The installation of a platform consists of transporting the various components of the platform to the installation site, positioning the platform on the site and assembling the various components into a stable structure in accordance with the design. The installation of a platform should be accomplished in such a manner that the platform will fulfill the intended design purpose.

5.2 Transportation

5.2.1 General

The movement of the platform components from a fabrication yard to an installation site presents a complex task which requires detailed planning. Basic considerations vary with reference to the type of platform to be transported. Included herein are items which should be considered.

5.2.2 Template-Type Platforms

5.2.2a. General. Consideration should be given to transporting the following items of a template-type platform: a jacket (welded tubular space frame), piling, superstructure and other miscellaneous items.

5.2.2b. Cargo or Launch Barge. An adequate number of seaworthy cargo barges should be provided. If the jacket portion of the platform is to be launched from a barge without the use of a derrick barge, the launch barge should be capable of this operation.

5.2.2c. Barge Stability. The various platform parts should be loaded on the barges in such a manner to insure a balanced and stable condition. Ballasting of the barge as required should be performed.

5.2.2d. Seafastening. Adequate ties for all platform components to prevent shifting while in transit should be designed and installed. These ties should be adequate for the severest weather that may be encountered during the voyage.

5.2.2e. Towing Vessels. The proper number of seagoing tugs with sufficient power and size to operate safely in any sea environment that may develop for each particular route or ocean traveled should be provided.

5.2.2f. Forces. Consideration should be given to the forces applied to the various platform components as they are lifted on and off the barges or as they are rolled on and launched off the barges. Localized loads on the barge structure should also be considered.

5.2.2g. Buoyancy and Flooding Systems. The buoyancy of any platform component to be launched should be determined to insure the unit will float. The flooding system, the buoyancy components and any necessary lifting connections should be designed to upright and land the structure safely.

5.2.3 Tower-Type Platform

5.2.3a. General. The tower-type platform consists of a tower substructure which is floated to the installation site and placed in position by selective

---

American Petroleum Institute, Dallas, 1984. By permission of API.

flooding. This substructure is also called a jacket. It has multiple piling and a superstructure. The movement considerations should include those specified for the template-type platform in addition to others listed herein.

5.2.3b. Water Tightness. The water tightness of the tower should be determined before towing commences.

5.2.3c. Flooding Controls. Consideration should be given to the location and accessibility of all controls for selective flooding and righting as well as the protection of the controls from environmental and operational hazards.

5.2.3d. Model Tests and Analysis. Model tests and detailed calculations should be considered for the prototype to determine towing and stability characteristics during towing and upending procedures.

5.2.4 Caisson-Type Platform The caisson-type platform, depending on the size, should include all applicable considerations specified above for both the template and tower-type platforms.

5.3 Removal of Jacket from Transport Barge

5.3.1 General

This section covers the removal of a template type platform jacket which has been transported to the installation site by a barge. Removal of the jacket from the barge is usually accomplished by either lifting with a derrick barge or launching.

5.3.2 Lifting Jacket

The rigging should be properly designed to lift safely the jacket off the barge and lower it into the water. Usually the slings are attached above the center of gravity of the jacket being lifted to avoid possible damage to the jacket and/or barge during the lifting process.

5.3.3 Launching Jacket

For those jackets which are to be launched a launching system should be provided considering the items listed below.

5.3.3a. Launch Barge. The launch barge should be equipped with launch ways, rocker arms, controlled ballast and de-watering system, and power unit (hydraulic ram, winch, etc.) to assist the jacket to slide down the ways.

5.3.3b. Loads. The jacket to be launched should be designed and fabricated to withstand the stresses caused by the launch. This may be done by either strengthening those members that might be overstressed by the launching operation or designing into the jacket a special truss, commonly referred to as a launch truss. A combination of the above two methods may be required.

5.3.3c. Flotation. A jacket which is to be launched should be water tight and buoyant. If upending is to be derrick barge assisted the launched structure should float in a position so that lifting slings from the derrick barge may be attached thereto and/or previously attached slings are exposed and accessible.

5.3.3d. Equipment. The derrick barge should be of sufficient size to change the position of the launched jacket from its floating position to its erected position, or to hold the launched jacket at the site until it can be righted by a controlled flooding system.

5.4 Erection

5.4.1 General

This section covers the placement and assembling of the platform so that the finished structure is ready for inspection prior to the initiation of its intended purpose.

5.4.1a. Placement and Assembly. Placement and assembling of the platform should be in accordance with instructions which indicate the installation procedure and pile make-up. Pile sections should be marked in a manner to facilitate installing the pile sections in proper sequence.

5.4.1b. Safety. Necessary measures should be employed to protect the construction, including observance of all safety regulations of the state and federal bodies having jurisdiction over the waters and area around the installation site. This should include, as required, the provision and maintenance of all necessary safety and navigational aids and other measures in observance of appropriate state and federal regulations.

5.4.2 Anchorage

Appropriate anchoring of the derrick and supply barges should be provided during the erection phase. Basic principles which should be considered are outlined herein.

5.4.2a. Anchor Lines. The length of anchor lines should be adequate for the water depth at the site.

5.4.2b. Anchors. Anchor sizes and shapes should be selected so that they will bite and hold in the ocean bottom at the site. This holding action should be sufficient to resist the strongest tides, currents and winds that may reasonably be expected to occur at the site during the erection phase.

5.4.2c. Orientation. Where it appears that the desired anchorage may not be totally possible, orientation of construction equipment should be such that, if the anchors slip, the derrick and supply barges will move away from the platform.

5.4.2d. Anchor Line Deployment. Where anchoring of derrick or supply barges is required within the field of the guyline system of a guyed tower, measures should be employed to prevent fouling or damage of the guylines.

5.4.3 Positioning

The term, "positioning," generally refers to the placement of the jacket on the installation site in preparation for the piling to be installed. This may require upending of those platform components which have been towed to the site or launched from a barge at the site. Generally, the upending process is accomplished by a combination of a derrick barge and controlled or selective flooding system. This upending phase requires advanced planning to predetermine the simultaneous lifting and controlled flooding steps necessary to set the structure on site. Closure devices, lifting connections, etc., should be provided where necessary. The flooding system should be designed to withstand the water pressures which will be encountered during the positioning process.

Where the jacket is to be installed over an existing well, the wellhead should be properly protected from damage through accidental contact with the substructure. Advance planning and preparation should be in such detail as to minimize hazards to the well and structure.

5.4.4 Guyline System Installation

Handling and erection of guyline system components offshore should employ equipment and procedures to minimize potential damage and installation problems.

5.4.4a. Guyline Handling Equipment. The design of equipment used to store, tension, and guide rope or strand should recognize minimum bending radius requirements. The handling equipment should be capable of supplying the necessary tensions to properly install the guylines. Special handling systems may be required to safely lower and position the clumpweights and anchors or anchor piles.

5.4.4b. Procedures. Maximum control of the guyline components should be a consideration in the development of installation procedures as design tolerances may require accurate positioning. Precautions should be taken to prevent fouling of the guylines. Elongation and rotation of guylines due to tensioning should be taken into account.

5.4.4c. Guyline Pretensioning. It may be desirable to preload the guylines to appropriate load levels in the installation phase. Accordingly, the tensioning equipment should be capable of supplying the specified pretension as well as any preload which may be required to seat the guying system. Prior to the completion of the installation phase, the guylines should be tensioned to the nominal levels within specified design tolerance.

5.4.5 Alignment and Tolerances

The degree of accuracy required to align and position a guyed tower jacket and guyline system is determined by design tolerances. Consideration should be given to the requirements for special position and alignment monitoring systems during the placement of the jacket, lead lines, clump-weights and anchors or anchor piles.

5.5 Pile Installation

5.5.1 General

Proper installation of piling, including conductor piles, is vital to the life and permanence of the platform and requires eac pile to be driven to or near design penetration, without damage, and for all field-made structural connections to be compatible with the design requirements.

5.5.2 Leveling Jacket

The jacket should be positioned at or near grade and leveled within acceptable tolerances before the piles are installed. Once level, care should be exercised to maintain grade and levelness of the jacket during the pile installation operation. Leveling of the jacket after pile installation has been initiated should be avoided, if possible, because of the high bending stresses which may result in the piles.

5.5.3 Jacket Column Closure and Pile Sleeve Closure

The closure device on the lower end of the jacket columns and pile sleeves, when required, should be designed to avoid interference with the installation of the piles.

5.5.4 Jacket Weight on Bottom

The soil loading at the base of the jacket can be critical prior to the installation of the permanent pile foundation. The load distribution on the soil at locations with a soft sea bottom should be considered for each combination of a factor of safety of 2.0 provided against a bearing failure. Steel stresses in jacket members that are affected by the soil loading should not be permitted to exceed allowable values, and the increase in soil loading resulting from waves of the maximum height anticipated during the installation period should be considered. Allowable soil loads and steel stresses may be increased by one-third when wave loading is included. In the event of rough seas or if the installation equipment must leave the site for other reasons before the jacket has been adequately secured with piles, the effective weight on bottom may require adjustment to minimize the possibility of jacket movement due to skidding, overturning, or soil failure.

5.5.5 Stabbing Guides

Add-on pile sections should be provided with guides to facilitate stabbing and alignment. A tight uniform fit by the guide should be provided for proper alignment. The guides should be capable of safely supporting the full weight of the add-on pile section prior to welding.

5.5.6 Lifting Eyes

When lifting eyes are used to facilitate the handling of the pile sections, the eyes should be designed, with due regard for impact, for the stresses developed during the initial pick-up of the section as well as those occurring during the stabbing of the section. When lifting eyes or weld-on lugs are used to support the initial pile sections from the top of the jacket, the entire hanging weight should be considered to be supported by a single eye or lug. The lifting eyes or support lugs should be removed by torch cutting ¼ inch (6.4 mm) from the pile surface and grinding smooth. Care should be exercised to ensure that any remaining protrusion does not prevent driving of the pile or cause damage to elements such as packers. If burned holes are used in lieu of lifting eyes, they should comply with the applicable requirements of this paragraph and consideration should be given to possible detrimental effects during hard driving.

5.5.7 Field Welds

The add-on pile sections should be carefully aligned and the bevel inspected to assure a full penetration weld can be obtained before welding is initiated. It may be necessary to open up the bevel by grinding or gouging. Welding should be in accordance with Section 4 of this Recommended Practice. Nondestructive inspection of the field welds, utilizing one or more of the methods referenced in Section 6, should be performed.

5.5.8 Obtaining Required Pile Penetration

The adequacy of the platform foundation depends upon each pile being driven to or near its design penetration. The driving of each pile should be carried to completion with as little interruption as possible to minimize the increased driving resistance which often develops during delays. It is often necessary to work one pile at a time during the driving of the last one or two sections to minimize "setup" time. Workable back-up hammers with leads should always be available when pile "setup" may be critical.

The fact that a pile has met refusal does not assure that it is capable of supporting the design load. Final blow count cannot be considered as assurance of the adequacy of piling. Continued driving beyond the defined refusal may be justified if it offers a reasonable chance of significantly improving the capability of the foundation. In some instances, when continued driving is not successful, the capacity of a pile can be improved by utilizing some method such as internal jetting or drilling to continue the advance of the pile. Any required remedial measures such as drilling, jetting, or continued driving beyond refusal should be approved by the design engineer.

5.5.9 Driven Pile Refusal

The definition of pile refusal is primarily for contractual purposes to define the point where pile driving with a particular hammer should be stopped and other methods instituted (such as drilling, jetting, or using a larger hammer) and to prevent damage to the pile and hammer. The definition of refusal should also be adapted to the individual soil characteristics anticipated for the specific location. Refusal should be defined for all hammer sizes to be used and is contingent upon the hammer being operated at the pressure and rate recommended by the manufacturer.

The exact definition of refusal for a particular installation should be defined in the installation contract. An example (to be used only in the event that no other provisions are included in the installation contract) of such a definition is:

Pile driving refusal is defined as the point where pile driving resistance exceeds either 300 blows per foot (.3 m) for five consecutive feet (1.5 m), or 800 blows for one foot (.3 m) of penetration. (This definition applies when the weight of the pile does not exceed four times the weight of the hammer ram. If the pile weight exceeds this, the above blow counts are increased proportionally, but in no case shall they exceed 800 blows for six inches [152 mm] of penetration).

If there has been a delay in pile driving operations for one hour or longer, the refusal criteria stated above shall not apply until the pile has been advanced at least one foot (.3 m) following the resumption of pile driving. However, in no case shall the blowcount exceed 800 blows for six inches (152 mm) of penetration.

5.5.10 Selection of Pile Hammer Size

When piles are to be installed by driving, the influence of the hammers to be used should be evaluated as a part of the design process as set forth in Par. 2.6.9. It is not unusual for alternate hammers to be proposed for use by the erector well after the design has been completed and re-evaluation by the designer may not be feasible. In such an event, justification for the use of an alternate hammer shall include calculation of stresses in the pile resulting therefrom as set out in Par. 2.6.9.

In lieu of an analytical solution for dynamic stress the following guidelines may be used:

Guideline Wall Thickness, In.

Pile Outside Diameter, in.	Hammer Size, Ft-Kips					
	36	60	120	180	300	500
24	1/2	1/2	7/8	—	—	—
30	9/16	9/16	11/16	—	—	—
36	5/8	5/8	5/8	7/8	—	—
42	11/16	11/16	11/16	3/4	1 1/4	—
48	3/4	3/4	3/4	3/4	1 1/8	1 3/4
60	7/8	7/8	7/8	7/8	7/8	1 3/8
72	—	—	1	1	1	1 1/8
84	—	—	—	1 1/8	1 1/8	1 1/8
96	—	—	—	1 1/4	1 1/4	1 1/4
108	—	—	—	—	1 3/8	1 3/8
120	—	—	—	—	1 1/2	1 1/2

Metric Table
Guideline Wall Thickness, mm

Pile Outside Diameter, mm.	Hammer Size, kJ					
	42	81	163	244	407	678
610	13	13	22	—	—	—
762	14	14	18	—	—	—
914	16	16	16	22	—	—
1067	18	18	18	19	32	—
1219	19	19	19	19	29	44
1524	22	22	22	22	22	35
1829	—	—	25	25	25	29
2134	—	—	—	29	29	29
2438	—	—	—	32	32	32
2743	—	—	—	—	35	35
3048	—	—	—	—	38	38

Values above the solid line based upon minimum pile area in square inches (mm^2) equal to 50% of the rated energy of the hammer in ft kips (23.8% in KJ). Values below line controlled by Par. 2.6.9f.

The preceding table is based on industry experience with up to 60 in. diameter piles and 300 ft-kip hammers.

When it is necessary to use a pile hammer to drive piles with less than the guideline wall thickness set out in the above table, or that determined by an analytical solution, the definition of refusal used should be reduced proportionally.

5.5.11 Drilled and Grouted Piles

Drilling the hole for drilled and grouted piles may be accomplished with or without drilling mud to facilitate maintaining an open hole. Drilling mud may be detrimental to the surface of some soils. If used, consideration should be given to flushing the mud with circulating water upon completion of drilling, provided the hole will remain open. Reverse circulation should normally be used to maintain sufficient flow for cutting removal. Drilling operations should be done carefully to maintain proper hole alignment and to minimize the possibility of hole collapse. The insert pile with an upset drill bit on its tip may be used as the drill string so that it can be left in place after completion of the hole.

Centralizers should be attached to the pile to provide a uniform annulus between the insert pile and the hole. A grouting shoe may be installed near the bottom of the pile to permit grouting of the annulus without grouting inside the pile. It may be necessary to tie down the pile to prevent flotation in the grout if a grouting shoe is used. The time before grouting the hole should be minimized in soils which may be affected by exposure to sea water. The quality of the grout should be tested at intervals during the grouting of each pile. Means should be provided for determining that the annulus is filled, as further discussed in Par. 5.5.14. Holes for closely positioned piles should not be open at the same time unless there is assurance that this will not be detrimental to pile capacity and that grout will not migrate during placement to an adjacent hole.

5.5.12 Belled Piles

In general, drilling of bells for belled piles should employ only reverse circulation methods. Drilling mud should be used where necessary to prevent caving and sloughing. The expander or under-reaming tool used should have a positive indicating device to verify that the tool has opened to the full width required. The shape of the bottom surface of the bell should be concave upward to facilitate later filling of the bell with tremie concrete.

To aid in concrete placement, longitudinal bars and spiral steel should be well spaced. Reinforcing steel may be bundled or grouped to provide larger openings for the flow of concrete. Special care should be taken to prevent undue congestion at the throat between the pile and bell where such congestion might trap laitance. Reinforcing steel cages or structural members should extend far enough into the pile to develop adequate transfer.

Concrete should be placed as tremie concrete, with concrete being ejected from the lower end of a pipe at the bottom of the bell, always discharging into fresh concrete. Concrete with aggregates ⅜-in. (10 mm) and less may be placed by direct pumping. Because of the long drop down the pile and the possibility of a vacuum forming with subsequent clogging, an air vent should be provided in the pipe near the top of the pile. To start placement, the pipe should have a steel plate closure with soft rubber gaskets in order to exclude water from the pipe. Care should be taken to prevent unbalanced fluid heads and a sudden discharge of concrete. The pile should be filled to a height above the design concrete level equal to 5% of the total volume of concrete placed so as to displace all laitance above the design level. Suitable means should be provided to indicate the level of the concrete in the pile. Concrete placement in the bell and adjoining section of the pile should be as continuous as possible.

5.5.13 Pile Installation Records

Throughout the pile driving operation, comprehensive driving and associated data should be recorded. The recorded data should include:

1. Platform and pile identification.
2. Penetration of pile under its own weight.
3. Penetration of pile under the weight of the hammer.
4. Blow counts throughout driving with hammer identification.
5. Unusual behavior of hammer or pile during driving.
6. Interruptions in driving, including "setup" time.

7. Lapsed time for driving each section.
8. Elevations of soil plug and internal water surface after driving.
9. Actual length of each pile section and cutoffs.
10. Pertinent data of a similar nature covering driving, drilling, grouting or concreting of grouted or belled piles.

5.5.14 Grouting Piles to Structures

If required by the design, the spaces between the piles and the surrounding structure should be carefully filled with grout using appropriate grouting equipment. The equipment should be capable of maintaining continuous grout flow until the annulus is filled. If the structure design does not require or permit grout to be returned to the surface, means should be provided to determine that the spaces have been filled as required. Such means might include but are not limited to underwater visual inspection, probing or detection devices.

5.6 Superstructure Installation

5.6.1 Alignment and Tolerances

After the piling has been driven and cut off to grade, the superstructure should then be set with proper care being exercised to insure proper alignment and elevation. The deck elevation shall not vary more than +3 in. (76 mm) from the design elevation shown in the drawing. The finished elevation of the deck shall be within ½ in. (13 mm) of level.

5.6.2 Appurtenances

Once the superstructure is installed, all stairways, handrails, and other similar appurtenances should be installed as specified.

5.7 Grounding of Installation Welding Equipment

5.7.1 General

Normal welding procedures use reverse polarity wherein the welding rod is positive (+) and the ground is negative (−). The current flow is positive to negative, and an adequate and properly placed ground wire is necessary to prevent stray currents, which, if uncontrolled, may cause severe corrosion damage. (See NACE RP-01-76, Sec. 7 Par. 7.3.).

5.7.2 Recommended Procedure

The welding machine should be located on and grounded to the structure whenever possible. When this is impossible or impractical, and the welding machine is located on the barge or vessel, both leads from the output of the welding machine should be run to the structure and the ground lead secured to the structure as close as practical to the area of welding. Under no conditions should the hull of the barge (or vessel) be used as a current path. The case or frame of the welding machine should be grounded to the hull to eliminate shock hazards to personnel.

The welding cables should be completely insulated to prevent stray currents. Damaged cables should not be allowed to hang in the water or to touch any part of the barge.

Grounding cable lugs should be tightly secured to grounding plates. The lug contact should be thoroughly cleaned to bare metal. The resistance of the connection should be a maximum of 125 microhms per connection, or the voltage drop across the connection should be a maximum of 62.5 millivolts for a current of 500 amperes. Use Ohm's Law ($V = IR$) for amperage other than 500 amperes.

The minimum cross-sectional area of the return ground cable should be one million circular mils per 1,000 amperes per 100 feet (645 circular mm per 1,000 amperes per 30.5 meters) of cable. One or more cables connected in parallel may be used to meet minimum cross-section requirements.

Note: 2/0 cable contains 133,392 circular mils (86 circular mm)

3/0 cable contains 169,519 circular mils (109 circular mm)

4/0 cable contains 212,594 circular mils (137 circular mm)

More than one ground cable of sufficient size is suggested to guard against a single return or ground becoming loose.

Connecting several welding machines to a common ground cable which is connected to the structure being welded will control stray currents if adequately sized and properly insulated from the barge or vessel containing welding machines.

5.7.3 Monitoring Remote Ground Efficiency

When welding is conducted using generators remote from a stucture, grounding efficiency can be monitored by simultaneously measuring the potential of the structure and barge or ship housing the welding generators. A change in potential reading from either indicates insufficient grounding.

Excerpts Concerning Grouted and Concreted Foundations

2.9.4 Cement Grout and Concrete

2.9.4a. Cement Grout. If required by the design, the space between the piles and the surrounding structure should be carefully filled with grout. Prior to installation, the compressive strength of the grout mix design should be confirmed on a representative number of laboratory specimens cured under conditions which simulate the field conditions. Laboratory test procedures should be in accordance with ASTM 109. The unconfined compressive strength of 28 day old grout specimens computed as described in ACI 214-77 but equating f'_c to f_{cu}, should not be less than either 2500 psi (17.25 MPa) or the specified design strength.

A representative number of specimens taken from random batches during grouting operations should be tested to confirm that the design grout strength has been achieved. Test procedures should be in accordance with ASTM 109. The specimens taken from the field should be subjected, until test, to a curing regime representative of the in situ curing conditions, i.e., underwater and with appropriate seawater salinity and temperature.

2.9.4b. Concrete. The concrete mix used in belled piles should be selected on the basis of shear strength, bond strength and workability for underwater placement including cohesiveness and flowability. The concrete mix may be made with aggregate and sand, or with sand only. The water-cement ratio should be less than 0.45. If aggregate is used, the aggregates should be small and rounded, the sand content should be 45% or greater, the cement content should be not less than 750 lb. per cubic yard (445 kg/m^3), and the workability as measured by the slump test should be 7 to 9 inches (180 to 230 mm). To obtain the properties required for proper placement, a suitable water-reducing and plasticizing admixture may be necessary.

4.1.6 Provisions for Grouted Pile to Sleeve Connections

Steel surfaces of piles and the structure, which are to be connected by grout, should be free of mill glaze, varnish, grease or any other material that would reduce the connection strength.

Centralizers should be used to maintain a uniform annulus or space between the pile and the surrounding structure. A minimum annulus width of 1½ in. (38 mm) should be provided where grout is the only means of load transfer. Adequate clearance between pile and sleeve should be provided, taking into account the shear keys' outstand dimension, h. Packers should be used as necessary to confine the grout. Proper means for the introduction of grout into the annulus should be provided so that the possibility of dilution of the grout or formation of voids in the grout will be minimized. The use of wipers or other means of minimizing mud intrusion into the spaces to be occupied by piles should be considered at sites having soft mud bottoms.

2.6.2 Pile Foundations

Types of pile foundations used to support offshore structures are as follows:

2.6.2a. Driven Piles. Open ended piles are commonly used in foundations for offshore platforms. These piles are usually driven into the sea-floor with impact hammers which use steam, diesel fuel, or hydraulic power as the source of energy. The pipe wall thickness should be adequate to resist axial and lateral loads as well as the stresses during pile driving. It is possible to predict approximately the stresses during pile driving using the principles of one-dimensional elastic stress wave transmission by carefully selecting the parameters that govern the behavior of soil, pile, cushions, capblock and hammer. For a more detailed study of these principles, refer to E. A. L. Smith's paper *Pile-Driving Analysis by the Wave Equation*. Transactions ASCE, Vol. 127, 1962, Part 1, Paper No. 3306, pp. 1145–1193. The above approach may also be used to optimize the pile-hammer-cushion and capblock with the aid of computer analyses (commonly known as the Wave Equation Analyses). The design penetration of driven piles should be determined in accordance with the principles outlined in Par. 2.6.3 through 2.6.6 and 2.6.8 rather than upon any correlation of pile capacity with the number of blows required to drive the pile a certain distance into the seafloor.

When hard driving is encountered before the pile reaches design penetration, one of the following procedures can be used to aid pile driving.

1. Plug Removal. The soil plug inside the pile is removed by jetting and air lifting or by drilling to reduce pile driving resistance. If plug removal results in inadequate pile capacities, the removed soil plug should be replaced by a grout or concrete plug having sufficient load-carrying capacity to replace that of the removed soil plug.
2. Soil Removal Below Pile Tip. Soil below the pile tip is removed either by drilling an undersized hole or by jetting and possibly air lifting. The drilling or jetting equipment is lowered through the pile which acts as the casing pipe for the operation. The effect on pile capacity of drilling an undersized hole is unpredictable unless there has been previous experience under similar conditions. Jetting below the pile tip should in general be avoided because of the unpredictability of the results.
3. Two-Stage Driven Piles. A first stage or outer pile is driven to a predetermined depth, the soil plug is removed, and a second stage or inner pile is driven inside the first stage pile. The annulus between the two piles is grouted to permit load transfer and develop composite action.

2.6.2b. Drilled and Grouted Piles. Drilled and grouted piles can be used in soils which will hold an open hole with or without drilling mud. Load transfer between grout and pile should be designed in accordance with Par. 2.8.2, 2.8.3, and 2.8.4. There are two types of drilled and grouted piles, as follows:

1. Single-Stage. For the single-staged, drilled and grouted pile, an oversized hole is drilled to the required penetraion, a pile is lowered into the hole and the annulus between the pile and the soil is grouted. This type pile can be installed only in soils which will hold an open hole to the surface. As an alternative method, the pile with expendable cutting tools attached to the tip can be used as part of the drill stem to avoid the time required to remove the drill bit and insert a pile.
2. Two-Stage. The two-staged, drilled and grouted pile consists of two concentrically placed piles grouted to become a composite section. A pile is driven to a penetration which has been determined to be achievable with the available equipment and below which an open hole can be maintained. This outer pile becomes the casing for the next operation which is to drill through it to the required penetration for the inner or "insert" pile. The insert pile is then lowered into the drilled hole and the annuli between the insert pile and the soil and between the two piles are grouted. Under certain soil conditions, the drilled hole is stopped above required penetration, and the insert pile is driven to required penetration. The diameter of the drilled hole should be at least 6 inches (150 mm) larger than th pile diameter.

2.6.2c. Belled Piles. Bells may be constructed at the tip of piles to give increased bearing and uplift capacity through direct bearing on the soil. Drilling of the bell is carried out through the pile by underreaming with an expander tool. A pilot hole may be drilled below the bell to act as a sump for unrecoverable cuttings. The bell and pile are filled with concrete of a height sufficient to develop necessary load transfer between the bell and the pile. Bells are connected to the pile to transfer full uplift and bearing loads using steel reinforcing such as structural members with adequate shear lugs, deformed reinforcement bars or pre-stressed tendons. Load transfer into the concrete should be designed in accordance with ACI 318. The steel reinforcing should be enclosed for their full length below the pile with spiral reinforcement meeting the requirements of ACI 318. Load transfer between the concrete and the pile should be designed in accordance with Par. 2.8.2, 2.8.3, and 2.8.4.

2.6.9 Pile Wall Thickness

2.6.9a. General. The wall thickness of the pile may vary along its length and may be controlled at a particular point by any one of several loading conditions or requirements which are discussed in the paragraphs below.

2.6.9b. Allowable Pile Stresses. The allowable pile stresses should be the same as those permitted by the AISC specification for a compact hot rolled section, giving due consideration to Par. 2.5.1 and 2.5.3. A rational analysis considering the restraints placed upon the pile by the structure and the soil should be used to determine the allowable stresses for the portion of the pile which is not laterally restrained by the soil. General column buckling of the portion of the pile below the mudline need not be considered unless the pile is believed to be laterally unsupported because of extremely low soil shear strengths, large computed lateral deflections, or for some other reason.

2.6.9c. Design Pile Stresses. The pile wall thickness in the vicinity of the mudline, and possibly at other points, is normally controlled by the combined axial load and bending moment which results from the design loading conditions for the platform. The moment curve for the pile may be computed with soil reactions determined in accordance with Par. 2.6.7, giving due consideration to possible soil removal by scour. It may be assumed that the axial load is removed from the pile by the soil at a rate equal to the ultimate soil-pile adhesion divided by the appropriate pile safety factor from Par. 2.6.3d. When lateral deflections associated with cyclic loads at or near the mudline are relatively large (e.g., exceeding y_c as defined in Par. 2.6.7c for soft clay), consideration should be given to reducing or neglecting the soil-pile adhesion through this zone.

2.6.9d. Stresses Due to Weight of Hammer during Hammer Placement. Each pile or conductor section on which a pile hammer (pile top drilling rig, etc.) will be placed should be checked for stresses due to placing the equipment. These loads may be the limiting factors in establishing maximum length of add-on sections. This is particularly true in cases where piling will be driven or drilled on a batter. The most frequent effects include: static bending, axial loads, and arresting lateral loads generated during initial hammer placement.

Experience indicates that reasonable protection from failure of the pile wall due to the above loads is provided if the static stresses are calculated as follows:

1. The pile projecting section should be considered as a freestanding column with a minimum effective length factor K of 2.1.
2. Bending moments and axial loads should be calculated using the full weight of the pile hammer, cap, and leads acting through the center of gravity of their combined masses, and the weight of the pile add-on section with due consideration to pile batter eccentricities. The bending moment so determined should not be less than that corresponding to a load equal to 2 percent of the combined weight of the hammer, cap, and leads applied at the pile head andperpendicular to its centerline.
3. No increase in AISC allowable stresses should be permitted.

2.6.9e. Stresses during Driving. Consideration should also be given to the stresses that occur in the free standing pile section during driving. A dynamic analysis may be used to determine the maximum dynamically induced stress level that occurs. In general, it may be assumed that column buckling will not occur as a result of the dynamic portion of the driving stresses. In the absence of reliable data regarding the maximum dynamically induced stresses, the static portion of the stress should be limited to one-half the yield strength of the pile material. The static stress during driving may be taken to be the stress resulting from the weight of the pile above the point of evaluation plus the pile hammer components actually supported by the pile during the hammer blows, including any bending stresses resulting therefrom. The pile hammers evaluated for use during driving should be noted by the designer on the installation drawings or specifications.

2.6.9f. Minimum Wall Thickness. The D/t ratio of the entire length of a pile should be small enough to preclude local buckling at stresses up to the yield strength of the pile material. Consideration should be given to the different loading situations occurring during the installation and the service life of a piling. For inservice conditions, and for those installation situations where normal pile-driving is anticipated or where piling installation will be by means other than driving, the limitations of Par. 2.5.2 should be considered to be the minimum requirements. For piles that are to be installed by driving where sustained hard driving (250 blows per foot [820 blows per meter] with the largest size hammer to be used) is anticipated, the minimum piling wall thickness used should not be less than

$$\left.\begin{array}{l} t = 0.25 + \dfrac{D}{100} \\ \text{Metric Formula} \\ t = 6.35 + \dfrac{D}{100} \end{array}\right\} \quad \ldots\ldots\ldots\ldots\ldots(2.6.9)$$

where:
t = wall thickness, in. (mm)
D = diameter, in. (mm)

Minimum wall thickness for normally used pile sizes should be as listed in the following table:

Minimum Pile Wall Thickness

Pile Diameter		Nominal Wall Thickness, t	
in.	mm	in.	mm
24	610	½	13
30	762	9/16	14
36	914	5/8	16
42	1067	11/16	17
48	1219	¾	19
60	1524	7/8	22
72	1829	1	25
84	2134	1⅛	28
96	2438	1¼	31
108	2743	1⅜	34
120	3048	1½	37

The preceding requirement for a lesser D/t ratio when hard driving is expected may be relaxed when it can be shown by past experience or by detailed analysis that the pile will not be damaged during its installation.

2.6.9g. Allowance for Underdrive and Overdrive. With piles having thickened sections at the mudline, consideration should be given to providing an extra length of heavy wall material in the vicinity of the mudline so the pile will not be overstressed at this point if the design penetration is not reached. The amount of underdrive allowance provided in the design will depend on the degree of uncertainty regarding the penetration that can be obtained. In some instances an overdrive allowance should be provided in a similar manner in the event an expected bearing stratum is not encountered at the anticipated depth.

2.6.9h. Driving Shoe. Provision of a driving shoe at the pile tip at least one diameter in length with a minimum wall thickness of 1.5 times the value established by Par. 2.6.9f should be considered. A driving head should be considered where hard driving is expected.

2.4. Installation Forces

2.4.1 General

Installation forces are those forces imposed upon the component parts of the structure during the operations of moving the components from their fabrication site to the offshore location, and installing the component parts to form the completed platform. Since installation forces involve the motion of heavy weights, the dynamic loading involved should be considered and the static forces increased by appropriate impact factors to arrive at adequate equivalent loads for design of the members affected. For those installation forces that are experienced only during transportation and launch, and which include environmental effects, basic allowable stresses for member design may be increased by ⅓ in keeping with provisions of 2.5.1b. Also see Section 5, "Installation," for comments complementary to this section.

2.4.2 Lifting Forces

2.4.2a. General. Lifting forces are imposed on the structure by erection lifts during the fabrication and installation stages of platform construction. The magnitude of such forces should be determined through the consideration of static and dynamic forces applied to the structure during lifting and from the action of the structure itself. Lifting forces on padeyes and on other members of the structure should include both vertical and horizontal components, the latter occurring when lift slings are other than vertical. Vertical forces on the lift should include buoyancy as well as forces imposed by the lifting equipment.

To compensate for any side loading on lifting eyes which may occur, in addition to the calculated horizontal and vertical components of the static load for the equilibrium lifting condition, lifting eyes and the connections to the supporting structural members should be designed for a horizontal force of 5% of the static sling load, applied simultaneously with the static sling load. This horizontal force should be applied perpendicular to the padeye at the center of the pinhole.

2.4.2b. Static Loads. When suspended, the lift will occupy a position such that the center of gravity of the lift and the centroid of all upward acting forces on the lift are in static equilibrium should be used to determine forces in the structure and in the slings. The movement of the lift as it is picked up and set down should be taken into account in determining critical combinations of vertical and horizontal forces at all points, including those to which lifting slings are attached.

2.4.2c. Dynamic Load Factors. For lifts where either the lifting derrick or the structure to be lifted is on a floating vessel, the selection of the design lifting forces should consider the impact from vessel motion. Load factors should be applied to the design forces as developed from considerations of Par. 2.4.2a and 2.4.2b.

For lifts to be made at open, exposed sea (i.e., off-shore locations), padeyes and other internal members (and both end connections) framing into the joint where the padeye is attached and transmitting lifting forces within the structure should be designed for a minimum load factor of 2.0 applied to the calculated static loads. All other structural members transmitting lifting forces should be designed using a minimum load factor of 1.35.

For other marine situations (i.e., loadout at sheltered locations), the selection of load factors should meet the expected local conditions but should not be less than a minimum of 1.5 and 1.15 for the two conditions previously listed.

2.4.2d. Allowable Stresses. The lift should be designed so that all structural steel members are proportioned for basic allowable stresses as specified in Par. 2.5.1. The AISC increase in allowable stresses for short-term loads should not be used. In addition, all critical structural connections and primary members should be designed to have adequate reserve strength to insure structural integrity during lifting.

2.4.2e. Effect of Tolerances. Fabrication tolerances and sling length tolerances both contribute to the distribution of forces and stresses in the lift system which are different from that normally used for conventional design purposes. The load factors recommended in Par. 2.4.2c are intended to apply to situations where fabrication tolerances do not exceed the requirements of Par. 4.1.5, and where the variation in length of slings does not exceed plus or minus ¼ of 1% of nominal sling length, or 1½ inches.

The total variation from the longest to the shortest sling should not be greater than ½ of 1% of the sling length, or 3 inches. If either fabrication tolerance or sling length tolerance exceeds these limits, a detailed analysis taking into account these tolerances should be performed to determine the redistribution of forces on both slings and structural members. This same type analysis should also be performed in any instances where it is anticipated that unusual deflections or particularly stiff structural systems may also affect load distribution.

2.4.2f. Slings, Shackles and Fittings. For normal offshore conditions, slings should be selected to have a factor of safety of 4 for the manufacturer's rated minimum breaking strength of the cable compared to static sling load. The static sling load should be the maximum load on any individual sling, as calculated in Par. 2.4.2a, b and e above, by taking into account all components of loading and the equilibrium position of the lift. This factor of safety should be increased when unusually severe conditions are anticipated, and may be reduced to a minimum of 3 for carefully controlled conditions.

Shackles and fittings should be selected so that the manufacturer's rated working load is equal to or greater than the static sling load, provided the manufacturer's specifications include a minimum factor of safety of 3 compared to the minimum breaking strength.

2.4.3 Loadout Forces

2.4.3a. Direct Lift. Lifting forces for a structure loaded out by direct lift onto the transportation barge should be evaluated only if the lifting arrangement differs from that to be used in the installation, since lifting in open water will impose more severe conditions.

2.4.3b. Horizontal Movement Onto Barge. Structures skidded onto transportation barges are subject to load conditions resulting from movement of the barge due to tidal fluctuations, nearby marine traffic and/or change in draft; and also from load conditions imposed by location, slope and/or settlement of supports at all stages of the skidding operation. Since movement is normally slow, impact need not be considered.

2.4.4 Transportation Forces

2.4.4a. General. Transportation forces acting on templates, towers, guyed towers and platform deck components should be considered in their design, whether transported on barges or self-floating. These forces result from the way in which the structure is supported, either by barge or buoyancy, and from the response of the tow to environmental conditions encountered enroute to the site. In the subsequent paragraphs, the structure and supporting barge and the self-floating tower are referred to as the tow.

2.4.4b. Environmental Criteria. The selection of environmental conditions to be used in determining the motions of the tow and the resulting gravitational and inertial forces acting on the tow should consider the following:

1. Previous experience along the tow route.
2. Exposure time and reliability of predicted "weather windows."
3. Accessibility of safe havens.
4. Seasonal weather system.
5. Appropriateness of the recurrence interval used in determining maximum design wind, wave and current conditions and considering the characteristics of the tow, such as size, structure, sensitivity and cost.

2.4.4c. Determination of Forces. The tow including the structure, sea fastenings and barge should be analyzed for the gravitational, inertial and hydrodynamic loads resulting from the application of the environmental criteria in Par. 2.4.4b. The analysis should be based on model basin test results or appropriate analytical methods. Beam, head and quartering wind and seas should be considered to determine maximum transportation forces in the tow structural elements. In the case of large barge-transported structures, the relative stiffnesses of the structure and barge is significant and should be considered in the structural analysis.

Where relative size of barge and jacket, magnitude of the sea states, and experience make such assumptions reasonable, tows may be analyzed based on gravitational and inertial forces resulting from the tow's rigid body motions using appropriate period and amplitude by combining roll with heave and pitch with heave.

2.4.4d. Other Considerations. Large jackets for templates and guyed towers will extend beyond the barge and will usually be subjected to submersion during tow. Submerged members should be investigated for slamming, buoyancy and collapse forces. Large buoyant overhanging members also may affect motions and should be considered. The effects on long slender members of wind-induced vortex shedding vibrations should be investigated. This condition may be avoided by the use of simple wire rope spoilers helically wrapped around the member.

For long transocean tows, repetitive member stresses may become significant to the fatigue life of certain member connections or details and should be investigated.

2.4.5 Launching Forces and Uprighting Forces

2.4.5a. Guyed Tower and Template Type. Guyed tower and template type structures which are transported by barge are usually launched at or near the installation location. The jacket is generally moved along ways, which terminate in rocker arms, on the deck of the barge. As the position of the jacket reaches a point of unstable equilibrium, the jacket rotates, causing the rocker arms at the end of the ways to rotate as the jacket continues to slide from the rocker arms. Forces supporting the jacket on the ways should be evaluated for the full travel of the jacket. Deflection of the rocker beam and the effect on loads throughout the jacket should be considered. In general, the most severe forces will occur at the instant rotation starts. Consideration should be given to the development of dynamically induced forces resulting from launching. Horizontal forces required to initiate movement of the jacket should also be evaluated. Consideration should be given to wind, wave, current and dynamic forces expected on the structure and barge during launching and uprighting.

2.4.5b. Tower Type. Tower type structures are generally launched from the fabrication yard to float with their own buoyancy for tow to the installation site. The last portion of such a tower leaving the launching ways may have localized forces imposed on it as the first portion of the tower to enter the water gains buoyancy and causes the tower to rotate from the slope of the ways. Forces should be evaluated for the full travel of the tower down the ways.

2.4.5c. Hook Load. Floating jackets for which lifting equipment is employed for turning to a vertical position should be designed to resist the gravitational and inertial forces required to upright the jacket.

2.4.5d. Submergence Pressures. The submerged, non-flooded or partially flooded members of the structure should be designed to resist pressure-induced hoop stresses during launching and uprighting.

A member may be exposed to different values of hydrostatic pressure during installation and while in place. The integrity of the member may be determined using the guidelines of Par. 2.5.4.

2.4.6 Installation Foundation Loads

2.4.6a. General. Calculated foundation loads during installation should be conservative enough to give reasonable assurance that the structure will remain at the planned elevation and attitude until piles can be installed. Reference should be made to appropriate paragraphs in Sections 2 and 6.

2.4.6b. Environmental Conditions. Consideration should be given to effects of anticipated storm conditions during this stage of installation.

2.4.6c. Structure Loads. Vertical and horizontal loads should be considered taking into account changes in configuration/exposure, construction equipment, and required additional ballast for stability during storms.

2.4.7 Hydrostatic Pressure

2.4.7a. General. Unflooded or partially flooded members of a structure should be able to withstand the hydrostatic pressure acting on them caused by their location below the water surface. A member may be exposed to different values of pressure during installation and while in place. The integrity of the member may be determined using the guidelines of Par. 2.5.4.

2.4.7b. Design Head. The hydrostatic pressure to be used in the calculations of Par. 2.5.4 should be determined from the design head, H_z, defined as follows:

$$H_z = z + \frac{H_w}{2}\left(\frac{\cosh[k(d-z)]}{\cosh kd}\right) \quad \ldots\ldots\ (2.4.6)$$

where:

z = depth below still water surface including tide, ft (m). z is positive measured downward from the still water surface. For installation, z should be the maximum submergence during launch or differential head during the upending sequence, plus a reasonable increase in head to account for structural weight tolerances and for deviations from the planned installation sequence.

H_w = wave height, ft (m)

$k = \dfrac{2\pi}{L}$ with L equal to the wave length, ft (m)

d = still water depth, ft (m)

C2.6.16 Installation and Removal of Shallow Foundations.

a. *Penetration of Shear Skirts.* Shear skirts can provide a significant resistance to penetration. This resistance, Q_d can be estimated as a function of depth by the following.

$$Q_d = Q_f + Q_p = fa_s + qA_p \quad \ldots\ldots\ (C2.6.16\text{-}1)$$

where

Q_f = skin friction resistance

Q_p = total end bearing

f = unit skin friction capacity

A_s = side surface area of skirt embedded at a particular penetration depth (including both sides)

q = unit end bearing pressure on the skirt

A_p = end area of skirt

The end bearing components can be estimated by bearing capacity formulae or alternatively by the direct use of cone penetrometer resistance corrected for shape difference. The side resistance can be determined by laboratory testing or other suitable experience. In most cases it is highly desirable to achieve full skirt penetration. This should be considered in selecting soil strength properties for use in analysis as low estimates of strength are non conservative in this case.

The foundation surface should be prepared in such a way to minimize high localized contact pressures. If this is not possible grout can be used between the structure foundation and soil to ensure intimate contact. In this case the grout must be designed so that its stiffness properties are similar to the soil.

In general water will be trapped within the shear skirt compartments. The penetration rate should be such that removal of the water can be accomplished without forcing it under the shear skirts and damaging the foundation. In some cases a pressure drawdown can be used to increase the penetration force however, an analysis should be carried out to insure that damage to the foundation will not result.

In assessing the penetration of shear skirts careful attention should be given to site conditions. An uneven seafloor, lateral soil strength variability, existence of boulders, etc. can give rise to uneven penetration and/or structural damage of skirts. In some cases site improvements may be required such as leveling the area by dredging or fill emplacement.

b. *Removal.* During removal suction forces will tend to develop on the foundation base and the tips of shear skirts. These forces can be substantial but can usually be overcome by sustained uplift forces or by introducing water into the base compartments to relieve the suction.

Appendix 4

Excerpts from Det Norske Veritas, Rules for Submarine Pipeline Systems

1.8 Documentation

1.8.2.8 Construction. The following information is to be submitted prior to start of construction.

Construction procedure specifications including installation, tie-ins and protection
Description of construction vessels and equipment
Specification for installation welding
Description of quality control system including specification for non-destructive testing
Specification for final surveys and tests

2.2 Pipeline Route

2.2.1 Location

2.2.1.1 The route should be selected with due regard to the probability of damages to the pipe and the consequences of a possible pipe rupture. Factors to take into consideration are:

population density
location of living quarters
ship traffic
fishing activity
offshore operations
unstable seabed
corrosivity of the environment

Known future operations in the vicinity of the route is to be taken into consideration.

Det Norske Veritas, Oslo, 1981. By permission of DNV.

2.2.2 Route Survey

2.2.2.1 A detailed route survey is to be performed to provide sufficient data for design and construction.

2.2.2.2 The route survey is to cover sufficient width and accuracy to permit the safe and proper installation and operation of the pipeline.

2.2.2.3 The accuracy needed may vary along the proposed route. A higher degree of accuracy is required in areas where other activities, obstructions or highly varied seabed topography or subsurface conditions may dictate more detailed investigations.

2.2.2.4 A proper investigation to reveal possible conflicts with existing or planned installations is to be performed. Examples of such installations are other submarine pipelines and communication cables.

2.2.2.5 The intended pipeline route is to be surveyed for wrecks and obstructions down to a depth exceeding that reached by the pipeline during installation, burial or operation.

2.2.2.6 The results of the survey are to be presented in an accurate route map indicating the location of the pipeline and related facilities and the seabed properties. See 2.2.4.

2.2.3 Bottom Topography

2.2.3.1 All topographical features influencing the stability and installation of the pipeline are to be covered by the route survey. The survey is at least to define:

obstructions in the form of rock outcrops, large boulders etc. that could require levelling or removal operations prior to pipeline installation topographical features that contains potentially unstable slopes, sand waves, deep valleys and erosion in form of scour patterns or material deposits.

2.2.4 Seabed Properties

2.2.4.1 All the geotechnical properties necessary for evaluating the effects of relevant loading conditions are to be determined for the subfloor deposits. This should include possible unstable deposits in the vicinity of the pipeline.

2.2.4.2 The geotechnical properties may be obtained through a combination of seismic survey, coring, in situ tests and borings with sampling.

Supplementary informations may be obtained from geological surveys, sea bottom topographical surveys, visual surveys, biological investigations, chemical examinations and laboratory testing on samples from borings.

Guidelines for site and laboratory investigations may be found in Veritas' Technical Note 302.

2.2.4.3 Special investigations of the subfloor deposits may be required to evaluate specific problems. Examples of such problems are:

ease of excavation and/or burial operations.
possibilities of flow slides or liquefaction as the result of repeated loadings.

2.3 Environmental Conditions

2.3.1 General

2.3.1.1 Possible effects of the various environmental actions are to be taken into account to the extent relevant to the situation considered.

2.3.2 Tide

2.3.2.1 Tides are to be taken into consideration when the water depth is a significant parameter, such as when determining wave loads on a riser, planning laying operations, determining maximum or minimum water pressures etc.

2.3.2.2 The assumed maximum tide is to include both astronomical tide and storm surge. Minimum tide estimates should be based on the astronomical tide and possible negative storm surge.

2.3.3 Wind

2.3.3.1 Direct action of wind is to be taken into consideration for slender risers. The possibility of vibrations of such risers excited by wind is to be considered. Special attention is to be paid to wind loads in the construction and transportation phases.

2.3.3.2 For risers the wind data used are in principle to be the same as those used for the design of the platform.

2.3.3.3 If the riser is positioned adjacent to other structural parts, possible effects due to disturbance in the flow field should be considered when determining the wind loads. Such effects may either be caused by an increase or reduction of the wind speed, or by dynamic excitations caused by vortexes shed from the adjacent structural parts.

2.3.4 Waves

2.3.4.1 The effect of waves is to be taken into consideration for both pipeline and riser. Examples of such effects are the action of wave forced on riser or on pipeline during installation or when resting on bottom (not buried). Examples of indirect effects are deformation of riser due to wave forces acting on the platform, and deformation of pipeline due to lay barge motions in waves.

Possible liquifaction and transportation of sea bed material is also to be considered.

2.3.4.2 If the riser is positioned adjacent to other structural parts, possible effects due to disturbance of the flow field should be considered when determining the wave loads. Such effects may either be caused by changes in the wave particle kinematics, or by dynamic excitation caused by vortexes shed from the adjacent structural parts.

2.3.4.3 For riser the wave data to be used are in principle to be the same as those used for the design of the platform.

2.3.4.4 For the assessment of wave conditions along the pipeline route a limited number of intervals may be assumed, each of which being characterized by water depth, bottom topography and other factors affecting the wave conditions.

2.3.5 Current

2.3.5.1 The effect of current is to be taken into consideration for both pipeline and riser.

2.3.5.2 The assumed current velocities are to include possible contributions from tidal current, wind induced currents, storm surge current, density current and possible other current phenomena. For near shore regions longshore current due to wave breaking should also be considered.

2.3.5.3 The tidal current may normally be determined from harmonic analyses of recorded data, while wind induced—storm surge and density currents may be determined either from statistical analyses of recorded data, or from numerical simulations.

Normally a wind induced surface current speed corresponding to 2 per cent of the 1 hour mean wind speed will be accepted.

2.3.5.4 In regions where bottom material may erode, special studies of the current conditions near the bottom including boundary layer effects may be required for onbottom stability calculations of pipelines.

2.3.5.5 For risers and for pipelines during laying reasonable assumptions should be made as to current velocity distribution over the depth. For risers this is normally to be the same as used for the platform.

2.3.6 Corrosivity

2.3.6.1 For the evaluation of the corrosion protection system the following properties, with seasonal variations of the sea water and soil along the route are to be considered:

- temperature
- salinity
- oxygen content
- pH-value
- resistivity
- current
- biological activity (sulfate-reducing bacteria etc.)

2.3.7 Ice

2.3.7.1 In case the installation is to be located in an area where ice may develop or drift, proper consideration of ice conditions and their possible effects on riser or pipeline is to be made. The ice conditions should be studied with particular attention to possible:

- ice forces on riser and on pipeline
- potential scour at pipeline location and contact with pipeline by floating ice
- ice problems during the installation operations

2.3.7.2 The description of ice conditions should preferably be in accordance with the «World Meterological Organization Sea-Ice Nomenclature».

2.3.8 Air and Sea Temperatures

2.3.8.1 Reasonably accurate air and sea temperature statistics are to be provided. These data are important for proper determination of design temperatures, possible thermal stresses, deformations, displacements, etc.

2.3.8.2 The period of observations on which the maximum and minimum air and sea temperature statistics are based, should preferably be several years.

2.3.9 Marine Growth

2.3.9.1 The effect of marine growth on riser and pipeline loads is to be considered, taking into account all biological and environmental factors relevant to the site in question.

2.3.9.2 For determination of the hydrodynamic loads special attention is to be paid to the effective diameter increase and the equivalent roughness of accumulated marine growth when determining the hydrodynamic coefficients.

2.4 Internal Pipe Conditions

2.4.1 Installation Conditions

2.4.1.1 A description of the internal conditions during storage, installation, and pressure testing is to be prepared. Of special concern is the duration of exposure to sea water and moist air, and whether inhibitors are to be used. See section 4, 5 and 6.

2.4.2 Operational Conditions

2.4.2.1 The physical and chemical composition of the product and the pressures and temperatures along the pipeline are to be specified.

2.4.2.2 Limits of temperatures and pressures, and

allowed concentrations of corrosive components for the product to be transported are to be specified. Of special concern is the content of:

sulphur compounds
water
chlorides
oxygen
carbon dioxide
hydrogen sulphide.

3.2.3 Functional Loads during Installation

3.2.3.1 The functional loads during installation may be grouped as

weight
pressure
installation forces.

3.2.3.2 If the buoyancy of the pipe is included in the term "weight", the longitudinal force due to pressure is to be added. If weight in air is used together with the actual pressure normal to the surface, the effect of pressure on the longitudinal force is automatically included in the result.

3.2.3.3 Installation forces are to include all forces acting on the pipe due to the installation operations. Typical installation forces are applied tension during laying and forces from the trenching machine if trenching is carried out after laying.

4.3 Pipeline/Riser during Installation

4.3.1 General

4.3.1.1 Strength considerations for the pipeline/risers during installation are to be made in order to determine how the pipeline/riser may be installed without suffering any damage which may impair the function or the safety of the completed line, or which may involve hazardous installation or repair work. See also Section 8.

4.3.1.2 If the installation analyses for a proposed pipeline/riser show that an acceptable set of installation parameters cannot be obtained with the installation equipment to be used, the pipeline/riser is to be modified.

4.3.1.3 The requirements of 4.3 apply also, as far as applicable, to repair operations.

4.3.1.4 Only those sections under 4.3.2, 4.3.3 and 4.3.4 found pertinent to the various installation techniques/phases should be considered.

4.3.1.5 Any installation phase/technique is to be checked. Such phases and techniques are:

Start of laying operation
Normal continuous laying
Pipe abandonment and retrieval
Termination of laying operation
Tow out
Bottom tow
Bottom pull
Spool on
Tie-in
Straightening
Trenching
Back fill

4.3.1.6 For any of the phases mentioned in 4.3.1.5 the pipeline/riser is to have the below required safety against the following forms of failure or damage

Excessive yielding; see 4.3.2.3–6.
Local buckling; see 4.3.2.7.
Fatigue effect; see 4.3.2.8.
Excessive damage to weight coating; see 4.3.2.9.

4.3.2.7 For installation methods involving a J or S shaped curve of the pipeline, N and M are to be determined by an appropriate method, suitable for the water depth, pipe stiffness and weight in question. Since the effect of the environmental loads is difficult to determine, the minimum required analyses are as follows:

(A) Loading condition (a) is to be analyzed in detail, and the formula of 4.3.2.6 is to be applied with a usage factor η of maximum 0.72.
(B) Loading condition b) is to be considered by evaluating the increase of M due to environmental loads on the basis of the assumed environmental conditions, the relevant characteristics of the installation equipment (particularly the laying vessel), and all available relevant experience. If there is reason to ex-

pect that the increase of M will exceed 33%, the assumed maximum M is to be inserted in the formula of 4.3.2.6, applying a usage factor of maximum 0.96.

(C) The particular effect of transverse forces acting on the pipe during laying, namely the change in direction of the pipe axis in the horizontal plane near the lift-off points, is to be specially considered.

4.3.4 Fatigue

4.3.4.2 When the bottom tow, bottom pull or the flotation method is used for installation of a pipeline, fatigue is considered to be a major effect and this effect should be paid special attention both through theoretical calculation and tests.

Section 8 Installation

8.1 General

8.1.1 Specifications

8.1.1.1 Installation of a pipeline system is to be carried out in accordance with written specifications, plans and drawings which are satisfying these Rules. The specifications are subject to approval by Veritas.

8.1.1.2 Welding procedures are to be specified as described in 8.5.2.

8.1.1.3 Field coating procedure is to be specified as described in 6.2.4.

8.1.1.4 NDT procedures are to be specified as described in Section 10.

8.1.1.5 A detailed quality control system has to be specified for all installation activities, see 1.4.4.2.

8.1.1.6 The installation specification is to give detailed information on parameters which have to be controlled in order to obtain the correct configuration of and stress levels in the affected portion of the pipeline. The range within which the parameters are allowed to vary is to be clearly stated, see 4.3.

8.1.1.7 Instrumentation systems used for measuring or controlling essential parameters during the installation operation are to be specified.

8.1.1.8 For a lay vessel the following should be included in the specification:

general lay-out drawings showing location of working stations, tension devices, stinger, supports, guides etc.

profile of ramp and stinger showing proposed pipeline configuration

brief description of the tension devices with information on pulling force, holding force and squeeze pressure

brief description of support and guides on lay-vessel and stinger, including information on possible horizontal and vertical adjustment

brief description of stinger including weight and buoyancy distribution and procedures for obtaining correct configuration

brief description of other systems or equipment essential for the installation operation.

8.1.1.9 For a riser installation the specification should include information such as:

description and general layout drawings of the riser showing location of supports, bends, flanges, etc.

detail drawings of riser supports, bends, flanges, spoolpieces, etc.

description and drawings of corrosion protection system.

description and specification of equipment essential for the installation

instrumentation systems used for measuring or controlling essential parameters during the installation operation

procedure specification covering all installation operations.

8.2 Pipeline Route

8.2.1 Route Survey

8.2.1.1 Adequate surveys are to be carried out prior to installation of the pipeline; see 2.2.

8.2.2 Seabed Preparation

8.2.2.1 Seabed preparation is to carried out in accordance with an approved specification.

The specification is to include information such as

- extent of preparation
- preparation methods and equipment
- inspection methods and equipment

8.3 Construction

8.3.1 Qualification

8.3.1.1 Construction has to be carried out by means of qualified personnel, procedures and equipment. The qualifications are to be proved prior to start of construction.

8.3.1.2 Welders and welding operators are to be qualified in accordance with 8.5.5.

8.3.1.3 Welding procedures are to be qualified in accordance with 8.5.3 and 8.5.8.

8.3.1.4 NDT procedures and operators are to be qualified in accordance with Section 10.

8.3.1.5 It may be required that installation vessels are surveyed prior to start of installation. This may include testing and calibration of equipment and instrumentation such as

- tension machines
- winches
- load cells
- depth gauges
- welding equipment

8.3.2 Handling and Storing

8.3.2.1 Pipes, fabricated sections and accessories are to be handled in a safe manner to prevent damange, and are to be adequately supported and protected during storage and transportation.

8.3.2.2 Pipes, prefabricated sections and accessories are to be inspected before installation. Damaged items are to be repaired to the satisfaction of the Surveyor or clearly marked and replaced; see 6.3.4 and 8.5.8.

8.3.2.3 Storing of pipes has to be carried out in such a way that the pipe is not being permanently deformed by its own weight or the weight of above layers of pipes. Special care should be taken for storing heavy coated anode joints.

8.3.3 Installation Operations

8.3.3.1 The installation of the pipeline system is to be carried out in accordance with approved procedures and in such a way that the pipe and coating will not be exposed to unacceptable strains/stresses or be damaged.

8.3.3.2 Mounting and application or riser supports are to be carried out so as to obtain the support conditions upon which the design calculations have been based.

8.3.3.3 Instrumentation systems used for measuring or controlling essential parameters are to be accessible for the Surveyor at any time.

8.3.3.4 Joining of pipes and subsequent non-destructive testing are to be carried out in accordance with 8.5 and 8.6 respectively.

Tie-ins of pipeline sections are to be carried out in accordance with 8.7.

8.3.3.5 Corrosion coating of field joints is to be carried out in accordance with 6.2.4.

8.3.3.6 Pipes which have suffered damage during abandon or retrieval operations are to be replaced or repaired to the satisfaction of the Surveyor.

Acceptance criteria for coating damages are to be worked out prior to start of laying.

8.3.3.7 Survey of the installed pipeline is required when there is reason to believe that damage has occurred, and that further laying may render later surveys and repairs difficult or impossible.

8.3.4 Pipeline and Cable Crossings

8.3.4.1 Crossing of pipeline and cables is to be carried out in accordance with an approved specification. Safety measures adopted to avoid damage on foreign installations or by other installations are to be specified.

8.3.4.2 The specification is to include information such as

- layout and profile of crossing
- auxiliary constructions or components including layers of separation
- methods and equipment adopted for installation
- inspection methods

8.3.3.3 Normally a minimum clearance distance of 0.3 m is to be maintained between the pipeline and other pipelines or cables.

8.3.5 Buckle Detection

8.3.6.1 In connection with pipelaying from vessel where pipe sections are joined onboard the vessel it may be required that continuous buckle detection is carried out during laying. In such cases the method of buckle detection is subject to approval. Normally a rigid disc is to be located within the pipe at a suitable distance behind the touch down point.

8.3.6.2 The diameter of the detector is to be chosen with due regard to pipeline inside diameter and tolerances on ovality, wall thickness, misalignment and height of internal weld bead.

The following formula may be used:

$$d = D - 2t - S$$

where
$S = 0.01\,D + 0.4\,t + 5\,l$
d = diameter of detector
D = nominal outer diameter of pipe
t = nominal wall thickness of pipe
l = 20% of t, max. 5 mm

8.4 Anchoring and Protection of Pipeline Systems

8.4.1 General

8.4.1.1 The pipeline system is to be protected and/or anchored against unacceptable loads and incidents such as:

- lateral axial movements
- impacts
- corrosion

8.4.1.2 Anchoring/protection of a pipeline system is to be carried out in accordance with an approved specification. The specification is at least to include

- definition of the final conditions
- description of methods and equipment
- description of means and instrumentation for control and inspection

Provisions for corrosion protection are covered in Section 6.

8.4.1.3 Measures for obtaining protection of risers and pipelines are outlined in 4.2.1.3 and 4.2.1.4.

8.5 Installation Welding

8.5.1 General

8.5.1.1 The schemes for installation welding described in this section have been based on current recognized practice. Other methods may also be used, but are then subject to special approval.

8.5.1.2 All installation welding is to be performed with equipment which has been proved reliable and suitable for field applications. Prequalification testing is to be performed for welding systems where previous field experience is limited, or the system will be used under new conditions.

8.5.1.3 Welding may be performed with the manual metal arc, the flux-cored arc, the gas metal arc or the tungsten inert gas metal arc process. Higher strength steels are to be welded with low hydrogen consumables unless special welding techniques are used ensuring an equal safety against cold cracking.

8.5.2 Welding Procedure Specification

8.5.2.1 A welding procedure specification is to be prepared for each procedure giving the following information:

Pipe material, standard grade and project specification.
Diameter and wall thickness.
Groove preparation and design.
Clamping device and line-up tolerances.
Welding process.
Welding consumable(s), trade name and recognized classification.
Electrode/wire diameter.
Shielding gas, mixture and flow rates.
Welding parameters, current, voltage, type of current and polarity, travel speed etc.
Welding position.
Welding direction.
Temporary backing and type (if any).
Number of passes.
Time lapse between passes.
Preheating and interpass temperatures.
Post weld heat treatment.

8.5.3 Qualification of the Welding Equipment and Welding Procedure

8.5.3.1 The selected type of welding equipment and the specified welding procedure is to be qualified prior to installation welding. The qualification test is to be carried out with the same or equivalent equipment as that to be used during installation. The test is normally to be performed on the yard or the vessel where the installation welding is to take place, and be conducted under representative conditions.

The test joints to be used for qualification testing are to be of sufficient length to give realistic restraint during welding. Pipes on the high side of the specified chemical composition are to be selected.

8.5.3.2 When manual welding is to be used, one complete test joint is to be made. For mechanized welding equipment, three consecutive complete test joints are to be made.

Each test joint is to be subject to visual examination, nondestructive tests and mechanical testing.

8.5.3.3 Non-destructive testing is normally to be radiography tested using X-rays. When the gas metal arc process is used, the test joints are also to be ultrasonic tested. Magnetic particle testing may be required in special cases.

Non-destructive testing is to be performed in accordance with Section 10 and the soundness of the test welds is to meet the acceptance limits given in this section.

8.5.3.4 The type and number of mechanical tests for each joint are given in Table 8.1. Sampling of test specimens, dimensions and method of testing are described in Appendix C.

The mechanical properties of the test welds are to meet the following requirements:

The ultimate tensile strength of the joint is to be at least equal to the specified ultimate tensile strength of the pipe material. When different steel grades are joined, the ultimate tensile strength of the joint is to be at least equal to the minimum specified ultimate tensile strength of the lower grade.

The guided bend tests are to disclose no defects exceeding 3 mm. Minor cracks, less than 6 mm, originating at the specimen edges may be disregarded if not associated with obvious defects.

The fracture surfaces of the nick break test specimens are to show complete fusion and penetration. Other defects exceeding the limits of Table 10.1 are not acceptable. "Fish eyes" may be disregarded unless associated with unacceptable amount of slag inclusions and porosity.

The average and single Charpy V-notch toughness at each position are not to be less than specified according to 5.2.7. When different steel grades are joined, a series of impact tests is to be performed in the heat affected zone on each side of the weld. The weld metal is then to meet the more stringent energy requirement.

The maximum hardness is not to exceed the limits given in 5.2.10 and 5.2.12 as applicable.

The macrosection is to show a sound weld merging smoothly into the pipe without defects as per the limits of Table 10.1.

8.5.3.5 Failure of a test specimen due to defective preparation may be disregarded, and is to be replaced by a new test specimen.

8.5.4 Essential Parameters for Welding Procedures

8.5.4.1 A qualified welding procedure remains valid as long as the essential parameters are kept within acceptable limits and production tests are performed regularly. When one or more variations outside the acceptable limits occur, the welding procedure is to be considered invalid, and is to be re-specified and qualified.

8.5.4.2 The essential parameters and the acceptable limits of variations are normally to be as described below. For special welding system other essential parameters and acceptable variation limits may have to be imposed.

Materials: A change from a lower strength grade to a higher, and any change in type, composition and processing significant for the weldability and the mechanical and properties of the weld. The C-content, alloy content, carbon equivalent and supply condition are to be specially considered.

Diameter: A change in diameter from one to another of the following ranges: OD $\leq$ 100 mm, 100 < OD $\leq$ 300 and OD > 300.

Thickness: A change outside the thickness interval 0.75 t to 1.5 t where t is the nominal thickness of the test joint.

Groove configuration: Any change(s) important for penetration, dilution and solidification pattern, i.e. groove type (V, U, Y, X) angles, root gap and root face are to be specially considered.

Welding process: Any change

Welding consumables: Any change of type, classification, diameter and brand as well as additions/omissions of powders, hot and cold wires.

Gas shielding: Any change of specified mixture, composition and flow rate range.

Welding position: A change to a principal position not being qualified according to Table 8.2.

Welding direction: A change from vertical down to vertical up or vice versa.

Current: Any change beyond ± 15% and from AC to DC.

Polarity: Any change.

Voltage: Any change beyond ± 10% except ± 5% for gas metal arc welding.

Travel speed: Any change beyond ± 10%.

Time lapse between root pass and first filler pass: Any delay significantly increasing the cold cracking risk.

Preheating: Any decrease.

Interpass temperature: Any significant change in the minimum and maximum interpass temperature limits.

Post weld heat treatment: Any change significantly affecting mechanical properties, the residual stress level, the corrosion resistance, i.e. the heating rate, cooling rate, temperature level and period, heating band and insulation width to be specially considered.

8.5.5 Qualification of Welders and Welding Operators

8.5.5.1 Qualification of welders and welding operators are generally to be as described in 7.2.3. For underwater welding, additional conditions apply, see 8.7.4.10.

8.5.5.2 Under special circumstances, qualification of welders may be based on visual examination and mechanical testing only, if so agreed by Veritas. In such cases bend testing and nick-break testing are to be carried out in accordance with Appendix C. Acceptance criteria for nick-breaks are, however, to be as follows:

The fractured surface is to show complete penetration and fusion. There is to be maximum one—1—gas pocket per cm^2, being less than 1.5 mm in extension. Only minor slag inclusions, with maximum depth 0.8 mm and with maximum length 3 mm spaced at least 12 mm, may be accepted. "Fish eyes" may be disregarded unless associated with significant number of slag inclusions and cluster porosity.

8.5.6 Welding and Workmanship

8.5.6.1 All installation welding is to be performed with qualified welding equipment, qualified welding procedures and type of equipment and by qualified welders/operators. The back lead of the welding equipment is to be correctly connected to avoid stray current giving arise to corrosion; see also 6.3.2.3. Identical welding units, either additional or replacement units, may be qualified by non-destructive testing of production welds.

8.5.6.2 The bevelled pipe ends are to be free from contamination by moisture, oil, grease, rust etc. which might affect the weld quality.

8.5.6.3 Internal or external line-up clamps are normally not to be removed before the first two passes are completed. When tack welds are necessary for alignment, these are only to be made in the weld groove using a qualified welding procedure. Defective tack welds are to be completely removed.

8.5.6.4 Welding is not to be discontinued before the joint has sufficient strength to avoid plastic yielding and cracking during pulling and handling. Prior to restarting after interruptions, preheating to the minimum specified preheating temperature is to be applied.

8.5.6.5 Supports, attachments, lifting devices etc. used for permanent positioning of risers and pipelines are normally to be welded to a doubler ring. Doubler rings for temporary use are to be clamped.

8.5.6.6 Permanent doubler rings are to be made as fully encircling sleeves and of materials satisfying the requirements for pressure parts; see 5.7. Longitudinal welds are to be made with a backing strip, avoiding penetration into the main pipe. The circumferential welds are to be continuous, and made in a manner minimizing the risk of root cracking and lamellar tearing.

8.5.7 Production Test

8.5.7.1 Production tests may be required during installation. The test is to be performed in a manner which, as far as possible, reproduces the actual welding, and is to cover welding of a sufficiently large pipe sector in a relevant position.

When production testing is required, half the

TABLE 8.1 Qualification of Girthwelding Procedure: Type and Number of Mechanical Tests for Each Joint

Test Joint		Number of Each Specified Test						
Wall Thickness (mm)	Outside Diameter (mm)	Transverse Weld Tensile	Root Bend	Face Bend	Side Bend	Nick Break[2]	Charpy V-Notch Samples[3,4,5,6]	Hardness and Macro
<12.5	<300	2	2	2	0	2	4	2
	>300	4	4	0	4[1]	4	4	2
>12.5	<300	2	0	0	4	2	4	2
	>300	4	0	0	8	4	4	2

Notes:

[1]Root and face bend tests may be used instead of side bends.

[2]Nick break tests may be omitted for manual metal arc welding to be performed above water.

[3]Impact testing is not required for t <5 mm.

[4]Each Charpy V-notch sample consists of 3 specimens.

[5]Impact testing is to be carried out with the V-notch positioned in the weld metal, on the fusion line, 2 mm from the fusion line and 5 mm from the fusion line.

[6]When more welding processes or more welding consumables are used, impact testing is normally to be carried out in the corresponding weld regions if the region tested cannot be considered representative for the complete weld.

number of tests specified in Table 8.1 are to be carried out. Impact test samples are to be located in the weld metal, and in the heat affected zone at the position which showed the lowest average energy absorption during the procedure qualification test, see 8.5.1.

8.5.8 Repair of Field Joints

8.5.8.1 Pipes and welds containing defects are to be repaired as described in 8.5.8.2 through 8.5.8.9.

8.5.8.2 Defects outside the weld are to be repaired by grinding only. If grinding reduces remaining wall thickness below the minimum specified thickness, the defective pipe section is to be cut out. Grinding is to be performed in a workmanlike manner, and with smooth transition into the pipe surface.

8.5.8.3 Defects in the weld may be repaired by grinding or welding. Repair welding specifications are to be prepared, and are to give the following information in addition to that relevant of 8.5.2.1.

Method for removal of defect.

Preparation of weld area.

Non-destructive tests for confirmation of defect removal.

Permissible minimum and maximum weld repair sizes.

8.5.8.4 The repair welding procedures are to be qualified. The qualification tests are to be made in a realistic manner simulating repair situations likely to occur, e.g.

Through thickness repair.

External repairs of undercuts with one stringer pass.

Inside root repair with one pass only.

Repeated weld repairs in same area.

The repair tests welds are to be made in the overhead through vertical position, using pipe with a chemical composition in the upper range of the specification.

8.5.8.5 The test weld covering through thickness repair is to be visually inspected, non-destructive tested and mechanical tested as required for the installation welding procedure, sec 8.5.3. The single

TABLE 8.2 Qualified Principal Welding Positions

Test Position	Applicable Welding Positions
1G	1G
2G	1G, 2G
5G	1G, 5G
2G + 5G	All
or 6G	All

pass test welds are to be visually inspected, magnetic particle examined and mechanical tested with two macro/hardness tests provided there is used the same welding consumables and parameters as for the major repairs.

8.5.8.6 Preheating is to be performed prior to repair welding. The minimum specified preheating/interpass temperature is to be maintained until the repair has been completed.

8.5.8.7 Long defects may require repair in several steps to avoid yielding and cracking. The maximum length of allowable repair step is to be calculated based on the maximum stresses in the joint during the repair operation. The repair length is to be at least approximately 100 mm even if the defect is of less extension.

8.5.8.8 Grinding is to be performed after arc air gouging to remove any carbon pick-up.

8.5.8.9 A joint may be repair welded twice in the same area. If the joint still contain defects, the complete joint is to be cut out unless special repair welding procedures simulating actual number of weld repairs have been qualified.

8.6 Visual Examination and Non-destructive Testing of Installation Welds

8.6.1 General

8.6.1.1 Installation welds including repairs made by grinding and welding are to be visual examined and non-destructive tested.

8.6.1.2 Non-destructive testing is to be performed in accordance with qualified procedures and qualified NDT-operators, see Section 10.

8.6.1.3 Inspection and NDT-records are to be made for each weld including any repair actions. The records are to be marked and identified in a suitable manner enabling tracebility to location of welds and the welding procedure(s) being used.

8.6.2 Visual Examination

8.6.2.1 Visual examination is to be carried out for all welds.

8.6.2.2 The finished welds and the pipe surfaces are to comply with the acceptance criteria specified in Table 10.1.

8.6.2.3 Welds which do not comply with Table 10.1 are to be repaired according to 8.5.8 or cut out.

8.6.3 Non-Destructive Testing

8.6.3.1 All installation welds are to be radiographed full length. Ultrasonic testing and magnetic particle testing may be required depending on the applied welding method.

8.6.3.2 Defects which exceed the acceptance limits in Table 10.1 are to be completely removed and repaired in accordance with 8.5.8. Magnetic particle testing is normally to be used to ensure complete removal of defects prior to repair welding.

8.6.3.3 Weld repairs are to be radiographed. This examination is to cover the repaired area and an additional length of 50 mm at each end of the repair weld.

8.6.3.4 Magnetic particle testing may replace radiography when the defect is located at the outside of the pipe, and is removed by grinding only.

8.7 Tie-Ins

8.7.1 General

8.7.1.1 Tie-ins between different portions of a pipeline, or between pipeline and riser, may be carried out by one of the following methods.

Mechanical connectors.
Welded connection on the lay vessel and subsequent lowering.
Underwater welding.

The choice of method is to be based on an evaluation of the conditions under which the tie-in is to be carried out and the service conditions under which the tie-in is to operate.

8.7.1.2 The tie-in operation is to be carried out in accordance with an approved tie-in specification.

8.7.1.3 Tie-in specification is to include:

description and specification of components which will be introduced as permanent parts of the pipeline.

calculation of stresses occurring during installation and operation.

procedure specifications covering all tie-in operations.

description and specification of equipment and instrumentation essential for the installation.

description and specification of methods of inspection and testing.

8.7.2 Mechanical Connectors

8.7.2.1 Mechanical connectors include flanges, couplings or other components adapting similar mechanical principles of obtaining strength and tightness.

8.7.2.2 An evaluation is to be carried out for loads and resulting stresses to which the components are subjected during installation and operation. Safety factors to be included to ensure an equivalent overall safety to that adopted for the adjacent pipeline.

8.7.3 Welded Tie-In on the Lay Vessel

8.7.3.1 Lifting and lowering of the pipeline during the tie-in operation are to be carried out so that induced stresses are within the allowable limits for pipeline or riser respectively during installation.

8.7.3.2 Suitable means for monitoring the configuration of the pipeline section are to be used.

8.7.3.3 Welding and inspection of the tie-in is, to be carried out in accordance with approved specifications; see 8.5 and 8.6.

8.7.4 Tie-In by Underwater Welding

8.7.4.1 Welding is to be carried out with a low hydrogen process in a chamber (habitat) from which the water has been displaced.

Other methods are subject to special approval.

8.7.4.2 Sealing devices are to be of a proven design and manufacture. Sealing pigs are to be pressure tested prior to installation into the pipeline sections unless this has been carried out at an earlier stage.

8.7.4.3 A detailed welding procedure specification is to be established, and is in addition to that specified in 8.5.2.1 to contain:

water depth.
pressure inside the chamber.
gas composition inside the chamber.
humidity level.
temperature fluctuations inside the chamber.

8.7.4.4 Storage and handling routines of welding consumables on the support vessel and in the welding chamber as well as the sealing and the transfer procedures to the welding chamber are to be specified.

8.7.4.5 The welding procedure is to be qualified under representative conditions in a suitable testing facility. The qualification test is to consist of minimum one complete joint for manual welding and minimum three joints for mechanized welding system.

The qualification program may be increased when the underwater welding will occur under conditions where previous experience is limited, or will be undertaken by a company with limited experience in this field.

8.7.4.6 The qualification test welds are to be inspected and tested as per 8.5.3 and comply with the requirements specified for the pipeline section in question.

8.7.4.7 Preheating to a suitable temperature is to be applied for moisture removal and hydrogen diffusion.

8.7.4.8 The essential parameters for underwater welding are those specified in 8.5.4.1 plus those given in 8.7.4.3. The acceptable variation limits are normally those specified in 8.5.4 plus the following:

Pressure inside chamber:	any increase
Gas composition inside chamber:	any change
Humidity:	any increase beyond specified range may be required

8.7.4.9 A confirmation test weld may be required made on location prior to starting the tie-in welding. The test weld is to be made on pipe coupons in the habitat under actual conditions. The coupons are to cover welding from the 6 o'clock to 9 o'clock region. Subject to acceptable visual inspection and radiography in accordance with 8.6 the tie-in welding may commence. Mechanical testing is to be performed as soon as possible. The number of mechanical tests is half that required for welding procedure qualification.

When the same welding habitat, equipment and welding procedure are used for consecutive tie-ins on the same pipeline under comparable conditions, further confirmation test welds are not required.

8.7.4.10 The tie-in weld is to be non-destructive

examined full length, as per 8.6 and comply with the applicable acceptance standard in Section 10.

8.7.4.11   Prior to qualification testing for underwater welding, the welder is to have passed a surface welding tests (see 7.2.3) and have relevant training for welding under pressure.

Qualification for underwater welding is to consist of at least one test weld made in a testing facility under representative conditions in accordance with the qualified underwater welding procedure. The test weld is to be visual inspected, radiographed and mechanical tested, see 7.2.3 and Appendix C.

8.8   Final Surveys and Tests

8.8.1   General

8.8.1.1   A final survey of the installed pipeline system is to be carried out in order to verify that the condition of the pipeline system satisfies the approved specification and the requirements of these Rules.

8.8.1.2   If the pipeline is to be buried or covered by other protection stabilization methods, surveys are normally required both before and after burial (covering) operations.

8.8.2   Survey of Installed Pipeline System

8.8.2.1   The final survey on the pipeline system is at least to provide the following information:

Detailed plot of the pipeline position

Thickness of cover or depth of trench (if applicable) and description of the state of rest along the route

Verification that the condition of weight coating or the anchoring system which provides for on-bottom stability is in accordance with the approved specification

Description of wreckage, debris or other objects which may affect the cathodic protection system or otherwise impair the pipeline

Description and location of damages to the pipeline, its coating or cathodic protection system

8.8.2.2   The final survey report of the installed riser is to verify that the riser, including supports, clamps, anchors, protection devices (e.g. fenders, casings, etc.) and corrosion protection system, are installed in accordance with approved drawings and specifications.

8.8.3   Survey of Corrosion Protection System

8.8.3.1   Inspection of the external coating of the pipeline system is required. Special attention should be given to the riser in the splash zone.

8.8.3.2   Spot measurements of the polarization along the pipeline may be required in areas with damaged coating. Special attention is to be paid to areas far from sacrificial anodes and areas with stress concentrations.

8.8.3.3   In areas where measurements indicate that cathodic protection has not been attained, some corrective action is to be arranged, e.g. mounting of additional sacrificial anodes, increasing current output from rectifiers, or application of protective coating.

8.8.3.4   The possibility of over-protection is to be investigated at locations where detrimental effects of over-protection may be suspected.

8.8.3.5   The possibility of stray currents are to be investigated by measurements and visual observations by qualified personnel.

8.8.4   Pressure Test

8.8.4.1   The pipeline system is to be pressure tested after installation. The testing is to be carried out in accordance with an approved procedure. A pipeline system may be tested in sections, e.g. between top of risers or between top of the riser and shore. When a pipeline is to be buried or covered, the pressure test is to be performed after such operation.

8.8.4.2   The test is normally to be carried out with liquid test medium.

8.8.4.3   The pressure test is to prove the strength and the tightness of the tested section. The minimum test pressure is to be 1.25 times the design pressure. Hoop stress in the pipe during testing is normally not to exceed 90 per cent of the minimum specified yield strength. Higher stresses will be considered in each case.

8.8.4.4   During pressurizing, added test liquid versus pressure is to be recorded in order to evaluate the amount of residual air in the test section.

8.8.4.5   After pressurizing, sufficient time has to

be allowed for stabilization of the pressure in the pipe section.

8.8.4.6 The holding time for pipeline sections is normally to be minimum 24 hours, after the pressure has stabilized. For shore lines and risers, 8 hours holding time may be accepted. For pipe section that can be 100% visually inspected the holding period is normally to be at least 2 hours.

8.8.4.7 Alternative pressure testing procedures may also be accepted. For guidance see Appendix E.

8.8.4.8 If the tested section bursts or leaks, the failure is to be corrected and the section retested.

8.8.4.9 Pressure testing of tie-in welds between already tested sections may in special cases be exempted provided the regular radiographic examination is extended with ultrasonic examination of other suitable methods. Monitoring may be required. The NDT procedures and operators are to be qualified for this testing, see Section 10.

8.8.5 Buckle Detection

8.8.5.1 Buckle detection is to be carried out by running a gauge pig (caliper pig) through each pipeline section after installation. When the pipeline is to be buried, the final buckle detection is to be performed after trenching.

8.8.6 Testing of Alarm and Shutdown Systems

8.8.6.1 It is the Owner's responsibility to protect the pipeline system against operational conditions for which the system is not designed.

8.8.6.2 Instrumentation for the safe operation of the pipeline system is to be tested according to generally recognized codes and the manufacturer's recommendations prior to start of operation.

8.8.6.3 Emergency shutdown systems are to be tested according to generally recognized codes prior to start of operation.

Appendix 5

Excerpts from Det Norske Veritas, Rules for the Design, Construction, and Inspection of Offshore Structures, Appendix H, "Marine Operations"

Section H1 Lifting

H1.1 General

H1.1.1 Introduction

H1.1.1.1 Marine Operations are dealt with in section 10 of the Rules. The intention of H1 is to present more detailed guidance. A worked design example for slinging arrangements is included in paragraph H1.9 to demonstrate a convenient method of applying the various factors.

H1.1.1.2 H1 covers well controlled one-hook inshore and offshore lifting of heavy objects where floating cranes or crane vessels are employed. Multi-hook lifts, which are not treated in detail, will normally require additional skew load factors and de-rating of the crane capacities.

H1.1.1.3 When a formal approval of the lifting operation is required, Veritas will upon request issue a "Lifting Declaration" when the attending Surveyor considers the preparations for the operations adequate.

H1.1.2 Documentation

H1.1.2.1 Before start of the lifting operation the documentation necessary for a complete description of the operation should be submitted for review. Such documentation includes: drawings, calculations, criteria for selection of slinging equipment and description of operation procedures.

H1.1.2.2 The documentation necessary to evaluate the operation against the design premises and operation procedures should be made available to the attending Surveyor.

H1.1.3 Assumptions

H1.1.3.1 It is the responsibility of the owner or contractor to establish a quality assurance program to ensure that the lifted object and lifting gear are designed and fabricated with adequate strength to survive the intended lifting operation(s).

H1.1.3.2 The lifting contractor should familiarize himself with the design assumptions for the lift and perform the operations in compliance with those assumptions. The operations should not be performed under more severe environmental conditions than those for which the objects involved are designed.

H1.1.3.3 Quality assurance documentation such as design assumptions, calculations, certificates, fabrication release note(s), weighing/test reports, etc. should be made available to the attending Surveyor.

H1.1.4 Survey

H1.1.4.1 The lifted object, including padeyes and lifting gear, should be made accessible for survey in due time before the lifting operation.

H1.1.4.2 In cases where the Surveyor finds that the properties of an item are not adequately documented, additional testing may be requested.

Det Norske Veritas, Oslo, 1977 (rpt. 1981). By permission of DNV.

H1.2 Loads

H1.2.1 Weight of the Lifted Object (W)

H1.2.1.1 The weight (W) is the weight of the object *as lifted*. It is recommended that this is determined by weighing. The weighing procedure should preferably also determine the position of the centre of gravity.

H1.2.1.2 The final weighing should take place when the fabrication of the object is completed. Provided the weighing is performed according to a reliable method, the weight (W) may be assumed to be equal to the weighed figure. If any weights are removed or installed after weighing, the weighed figure should be corrected accordingly, or the weighing repeated.

H1.2.1.3 During the design stage, the weight should be calculated on the basis of an accurate weight take-off. It is recommended that the weight (W) is taken to equal the calculated weight multiplied by a weight growth factor of not less than 1.1 to allow for weight increases normally experienced during design and fabrication.

H1.2.2 Dynamic Amplification Factor (DAF)

H1.2.2.1 The dynamic amplification factor (DAF) is a factor accounting for global dynamic effects normally experienced.

H1.2.2.2 The actual DAF a lift will be exposed to will be significantly influenced by a number of factors, in particular the environmental conditions. The DAF given below should be considered as minimum factors which may be used provided the lifting operation will not take place under adverse conditions.

Weight of the Lifted Object (W)	< 100 t	100–1000 t	> 1000 t
DAF Offshore	1.3	1.2	1.15
DAF Inshore	1.15	1.1	1.05

H1.2.3 Weight of Rigging (RW)

H1.2.3.1 The weight of rigging (RW) is the total weight of rigging such as: shackles, slings, spreader bars or frames, etc.

H1.2.4 Special Loads (SPL)

H1.2.4.1 When appropriate, allowances for special loads should be made.
Special Loads (SPL) are loads such as: tugger line forces, guide forces, wind forces, hydrodynamic and hydrostatic forces, etc.

H1.2.5 Skew Load Factor (SKL)

H1.2.5.1 Skew load is the extra loading on slings caused by the effect of inaccurate sling lengths and other uncertainties with respect to force distribution in the slinging arrangement.

H1.2.5.2 For slinging arrangements with tolerances within the limits specified in H1.4.4.2, the following skew load factors (SKL) are normally applicable:

1.0: for statically determinant 1, 2, 3 and 4 points lifts.

1.25: for statically indeterminant 4 points lifts.

H1.2.5.3 Higher skew load factors (SKL) should be applied for lifts with excessive fabrication tolerances and for lifts which are particularly sensitive with respect to sling force distribution. Examples of such lifts are:

Lifts with more than 4 lifting points.
Lifts with 3 or more lifting points in the same vertical plane.
Some multi-hook lifts.

H1.3 Loadcases and Analysis of Forces

H1.3.1 General

H1.3.1.1 A lifting operation does not represent one well defined loadcase, but a sequence of different loadcases. Uncertainties with respect to internal force distribution, possible accidental loads, etc., will introduce further complications.

H1.3.1.2 The Designer should in principle consider the entire lifting sequence step-by-step and identify the most critical (dynamic) loadcase for each specific member. However, for most conventional one hook lifts, the entire sequence can be adequately covered by the basic loadcases described in H1.3.2 and the additional loadcases described in

H1.3.3. Multi-hook lifts, upending operations, etc., will require special consideration.

H1.3.2   Basic Loadcase and Force Distribution

H1.3.2.1   The dynamic hook load (DHL) can normally be expressed as:

$$DHL = DAF(W + RW) + SPL$$

where
DAF = Dynamic amplification factor (see H1.2.2)
W   = Weight of the lifted object (see H1.2.1)
RW = Weight of rigging equipment (see H1.2.3)
SPL = Special loads (see H1.2.4)

H1.3.2.2   The force distribution in a lift can normally be calculated as a quasi static loadcase by applying DHL at the hook position and associate mass and possible special loads to each element.

H1.3.2.3   For lifts with SKL equal to 1.0 (see H1.2.5), the dynamic forces calculated according to H1.3.2.2 can normally be used as "maximum dynamic loads" for design purposes as described in H1.4, H1.5, and H1.6.

H1.3.2.4   For lifts with SKL exceeding 1.0 (see H1.2.5) the ideal force distribution calculated according to H1.3.2.2 must be adjusted to account for uncertainties in the internal force distribution in the slinging arrangement.

For a conventional four sling lift, the following skew load cases should normally be considered:

(a) The force distribution calculated according to H1.3.2.2 modified by multiplying the forces in two slings positioned diagonally opposite to each other with the skew load factor. The forces in the remaining two slings are to be determined by (quasi) static equilibrium.

(b) As (a), but with the additional skew load on the other pair of slings.

H1.3.3   Additional Loadcases

H1.3.3.1   Members which may be exposed to loads not adequately covered under H1.3.2 should be identified and designed accordingly. Bumpers and guides should be able to absorb a relevant amount of energy without causing operational problems and without damaging vital members.

H1.3.3.2   If tugger lines are attached to the lifted object, the attachment points are to have adequate structural strength to withstand the maximum forces which can be imposed by the tugger lines.

H1.3.3.3   Lifting points (padeyes) are normally to be positioned so that the sling design force acts in plane with the main padeye plate. Lifting points and their attachments to the structure should be designed for the maximum in plane loads plus possible lateral load.

A lateral load acting simultaneously with the in plane load should normally not be taken as less than 3% of the maximum in-plan load. If feasible, a plastic consideration could be adopted.

H1.4   Slings and Grommets

H1.4.1   Breaking Strength

H1.4.1.1   The minimum breaking strength of steel ropes used for fabrication of slings and grommets is to be determined by testing the entire rope, or a part of the rope, to destruction. When the minimum breaking strength of ropes are calculated based on the tested strength of a fraction of the rope, appropriate spinning loss factors (or lay factors) should be applied.

For grommets with unspliced core, the strength of the core is not to be included when calculating the minimum breaking strength.

H1.4.1.2   Minimum required breaking strength is the nominal safety factor given in H1.4.2 multiplied by the maximum dynamic sling force calculated according to H1.3.2.

H1.4.2   Nominal Safety Factor

H1.4.2.1   For lifts equal to or exceeding 50 tonnes the nominal safety factor is not to be less than 3.3. For lifts below 40 tonnes the nominal safety factor is not to be less than 4.0. For lifts between 40 and 50 tonnes the nominal safety factor is to be determined by linear interpolation between 4.0 and 3.3.

H1.4.2.2   The nominal safety factor given above contains an allowance for strength reductions of the ropes due to splicing and bending. If the minimum breaking strength of a sling or grommet is reduced more than 25% compared with H1.4.1.1 due to splicing or bending, the capacity is to be de-rated accordingly.

H1.4.3 Bending of Slings and Grommets

H1.4.3.1 The strength reduction of a sling or grommet due to bending will normally be within acceptable limits (no de-rating required) provided the limitations given below are complied with.

H1.4.3.2 The eye of a single part cable laid sling should not be bent round a diameter less than the nominal diameter of the sling. However, in order to maintain the sling in good condition, it is normally recommended to avoid bending round a diameter less than two times the diameter of the sling. No other part of a single part cable laid sling shall be bent around a diameter of less than 4 times the nominal diameter of the sling.

H1.4.3.3 No eye or other part of a cable laid grommet should be bent round a diameter of less than 6 times the nominal diameter of the grommet.

H1.4.3.4 Bending in way of splices should be avoided.

H1.4.4 Manufacturing

H1.4.4.1 The manufacturing of slings and grommets should only be performed by a recognized steel wire rope manufacturer. The rope construction should be well suited for the intended use and comply with recognized codes or standards e.g. International Standard ISO 2408, and for heavy cable laid ropes:

Guidance Note PM 20: "Cable laid slings and grommets" from the British Health and Safety Executive.

H1.4.4.2 The length of slings or grommets should normally be within tolerances of plus or minus 0.25 percent of their nominal length. During measuring, the sling or grommet should be fully supported and adequately tensioned. Matched slings should be measured under similar conditions.

H1.4.5 Certification

H1.4.5.1 For each sling and grommet a Maker's Certificate should be provided. For slings or grommets used with a nominal safety factor of less than 4 (see H1.4.2), a certificate issued by a recognized Certifying Body is normally required.

H1.4.5.2 The Sling Certificate or Grommet Certificate should contain the following minimum information:

Certificate number
Sling identification code
Name of manufacturer
Date of manufacture
Sling or grommet diameter and length
Type of construction
Certificate no. for unit rope (Certificate to be enclosed)
Calculated minimum breaking load of cable laid rope
Calculated safe working load (SWL) of sling or grommet.

H1.4.5.3 Each sling or grommet should be clearly identified with reference to the corresponding certificate. The safe working load (SWL) as specified in the certificate should be clearly marked on the sling or grommet.

H1.4.5.4 For slings and grommets with a nominal safety factor of less than 4, the certificate should be endorsed after each lift by the Maker or a competent person confirming that the sling is in "as new" condition. Otherwise, if a reduction in strength is suspected, a relevant de-rating should be noted on the certificate or the sling should not be used.

H1.5 Shackles

H1.5.1 Safe Working Load (SWL)

H1.5.1.1 As a reference for the strength of a shackle, the safe working load (SWL) is used. SWL is normally determined by the maker or a Certifying Body.

For lifting purposes, only shackles with breaking strength minimum four times SWL are to be used.

H1.5.1.2 Shackles are normally to be selected with SWL equal to or greater than: the maximum dynamic shackle force including possible skew load (see H1.3.2).

However, big shackles (700 t SWL) may be selected based on a breaking strength consideration. The breaking strength consideraion will normally permit 21% higher load in the shackle than the SWL philosophy (see H1.4.2.1 and H1.5.1.1).

H1.5.2 Design Considerations

H1.5.2.1 Shackles are designed and load-rated to support centreline loading of the shackle. Other load conditions should be avoided.

H1.5.2.2 Shackle dimensions should be selected with due regard to bending radii of slings and grommets, see H1.4.3.

H1.5.3 Manufacturing and Testing

H1.5.3.1 The manufacturing and testing of shackles to be used for lifting should be carried out according to sound practice and in accordance with a recognized code or standard, or according to specifications approved by Veritas.

H1.5.3.2 The shackle material should be well suited for the intended use. Strain ageing resistant material is to be used. Special attention should be paid to obtain adequate resistance against brittle fracture particularly for shackles used in a cold environment; see 6.1.5 (R).

H1.5.3.3 Each individual shackle is to be proof loaded.

For shackles with SWL less than or equal to 25 t, the proofload should not be less that two times SWL.

For shackles with SWL between 25 and 180 t, the proofload should not be less than:

$(1.22 \times SWL) + 20$ t.

For shackles with SWL equal to or greater than 180 t, the proofload should not be less than 1.33 times SWL.

H1.5.3.4 An inspection after the proof loading should not reveal any geometrical deformations, cracks or other defects.

H1.5.3.5 After each use, or before subsequent use, each shackle should be inspected. The type of inspection will be at the discretion of the attending Surveyor but will normally consist of visual inspection and crack tests. Additional proxofloading may be requested for old shackles.

H1.5.4 Certification of Shackles

H1.5.4.1 A makers certificate and a proofloading certificate signed by a recognized Certifying Body should be provided for each shackle.

H1.5.4.2 A shackle Certificate should normally contain the following minimum information:

Certificate identification code
Shackle identification code
Name of manufacturer
Date of manufacture
Manufacturing method/Material information
Reference code, standard or specification
Minimum ultimate strength
Proof load
Safe Working Load

H1.5.4.3 Each shackle should be clearly identified with reference to the corresponding certificate. The safe working load as specified in the certificate should be clearly marked on the shackle.

H1.6 Structures

H1.6.1 General

H1.6.1.1 Loadcases and analysis of forces taking into account global dynamic effects for lifts are described in H1.3. However, additional local dynamic effects such as shockloads in padeyes and similar are included in the factors given in H1.6.2 and H1.6.3. Application of the factors for a typical module structure is shown in H1.9.3.1.

H1.6.2 Design According to Section 4 of the Rules.

H1.6.2.1 The "ordinary" load condition in 4.4.4.3 (R) will be governing. The factors given below includes the loadfactors in 4.4.4.3 (R) as well as an allowance for local shock loads and can normally be applied directly for design purposes.

Item	Design Factor
Spreader frames, spreader beams, etc.	1.75
Lifting points (padeyes) and their attachments to the structure	1.75
Load transferring members supporting the lifting points	1.5
Other members of the lifted object	1.3

H1.6.2.2 The material factor for steel is 1.15 for elastic design, and 1.30 for plastic design.

H1.6.3 Design According to AISC

H1.6.3.1 The loads calculated according to H1.3.2, should be multiplied with the following local factors.

Item	Local Factor
Spreader frames, spreader beams, etc.	1.3
Lifting points (padeyes) and their attachments to the structure	1.3
Load transferring members supporting the lifting points	1.15
Other members of the lifted object	1.0

H1.6.3.2 When the loads calculated according to H1.6.3.1 are applied, the stresses should be within the limits specified by AISC without increase.

H1.6.4 Design Considerations for Lifting Points (Padeyes)

H1.6.4.1 Padeye plates should be orientated in such a direction that the possibility for out-of-plane loading of the padeye plate and on the shackle is minimized. Designs which may fail as a result of a moderate deviation in slingforce direction are to be avoided.

H1.6.4.2 It is normally recommended that padeyes are designed with the main connections in shear rather than tension. High tension loads in the thickness direction of steel materials should be avoided. (Example: Padeye plates should be slotted through horizontal flanges and welded directly to vertical web plates.) If it is not practical to avoid tension loads in the thickness direction, it is recommended that materials with guaranteed through thickness properties are used.

H1.6.5 Fabrication and Inspection of Lifting Points (Padeyes), Plate Shackles, Spreader Beams, etc.

H1.6.5.1 Lifting points, plate shackles, spreader beams, etc. should normally comply with the requirements for primary steel. See 6.1 (R).

H1.6.5.2 Fabrication (welding) should be in accordance with 6.4 (R).

H1.6.5.3 Materials used in lifting points and welds in connection with lifting points should be 100% non-destructively tested, in each case applying the most suitable method. See 6.5 (R).

H1.6.5.4 Spreader beams and frames should be inspected and tested according to 6.5 (R), as primary structural steel. For lifting points, see H1.6.5.3.

H1.6.5.5 Relative padeye positions and fabrication tolerances of plate shackles, spreader beams, etc., are to be measured. Tolerances which will result in excessive skew load are to be avoided (see H1.2.5.2), or an increased skew load factor (SKL) is to be applied, see H1.2.5.3.

H1.6.6 Design Considerations for Seafastening

H1.6.6.1 Design of seafastening for transportation is not covered in H1. However, the following recommendation should be observed with respect to lifting:

The seafastening design should allow for easy release and provide adequate support and horizontal restraint until the object can be lifted clear of the barge.

H1.7 Crane and Crane Vessel

H1.7.1 General

H1.7.1.1 The crane, crane vessel, and all associated equipment should be in good condition, properly manned and fit for performing the intended operations with an adequate safety margin.

H1.7.1.2 Normally the crane vessel should be in possession of the following certificates, which should be made available to the attending Surveyor upon request:

(a) Certificate of Registry.
(b) Certificate of Classification.
(c) Safety Construction Certificate.
(d) Certificate of International Load Line.
(e) Safety Equipment Certificate.

H1.7.1.3 Hydrostatic stability data should be available onboard. Adequate hydrostatic stability during the lifting operation should be verified.

H1.7.1.4 With respect to the crane the following documentation should normally be available:

(a) Certificate of Classification or makers certificate.
(b) Crane test and installation report issued by a recognized authority.
(c) Latest annual survey report.
(d) Lift record for preceeding operations.

(e) Load-radius curves for static and dynamic lifting conditions.
(f) Instructions for crane operation including limiting parameters for crane operation (windspeed, roll/pitch angles etc.).

H1.7.1.5 The crane should be equipped with a reliable load monitoring system sufficiently accurate to show cyclic dynamic loads.

H1.8 Operations

H1.8.1 Mooring/Anchoring

H1.8.1.1 Before positioning of a crane vessel for lifting, a mooring/anchoring plan should be prepared. It should be checked that the holding capacity of the mooring/anchoring arrangement is sufficient to hold the vessel in position during the lifting operations.

H1.8.1.2 The mooring/anchoring plan should describe the operation position as well as the stand off position and the procedure for moving clear of the structure or platform in question.

H1.8.2 Lifting

H1.8.2.1 The intended lifting operation is to be described in an operation manual.

H1.8.2.2 The chain of command and individual responsibilities of the personnel participating in the lifting procedures should be established in writing prior to the operation.

H1.8.2.3 Before starting the lifting procedure, or cutting the seafastening beyond a point of no return, the person responsible for the decision to proceed with the operations should be satisfied that the following conditions are complied with:

(a) The environmental conditions, including the forecasts are such that the operations can be completed in a well controlled manner and in accordance with the design assumptions for the objects involved.
(b) The personnel have been properly briefed and are standing by.
(c) The crane and all equipment necessary for the operations are correctly rigged and ready to be used.
(d) Obstacles which may unduly delay the operations have been removed.

H1.8.2.4 During the operations a detailed log should be kept by the lifting contractor. The environmental conditions, motion response (relative hook movement), crane load and the sequence of events are to be recorded.

H1.9 Design Examples

H1.9.1 Statically Indeterminate Four Sling Lifts

H1.9.1.1 Problem: A module with weighed weight W is to be lifted from a barge to a rig. Special loads (SPL) is zero. A conventional, redundant, statically indeterminate, 4-sling lifting arrangement is used. Tolerances of the slinging arrangement is within 0.25%. Select slings and shackles.

H1.9.1.2 Static analysis of forces:

(a) Calculate the centre of gravity (COG).
(b) The equilibrium position of the hook will be on the vertical line through the COG when SPL is zero. Select a desired hook position and calculate sling lengths and angles.
(c) Calculate ideal static force distribution at the padeyes. (Skewload and DAF ignored).

The static force at one position in the slinging arrangement is the force at the padeye plus the weight of the slinging arrangement "below" the point in question.

H1.9.1.3 Shackles can normally be selected with SWL equal to or greater than the ideal static force multiplied with the factor given below.

Weight of the Lifted Object	< 100 t	100–1000 t	> 1000 t
Offshore	(1.3 × 1.25 =) 1.62	(1.2 × 1.25 =) 1.5	(1.15 × 1.25 =) 1.44
Inshore	(1.15 × 1.25 =) 1.44	(1.1 × 1.25 =) 1.37	(1.05 × 1.25 =) 1.31

Slings can be selected with breaking strength equal to or greater than the ideal static force multiplied with the factor given in Table H1.1.

H1.9.2 Statically Determinate Lift (1, 2 or 3 Slings)

H1.9.2.1 For conventional lifts with statically determinate slinging arrangements, i.e. lifts with SKL 1.0, the same general procedure as described in H1.9.1 can be used, but the following factors will apply:

Shackles:

Weight of the Lifted Object	< 100 t	100–1000 t	> 1000 t
Offshore	1.3	1.2	1.15
Inshore	1.15	1.1	1.05

Slings: See Table H1.2.

TABLE H1.1 Sling Factors with Skewload (SKL = 1.25 Included)

Weight of the Lifted Object	< 50 t	50–100 t	100–1000 t	> 1000 t
Offshore	(1.3 × 1.25 × 4 =) 6.5	(1.3 × 1.25 × 3.3 =) 5.36	(1.2 × 1.25 × 3.3 =) 4.95	(1.15 × 1.25 × 3.3 =) 4.74
Inshore	(1.15 × 1.25 × 4 =) 5.75	(1.15 × 1.25 × 3.3 =) 4.74	(1.1 × 1.25 × 3.3 =) 4.54	(1.05 × 1.25 × 3.3 =) 4.33

TABLE H1.2 Sling Factors without Skewload (SKL = 1.0)

Weight of the Lifted Object	< 50 t	50–100 t	100–1000 t	> 1000 t
Offshore	(1.3 × 4 =) 5.2	(1.3 × 3.3 =) 4.29	(1.2 × 3.3 =) 3.96	(1.15 × 3.3 =) 3.8
Inshore	(1.15 × 4 =) 4.6	(1.15 × 3.3 =) 3.80	(1.1 × 3.3 =) 3.63	(1.05 × 3.3 =) 3.46

Index

Above-bottom pull (pipelines), 357, 358
Access, for constructibility, 417, 418
Air cushion, 116
Airlift excavation, 154
Anchor(s), 119–121
    clump, 121
    drag embedment, 120, 121
    gravity, 121
    handling boat, 110–111
    pile, 121
    propellant, 122
    suction, 122
Anchorage, of derrick (API–RP2A), 510
Anchoring, 483–487, 490, 491
    of pipelines (DNV), 529
API-RP2A, extracts from, 508–522
Arctic:
    constructibility, 474
    deployment, 463–465
    dredging, 449, 455
    ecology, 449, 450
    earthwork, 450–453
    icebreaking, 450, 451
    installation, 465–470
    logistics, 450, 451
    marine structures, 440–475
    pipelines, 474–476
    removals, 469
    seafloor, 446–448
    slope protection, 452, 453
    storms, 449
    towing, 450, 451, 463–465
Arctic structures, 457–470
    caisson-retained islands, 457, 458
    concrete, 458–462
    gravity base, 458–462
    installation, 465–470
    removal, 469
    steel, 457
    towing, 463–465
Articulated columns, 297–307
Articulated mattress, 332
Asphaltic materials, 76, 77
Assembly afloat, 411, 412
Atmospheric icing, 445
Aurora borealis, 446

Ballasting, of GBS, 251
Barges, 85–89
    bury, 112
    crane, 89–93
    derrick, 93–96
    jack-up, 99–101
    launch, 87, 88, 101–103, 410–411
    pipelaying, 107–110, 336–350
    reel, 112
    semi-submersible, 96–98
    sheer legs, 90–93
    submergence, 87, 88
    submersible, 88
Basin, construction, 408, 409
Bearing on seafloor, 43
Belled footings, 194, 195, 196, 394, 513–515
"Billy Pugh" net, 133
Bituminous materials, 76
Blasting (underwater), 156, 157, 158, 159
Boats:
    anchor handling, 110–111
    crew, 112
    supply, 110, 111
    tow, 110, 111
Bolts, high strength, 60
Bottom-pull, of pipelines, 353–357
Boulders, ice-rafted, 40
Breasting dolphins, 280, 283, 284
Bridge piers (offshore), 319–322
Buckle detection, 529
Bumper piles, 315
Bundled pipes, 360
Buoyancy:
    of structures, 12–14, 80
    tanks, 314
    temporary, 147, 148
Burial, of pipelines, 361–367
Bury barge, 112

Cables:
    arrays, 322–324
    crossing with pipelines, 528
    laying, 373–375
Caissons, for offshore terminals, 286
Calcareous soils, 40
Cap rock, 44

Cemented soils, 44
Cerveza (installation), 231, 233, 234, 235
Clamshell:
   dredge, 105–107, 154
   dredging, 154
Clathrates, 447
Clays:
   consolidation of, 44
   and muds, 42
   scour of, 44
   weak, 41
Coatings:
   for concrete, 70
   for steel, 63
Cobbles, on seafloor, 47
Coflexip pipelines, 369
Cognac, installation, 228, 229, 230, 231, 232, 233
Concrete:
   abrasion resistance, 65
   for bells, 515
   bond, 65
   construction joints, 70
   creep, 65
   curing, 67
   forming, 70
   freeze-thaw durability, 65
   heat of hydration, 65
   hybrid concrete-steel, 72-75
   mixes, 64–66
   penetrations, 396–397
   permeability, 64
   placement, 66, 67
   platforms, 236–278, 481–507
   precast, 72
   properties, 64
   pumped, 141
   reinforcement of, 67, 68
   repairs, 374–375, 506, 507
   slip forms, 70
   structural, 64
   sulfate resistance, 65
   tolerances, 71
Conductor installation, 222, 224, 272
Consolidation:
   of underwater fills, 161, 162, 163
   of weak soils, 163
Constructibility, 404–426
   in Arctic, 474
Construction:
   basin, 408–409
   facilities, 408–409
   principles, 408
   procedures, 412–417
   stages, 405–408
   platforms, 237, 498–501
Construction afloat, 493, 501
Contingency:
   planning, 422, 423
   for towing, 117
Contractual relationships, 4, 5
Coral, 44

Corrosion, protection:
   for embedments, 70
   for prestressing steel, 69
   for reinforcing steel, 68
   for steel, 63
Crane barges, 89–93
Crew boat, 112
Currents, effects of, 17, 18, 19, 20
Cylinder piles, 288, 289

Damage control, 83, 84, 85
Deck:
   construction, 255, 256
   installation, 224, 225
   mating, 261–263
   transport, 256
Decompression, of divers, 135
Deep sea:
   construction methods, 427–439
   material properties, 429–434
   phenomena, 428
   seawater properties, 429
   techniques, 428, 429
Densification of underwater fills, 161, 162
Deployment, seafloor, 322–324
Derrick barge, 93–96
Det Norske Veritas, excerpts from:
   "Marine Operations," 537–544
   Rules for Offshore Platforms, 481–496
   Rules for Submarine Pipelines, 523–536
Dikes, underwater, 159, 160, 161
Diving, 133–138
   "bounce," 135
   decompression, 135
   saturation, 135
   systems, 138
Draft, 80, 81
Dredge, 103–107
   clamshell, 105–107
   hydraulic, 105
   offshore, 103–107
   trailer-suction, 103–105
Dredging, 44, 153–159
   in deep water, 154, 155, 428–439
   effects of, 51
   hard material, 156–159
   rock, 156–159
   soft soils, 153–155
   unsuitable soils, 152–156
Drilled and grouted piles, 513
Drilling vessel, 111, 112
Durability, in Arctic, 473, 474

Earthquakes, effects of, 37
Ecological considerations, 50–53
Eductor:
   dredging, 455
   system, 105, 154
Embankments, underwater, 159–161
Embedments, 69, 70
Erection of jacket, 509, 510

Erosion protection, 164, 165, 166
Existing structures, protection of, 53, 54
Explosives, underwater, 156–158

Feedback, constructibility, 421, 422
Fender units, 285
Fiber, reinforced glass pipelines, 368, 369
Fills:
   densification, 161–163
   underwater, 159–161
   vibration, 162
Filter fabrics, 76
F.I.P. recommendations (4th edition), extracts of, 497–507
Flexible pipelines, 369
Flotation:
   of pipelines, 358, 359
   underwater, 359
   units, 115
Float-out, 243, 500
Fog:
   Arctic, 445
   effects of, 29, 30
Foundations, 504, 505
Freeboard, 80, 81
Freezing, of soils, 452, 453
FRG pipe, 368, 369
Frost heave, 453
Future of offshore construction, 477

Giant modules, 379, 380
Glacial till, 40
Gravel deposits, 47
Gravity base structures (GBS):
   foundation strengthening, 275–277
   installation, 268–270, 497, 498
   platforms, 236–278, 497, 507
   penetration, 274
   removal, 278, 401–403
   strengthening, 275–277, 395–397
   structures, 236–278
Grounding, of welding equipment, 514
Grout:
   annulus, 143
   underbase, 142
Grouted connections, 515, 516
Grouting:
   of seafloor soils, 163, 164
   underbase, 332–334
Grout-intruded aggregate, 141
Guyed towers, 306–310, 510
Guyline installation, 510

Handling heavy loads, 127–132
Heather platform, piling, 185
Heat of hydration, 65
Heavy lifts, 127–132, 518, 519, 538–544
   of decks, 225, 226
"Hi-Deck," 226, 380
Hondo platform:
   installation, 227
   piling, 184, 185

Honshu-Shikohu bridge, 320–322
Hookup:
   for GBS, 263
   of topsides, 378, 379
Hutton TLP (platform), 312, 313
Hydraulic dredge, 105
Hydrostatic pressure, 12, 14, 521

Icebergs, 30, 35, 36, 37, 440, 445
Ice islands:
   artificial, 456, 457
   natural, 33
Icing, atmospheric, 29, 30
Ice management, 470–472
Ice scour, of seafloor, 447
Ice structures, 456, 457
Ice surveys, 470–472
Increasing pile capacity:
   axial, 391–394
   lateral, 394–395
Inflatable buoys, 403
Insert piles, 188, 189, 391, 392
Installation:
   of GBS, 268–270, 497, 498
   of jacket, 221–224
      examples, 227–235
      over template, 221
Instruction sheets, for construction, 424, 425
Intakes, 369–372
Interference, by vessels, 54

Jacketing down, for launching, 410
Jackets, 199–235, 283
   erection, 510
   installation, 521
   launching, 509
   leveling, 511
   lifting, 509
   positioning, 510
   removal, 399
   transport, 508–510
   weight on bottom, 511
Jack-up construction:
   barge, 99, 100, 101
   foundations, 491
Jet sled, 112, 155, 363–367
Jointing afloat, 411, 412

Kevlar, 76
Khazzan Dubai (offshore oil storage), 316

Lateral resistance, of piles, 170, 177
Launch barge, 101, 102, 103, 201
Launching:
   from barge, 87, 88, 209–217, 494, 509, 520, 521
   from ways, 409
Lay barge, 338, 351
Lena platform, 310
Leveling:
   jacket, 511
   of seafloor, 151, 152

Life saving gear, 85
Lifting:
  heavy, 518, 519, 538–544
  of jacket, 209, 210, 509
  offshore, 495, 496
Lightning, effects of, 29, 30
Liquefaction, prevention of, 114
Load out, 199, 201, 202, 204, 205, 493, 519, 520
Lowering loads into water, 129, 132

Manuals, for construction, 423, 424
Marine operations, 113–148, 491–496
Marine organisms and bacteria, 15
Match casting, 246–248
Mating, 327, 328
  of column and base, 306, 307
  of jacket sections, 227, 230
  underwater, 147
Maui A piling, 186
Metacentric height, 81–83
Methane gas, 42, 447
Mock-ups, 415
Module:
  erection, 376–377
  giant, 379, 380
Moored buoys, 322–324
Mooring, 119–127, 282, 483–487
  buoys, 122
  chain, 126
  dynamic, 127
  of GBS, 246–251
  lines, 120, 124–126
  for mating, 123
  shallow water, 123
  storm, 282
  survival, 124, 282
Motions, of floating structures and equipment, 78, 79
Mud, 42
Mudmats, 221, 222
Mudslide:
  conductor protection, 224
  jacket penetration, 222

Navigation, offshore, 143–147
Noise, ecological disturbance, 51

Ocean thermal energy systems, 326–328
Obstruction removal (from seafloor), 151, 152
Ocean:
  intakes, 369–372
  outfalls, 369, 372
Offshore terminals, 279
Oil spill, effects and constraints, 52
Oosterschelde storm surge barrier, 329–335
Oozes, seafloor, 47
OTEC, 326–328
Outfalls, 369–372

Piles:
  anchoring, 189
  belled, 194–196, 394, 513–515
  capacity, 196–198, 391–395
  connection to jackets, 223
  drilled and grouted, 191–194
  drilling, 188–196
  driving, 43
  driving refusal, 178, 179
  driving underwater, 182
  end closure, 181–183
  end plugs, 383
  fabrication, 170, 171
  freezing, 197
  field splicing, 176
  grouting, 183, 184, 393
  hammers, 174–180
  installation, 172–176, 222, 227, 229, 279, 280, 310, 495, 510–514, 515–518
    by dredging, 196
    by drilling, 188–194
    examples, 184–186
    by weighting, 196
  improving capacity of, 196, 197, 198
  improving lateral capacity of, 197
  increasing penetration, 186–188
  insert, 188, 189, 391–393
  jetting, 186, 187
  load transfer, 183, 184
  minimum wall, 175, 180
  for offshore terminals, 190, 279, 280
  penetration, increasing, 186–188
  prestressed concrete, 189, 190
  removal, 399–400
  tip reinforcement, 179–182
  transportation, 171, 172
Pingos:
  subsea, 447
  underwater, 42
Pin piles, installation, 222
Pipelaying:
  barge, 107–110
  equipment, 107–110
  operations, 107–110, 336–350
Pipelines:
  anchoring, 529
  Arctic, 474–476
  concrete, 372
  crossings, 528
  flexible, 368, 369
  under ice, 360, 361
  polyethylene, 368
  repairs, 387–389
  route, 523–527
  submarine, 336–367
  support, 367
  surveys, 535
  tests, 535, 536
  tie-ins, 533, 534
  trenching, 363, 367
  welding, 529, 533
Piping, of seafloor soils, 49

Plastic materials, 75
Platforms, removal of, 398–403
Polyethylene:
    materials, 75, 76
    pipelines, 368
Positioning:
    of jacket (API-RP2A), 510
    of structures, 494, 495
Pressure ridges, 31, 32
Prestressing:
    operations, 69
    tendons and accessories, 68, 69
Principles of construction, 408
Properties of materials, deep sea, 429–434
Pumping concrete and grout, 141
Padeyes, 130, 131
Painting of steel, 63
Penetration:
    of GBS, 275
    into seafloor, 44
    of skirts, 275, 521, 522
Permafrost subsea, 446, 447
Personnel transfer, 132, 133

Quality control, 420

Rain, effects of, 29, 30
Reel barge, 112, 351–353
Regulation and verification, 5
Reinforcing:
    corrosion prevention, 68
    steel, 67
Removal:
    of GBS, 401–403, 522
    of jackets, 399–401
    of piled structures, 399–401
    of piles, 399–400
    of platform, 398–403
Repairs:
    concrete structures, 385, 506, 507
    fire damage, 387
    foundations, 386, 387
    pipeline, 387–389
    steel bracing, 382–384
Riser installation, 347–349
Risk and reliability, in construction, 425, 426
Rock:
    material properties, 76, 77
    outcrops, 46
    placement underwater, 160
Rolling in, for launching, 410
ROV's, 137

Safety, 421
Salvage, of platforms, 398–403
Sampling, of soils, 44
Sand:
    calcareous, 40
    dense, 39
    flow methods, 271
    jacketing, launching, 409, 410
    piles, 455
    unconsolidated, 45
    waves, 38, 46
Scour, 18, 19
    of clays, 44
    by ice, 38
    protection, 164–166, 223, 294
Screeding, of seafloor, 151, 152
S-curve, pipelines, 360
Seafloor densification, 151, 152
    deployment, 322–324
    dredging, 151, 152
    erosion protection, 151, 152
    improvements, 149
    leveling, 151, 152
    liquefaction prevention, 151, 152
    modifications, 149
    obstruction removal, 151, 152
    survey, 150, 151
    templates, 313, 314
Sea ice, 30, 31, 32, 33, 34, 440–445
Seawater:
    chemistry, 15
    deep sea, 429
Seismicity, 447, 448
Self-floating, jacket, 216
Semi-submersible barge, 96–98
Shackles (DNV), 540, 541
Sheer-legs barge, 90, 91
Shore crossings, 349, 361, 362, 366, 367
Side launching, 215, 216
Silts:
    overconsolidated, 41
    weak, 41
Single-point moorings, 292-299
Slings, 128, 129, 132, 539, 540
    factors, 543, 544
Slope protection, 452, 453
Slopes, stable, 42, 43, 45, 46
Slumping, of seafloor, 47, 48
Snow, effects of, 29, 30
Spray, effects of, 29, 30
Spud piles, 273, 274
Stability, 81, 82, 83, 481, 482, 487, 490, 492
    during tow, 209
Stages, of construction, 405, 408
Standards and rules, 9, 10, 11
Steel:
    coatings, 63
    concrete hybrid, 72–75
    corrosion protection, 63
    erection, 61, 62
    fabrication, 57, 59
    jackets, 199–224
    low temperature, 85, 473
    material specifications, 55
    painting, 63
    structures, 55–64
    welding, 57–60

Storm:
   moorings, 282
   surges, 28, 29, 449
Strengthening:
   existing platforms, 390–397
   seafloor, 163, 164
   steel bracing, 390, 391
   underwater fills, 163
Subaqueous tunnels, 324–326
Sub-base, for GBS, 277
Submarine:
   pipelines, 292, 336–367, 525–536
   work, 137
Submergence, of structures, 260, 494, 495
Submersible barge, 88
Submersibles, 137
Subsea:
   production, 439
   storage, 315–319
   templates, 313–317
Subsurface pull, of pipelines, 359
Sunken tunnel segments, 324–332
Superstructure, installation, 514
Supply boat, 110, 111
Survey:
   channel, 115
   control, 419, 420
   offshore, 143–147
   seafloor, 150, 151
Strudl-scour, 447
Swells, 20, 21, 24, 25
Syntatic foam, 403
Synthetic materials, 75

Temperature effects, 13
Temporary buoyancy, 269, 314, 400–403
Tensioner (pipeline), 339, 341, 347
Tension-leg platform, 310–313
Thermo-karst, 448
Thistle platform, 185
Tide surges, 28, 29
Tie-downs, 199, 204, 205, 206
Tie-ins, pipeline, 533, 534
Titanium, 76
TLP, 310–313
Tolerances:
   for concrete fabrication, 72
   for constructibility, 418, 419
   for steel erection, 61
Topsides installation, 376–380
Tow boat, 110–111
Towed structure, stability requirements, 119
Towing, 113–119, 497, 501–504
   of GBS, 253, 254, 264–268
   in ice, 117
Towline attachments, 113
Toxic chemicals, 52
Tractor, sea-floor crawler, 112
Trailer-suction dredge, 103–105, 153
Transport, of jackets, 207, 493, 508, 509, 520
Tremie concrete, 137–141
Trenching of pipelines, 363–367
Trestle construction, 290, 291
Tsunamis, effect of, 37
Tubes, 324–326
Turbidity:
   currents, 47
   particles, 50, 51

Underbase:
   grout, 270, 271, 293, 294
   sand fill, 271
Underwater:
   admixtures, 139, 141
   concreting, 137–142
   mixes, 139
Underwater grouting, 137, 141–143
Underwater inspection, 332, 333
Underwater mating, 147
Underwater oil storage, 315–319
Underwater repairs, 381–389
Underwater tools, 134, 136, 137
Underwater work systems, 133–137
Upending, of jacket, 217–221, 494

Vortex shedding, 18

Watertight integrity, 490
Waves, 20–25
Welding, 57–60
   grounding of, 514
   wet, 136
Well protectors, Arctic, 462
White-out, 29, 30, 445
Winds, 25–28